Image Modeling

Contributors

Narendra Ahuja
Harry G. Barrow
P. C. Chen
F. Cohen
Richard W. Conners
D. B. Cooper
Larry S. Davis
H. Elliott
Martin A. Fischler
H. Freeman
B. Roy Frieden
R. W. Fries
K. S. Fu
Robert M. Haralick
Charles A. Harlow
Martin Hassner
Gabor T. Herman
Thomas S. Huang
B. R. Hunt
Laveen N. Kanal
R. L. Kashyap
Donald E. McClure
R. E. Miles
Amar Mitiche
J. W. Modestino
T. Pavlidis
L. Reiss
J. A. Saghri
Bruce Schachter
J. Serra
Jack Sklansky
P. Symosek
Jay M. Tenenbaum
Demetri Terzopoulos
Julius T. Tou
A. L. Vickers
Steven W. Zucker

Image Modeling

Edited by AZRIEL ROSENFELD

Computer Vision Laboratory
Computer Science Center
University of Maryland
College Park, Maryland

ACADEMIC PRESS, INC.
(Harcourt Brace Jovanovich, Publishers)
Orlando San Diego New York London
Toronto Montreal Sydney Tokyo

ACADEMIC PRESS, INC.
Orlando, Florida 32887

United Kingdom Edition published by
ACADEMIC PRESS, INC. (LONDON) LTD.
24/28 Oval Road, London NW1 7DX

Library of Congress Cataloging in Publication Data
Main entry under title:

Image modeling.

Papers originally presented at a workshop held in Rosemont, Ill., Aug. 6-7, 1979.
Contents: Mosaic models for textures / N. Ahuja -- Image segmentation as an estimation problem / P. C. Chen and T. Pavlidis -- Toward a structural textual analyzer based on statistical methods / Richard W. Conners and Charles A. Harlow -- [etc.]
1. Computer graphics--Congresses. 2. Image processing --Congresses. I. Rosenfeld, Azriel, Date.
T385.I45 621.3819'598 81-3562
ISBN 0-12-597320-9 AACR2

PRINTED IN THE UNITED STATES OF AMERICA

85 86 87 88 9 8 7 6 5 4 3 2

Contents

List of Contributors

Numbers in parentheses indicate the pages on which the authors' contributions begin.

NARENDRA AHUJA (1), Coordinated Science Laboratory, University of Illinois, Urbana, Illinois 61801

HARRY G. BARROW* (371), Artificial Intelligence Center, SRI International, Menlo Park, California 94025

P. C. CHEN† (9), Department of Electrical Engineering and Computer Science, Princeton University, Princeton, New Jersey 08540

F. COHEN (63), Division of Engineering, Brown University, Providence, Rhode Island 02912

RICHARD W. CONNERS (29), Department of Electrical Engineering, Louisiana State University, Baton Rouge, Louisiana 70803

D. B. COOPER (63), Division of Engineering, Brown University, Providence, Rhode Island 02912

LARRY S. DAVIS (95), Computer Sciences Department, The University of Texas at Austin, Austin, Texas 78712

H. ELLIOTT (63), Department of Electrical Engineering, Colorado State University, Fort Collins, Colorado 80523

MARTIN A. FISCHLER (371), Artificial Intelligence Center, SRI International, Menlo Park, California 94025

H. FREEMAN (111), Rensselaer Polytechnic Institute, Troy, New York 12181

B. ROY FRIEDEN (133), Optical Sciences Center, University of Arizona, Tucson, Arizona 85721

R. W. FRIES‡ (301), Electrical, Computer and Systems Engineering Department, Rensselaer Polytechnic Institute, Troy, New York 12181

K. S. FU (153), School of Electrical Engineering, Purdue University, West Lafayette, Indiana 47907

ROBERT M. HARALICK (171), Department of Electrical Engineering, and Department of Computer Science, Virginia Polytechnic Institute and State University, Blacksburg, Virginia 24061

CHARLES A. HARLOW (29), Department of Electrical Engineering, Louisiana State University, Baton Rouge, Louisiana 70803

MARTIN HASSNER (185), School of Engineering, University of California, Irvine, Irvine, California 92717

* Present address: Fairchild, Artificial Intelligence Research Laboratory, 4001 Miranda Avenue M/S 30-888, Palo Alto, California 94304.

† Present address: P. O. Box 2189, Exxon Production Research Company, Houston, Texas 77001.

‡ Present address: PAR Technology Corp., Rome, N.Y. 13440.

GABOR T. HERMAN* (199), Medical Image Processing Group, Department of Computer Science, State University of New York at Buffalo, 4226 Ridge Lea Road, Amherst, New York 14226

THOMAS S. HUANG† (215), School of Electrical Engineering, Purdue University, West Lafayette, Indiana 47907

B. R. HUNT (225), Systems Engineering Department and Optical Sciences Center, University of Arizona, Tucson, Arizona 85721

LAVEEN N. KANAL (239), Laboratory for Pattern Analysis, Department of Computer Science, University of Maryland, College Park, Maryland 20742

R. L. KASHYAP (245), School of Electrical Engineering, Purdue University, West Lafayette, Indiana 47907

DONALD E. McCLURE (259), Division of Applied Mathematics, Brown University, Providence, Rhode Island 02912

R. E. MILES (277), Department of Statistics, Institute of Advanced Studies, Australian National University, P. O. Box 4, Canberra, Australia Capital Territory 2600, Australia

AMAR MITICHE (95), Computer Sciences Department, The University of Texas at Austin, Austin, Texas 78712

J. W. MODESTINO (301), Electrical, Computer, and Systems Engineering Department, Rensselaer Polytechnic Institute, Troy, New York 12181

T. PAVLIDIS‡ (9), Department of Electrical Engineering and Computer Science, Princeton University, Princeton, New Jersey 08540

L. REISS (63), Department of Computer Science, Brown University, Providence, Rhode Island 02912

J. A. SAGHRI (111), Rensselaer Polytechnic Institute, Troy, New York 12181

BRUCE SCHACHTER§ (327), General Electric Company, P. O. Box 2500, Daytona Beach, Florida 32015

J. SERRA (343), Centre de Morphologie Mathématique, Ecole des Mines de Paris, 35 Rue Saint-Honoré, 77305 Fontainebleau, France

JACK SKLANSKY (185), School of Engineering, University of California, Irvine, Irvine, California 92717

P. SYMOSEK (63), Division of Engineering, Brown University, Providence, Rhode Island 02912

JAY M. TENENBAUM¶ (371), Artificial Intelligence Center, SRI International, Menlo Park, California 94025

DEMETRI TERZOPOULOS (423), Computer Vision and Graphics Laboratory, Department of Electrical Engineering, McGill University, Montreal, Quebec, Canada

* Present address: Medical Imaging Section, Department of Radiology, Hospital of the University of Pennsylvania, 3400 Spruce Street, Philadelphia, Pennsylvania 19104.

† Present address: Coordinated Science Laboratory, University of Illinois, Urbana, Illinois 61801

‡ Present address: Bell Laboratories, Murray Hill, New Jersey 07974.

§ Present address: Westinghouse Defense and Electronics Systems Center, Box 746, Mail Stop 451, Baltimore, Maryland 21203.

¶ Present address: Fairchild, Artificial Intelligence Research Laboratory, 4001 Miranda Avenue M/S 30-888, Palo Alto, California 94304.

JULIUS T. TOU (391), Center for Information Research, University of Florida, Gainesville, Florida 32611

A. L. VICKERS (301), Electrical, Computer, and Systems Engineering Department, Rensselaer Polytechnic Institute, Troy, New York 12181

STEVEN W. ZUCKER (423), Computer Vision and Graphics Laboratory, Department of Electrical Engineering, McGill University, Montreal, Quebec, Canada

Preface

It has long been recognized in the field of image processing that the design of processing operations should be based on a model for the ensemble of images to be processed. This realization is becoming increasingly prevalent in the field of image analysis as well. Unfortunately, it is difficult to formulate realistic models for real-world classes of images; but progress is being made on a number of fronts, including models based on Markov processes, random fields, random mosaics, and stochastic grammars, among others. At the same time, analogous models are being developed in fields outside image processing, including stereology, mathematical morphology, integral geometry, statistical ecology, and theoretical geography. It is hoped that this volume, by focusing attention on the field of image modeling, will serve to stimulate further work on the subject, and will promote communication between researchers in image processing and analysis and those in other disciplines.

The papers in this volume were presented at a workshop on image modeling in Rosemont, Illinois on August 6–7, 1979. The workshop was sponsored by the National Science Foundation under Grant MCS-79-04414, and by the Office of Naval Research under Contract N00014-79-M-0070; their support is gratefully acknowledged. Three of the papers presented at the workshop are not included in this volume: B. Julesz, Differences between attentive (figure) and preattentive (ground) perception; W. K. Pratt and O. D. Faugeras, A Stochastic texture field model; W. R. Tobler, Generalization of image processing and modeling concepts to polygonal geographical data sets. All but the first of the papers in this book appeared in Volume 12 of the journal *Computer Graphics and Image Processing.*

Mosaic Models for Textures

NARENDRA AHUJA

Coordinated Science Laboratory and Department of Electrical Engineering, University of Illinois, Urbana, Illinois 61801

Traditionally the models of image texture have been classified as statistical or structural [15, 29, 30]. However, in [6, 9] we have suggested a classification of image models into *pixel-based* and *region-based* models, which we believe is more useful. The pixel-based models view individual pixels as the primitives of the texture. Specification of the characteristics of the spatial distribution of pixel properties constitutes the texture description [15, 28]. The region-based models conceive of a texture as an arrangement of a set of spatial subpatterns according to certain placement rules [30, 34]. Both the subpatterns and their placement may be characterized statistically.

Most of the models used in the past are pixel-based. These models have been proposed for images representing a variety of natural phenomena, including ocean waves and the earth's surface. However, for many images the region-based models appear to be more natural [1, 5, 37] than the pixel-based models, although relatively little research has been done on their development [9, 14]. In this paper we shall discuss a specific class of region-based models known as mosaic models, and shall review the work done on these models and their application to modeling textures.

1. MOSAIC MODELS

Mosaic models are defined in terms of planar random pattern generation processes. The characteristics of the patterns generated by a given process may be obtained from the definition of the process. These properties then determine the class of images for which the corresponding model is suitable. A variety of processes may be used to define mosaic models. We describe below briefly two classes of such processes that we have considered in our work. For details see [1–4, 8].

1.1. Cell Structure Models

Cell structure mosaics are constructed in two steps:

(a) Tessellate a planar region into cells. We shall consider only tessellations composed of bounded convex polygons.

(b) Independently assign one of m colors $c_1, c_2, \ldots, c_m$ to each cell according to a fixed set of probabilities

$$p_1, \ldots, p_m, \qquad \sum_{i=1}^{m} p_i = 1.$$

ISBN 0-12-597320-9

Let $P_{ij}(d)$ denote the probability that one end of a randomly dropped needle of length d falls on color c_i given that the other end is in a region of color c_j. Let $W(d)$ be the probability that a randomly dropped needle of length d falls completely within a cell. Then it can be shown that

$$P_{ij}(d) = p_i(1 - W(d)) + \delta_{ij}W(d),$$

where δ is the Kronecker function.

Given the coloring process in step (b), the cell structure models form a family whose members differ in the manner in which the plane is tessellated. We shall now describe some members of this family that we have used, starting from the three regular tessellations and progressing toward some random ones.

1.1.1. Square model. This is an example of a cell structure model where the cells are of a uniform size. A square (checkerboard) model can be formed by the following procedure. First, choose the origin of an x–y coordinate system on the plane with uniform probability density. Then tessellate the plane into square cells of side length b. Next, this "checkerboard" is rotated by an angle chosen with uniform probability from the interval $(0, 2\pi)$. The cells are now independently assigned one of the m tile types.

Modestino *et al.* [24, 26] have considered tessellations of the plane into rectangles and parallelograms. The lengths of the sides of the rectangles or the parallelograms are determined by two independent renewal processes defined along a pair of axes.

1.1.2. Hexagonal model. This model uses a network of identical hexagons to tessellate the plane. The hexagons can be oriented at any angle to the axes.

1.1.3. Triangular model. This is similar to (1) and (2) except that a triangular tessellation of the plane is used.

All three regular tessellations described above can be viewed as the result of a growth process from a set of nuclei placed at the points of an appropriate regular lattice. Assume all the nuclei start growing simultaneously along a circular frontier, at any given instant. At some later time the circles centered at neighboring lattice points come into contact. As the cells continue to grow, these points of contact become the midpoints of growing straight line segments along which the growth frontiers meet and the growth is stopped. Finally, the grown line segments form the sides of polygons that have the original nuclei as their centers. Expressions for $W(d)$ for these tessellations are known [1, 8, 17, 31, 32].

An interesting special case arises when we consider cells of unit area. Then the resulting mosaic is the realization of a random lattice point process defined by the coloring process. We shall now describe some random cell structure models.

1.1.4. Poisson line model. Consider a system of intersecting lines in the plane with random positions and orientations. Such a system when derived by the following Poisson process possesses fundamental properties of homogeneity and isotropy. A Poisson process of intensity τ/π determines points (θ, ρ) in the infinite rectangular strip $[0 \leq \theta < \pi, -\infty < \rho < \infty]$. Each of these points can be used to construct a line in the plane of the form $x \cos\theta + y \sin\theta - \rho = 0$, where ρ is the distance between the line and an arbitrarily chosen origin. This process is used to tessellate the plane into convex cells.

[1, 32] list some important characteristics of the Poisson line tessellation, such as the expected cell area, expected cell perimeter, expected number of cells meeting at a vertex, and expected total line length per unit area. A detailed discussion can be found in [19, 20, 36].

1.1.5. Voronoi model. This model is based upon a tessellation that is the result of a growth process similar to that used for the regular cell structure models described earlier except that the growth now starts at randomly located points. Each of these points spreads out to occupy a "Dirichlet cell" [13, 21, 22] consisting of all the points that are nearer to it than to any other nucleus. The random initial arrangement of the nuclei may result in cell edges with any of infinitely many slopes, and therefore, a random tessellation. The cells are then independently colored as usual to obtain a *Voronoi mosaic.* [1, 32] present some properties of the Voronoi tessellation. For details, see [13, 21, 22].

1.1.6. Delaunay model. The Delaunay tessellation is closely related to the Voronoi tessellation. *Delaunay triangles* [21–23] can be constructed in the Voronoi tessellation by joining all pairs of nuclei whose corresponding Voronoi polygons share an edge. Thus the vertices of Voronoi polygons are the circumcenters of the Delaunay triangles. The properties of Delaunay tessellations are discussed in [21–23].

1.2. Coverage Models

Coverage or "bombing" models constitute the second class of mosaic models that we have considered. A *coverage mosaic* is obtained by a random arrangement of a set of geometric figures ("bombs") in the plane.

We shall first define the class of binary coverage models. Consider a geometric figure in the plane and identify it by (i) the location of some distinguished point in the figure, e.g., its center of gravity, hereafter called the center of the figure, and (ii) the orientation of some distinguished line in the figure, e.g., its principle axis of inertia. Let a point process drop points on the plane, and let each point represent the center of a figure. If the points are replaced by their corresponding figures, the plane is partitioned into foreground (covered by the figures) and background.

A *multicolored coverage mosaic* is obtained by considering figures of more than one color. The color of a given figure is randomly chosen from a known vector of colors $\mathbf{c} = (c_1, c_2, \ldots, c_m)$ according to a predetermined probability vector $\mathbf{p} = (p_1, p_2, \ldots, p_m)$. Let c_0 denote the background color. Since, in general, the figures overlap, we must have a rule to determine the colors of the regions that are covered by figures of more than one color. We shall give one example of such a rule. Let us view the point process as dropping the centers sequentially in time. Each time a new point falls, the area covered by the associated figure is colored with the color of that figure irrespective of whether any part of the area has already been included in any of the previously fallen figures. The color of a region in the final pattern is thus determined by the color of the latest figure that covered it. (Note that we could just as well have allowed a figure to cover only an area not included in any of the previous figures.)

As in the case of the cell structure models, $P_{ij}(d)$ denotes the probability that one of the ends of a randomly dropped needle of length d falls in a region of color c_i given that the other end is in a region of color c_j, $0 \leq i, j \leq m$, where c_0 denotes the color of the background, the region not occupied by any of the figures. Some general properties of coverage models are discussed in [1, 12, 35].

2. PROPERTIES OF MOSAIC MODELS

A major part of our past effort has been devoted to relating properties of the patterns generated by mosaic models to the parameters occurring in their definitions. These results have then been used to fit the models described in Sections 1.1 and 1.2 to real textures. We now summarize the past work.

2.1. Geometric Properties of Components in Cell Structure Mosaics

Ahuja [1, 2] presents a detailed analysis of the geometric properties of components in the cell structure mosaics. To avoid the numerous details, we shall present here only a qualitative description of the basic approaches involved. A concise but more illustrative discussion appears in [5]. Some experimental results are presented in [7, 10].

To estimate the expected component area in a regular cell structure mosaic, let us first consider the colored regular lattice defined by cell centers, each having the same color as its cell. The expected number of points in a component of this lattice is obtained by viewing the component as a stack of overlapping identically colored runs in succeeding rows, formed as a result of a one-dimensional row-incremental Markov growth process. The statistics of the within-row components, or runs, are easy to obtain. The expected number of cells in a component of a regular mosaic is the same as the expected number of points in a component of the regular lattice. The expected area of the mosaic component is then obtained by using the known cell area. For the random models, the cell centers do not form a regular lattice. However, the expected number of neighbors of a cell and the expected number of cells meeting at a vertex are fixed for a given tessellation. A conjecture is presented that suggests that the expected area of a component in a random mosaic can be approximated by the expected area of a component in a regular mosaic that has the same cell area and number of cell neighbors as the corresponding expected values in the random mosaic.

The expected perimeter of a component is estimated in terms of the expected number of sides of a cell in the component that belong to the component border. Expected component perimeter follows from the known expected perimeter of a cell, the expected number of sides of a cell, and the expected number of cells in a component obtained as described above.

The problem of estimating the expected width of a component, i.e., the expected length of intercept on an arbitrary component due to a randomly located and oriented line transect, is also considered in [1, 2]. The probability that the number of cells along the intercept is n can be determined easily. Given the orientation of the transect, the total length of the intercept in a regular tessellation can then be expressed in terms of the cell size. For the random tessellations, the orientation of the transect need not be known, since the intercept length is independent of the direction in which it is measured. The expected intercept length can be found by considering different values of n.

2.2. Geometric Properties of Components in Coverage Mosaics

Estimation of the expected area, expected perimeter, and expected width of a component in a coverage mosaic is discussed in detail in [1, 3]. Here we shall briefly outline the approaches used without giving any mathematical details.

The computation of the expected component area is very similar to that for the cell structure models. A component is viewed as resulting from stacking of overlapping runs of figure centers. A run of centers is defined as the sequence of those successive centers within a row whose corresponding figures overlap. A run in a given row may overlap with a run in a distant row if the figures are sufficiently large. The expected total number of components in a given image is derived from a Markov formulation of the component growth process. The expected total area covered by the figures is easy to obtain in terms of the probability that an arbitrary point is isolated. These two results together provide the expected component area.

The estimation of the expected perimeter makes use of the estimate of the expected total length of that part of the border of a figure that is not covered by any other figure. This latter estimate can be made in terms of the expected number of uncovered segments along the border of a figure and the expected length of one such segment. Exact formulas are obtained for the Euclidean plane mosaics, but results for the grid case are approximate. In multicolored coverage patterns the perimeter is computed from borders between different colors and the background and between different colors. It is easy to see that the former is the same as in binary coverage patterns, where all bombs have the same color. Different colors share this border with the background according to their stationary probabilities. Similarly, the expected length of the border between a given color and other colors is the difference of its expected lengths of border with the background when the figures with the other colors are not dropped and when they are dropped. Different colors share this border according to their stationary probabilities.

Computation of expected width of a component is relatively more complex for coverage models. The intercept of a component along a transect consists of smaller intercepts due to many overlapping figures. The distribution of the length of each of these smaller intercepts can be obtained. The component intercept can then be interpreted as formed by a renewal process where the ends of the smaller intercepts define the renewal "times". The expected length of the intercept is given by the renewal equation. This approach, however, requires that the figures used be convex.

2.3. Spatial Correlation in Mosaics

We shall now review those properties of the patterns generated by mosaic models that involve relationships between the gray levels (or colors, etc.) at a pair of points at a given distance and orientation. Once again, we keep the description nonmathematical for brevity. For details, see [1, 4] and the other references cited below.

The joint probability density function for a pair of points in a cell structure mosaic can be expressed in terms of the probability that the two points belong to the same cell. For points chosen at a random orientation, this latter probability is only a function of the distance d between them, and was denoted $W(d)$ earlier. For the regular cell structure models and the Poisson line model, the analytic expressions for $W(d)$ are known. For the occupancy model, it has been shown [17] to involve the solution of a complicated double integral. Ahuja [1, 4] has empirically estimated $W(d)$ for the occupancy and the Delaunay models. Since then, Moore [27] also has conducted experiments with the occupancy model, and has estimated $W(d)$ for that model. For the coverage models, computation of the joint probability density involves point containment properties of certain regions determined by the figures involved and the separation and orientation of the points. For the multicolored coverage models, one has to consider further the cases in which these regions may have different colors.

The joint probability density function can be used to derive many joint pixel properties. The autocorrelation function is a commonly used second-order statistic. For cell structure models, it is the same as the function $W(d)$, and therefore is known for all of the models we have considered. Modestino *et al.* [24] present an integral for the autocorrelation function for their generalized checkerboard model where the cell sides have exponentially distributed lengths. They also present the corresponding expression for the power spectral density. The second-order properties of the parallelogram tessellation model are given in [25]. We may note here that Modestino *et al.* assign normally distributed gray levels to the cells such that the gray levels of adjacent cells are correlated. This is in contrast to the process described in Section 1.1, in which the gray levels of the cells are independent. For the coverage models, the autocorrelation function is obtained by a straightforward application of its definition in conjunction with the known joint probability density function.

The variogram [16, 18], the expected squared difference between the colors of a randomly chosen pair of points, is another useful second-order property, similar to the autocorrelation function. The joint probability density functions for point pairs are used to obtain the variograms for the individual models.

The gradient density is a useful measure of the spatial variation of color in Euclidean plane patterns. For grid patterns generated by mosaic models, Ahuja [1, 4] relates the digital edge density (analogous to the gradient density) to the perimeter results for the Euclidean plane patterns. The orientation distribution of the edges is known from the underlying tessellation (cell structure models) or the shapes of the figures (coverage models). Approximate responses of several digital edge operators, such as horizontal, vertical, and Roberts, when applied to mosaics, are given.

2.4. *Fitting Mosaic Models to Textures*

In [33] some preliminary experiments on fitting mosaic models to real textures are described. Predicted variograms were computed for two models, checkerboard and Poisson line, and were fitted to the actual variograms of ten texture samples from Brodatz's album [11]. These textures were also thresholded, and the average component width was computed. This width agreed very closely with the width predicted by the better fitting model in each case.

Some further experiments on mosaic model fitting are reported in [7, 10]. Samples of four Brodatz textures (wool, raffia, sand, and grass) [11] and three terrain textures were segmented, and average component area and perimeter were computed. Values predicted by six cell structure models (checkerboard, hexagonal, triangular, Poisson line, occupancy, and Delaunay) were also computed. (Predictions were also made for the square bombing model, but they were very poor in all cases.) For each texture, the model parameters were adjusted to make the area predictions match the observed values, and the resulting errors in predicted perimeter were tabulated; and vice versa. The minimum area error and minimum perimeter error models for each texture were the same in nearly all cases, and were consistent from sample to sample for nearly all the textures.

REFERENCES

1. N. Ahuja, Mosaic models for image analysis and synthesis, Ph.D. dissertation, Department of Computer Science, University of Maryland, College Park, Maryland, 1979.
2. N. Ahuja, Mosaic models for images, 1: geometric properties of components in cell structure mosaics, *Inform. Sci.* **23,** 1981, 69–104.
3. N. Ahuja, Mosaic models for images, 2: geometric properties of components in coverage mosaics, *Inform. Sci.* **23,** 1981, 159–200.
4. N. Ahuja, Mosaic models for images, 3: spatial correlation in mosaics, *Inform. Sci.* **24,** to appear.
5. N. Ahuja and A. Rosenfeld, Mosaic models for textures, *IEEE Trans. Pattern Analysis Machine Intelligence* **3,** 1981, 1–11.
6. N. Ahuja and A. Rosenfeld, Image models, in *Handbook of Statistics,* Vol. 2 (P. R. Krishnaiah, Ed.), North-Holland, New York, to be published.
7. N. Ahuja and A. Rosenfeld, Fitting mosaic models to textures, in *Image Texture Analysis* (R. M. Haralick, Ed.), Plenum, New York, to be published.
8. N. Ahuja and B. Schachter, *Pattern Models,* Wiley, New York, to be published.
9. N. Ahuja and B. Schachter, Image models, *Comput. Surveys,* to appear.
10. N. Ahuja, T. Dubitzki, and A. Rosenfeld, Some experiments with mosaic models for images, *IEEE Trans. Systems, Man, Cybernet.* **SMC-10,** 1980, 744–749.
11. P. Brodatz, *Textures: A Photographic Album for Artists and Designers,* Dover, New York, 1966.
12. D. Dufour, Intersections of random convex regions, Stanford University, Dept. of Statistics, T.R. 202, 1973.
13. E. N. Gilbert, Random subdivisions of space into crystals, *Ann. Math. Stat.* **33,** 1962, pp. 958–972.
14. R. M. Haralick, Statistical and structural approaches to texture, in *Proc. 4th Int. Joint Conf. Pattern Recognition, November 1978,* pp. 45–69.
15. J. K. Hawkins, Textural properties for pattern recognition, in *Picture Processing and Psychopictorics* (B. S. Lipkin and A. Rosenfeld, Eds.), pp. 347–370, Academic Press, New York, 1970.
16. C. Huijbregts, Regionalized variables and quantitative analysis of spatial data, in *Display and Analysis of Spatial Data* (J. Davis and M. McCullagh, Eds.), pp. 38–51, Wiley, New York, 1975.

17. B. Matern, Spatial variation, *Medd Statens Skogsforskningsinstit., Stockholm* **36,** 1960, 5.
18. G. Matheron, The theory of regionalized variables and its applications, *Cahiers Centre Morphologie Math. Fontainbleau* **5,** 1971.
19. R. E. Miles, The various aggregates of random polygons determined by random lines in a plane, *Adv. in Math.* **10,** 1973, 256–290.
20. R. E. Miles, Random polygons determined by random lines in the plane, *Proc. Nat. Acad. Sci. USA* **52,** 1969, 901–907, 1157–1160.
21. R. E. Miles, On the homogeneous planar Poisson point process, *Math. Biosci.* **6,** 1970, 85–127.
22. R. E. Miles, The random division of space, *Supp. Adv. Appl. Probl.*, 1972, 243–266.
23. R. E. Miles, Probability distribution of a network of triangles, *SIAM Rev.* **11,** 1969, 399–402.
24. J. W. Modestino and R. W. Fries, Stochastic models for images and applications, in *Pattern Recognition and Signal Processing* (C. H. Chen, Ed.), pp. 225–249, Sijthoff and Noordhoff, Alphen aan den Rijn, The Netherlands, 1978.
25. J. W. Modestino and R. W. Fries, Construction and properties of a useful two-dimensional random field, *IEEE Trans. Inform. Theory* **IT-26,** 1980, 44–50.
26. J. W. Modestino, R. W. Fries, and D. G. Daut, A generalization of the two-dimensional random checkerboard process, *J. Opt. Soc. Amer.* **69,** 1979, 897–906.
27. M. Moore, The transition probability function for the occupancy model, Ecole Polytechnique Mathematics T.R. 40, October 1978.
28. J. L. Muerle, Some thoughts on texture discrimination by computer, in *Picture Processing and Psychopictorics* (B. S. Lipkin and A. Rosenfeld, Eds.), pp. 347–370, Academic Press, New York, 1970.
29. R. M. Pickett, Visual analysis of texture in the detection and recognition of objects, in *Picture Processing and Psychopictorics* (B. S. Lipkin and A. Rosenfeld, Eds.), pp. 289–308, Academic Press, New York, 1970.
30. A. Rosenfeld and B. S. Lipkin, Texture synthesis, in *Picture Processing and Psychopictorics* (B. S. Lipkin and A. Rosenfeld, Eds.), pp. 309–322, Academic Press, New York, 1970.
31. L. A. Santalo, *Integral Geometry and Geometric Probability*, Addison-Wesley, Reading, Massachusetts, 1976.
32. B. Schachter and N. Ahuja, Random pattern generation processes, *Computer Graphics Image Processing* **10,** 1979, 95–114.
33. B. Schachter, A. Rosenfeld, and L. S. Davis, Random mosaic models for textures, *IEEE Trans. Systems, Man, Cybernet.* **SMC-8,** 1978, 694–702.
34. J. Serra and G. Verchery, Mathematical morphology applied to fibre composite materials, *Film Sci. Technol.* **6,** 1973, 141–158.
35. P. Switzer, Reconstructing Patterns from sample data, *Ann. Math. Stat.* **38,** 1967, 138–154.
36. P. Switzer, A random set process in the plane with a Markovian property, *Ann. Math. Stat.* **36,** 1965, 1859–1863.
37. S. Zucker, Toward a model of texture, *Computer Graphics Image Processing* **5,** 1976, 190–202.

Image Segmentation as an Estimation Problem *

P. C. CHEN† AND T. PAVLIDIS‡

Department of Electrical Engineering and Computer Science, Princeton University, Princeton, New Jersey 08540

Picture segmentation is expressed as a sequence of decision problems within the framework of a split-and-merge algorithm. First regions of an arbitrary initial segmentation are tested for uniformity and if not uniform they are subdivided into smaller regions, or set aside if their size is below a given threshold. Next regions classified as uniform are subject to a cluster analysis to identify similar types which are merged. At this point there exist reliable estimates of the parameters of the random field of each type of region and they are used to classify some of the remaining small regions. Any regions remaining after this step are considered part of a boundary ambiguity zone. The location of the boundary is estimated then by interpolation between the existing uniform regions. Experimental results on artificial pictures are also included.

1. INTRODUCTION

Signal processing in the time domain usually assumes that the processes under consideration are stationary. The extension of the classical methodology to picture processing is faced with the problem of a more prominent nonstationarity. Most pictures have well-defined regions of very distinct properties. Therefore, the problem of *image segmentation* has received considerable attention in the literature [1, 2]. Two types of methodologies are used widely in attempts to solve this problem. Edge detection searches for parts of the picture where a transition occurs from one "uniform" region to another. Region growing starts from small regions which are uniform and expands them as far as possible without violating their uniformity. Most of the literature dealing with these topics is centered on heuristic techniques and expressions of this problem in terms of estimation theory have been rather sketchy [2].

The goal of this paper is to present some preliminary results dealing with picture segmentation as an estimation problem. For simplicity we assume at first that the picture contains only two types of regions (e.g., two colors or two kinds of texture). A generalization is straightforward, except that the mathematical notation becomes more complicated.

* Research supported by NSF Grant ENG76-16808. An extended summary of this paper was presented at the IEEE Control and Decision Conference, Dec. 12–14, 1979.

† Present address: P. O. Box 2189, Exxon Production Research Company, Houston, Texas 77001.

‡ Present address: Bell Laboratories, Murray Hill, New Jersey 07974.

ISBN 0-12-597320-9

Let R be a connected region of the plane where the picture is defined. We have three possible hypotheses: H_0, the region is type I; H_1, the region is type II; H_2, the region has parts of both types. As a rule the larger the region area, the higher is the confidence level of decision regarding the three alternatives, but also the more likely is the occurrence of the third alternative. Edge detection can be seen as an estimation strategy where the size of the regions tested is very small so that if hypothesis H_2 is accepted a segment of the boundary between regions can be readily determined. Region growing in its simplest forms also starts with very small regions but more sophisticated versions use larger regions. One particular method, the split-and-merge algorithm [1], starts with large regions and if H_2 is true it subdivides them into smaller regions and tests them again. If H_0 (or H_1) is true on two adjacent regions these are merged. The method proceeds recursively and its implementation is described in detail elsewhere [1, 3]. However, as the size of the regions decreases, the confidence level of the decision drops and for this reason the method has been supplemented with heuristic criteria. This last process is commonly referred to as "small region elimination" and it is shared by all region-growing techniques [3–5].

Therefore, both region growing and edge detection are faced with the small region problem. One inherent disadvantage of edge detection is that it is always forced to make a decision. Even some of the more sophisticated detectors (e.g., [6]), which make a locally optimal decision, are not immune from this problem. For this reason various heuristic techniques have been suggested for postprocessing the results of edge detection [7–10].

We proceed now to analyze these decision processes and suggest certain solutions to the problem. In this paper we assume some rather simple statistical models for pictorial data which are more tractable for analysis. The use of more realistic distributions will be the subject of a future paper. We demonstrate the effectiveness of schemes based on estimation theory by providing examples of implementation on artificial pictures.

2. OVERALL STRATEGY

In general even though we know that only two types of regions exist, we do not know their statistical properties. Thus testing for the alternative hypotheses H_0, H_1, and H_2, one cannot assume prior knowledge of the parameters. However, after a number of "uniform" regions of the picture have been identified, these parameters may become known and may be used to facilitate further segmentation. In this sense, a region-growing-type strategy has the advantage over edge detection that it may estimate these parameters before approaching the neighborhood of the boundary.

Initially, we cannot distinguish between hypotheses H_0 and H_1 because we do not know the parameters at all. We can only test "uniformity" (hypotheses H_0 and H_1) against "nonuniformity" (hypothesis H_2). We divide the picture into a number of blocks. Each one of them is considered as a sample used to test the hypothesis. If the sample satisfies the "nonuniformity" hypothesis then it is divided into smaller samples. If two adjacent samples satisfy the "uniformity" test, then they are merged together into a single sample. At the end of this

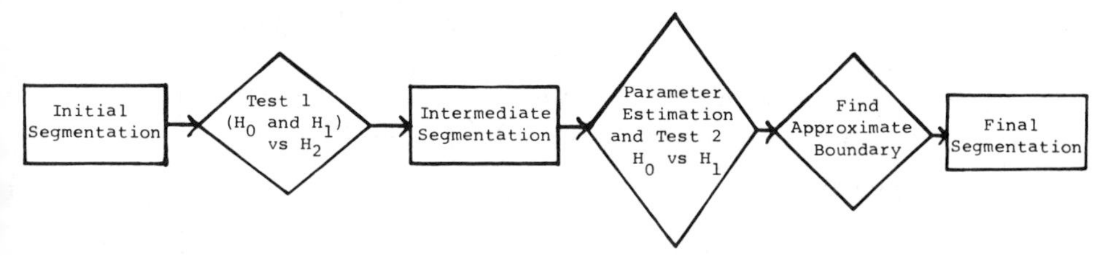

FIG. 1. Illustration of the overall strategy for segmentation.

process the whole picture is separated into many variable-size samples. Each sample corresponds to a "uniform" region. From the large samples, the statistical parameters characterizing the type I or type II regions can be estimated precisely. Then the estimated parameters are adopted to test the small-size samples adjacent to the large samples as to whether or not they are of type I or type II. If the test is satisfied, the large-size sample is enlarged by merging into it the small regions. During the test, the boundary is gradually approached.

The exact boundary cannot be reached because we cannot have accurate estimation with small-size samples. In the following, we show that there exists an intrinsic boundary region, separating the type I and type II regions. Then a curve-fitting method is applied to find an approximate boundary. The whole process is diagrammed as Fig. 1.

3. REGION TESTING

To implement the above test, we assume that the picture is a two-dimensional discrete random field which is a collection of random variables [11]. Each random variable, supposed to have a Gaussian distribution, denotes the brightness at the points of the picture. The random field is separated into two parts. In part 1, the random variables have mean m_1 and variance σ_1^2 and, in part 2, they have mean m_2 and variance σ_2^2. We say part 1 consists of type I regions and part 2 of type II regions. We also assume the random variables to be stochastically independent. In the following, we use the mean estimator and variance estimator to evaluate the statistical properties of the regions, and the split-and-merge operation for the post-testing actions.

A. Parameter Estimators

Let R_j be a connected region of type I (for $j = 1$) or type II (for $j = 2$) which is formed by the set of random variables $\{x_{ij}, i = 1, 2, \ldots, n\}$. The following estimators for R_j are used for hypothesis testing and parameter estimation.

Mean estimator:

$$\bar{x}_j = \sum_{i=1}^{n} x_{ij}/n, \qquad j = 1, 2. \tag{1}$$

Since $\bar{x}_j$ is a linear combination of Gaussian-distributed random variables, $\bar{x}_j$ is a Gaussian-distributed random variable. $\bar{x}_j$ has the mean $E\{\bar{x}_j\} = m_j$ and variance $E\{(\bar{x}_j - m_j)^2\} = \sigma_j^2/n$ [12], where n can be thought of as the region size of R_j.

In the case $m_1 \approx m_2$, the mean estimator is not able to discriminate the type I region from the type II and the variance estimator must be used.

Variance estimator:

$$\bar{\nu}_j = \sum_{i=1}^{n} (x_{ij} - m_j)^2/n, \qquad j = 1, 2. \tag{2}$$

Since x_{ij}, $i = 1, 2, \ldots, n$, are Gaussian-distributed, $n\bar{\nu}_j/\sigma_j^2$ has a chi-square distribution with n degrees of freedom. $\bar{\nu}_j$ has mean $E\{\bar{\nu}_j\} = \sigma_j^2$ and variance $E\{(\bar{\nu}_j - \sigma_j^2)^2\} = (2/n)\sigma_j^4$. When n is large, the central limit theorem implies that $\bar{\nu}_j$ can be approximated by a Gaussian-distributed random variable [12]. Notice that the variance of both estimators is inversely proportional to the region size n.

The merit of an estimator is obviously the confidence it can provide. Suppose we are given an interval $(m_j - \epsilon, m_j + \epsilon)$ where $\epsilon > 0$ and consider the mean estimator. We wish to know the probability with which the estimator $\bar{x}_j$ will fall in the interval. If the probability is high, we put strong confidence on $\bar{x}_j$. Let us define the confidence level, C_j, of the estimator $\bar{x}_j$ as

$$C_j = \text{prob}\,\{m_j - \epsilon \leqslant \bar{x}_j \leqslant m_j + \epsilon\}.$$

This can be rewritten as

$$C_j = \text{prob}\left\{\frac{-\epsilon n^{\frac{1}{2}}}{\sigma_j} \leqslant \frac{n^{\frac{1}{2}}(\bar{x}_j - m_j)}{\sigma_j} \leqslant \frac{\epsilon n^{\frac{1}{2}}}{\sigma_j}\right\}. \tag{3}$$

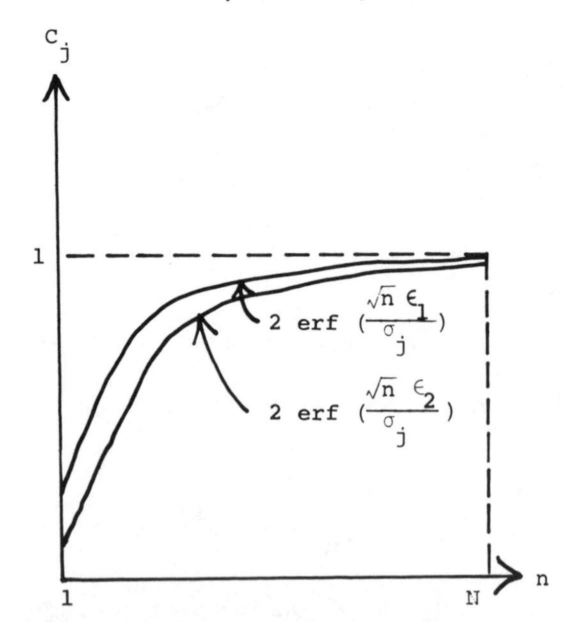

FIG. 2. Confidence level as function of size and threshold $\epsilon(\epsilon_1 > \epsilon_2)$.

Since $\bar{x}_j$ is Gaussian-distributed with mean m_j and variance σ_j^2, $n^{\frac{1}{2}}(\bar{x}_j - m_j)/\sigma_j$ has a Gaussian distribution with mean 0 and variance 1. C_j can be expressed in terms of the error function, erf (.):

$$C_j = 2 \operatorname{erf}\left(\frac{n^{\frac{1}{2}}\epsilon}{\sigma_j}\right) \tag{4}$$

For fixed σ_j, C_j is an increasing function of n and ϵ, diagrammed in Fig. 2.

For the variance estimator, the confidence level

$$\begin{aligned} C_j &= \text{prob}\,\{\sigma_j^2 - \epsilon \leqslant \bar{v}_j \leqslant \sigma_j^2 + \epsilon\} \\ &= \text{prob}\left\{n\left(1 - \frac{\epsilon}{\sigma_j^2}\right) \leqslant \frac{n\bar{v}_j}{\sigma_j^2} \leqslant n\left(1 + \frac{\epsilon}{\sigma_j^2}\right)\right\} \\ &= \int_{n(1-(\epsilon/\sigma_j^2))}^{n(1+(\epsilon/\sigma_j^2))} \frac{1}{\Gamma(n/2)2^{n/2}} x^{n/2-1}e^{-x/2}dx, \end{aligned} \tag{5}$$

where $\Gamma(.)$ is the Gamma function. For fixed σ_j, C_j is an increasing function of n and ϵ.

B. Segmentation

In this section, we describe the test of H_2 versus H_0 and H_1 (test 1) and then H_0 versus H_1 (test 2). The main purpose of the first test is to form the large uniform regions which can be the cores for the type I or type II regions. Before designing this test, we must explain under what conditions the region is called "uniform."

For example, if we take two samples x_1, x_2 which are from the same type of region and have the same probability distribution, their union should be called uniform. The mean for the two samples is $(x_1 + x_2)/2$. Consider the difference

$$x_2 - \frac{x_1 + x_2}{2}. \tag{6}$$

The mean for it is

$$E\left\{x_2 - \frac{x_1 + x_2}{2}\right\} = 0 \tag{7}$$

and the variance is

$$E\left\{\left(x_2 - \frac{x_1 + x_2}{2}\right)^2\right\} = \frac{\sigma_1^2}{2} \quad \left(\text{or } \frac{\sigma_2^2}{2}\right). \tag{8}$$

When two samples x_1 and x_2 are of different type, say x_1 is of type I and x_2 is of type II, their union should be called nonuniform. The mean for $x_2 - (x_1 + x_2)/2$ is

$$E\left\{x_2 - \frac{x_1 + x_2}{2}\right\} = \frac{m_2 - m_1}{2} \tag{9}$$

and the variance is

$$E\left\{\left(x_2 - \frac{x_1 + x_2}{2} - \frac{m_2 - m_1}{2}\right)^2\right\} = \frac{\sigma_1^2 + \sigma_2^2}{4}. \tag{10}$$

Suppose we are given a threshold $\epsilon > 0$. If the sample difference value $x_2 - (x_1 + x_2)/2$ falls in the interval $(-\epsilon, \epsilon)$, then we say that $\{x_1, x_2\}$ is uniform; otherwise, that it is nonuniform. The union of two samples of type I is considered as uniform with probability

$$p_1 = \int_{-\epsilon}^{\epsilon} \frac{1}{\pi^{\frac{1}{2}}\sigma_1} \exp\left(-\frac{x^2}{\sigma_1^2}\right) dx \tag{11}$$

and if two samples are of different types, the union is considered as uniform with probability

$$p_2 = \int_{-\epsilon}^{\epsilon} \frac{1}{(2\pi)^{\frac{1}{2}}(\sigma_1^2 + \sigma_2^2)^{\frac{1}{2}}/2} \exp\left\{-\left(\left(x - \frac{m_2 - m_1}{2}\right)^2 \Big/ \frac{\sigma_1^2 + \sigma_2^2}{4}\right)\right\} dx \tag{12}$$

If $m_2 - m_1$ is large and σ_1^2 and σ_2^2 are small, then $p_1 \gg p_2$. The same is true for $x_1 - (x_1 + x_2)/2$, since $x_1 - (x_1 + x_2)/2 = -(x_2 - (x_1 + x_2)/2)$. Hence we can detect the region's uniformity with very high confidence.

We implement test 1 with the split-and-merge algorithm on the pyramid data structure, as detailed in [1, 3]. This structure subdivides a picture in a sequence of successively finer square grids. A square of a given size is said to be uniform if the estimated mean for it differs by less than ϵ from the estimated mean for each one of the four subsquares into which is divided. Otherwise split it into the four subsquares. A possible error occurs when a square of a single type is split into four, but it can be corrected during test 2. However, when squares of mixed type are considered as uniform, then there is no error recovery. To avoid this, it is better to set a high threshold. When $m_1 \approx m_2$, $\sigma_1^2 \neq \sigma_2^2$, instead of the mean estimator, one should use the variance estimator.

Consider a square; the estimated mean for it is

$$\bar{x}_j = \sum_{i=1}^{4n} x_{ij}/4n, \qquad j = 1 \text{ or } 2, \tag{13}$$

where $\{x_{ij}, i = kn + 1, kn + 2, \ldots, kn + n\}$, $k = 0, 1, 2, 3$, are the sample points from the four subsquares. The random variables, Y_k, $k = 0, 1, 2, 3$, which denote the differences of estimated means for the square and the individual subsquares, are

$$Y_k = \bar{x}_j - \left(\sum_{i=kn+1}^{kn+n} x_{ij}/n\right), \qquad k = 0, 1, 2, 3, \tag{14}$$

$$E\{Y_k\} = 0, \qquad k = 0, 1, 2, 3, \tag{15}$$

$$E\{Y_k^2\} = \frac{3\sigma_j^2}{4n}, \tag{16}$$

$$E\{Y_h Y_k\} = -\frac{\sigma_j^2}{4n}, \qquad h \neq k, \quad h, k = 0, 1, 2, 3. \tag{17}$$

Since x_{ij}, $i = 1, 2, \ldots, 4n$ are Gaussian-distributed, $\{Y_k, k = 0, 1, 2\}$ has a multivariate Gaussian distribution with zero mean vector and covariance matrix V_n:

$$V_n = \begin{bmatrix} \frac{3\sigma_j^2}{4n} & \frac{-\sigma_j^2}{4n} & \frac{-\sigma_j^2}{4n} \\ \frac{-\sigma_j^2}{4n} & \frac{3\sigma_j^2}{4n} & \frac{-\sigma_j^2}{4n} \\ \frac{-\sigma_j^2}{4n} & \frac{-\sigma_j^2}{4n} & \frac{3\sigma_j^2}{4n} \end{bmatrix} = \frac{\sigma_j^2}{4n} \begin{bmatrix} 3 & -1 & -1 \\ -1 & 3 & -1 \\ -1 & -1 & 3 \end{bmatrix}. \tag{18}$$

Note that $Y_3 = -(Y_0 + Y_1 + Y_2)$ depends statistically on Y_0, Y_1, and Y_2. Let $Y = [y_0\ y_1\ y_2]$. Compute

$$|V_n|^{\frac{1}{2}} = \frac{1}{2} \frac{\sigma_j^3}{n^{\frac{3}{2}}}, \tag{19}$$

$$V_n^{-1} = \frac{n}{\sigma_j^2} \begin{bmatrix} 2 & 1 & 1 \\ 1 & 2 & 1 \\ 1 & 1 & 2 \end{bmatrix}. \tag{20}$$

Let $S = \{(y_0, y_1, y_2) \mid -\epsilon \leqslant y_0, y_1, y_2 \leqslant \epsilon \text{ and } -\epsilon \leqslant y_0 + y_1 + y_2 \leqslant \epsilon\}$. If a square is of a single type, then it will be considered as a uniform square with probability

$$p' = \iiint_S \frac{1}{(2\pi)^{\frac{3}{2}} |V_n|^{\frac{1}{2}}} \exp\{-\tfrac{1}{2} \bar{Y} V_n^{-1} \bar{Y}^T\} dy_0 dy_1 dy_2. \tag{21}$$

Each entry of the covariance matrix is inversely proportional to the square

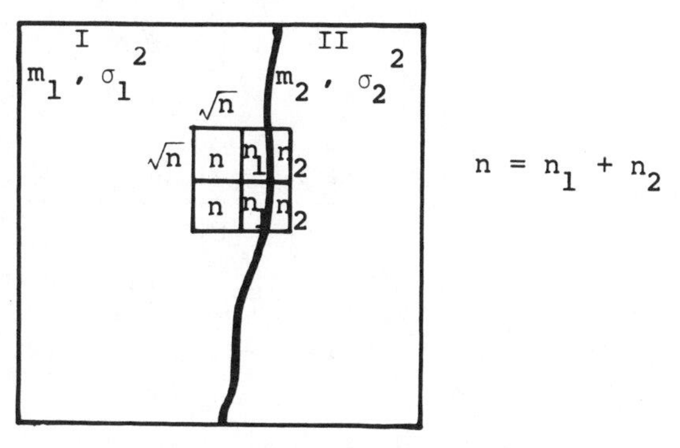

FIG. 3. Illustration of the quantities used for deriving Eq. (34).

size $4n$. p' will increase with n. Hence, for large n, we have high confidence of recognizing a uniform square.

Now let us consider a simple case which is a typical mixed type square shown in Fig. 3:

$$\bar{x} = \left(\sum_{i=1}^{2n+2n_1} x_{i1} + \sum_{i=1}^{2n_2} x_{i2} \right) / 4n. \tag{22}$$

Let

$$Y_0 = \bar{x} - \left(\sum_{i=1}^{n} x_{i1}/n \right), \tag{23}$$

$$Y_1 = \bar{x} - \left(\sum_{i=n+1}^{n+n_1} x_{i1} + \sum_{i=1}^{n_2} x_{i2} \right) / n, \tag{24}$$

$$Y_2 = \bar{x} - \left(\sum_{i=n+n_1+1}^{n+2n_1} x_{i1} + \sum_{i=n_2+1}^{2n_2} x_{i2} \right) / n, \tag{25}$$

$$Y_3 = \bar{x} - \left(\sum_{i=n+2n_1+1}^{2n+2n_1} x_{i1}/n \right). \tag{26}$$

Obviously, $Y_0 + Y_1 + Y_2 + Y_3 = 0$

$$E\{Y_k\} = \frac{2n_2(m_2 - m_1)}{4n}, \qquad k = 0, 3. \tag{27}$$

$$E\{Y_k\} = \frac{2n_2(m_1 - m_2)}{4n}, \qquad k = 1, 2. \tag{28}$$

Hence the mean vector, $\bar{m}'$, for $\{Y_k, k = 0, 1, 2\}$ is $(2n_2(m_2 - m_1)/4n)$ $[1 \ -1 \ -1]$ and Y_3 is statistically dependent upon Y_0, Y_1, and Y_2.

$$E\{(Y_k - E\{Y_k\})^2\} = \frac{12n\sigma_1^2 + 2n_2(\sigma_2^2 - \sigma_1^2)}{16n^2}, \qquad k = 0, 3, \tag{29}$$

$$= \frac{12n\sigma_1^2 + 10n_2(\sigma_2^2 - \sigma_1^2)}{16n^2}, \qquad k = 1, 2, \tag{30}$$

$$E\{(Y_h - E\{Y_k\})(Y_k - E\{Y_k\})\} = \frac{-4n\sigma_1^2 - 2n_2(\sigma_2^2 - \sigma_1^2)}{16n^2},$$

$$h = 0, 3, \quad k = 1, 2 \quad \text{or} \quad h = 1, 2, \quad k = 0, 3, \tag{31}$$

$$= \frac{-4n\sigma_1^2 + 2n_2(\sigma_2^2 - \sigma_1^2)}{16n^2},$$

$$h = 0, \quad k = 3 \quad \text{or} \quad h = 3, \quad k = 0, \tag{32}$$

$$= \frac{-4n\sigma_1^2 - 6n_2(\sigma_2^2 - \sigma_1^2)}{16n^2},$$

$$h = 1, \quad k = 2 \quad \text{or} \quad h = 2, \quad k = 1. \tag{33}$$

The covariance matrix for $\{Y_k, k = 0, 1, 2\}$ is

$$V'_n = \frac{\sigma_1^2}{4n}\begin{bmatrix} 3 & -1 & -1 \\ -1 & 3 & -1 \\ -1 & -1 & 3 \end{bmatrix} + \frac{n_2(\sigma_2^2 - \sigma_1^2)}{8n^2}\begin{bmatrix} 1 & -1 & -1 \\ -1 & 5 & -3 \\ -1 & -3 & 5 \end{bmatrix}. \tag{34}$$

To simplify the analysis, suppose $\sigma_1^2 \approx \sigma_2^2$. Hence $V'_n \approx V_n$. If we have a square such as shown in Fig. 3, it will be considered uniform with probability

$$p'' = \iiint_S \frac{1}{2\pi^{\frac{3}{2}}|V'_n|^{\frac{1}{2}}} \exp\{-\tfrac{1}{2}(\bar{Y} - \bar{m}')V'^{-1}_n(\bar{Y} - \bar{m}')^T\} dy_0 dy_1 dy_2. \tag{35}$$

Note that $n_2/n^{\frac{1}{2}}$ denotes the error quantity in Fig. 3. When $n_2/n^{\frac{1}{2}}$ gets smaller, $\bar{m}'$ is very close to zero, and p'' gets larger. That is, if n is fixed, for smaller n_2, we are more likely to have an error. If this error does happen for some n_2 and some $n = n'$ then the error will propagate along the merging sequence to the segmentation.

Besides the error mentioned above, there is another kind of error, shown in Fig. 4, from test 1. This error occurs because the square is located just in a position in which the error is evenly distributed among its four subsquares. The occurrences of this error are rare and it usually does not have the propagating property when the error is large.

After test 1, generally, large squares are formed in the areas of both types and small squares aggregate along the boundary area. We select the largest squares from the final result of test 1. Then the parameters are estimated; i.e., the mean and variance, for the largest squares. From these largest squares, we can find two clusters of parameters which characterize the type I and type II regions. From each cluster, one region is selected as the representative. If two largest

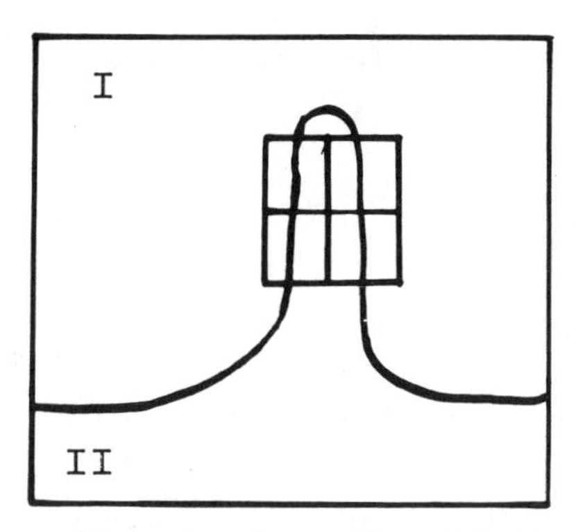

FIG. 4. Illustration of a possible merging error.

squares are adjacent and their parameters are in the same cluster, we consider them as a single region.

Now we have the parameters and proceed to implement test 2. In this test, we apply the Neyman–Pearson approach to the remaining small squares.

Assume the representative means and variances for both types are $\bar{x}_1$, $\bar{x}_2$, $\bar{\nu}_1^2$, and $\bar{\nu}_2^2$. Consider now a square of size q. Let $\{x_j, i = 1, 2, \ldots, q\}$ be the sample points in the square. If H_0 is true for the square, then the square should have the joint probability density function

$$f_0 = \left(\frac{1}{2\pi\bar{\nu}_1^2}\right)^{q/2} \exp\{-(\sum_{i=1}^{q} (x_i - \bar{x}_1)^2/2\bar{\nu}_1^2)\}. \tag{36}$$

If H_1 is true, the joint probability density function is

$$f_1 = \left(\frac{1}{2\pi\bar{\nu}_2^2}\right)^{q/2} \exp\{-(\sum_{i=1}^{q} (x_i - \bar{x}_2)^2/2\bar{\nu}_2^2)\}. \tag{37}$$

The likelihood ratio f_0/f_1 is used for the decision. Assume an error tolerance, α, is given which is a probability value, indicating the upper bound for misjoining.

We use the term "case A" when $\bar{x}_2 > \bar{x}_1$, $\bar{\nu}_1^2 \approx \bar{\nu}_2^2 = \bar{\nu}^2$. The likelihood ratio is

$$\frac{f_0}{f_1} = \exp\left\{(\sum_{i=1}^{q} (\bar{x}_1 - \bar{x}_2)\bar{x}_i/2\bar{\nu}^2) + \frac{q(\bar{x}_2^2 - \bar{x}_1^2)}{2\bar{\nu}^2}\right\}. \tag{38}$$

If $\bar{x}_1 \approx \bar{x}_2 = \bar{x}$, $\bar{\nu}_2^2 > \bar{\nu}_1^2$, then we shall refer to it as "case B."

$$\frac{f_0}{f_1} = \left(\frac{\bar{\nu}_2}{\bar{\nu}_1}\right)^q \exp\left\{\sum_{i=1}^{q} (x_i - \bar{x})^2 \frac{\bar{\nu}_1^2 - \bar{\nu}_2^2}{2\bar{\nu}_1^2\bar{\nu}_2^2}\right\}. \tag{39}$$

Suppose that if $f_0/f_1 \leqslant k$ $(k > 0)$ the hypothesis H_0 is rejected, and otherwise accepted. We simplify the expression for the likelihood ratio as follows:

For case A,

$$(\sum_{i=1}^{q} x_i/q) \geqslant k', \tag{40}$$

where

$$k' = \frac{q(\bar{x}_2^2 - \bar{x}_1^2) - 2\bar{\nu}^2 \ln k}{q(\bar{x}_2 - \bar{x}_1)} \tag{41}$$

is the decision threshold

If H_0 is true,

$$\sum_{i=1}^{q} x_i/q$$

has a Gaussian distribution with mean $\bar{x}_1$ and variance $\bar{\nu}^2/q$, and we can select k' in terms of the formula

$$\text{prob}\{(\sum_{i=1}^{q} x_i/q) \geqslant k'\} = \alpha. \tag{42}$$

If the estimated value

$$\left(\sum_{i=1}^{q} x_i/q\right) < k'$$

then the square would be considered as of type I with error probability less than α.

Since

$$\text{prob}\left\{\left(\sum_{i=1}^{q} x_i/q\right) \geqslant k'\right\} = \text{prob}\left\{\left(q^{\frac{1}{2}}\left(\sum_{i=1}^{q} \frac{x_i}{q} - \bar{x}_1\right)/\bar{\nu}\right) \geqslant \frac{k' - \bar{x}_1}{\bar{\nu}} q^{\frac{1}{2}}\right\}$$

and

$$q^{\frac{1}{2}}\left(\sum_{i=1}^{q} (x_i/q) - \bar{x}_1\right)/\bar{\nu}$$

has a Gaussian distribution with mean 0 and variance 1, for fixed α,

$$\frac{k' - \bar{x}_1}{\bar{\nu}} q^{\frac{1}{2}} \tag{43}$$

is constant. Let

$$k = \frac{k' - \bar{x}_1}{\bar{\nu}} q^{\frac{1}{2}} \qquad (k \text{ is constant}),$$

$$k' - \bar{x}_1 = \frac{k\bar{\nu}}{q^{\frac{1}{2}}}. \tag{44}$$

$k' - \bar{x}_1$ is the distance between the decision point and the estimated mean which is inversely proportional to $q^{\frac{1}{2}}$.

For case B, the likelihood ratio expression is simplified as

$$\frac{1}{q} \sum_{i=1}^{q} (x_i - \bar{x})^2 \geqslant k'', \tag{45}$$

where

$$k'' = \ln\left\{k\left(\frac{\bar{\nu}_1}{\bar{\nu}_2}\right)^q\right\} \frac{2\bar{\nu}_1^2\bar{\nu}_2^2}{q(\bar{\nu}_2^2 - \bar{\nu}_1^2)} \tag{46}$$

depends on k.

If H_0 is true,

$$\sum_{i=1}^{q} (x_i - \bar{x})^2/\bar{\nu}_1^2 \tag{47}$$

has a chi-square distribution with 1 degree of freedom. The decision point, k'', is chosen from the formula

$$\text{prob}\left\{\frac{1}{q} \sum_{i=1}^{q} (x_i - \bar{x})^2 \geqslant k''\right\} = \alpha. \tag{48}$$

If the estimated value

$$(1/q) \sum_{i=1}^{q} (x_i - \bar{x})^2$$

is less than k'', then the square is decided to be of type I with error probability less than α. From the chi-square distribution table, we know that $k'' - \bar{\nu}_1^2$ is decreasing with q. Hence we conclude that $k' - \bar{x}_1$ (or $k'' - \bar{\nu}_1^2$) for the test varies with the square size and should not be a constant as usual. For simplicity, we always assume $k' - \bar{x}_1$ (or $k'' - \bar{\nu}_1^2$) is a constant which is not greater than $(\bar{x}_2 - \bar{x}_1)/2$ (or $(\bar{\nu}_2^2 - \bar{\nu}_1^2)/2$).

If we force all small regions into one of the two types, then various small isolated regions will be formed along the boundary, because we misrecognize them with high probability. Actually we obtain only a rather poor estimate of the boundary. If we do not force all small regions to be classified, then the unclassified small regions form an area along the boundary. We shall show next that this is not necessarily an undesirable result.

4. EVALUATION OF INTRINSIC BOUNDARY AMBIGUITY

In this section, we show that regardless of the method used it is not possible to determine an exact boundary, or, in other words, that there exists an intrinsic ambiguity in its location. For simplicity we assume a configuration as shown in Fig. 5, and consider an estimation process using the little square (of course, we can use other shapes, the square is convenient for computation), containing n sample points.

A. *For the case* $m_1 \neq m_2$ and $\sigma_1^2 \neq \sigma_2^2$

The mean estimator is

$$\bar{x}_j = \sum_{i=1}^{n} x_{ij}/n, \qquad j = 1, 2. \tag{49}$$

The mean of $\bar{x}_j$ is m_j and the variance is σ_j^2/n.

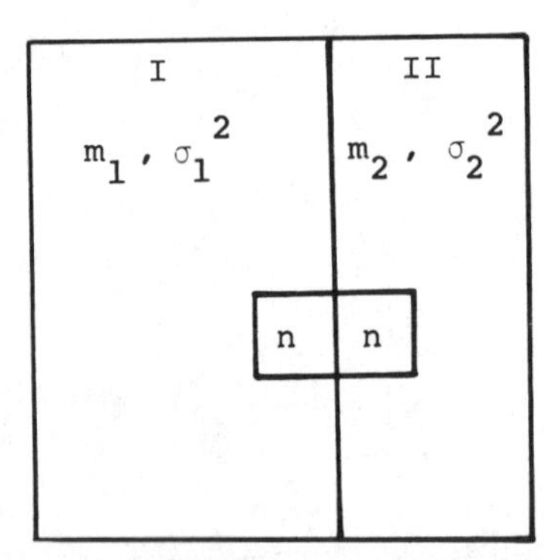

FIG. 5. Illustration of the quantities used for deriving Eqs. (49) and (50).

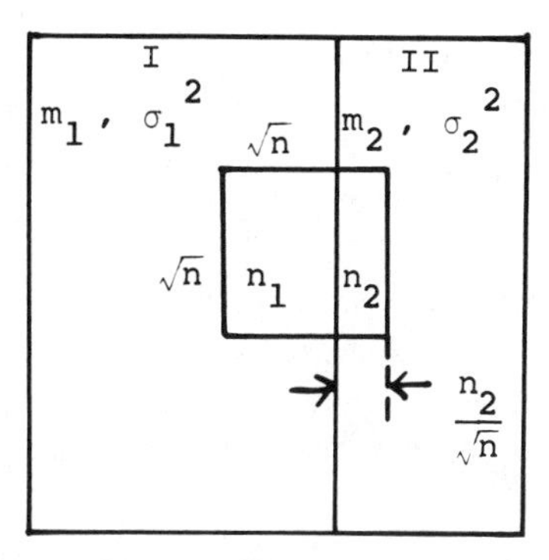

FIG. 6. Illustration of the quantities used for deriving Eqs. (51)–(60).

The probability, p, of the event $\{\bar{x}_1 \geqslant m_1 + s\sigma_1/n^{\frac{1}{2}}\}$

$$= \int_{(m_1+(s\sigma_1/n^{\frac{1}{2}}))}^{\infty} \frac{n^{\frac{1}{2}}}{2^{\frac{1}{2}}\pi\sigma_1} \exp\left\{-\frac{(x-m_1)^2}{2\sigma_1^2/n}\right\} dx \tag{50}$$

and the probability of the event $\{\bar{x}_2 \leqslant m_2 - (s\sigma_2/n^{\frac{1}{2}})\} = p$, where s is a positive number which is used to determine the confidence level.

Consider a square crossing the boundary line as shown in Fig. 6 and assume $m_2 > m_1$. In this square, there are n_1 points in region I and n_2 points in region II and $n = n_1 + n_2$.

The mean estimator in this case is

$$\bar{x} = (\sum_{i=1}^{n} x_{i1} + \sum_{i=1}^{n_2} x_{i2})/n, \tag{51}$$

$$E\{\bar{x}\} = \frac{n_1 m_1 + n_2 m_2}{n} = m_1 + \frac{n_2}{n}(m_2 - m_1), \tag{52}$$

$$E\{(\bar{x} - E\{\bar{x}\})^2\} = \frac{n_1\sigma_1^2 + n_2\sigma_2^2}{n^2}. \tag{53}$$

If the mean evaluated from the real data is equal to $m_1 + (s\sigma_1/n^{\frac{1}{2}})$, it is most likely to regard it as a square with n_1 points in region of type I and n_2 points in region of type II.

Hence

$$m_1 + \frac{s\sigma_1}{n^{\frac{1}{2}}} = m_1 + \frac{n_2}{n}(m_2 - m_1)$$

or

$$\frac{n_2}{n^{\frac{1}{2}}} = \frac{s\sigma_1}{m_2 - m_1}. \tag{54}$$

$n_2/n^{\frac{1}{2}}$ denotes the distance from the estimated boundary to the real boundary as shown in Fig. 6.

$$\frac{n_2}{n^{\frac{1}{2}}} \geqslant \frac{s\sigma_1}{m_2 - m_1} \text{ will happen with probability } p.$$

In a similar way, $n_1/n^{\frac{1}{2}} \geqslant s\sigma_2/(m_2 - m_1)$ will happen with probability p. Hence we find a region with width

$$\frac{n_1}{n^{\frac{1}{2}}} + \frac{n_2}{n^{\frac{1}{2}}} = \frac{s(\sigma_1 + \sigma_2)}{m_2 - m_1} \tag{55}$$

in which the boundary is located with probability p. With fixed s and p, for various m_1, m_2, σ_1^2, and σ_2^2, the region width is proportional to the sum of variances and inversely to the difference of the means. Therefore, the quantity $s(\sigma_1 + \sigma_2)/(m_2 - m_1)$ can be said to represent the intrinsic ambiguity of the boundary location.

B. For the case $m_1 \approx m_2$, $\sigma_1^2 \neq \sigma_2^2$

Consider Fig. 5 again. Assume $\sigma_1^2 < \sigma_2^2$, $m_1 \approx m_2 = m$. The probability distribution for $Y_{ij} = (x_{ij} - m)^2$ is

$$\begin{aligned} &\frac{1}{(2\pi y)^{\frac{1}{2}}\sigma_j} \exp\left\{-\frac{y}{2\sigma_j^2}\right\} && \text{for } y > 0 \\ &\text{and zero} && \text{for } y \leqslant 0 \end{aligned} \tag{56}$$

The variance estimator

$$\bar{\nu}_j = \sum_{i=1}^{n} Y_{ij}/n$$

approaches a Gaussian-distributed random variable when n is large. This is due to the central limit theorem.

With the same procedure as for case A, we can estimate the intrinsic ambiguity boundary region width. Consider Fig. 6 with $m_1 \approx m_2 = m$.

Variance estimator:

$$\bar{\nu} = \left(\sum_{i=1}^{n_1} (x_{i1} - m)^2 + \sum_{i=1}^{n_2} (x_{i2} - m)^2\right)/n, \tag{57}$$

$$E\{\bar{\nu}\} = \frac{1}{n}(n_1\sigma_1^2 + n_2\sigma_2^2) = \sigma_1^2 + \frac{n_2}{n}(\sigma_2^2 - \sigma_1^2), \tag{58}$$

$$E\{(\bar{\nu} - E\{\bar{\nu}\})^2\} = \frac{1}{n^2}(n_1\sigma_1^4 + n_2\sigma_2^4). \tag{59}$$

In the same way as for case A, equate

$$\sigma_1^2 + s\left(\frac{2}{n}\right)^{\frac{1}{2}}\sigma_1^2 = \sigma_1^2 + \frac{n_2}{n}(\sigma_2^2 - \sigma_1^2),$$

$$\frac{n_2}{n^{\frac{1}{2}}} = \frac{s2^{\frac{1}{2}}\sigma_1^2}{\sigma_2^2 - \sigma_1^2}. \tag{60}$$

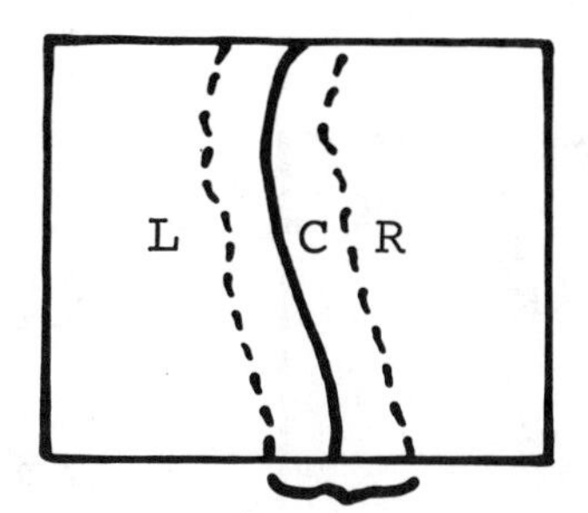

FIG. 7. Illustration of the ambiguous region and the central line.

Similarly, we can find

$$\frac{n_1}{n^{\frac{1}{2}}} = \frac{s2^{\frac{1}{2}}\sigma_2^{\,2}}{\sigma_2^{\,2} - \sigma_1^{\,2}}.$$

Hence, in case B, the intrinsic ambiguity boundary width is

$$\frac{s2^{\frac{1}{2}}(\sigma_1^{\,2} + \sigma_2^{\,2})}{\sigma_2^{\,2} - \sigma_1^{\,2}}. \tag{61}$$

5. APPROXIMATE BOUNDARY

We have seen that our method leaves an area of the picture which is unlabeled. (This also happens for other region growing schemes.) However, we are faced with an easier problem. During the initial process the exact values of the mean and variance are unknown and must be estimated at the same time as one decides whether a region is uniform. Now, however, the values are known so that the confidence level of the decision is higher for a given region size. One can also use certain geometric considerations.

In particular, we shall use the central line of the ambiguous region as an initial estimation for the boundary.

Consider the left limit of the region as the set L and right limit as the set R. The central line C is a set

$$C = \{(i, j) \mid d((i, j), L) = d((i, j), R)\}, \tag{62}$$

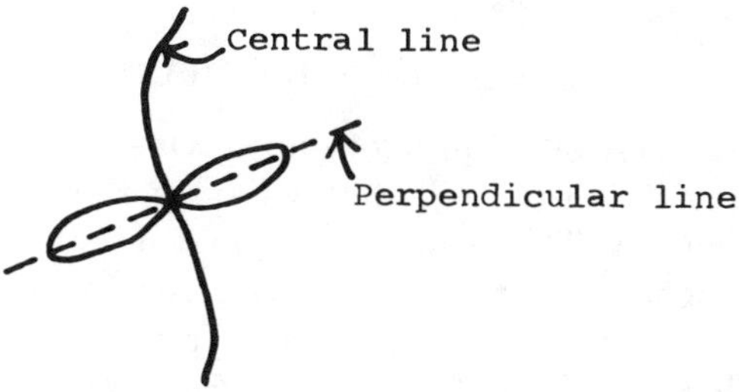

FIG. 8. Configuration for evaluating the mean (or variance) difference.

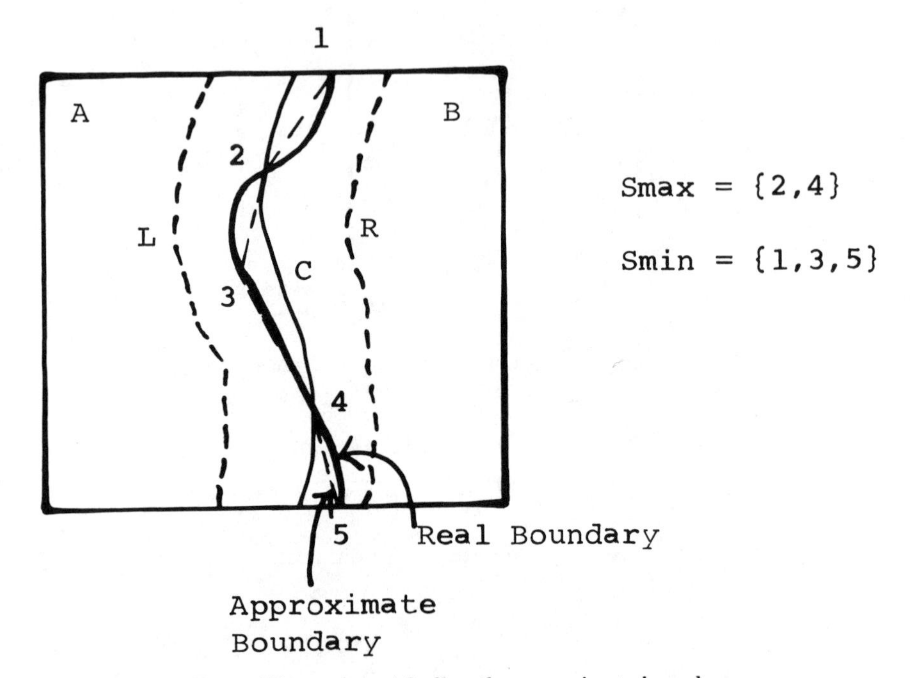

FIG. 9. Illustration of finding the approximate boundary.

where $d((i, j), L)$ or $d((i, j), R)$ is the minimum distance between (i, j) and the set L or R. (i, j) is a point in the picture region as in Fig. 7.

For each point on the central line, find a line perpendicular to the central line. From both sides of the central line on the perpendicular line, take a finite string of points as shown in Fig. 8, estimate their mean values and take the absolute value of their difference. Then we collect all the points, S_{max} (S_{min}), along the central line such that the absolute mean difference is locally maximum (minimum).

For each point in S_{min}, move it along its perpendicular line within the ambiguous region to find a point whose estimated absolute mean value difference is maximal along the perpendicular line. Connect this point with its two neighboring points in S_{max}. In this way, we construct an approximate boundary as in Fig. 9.

This last method is somewhat related to the approach of Nagao and Matsuyama [9] where they look for an edge along the direction of maximum variance around a pixel.

6. EXPERIMENTAL RESULTS

The above analysis leaves two questions open. One is how well the model fits real images. The second is how well the method performs, assuming that the statistical model is correct. The answer to the first question is still open because of the wide variety of natural images. Our next goal is to study more realistic models, which assume interdependence among pixels and other distributions besides Gaussian. However, in order to see whether such an effort is worthwhile one must answer the second question. For this purpose we generated a sequence

FIG. 10. Artificial image containing two random fields with different means.

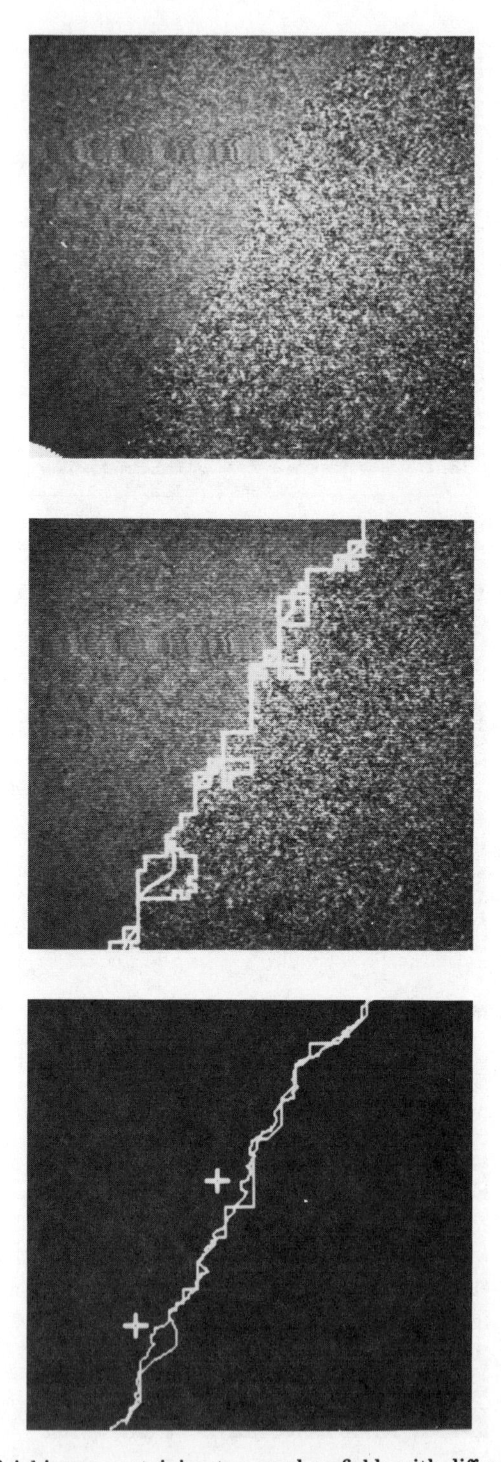

FIG. 11. Artificial image containing two random fields with different variances.

of images using the current model and then applied the algorithm to them. Figure 10a shows regions with different means (122 and 116) and the same variance 156, and Fig. 11a shows regions with the same mean (126) and different variances (59 and 285). The maximum range of values of the pixels displayed is 0–255 and each picture had 256 $\times$ 256 pixels. Figures 10b and 11b shows the unlabeled zones plus the "central" line used as the boundary whenever there was a nonzero width. The staircase appearance is due to the use of squares as test regions. Figures 10c and 11c show the estimated boundary overlayed on the original (exact) boundary. The +'s in Fig. 11c mark places of significant difference between the two. An obvious and easy improvement of the result is to interpolate a smooth curve (e.g., a spline) on the estimated boundary points.

The smallest region used in test 1 was 4 $\times$ 4, i.e., 16 pixels. The threshold ϵ was 2 for the means and 67 for the variance and (Eq. (5)) 0.48 for the different variance case. The initial segmentation used 16 $\times$ 16 squares where the confidence level is virtually 1.00 for the mean and 0.98 for the variance. The intrinsic boundary width according to Eq. (54) was about 8.

7. CONCLUSIONS

Hueckel [6] and Yakimovsky [2] have utilized local statistical properties to find the boundary line. Such properties are very sensitive to the local samples and the number of samples on the local pass, especially in textured pictures. As illustrated above, the boundary line is actually located within an ambiguous zone. Without global information, such methods lose the edge-tracing direction and create open cracks or many uncertain small regions. The traced boundary does not fit the real one very well.

In the split-and-merge algorithms, global information can be gradually aggregated from the merging operation, and local properties are utilized in the splitting operation (as in Section 2). The global information can direct the algorithm to proceed. It is the statistical property that causes the ambiguous boundary region. The algorithm reaches the ambiguous region, then uses the curve-fitting method to find the approximate boundary.

Our examples illustrate that if the picture fits the model then an estimation-theory-based method can find the boundary successfully, even where visual discrimination is very hard.

REFERENCES

1. S. L. Horowitz and T. Pavlidis, Picture segmentation by a tree traversal algorithm, *J. Assoc Comput. Mach.* **23**, Apr. 1976, 368–388.
2. Y. Yakimovsky, Boundary and object detection in real world images, *J. Assoc. Comput. Mach.* **23**, Oct. 1976, 599–618.
3. T. Pavlidis, *Structural Pattern Recognition.* Springer-Verlag, Berlin, Heidelberg, 1977.
4. R. A. Jarvis, Interactive image segmentation: Line, region, and semantic structure, in *Data Structures, Computer Graphics, and Pattern Recognition*, Academic Press, New York, 1977.
5. R. Ohlander, K. Price, and D. R. Reddy, Picture segmentation using a recursive region splitting method, *Computer Graphics Image Processing* **8**, 1978, 313–333.
6. M. H. Hueckel, A local visual operator which recognizes edges and lines, *J. Assoc. Comput.* Mach. **20**, Oct. 1973, 634–647.

7. A. K. Griffith, Edge detection in simple scenes using a priori information, *IEEE Trans. Computers* **C-22**, Apr. 1973, 371–381.
8. L. S. Davis, A survey of edge detection techniques, *Computer Graphics Image Processing* **4**, 1975, 248–270.
9. M. Nagao and T. Matsuyama, Edge preserving smoothing, in *Proceedings of Fourth International Joint Conference on Pattern Recognition*, Kyoto, Japan, pp. 518–520, No. 1978.
10. J. W. Modestino and R. W. Fries, Edge detection in noisy images using recursive digital filtering, *Computer Graphics Image Processing* **6**, 1977, 409–433.
11. E. Wong, Recursive causal filtering for two-dimensional random fields, *IEEE Trans. Inform. Theory* **IT-24**, Jan. 1978, 50–59.
12. C. H. Chen, *Statistical Pattern Recognition*. Spartan, New Jersey, 1973.

Toward a Structural Textural Analyzer Based on Statistical Methods *

RICHARD W. CONNERS AND CHARLES A. HARLOW

*Department of Electrical Engineering, Louisiana State University,
Baton Rouge, Louisiana 70803*

This paper reports investigations aimed at developing a feature set for the Spatial Gray Level Dependence Method (SGLDM) which measures visually perceivable qualities of textures. In particular it will be shown that the inertia measure commonly used with the SGLDM can be used to characterize the placement rules and the unit pattern of periodic textures. In this way one may formulate a structural approach to texture analysis based on the statistical SGLDM. To mathematically verify that the features used with the SGLDM can be used to characterize the unit pattern and placement rules of a periodic texture, a mathematical tiling theory model is proposed. This model allows one to develop the mathematical machinery necessary to prove the result. In a companion paper other features which measure visually perceivable qualities of patterns will be developed. The reason for concentrating on the SGLDM for developing such a feature set is predicated on perceptual psychology experiments and comparison studies of various texture algorithms. All of these studies indicate that second-order probabilities of the type measured by the spatial gray level dependence matrices are important in human texture discrimination and that these matrices contain more important texture-context information than the intermediate matrices of other statistical texture analysis algorithms.

1. INTRODUCTION

The aim of the work reported here is to develop a texture analysis algorithm which will model the primitive mechanisms of human texture perception. While this aim might seem to be the antithesis of current efforts by a number of investigators to model the underlying random fields which represent textures [1–6], the overall objective is basically the same, that is, to develop a set of features which are in some sense meaningful. "Meaningful" in the context of this work means that each feature in the feature set measures a visually perceivable quality of texture patterns. It is also desired that the resulting algorithm be a powerful one. This necessarily implies that the algorithm can be used on deterministic as well as stochastic textures. Hence, the choice was made to concentrate primarily on statistical analysis procedures.

Unfortunately, most if not all statistical texture analysis algorithms share a

* This research supported in part by AFOSR Contract F49620-79-C-0042 and NSF Grant ENG 79-5154.

ISBN 0-12-597320-9

common malady: they do not have features which correspond to any meaningful qualities of textural patterns. In particular, they do not have features which correspond to visual qualities of patterns. This, in turn, poses many problems for the image analysis practitioner trying to solve an applications problem. First, it forces him to place complete reliance on statistical measure selection and pattern recognition schemes. This reliance necessitates the use of large data bases, in order for the study to be meaningful. Secondly, even after the investigator has obtained the results of the application study, they are difficult, if not impossible, to interpret. One can never be sure precisely what properties of the textural patterns are being used to do the discrimination. This tends to make the results less palatable for the experts in applications area. Finally, the lack of visually meaningful features precludes the full utilization of the experience of highly trained experts in the applications area. These experts are used merely to classify the data base so that supervised learning procedures can be used to train the classifier. One cannot easily incorporate the experience of the experts in the very important and theoretically troublesome measurement selection process.

The basic premise of this research is that the Spatial Gray Level Dependence Method (SGLDM) is the most powerful statistical texture analysis algorithm. By this we mean that the intermediate matrices, namely, the spatial gray level dependence matrices, contain more important texture-context information than the intermediate matrices of any other texture analysis algorithm. Consequently, the research activities are directed toward finding a set of features defined on the spatial gray level dependence matrices which measure visual qualities of patterns. While it has not been conclusively proven that the SGLDM is the most powerful statistical texture analysis algorithm currently available there is at least strong supporting evidence. First, experiments on human texture perception indicate that second-order probabilities of the form measured by the spatial gray level dependence matrices play an important role in human texture discrimination [10–12]. And even though Julesz *et al.* [13–15] and Gagalowicz [16] have found some counterexamples to the well-known Julesz conjecture, both authors quickly point out that the preponderance of human texture discrimination still seems dependent on second-order probabilities. Secondly, the SGLDM has been used successfully on number of real-world texture analysis problems. References [9, 10, 18–20] describe just a few. This seemingly widespread utility must be considered at least somewhat indicative of the algorithm's power. Finally, two comparison studies, one done by Weszka *et al.* [21] and the other conducted by the authors [22], have both shown that the SGLDM is more powerful than the Power Spectral Method [23], the Gray Level Run Length Method [24], or the Gray Level Difference Method [21].

The results of this paper indicate that the Spatial Gray Level Dependence Method can be used as a basis for a structural analyzer for texture. That is, a feature defined on the spatial gray level dependence matrices is shown to characterize the unit patterns of repetitive textures and also to characterize the placement rules of these unit patterns.

An important characteristic of the structural analysis procedure based on the

SGLDM is that texture analysis begins to take on a hierarchical structure. In particular the methodologies employed indicate that texture analysis is at least a two-level process. The first level involves pure discrimination based on primitive features. The second and possibly higher levels involve the explanation of this discrimination. It is important to note that the hierarchical structure indicated was not a design guideline forced into the framework of the structural analyzer described here but rather resulted naturally as the system evolved. The hierarchy suggested seems reasonable based on human experience and seems to provide a certain esthetic appeal to the statistical procedures used.

Also during the development, a mathematical tiling theory model for texture will be developed. This model is similar to the Zucker model [25] but it is of utility in that it allows one to develop some needed mathematical machinery based on tiling theory.

The initial motivation for this work came from a desire to find features based on the SGLDM which would gauge two image qualities believed to be important in human perception. These are

1. Periodicity detection (Mach [26]) and the texture gradient (Gibson [27]).
2. Uniformity and proximity (Wertheimer [28])—also called connectivity detection (Julesz [10]).

For purposes of this study periodicity detection and texture gradient detection are put together as one perceptual entity. The reason behind this combination is that these two concepts are obviously very closely related.

While developing features which would measure these two perceptual quantities was the initial motivation, the results in some sense transcend these objectives. That is, in performing the analysis it became apparent that the SGLDM could be used to develop a statistically based structural analyzer. Also the hierarchy mentioned previously became apparent.

It should also be mentioned that the concept of developing a texture algorithm which measures visually perceivable characteristics of patterns is not new. Tamura *et al.* reported investigations toward these ends in [27]. However, the difference between this approach and Tamura's is that we are attempting to define features based on the SGLDM. Tamura, on the other hand, did not constrain his features to the SGLDM.

This paper concentrates on showing that the inertia measure can be used to detect the periodicity of textural patterns and determine the texture gradient. A companion paper will investigate features defined on the SGLDM which will measure the concepts of uniformity and proximity.

We begin the analysis with some background information.

2. BACKGROUND

A. A Brief Description of the SGLDM

Computationally the heart of the Spatial Gray Level Dependence Method [8, 9] is the spatial gray level dependence matrices, $S_\theta(d) = [s_\theta(i, j|d)]$. An element $s_\theta(i, j|d)$ of the matrix $S_\theta(d)$ represents the estimated probability of going from gray level i to gray level j given that the intersample spacing distance is d and the angular direction is θ.

In this analysis we do not force the matrix $S_\theta(d)$ to be symmetric as is commonly done by many investigators. The reason for not forcing the spatial gray level dependence matrices to be symmetric is discussed in [22]. Consequently, for example, $S_0(d)$ may not equal $S_{180}(d)$.

Classically, from each of $S_\theta(d)$ matrices, five features have been computed. These are

1. Energy

$$\mathcal{E}(S_\theta(d)) = \sum_{i=0}^{N_G-1} \sum_{i=0}^{N_G-1} [s_\theta(i, j|d)]^2; \tag{1}$$

2. Entropy

$$H(S_\theta(d)) = -\sum_{i=0}^{N_G-1} \sum_{j=0}^{N_G-1} s_\theta(i, j|d) \log s_\theta(i, j|d); \tag{2}$$

3. Correlation

$$C(S_\theta(d)) = \sum_{i=0}^{N_G-1} \sum_{j=0}^{N_G-1} (i - \mu_x)(j - \mu_y) s_\theta(i, j|d)/\sigma_x \sigma_y; \tag{3}$$

4. Local homogeneity

$$L(S_\theta(d)) = \sum_{i=0}^{N_G-1} \sum_{j=0}^{N_G-1} \frac{1}{1 + (i - j)^2} s_\theta(i, j|d); \tag{4}$$

5. Inertia

$$I(S_\theta(d)) = \sum_{i=0}^{N_G-1} \sum_{j=0}^{N_G-1} (i - j)^2 s_\theta(i, j|d); \tag{5}$$

where N_G is the number of gray levels in the picture from which the spatial gray level dependence matrix $S_\theta(d)$ was computed and where

$$\mu_x = \sum_{i=0}^{N_G-1} i \sum_{j=0}^{N_G-1} s_\theta(i, j|d),$$

$$\mu_y = \sum_{j=0}^{N_G-1} j \sum_{i=0}^{N_G-1} s_\theta(i, j|d),$$

$$\sigma_x^2 = \sum_{i=0}^{N_G-1} (i - \mu_x)^2 \sum_{j=0}^{N_G-1} s_\theta(i, j|d),$$

and

$$\sigma_y^2 = \sum_{j=0}^{N_G-1} (j - \mu_y)^2 \sum_{i=0}^{N_G-1} s_\theta(i, j|d).$$

An observation should be made with regard to the features defined in Eqs. (1)–(5). It was shown in the comparison study in [22] that these five features do not contain all the important texture-context information present in the

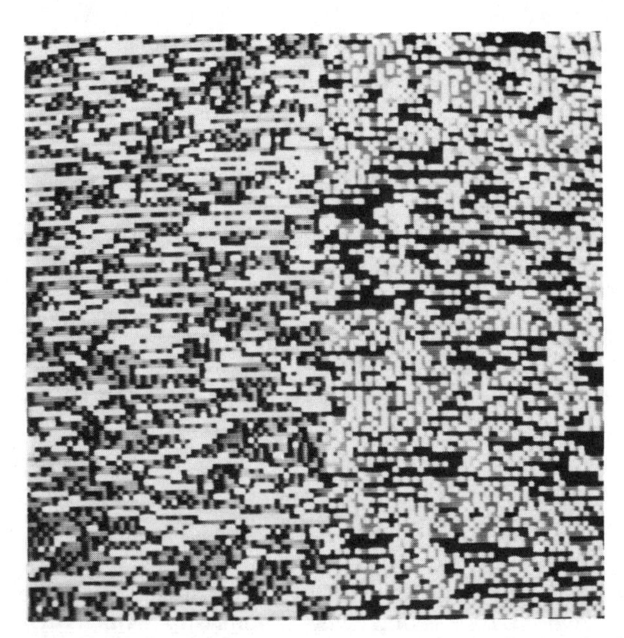

FIG. 1. Two visually distinct textures which cannot be discriminated by the five features commonly used with the SGLDM but which can be discriminated by information contained in the spatial gray level dependence matrices.

matrix $S_\theta(d)$. This statement is based on the fact that there exists a pair of visually distinct textures which have identical feature values for the above features but which have different corresponding spatial gray level dependence matrices. In particular, $S_0^{(1)}(d) \neq S_0^{(2)}(d)$ for $d = 1, 2, \ldots,$ where $S_0^{(1)}(\cdot)$ are the matrices computed from one texture and $S_0^{(2)}(\cdot)$ are the matrices computed from the other texture in the texture pair. Figure 1 shows one such visually distinct texture pair.

It was precisely this observation that motivated the search for a new feature set for the SGLDM.

B. *Structural Approaches to Texture Analysis*

Structural approaches to texture analysis are based on the concepts of unit patterns and well-defined placement rules. The basic steps associated with the implementation of these approaches are:

1. locate the unit patterns of the texture;
2. extract features which characterize the unit patterns;
3. extract features which characterize the placement rules of the unit patterns.

While structural approaches have an esthetic appeal because they correspond to our intuitive idea of what comprises a texture, there are many problems associated with the implementation of these approaches. First there is the problem of locating the unit patterns. This is complicated by the fact that the unit patterns may be composed of subpatterns, and these subpatterns may in turn be composed of still smaller subpatterns, etc. A further complication is the selection of the features that will be used to characterize the unit patterns and the placement rules. In this respect structural approaches share a common problem with statistical approaches. That is, the selection of the actual features used in the classification are largely heuristic.

Because of the complications associated with the implementation of structural analyzers it seems appropriate to briefly review the literature in an effort to determine how various investigators have approached the above mentioned problems.

First consider the structural analyzer proposed by Tsuji and Tomita [30]. In the formulation of this analyzer the problem of finding the unit patterns, or as they are called in [30] atomic regions, is largely ignored in that it was assumed the unit patterns could be found using a simple thresholding method. The measures used to characterize the unit patterns were a series of i, jth order central moments $M_{i,j}$. The specific moments used were M_{20}, M_{02}, M_{21}, and M_{11}. The placement rules for the various unit patterns were in part measured by a density descriptor. To define the density descriptor, one must first define the distance between atomic regions in a group. This distance is defined to be the distance between their centers of gravity. Then, an atomic region A, in a given set S, has a density descriptor D_S, whose value is the minimum distance from A to other atomic regions in S. Because of the simple procedure used to find the unit patterns and the single measure used to define the placement rules it would appear that this analyzer is very limited as to the class of textures it can handle.

Next consider the structural analyzer of Zucker *et al.* [31]. The purpose of this analyzer is to generalize the approach taken by Tsuji and Tomita [30] to natural textures. This analyzer considered only various size "spot" unit patterns and found them using a spot detector. Since only various size spot unit patterns were considered, no measures were extracted to characterize these unit patterns. Rather, only the number of "spots" of the various sizes found present in the texture were recorded. This analyzer had no mechanism for characterizing the placement rules of the "spots." The limitation of the unit patterns to spots and the fact that the placement rules were not considered would seem to limit the generality of the approach.

Another structural analyzer which has been proposed is the one of Ehrich and Foith [32, 33]. The unit patterns considered by this analyzer were regions centered about a local maximum which is bounded on all sides by local minima. These fundamental patterns are referred to as "peaks." The measures used to characterize the peaks are absolute peak height, relative peak height, and the area of the region where the peak is located. This region is assumed to be circular. This analyzer, again, does not characterize the placement rules of the unit patterns. This method also seems somewhat limited because there is no direct measurement of the placement rules of the unit patterns and there is no indication that the peak unit patterns correspond to perceived properties of texture.

The final structural analyzer to be discussed is the one of Lu and Fu [34, 35]. The unit patterns of this analyzer consist of square windowed regions of the texture. The selection of the size of the windowed regions to be used is not computed but rather is assumed to be known. A tree grammar is used to characterize the resulting unit patterns and another tree grammar is used to characterize the placement rules. A difficulty with this analyzer is that its computational complexity is heavily dependent on the window size used and, as yet, the authors have not suggested a mechanism for computing an optimal size window. The application of the method to texture patterns which have several gray levels remains to be demonstrated. Also there is a possible difficulty in determining the grammar rules in the actual analysis process.

3. A MODEL FOR TEXTURE BASED ON MATHEMATICAL TILING

In this section a tiling theory model for texture will be introduced. The reason for formulating this model is that it provides a mathematical formalism for the problem of locating the unit patterns and defining the placement rules for repetitive textural patterns. In particular it allows one to use an important theorem from mathematical tiling theory to attack the problem of defining the unit patterns of periodic textures.

Before presenting the model some important definitions and theorems form mathematical tiling theory will be given.

A. Mathematical Tiling Theory: Basic Concepts

The definitions and theorems in this subsection were taken from Grünbaum and Shephard's forthcoming book on tiling theory [36].

DEFINITION 1. A *plane tiling* is a family of closed topological disks, $\mathcal{T} = \{T_i | i = 1, 2, \ldots\}$ which covers the Euclidean plane without holes or overlaps.

The requirement that each tile T_i be a closed topological disk guarantees that T_i is closed, bounded, connected, and simply connected. In other words, the boundary of each T_i, i.e., $\dot{T}_i = \mathrm{Int}(T_i)$ where $\mathrm{Int}(T_i)$ is the interior of T_i, is a simple closed curve. That is a curve whose end points join up to form a "loop" which has no crossings or branches. The requirement that the tiling $\mathcal{T}$ covers the plane without gaps or overlaps means that the union of the sets T_i is E^2 the whole Euclidean plane and the interiors of the sets T_i are pairwise disjoint.

From Definition 1 it follows that the intersection of any finite set of two or more distinct tiles of $\mathcal{T}$ has area zero. Such an intersection may be empty or consist of a finite set of isolated points and arcs. This leads us to the following definitions.

DEFINITION 2. Let $\mathcal{V}$ be a finite intersection of two or more distinct tiles of $\mathcal{T}$. If $\mathcal{V}$ is a finite set of isolated points these points are called *vertices* of the tiling.

DEFINITION 3. Let $\mathcal{E}$ be a finite intersection of two or more distinct tiles of $\mathcal{T}$. If $\mathcal{E}$ is a finite set of isolated arcs these arcs are referred to as *edges* of the tiling.

It can be easily shown that an edge connects two vertices. These vertices are called the endpoints of the edge. Similarly it can be shown that every vertex is the endpoint of an edge.

DEFINITION 4. A function $\sigma: E^2 \to E^2$ is called an *isometry* or *congruence transformation* if it maps the Euclidean plane onto itself and if the function preserves distance. That is, if $\bar{x}$ and $\bar{y}$ are points in E^2 then $\|\bar{x} - \bar{y}\| = \|\sigma(\bar{x}) - \sigma(\bar{y})\|$.

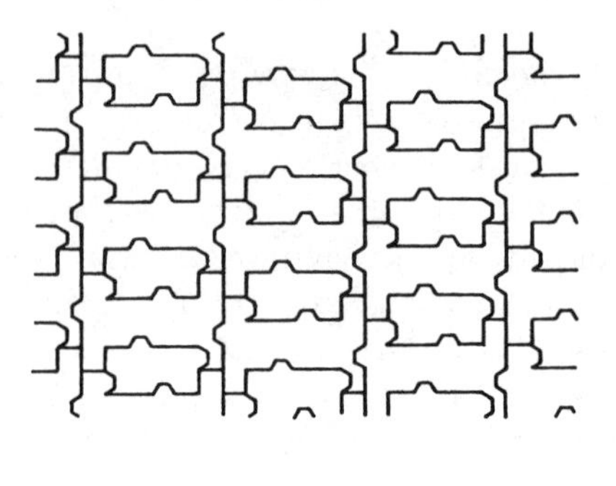

FIG. 2. Two ways the plane can be tiled using the same set of prototiles.

THEOREM 1. *Every isometry is of one of four types:*

(*i*) *Rotation about a point O through a given angle θ.*
(*ii*) *Translation in a given direction through a given distance.*
(*iii*) *Reflection about a given line L called the line of reflection.*
(*iv*) *Glide Reflection in which reflection about a line L is followed by a translation through a given distance D parallel to L.*

Isometries of types (i) and (ii) are usually called *direct* because if the points ABC form the vertices of a triangle named in a clockwise direction, then the same is true of the images under an isometry of either type (i) or type (ii). Isometries of types (iii) and (iv) are called *indirect* or *reflective* because the images of the points ABC will form a triangle named in a counterclockwise direction.

DEFINITION 5. Two tilings $\mathcal{T}_1 = \{T_i^1 | i = 1, 2, \ldots\}$ and $\mathcal{T}_2 = \{T_i^2 | i = 1, 2, \ldots\}$ are said to be equal if there exists an isometry σ and a positive real number α such that for each i $\alpha\sigma(T_i^1) = T_j^2$ for some j.

Definition 6. Let $\mathcal{S}$ be a set of tiles. If there exists a tiling $\mathcal{T}$ such that every tile in $\mathcal{T}$ is congruent either directly or reflectively to a tile in $\mathcal{S}$ then $\mathcal{S}$ is called the *set of prototiles* of $\mathcal{T}$ and each tile in $\mathcal{S}$ is called a *prototile*. In such cases it is said that $\mathcal{S}$ *admits* the tiling $\mathcal{T}$.

Two points need to be made about prototiles. First, not every set $\mathcal{S}$ of tiles will admit a tiling of the plane. For example if the set $\mathcal{S}$ consist of just one tile, a regular octagon, then no tiling of the plane is possible. Secondly, there are instances when a set of prototiles will admit more than one tiling of the plane. Figure 2 shows two ways a plane can be tiled using the same set of prototiles.

DEFINITION 7. A tiling $\mathcal{T} = \{T_i | i = 1, 2, \ldots\}$ is a monohedral tiling if there exists a single prototile which admits $\mathcal{T}$.

Initial investigations in tiling theory concerned the examination of prototiles which were polygons. To limit the number of tiling possibilities a special, very natural tiling was defined, the so-called edge-to-edge tiling.

DEFINITION 8. An edge-to-edge tiling of polygons is one in which the intersection of any two distinct tiles is

(i) equal to the empty set ϕ;

(ii) contains precisely one point which is a vertex of all the polygons of which it is an element; or

(iii) is an edge of both the tiles being intersected.

Figure 3 shows an edge-to-edge tiling by congruent parallelograms. Such a tiling is possible for any parallelogram $\mathcal{P}$ and will be called the *parallelogram tiling* for $\mathcal{P}$.

DEFINITION 9. An isometry σ is a *symmetry* of the tiling $\mathcal{T}$ if it maps every tile of $\mathcal{T}$ onto a tile of $\mathcal{T}$.

For any tiling $\mathcal{T}$, let $\mathcal{S}(\mathcal{T})$ be the group of symmetries of $\mathcal{T}$. It is possible for $\mathcal{S}(\mathcal{T})$ to consist of only the identity isometry.

DEFINITION 10. If a tiling $\mathcal{T}$ admits any symmetry in addition to the identity symmetry then $\mathcal{T}$ is called *symmetric*.

DEFINITION 11. If the symmetry group of a tiling $\mathcal{T}$ contains at least two translations in nonparallel directions then $\mathcal{T}$ is said to be *periodic*.

Let the nonparallel translation be represented by the vectors $\bar{a}$ and $\bar{b}$. Then $\mathcal{S}(\mathcal{T})$ contains all the translations $n\bar{a} + m\bar{b}$ where n and m are integers. Starting from any fixed point $\mathcal{O}$ the set of images of $\mathcal{O}$ under the set of translations $\{n\bar{a} + m\bar{b} | m$ and n integers$\}$ form a lattice. A lattice may be thought of as consisting of the vertices of a parallelogram tiling.

This leads one to the following very important theorem.

THEOREM 2. *If $\mathcal{T}$ is a periodic tiling then $\mathcal{T}$ has an associated lattice and this lattice may be regarded as the vertices of a parallelogram tiling $\mathcal{T}_{\mathcal{P}}$. Further, if the "pattern" formed by the tiles of $\mathcal{T}$, i.e., the edges and vertices, that are incident with one of the parallelograms of $\mathcal{T}_{\mathcal{P}}$ are known, then the rest of $\mathcal{T}$ can be constructed by repeating this pattern in every parallelogram of $\mathcal{T}_{\mathcal{P}}$.*

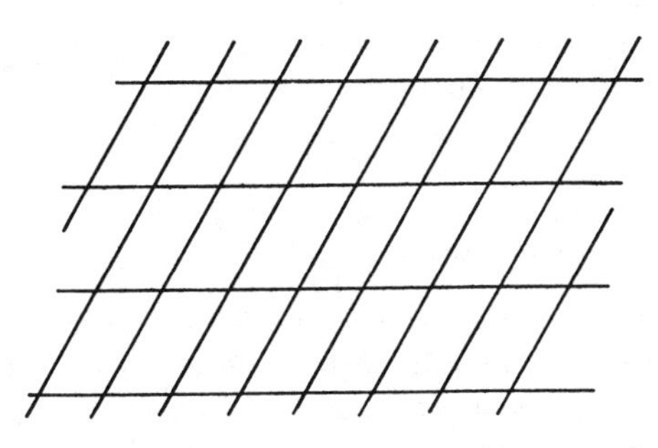

FIG. 3. The parallelogram tiling of the parallelogram prototile $\mathcal{P}$.

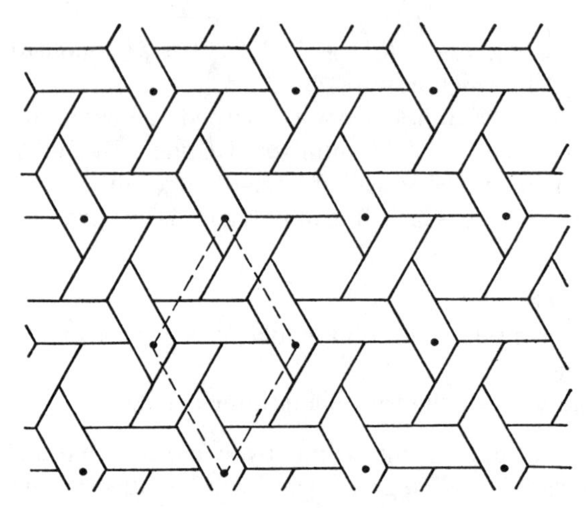

FIG. 4. A period parallelogram of a periodic tiling.

DEFINITION 12. The tiles of the parallelogram tiling $\mathcal{T}_{\mathcal{P}}$ are known as *period parallelograms.*

A typical periodic tiling $\mathcal{T}$ is shown in Fig. 4. In this figure the edges of tiles of the tiling $\mathcal{T}$ are shown as black lines; a set of lattice points of this periodic tiling are shown as black dots; and a period parallelogram prototile $\mathcal{P}$ of the tiling $\mathcal{T}$ is indicated by the dashed lines. The importance of the Theorem 2 stems from the fact that if one were to replicate the pattern formed by the edges and vertices of the tiles of $\mathcal{T}$ incident with $\mathcal{P}$ on every tile of the parallelogram tiling $\mathcal{T}_{\mathcal{P}}$ defined by $\mathcal{P}$, then the pattern generated would be the same as the pattern created by the tiles of $\mathcal{T}$. This statement should immediately suggest that Theorem 2 could be of use in doing structural analysis. In the next sections the utility of Theorem 2 will be fully developed.

Other examples of period parallelograms of periodic tilings are shown in Fig. 5.

B. A Tiling Model for Texture

The purpose of this model is to provide a theoretical basis for the statistical structure analyzer developed here. However, the model is within itself of some interest because it allows one to formalize some basic intuitive concepts and in turn precisely state some fundamental problems of analysis.

The model described is similar to the Zucker model for texture [25] and, in part, it was even suggested by Zucker in his formulation. This model has the same basic components as the Zucker model. These are:

(1) primitives,
(2) ideal textures,
(3) observable surface textures.

The interrelationships of these basic components are also the same. These interrelationships are shown in Fig. 6. However, here the similarity ends. This model is devoid of the perceptual parameters Zucker incorporated in his model. The idea was to create a purely mathematical model.

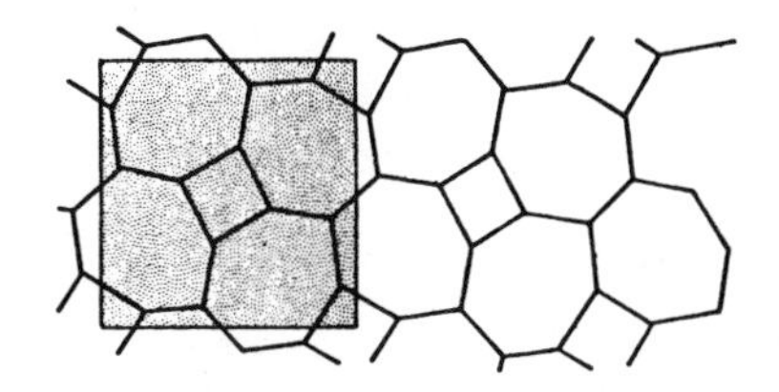

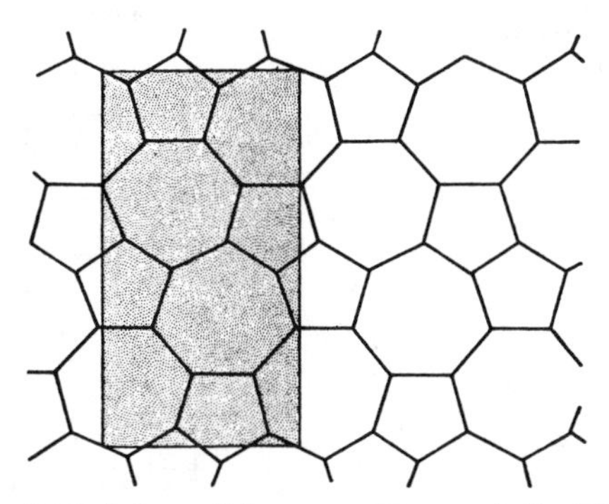

FIG. 5. Period parallelograms of two periodic tilings.

1. Primitives

DEFINITION 13. A primitive is a set $\{T, f\}$ where T is a tile and f is a function which maps the points of T into a bounded subset of the nonnegative real numbers.

One may think of f as a painting function which allows one to put a pattern on the tile T.

As was pointed out in the comments immediately following Definition 6, not every prototile T or set of prototiles will admit a tiling of the plane. Consequently, the following definition is needed.

DEFINITION 14. A set $\mathcal{S}_p = \{\{T_1, f_1\}, \{T_2, f_2\}, \ldots, \{T_n, f_n\}, \ldots\}$ is an admissible set of primitives if the set $\{T_1, T_2, \ldots, T_n, \ldots\}$ of prototiles admits at least one tiling of the plane.

One can now formally state a definition for the intuitive concept of a unit pattern.

DEFINITION 15. If $\mathcal{S}_p$ is an admissible set of primitives then an element $\{T_i, f_i\}$ of $\mathcal{S}_p$ is called a unit pattern.

2. Ideal Texture

DEFINITION 16. A placement rule of a tile T is an isometry which maps T onto E^2.

Given a set of prototiles $\mathcal{S}_p$ not every set of placement rules can be used to tile in the plane using $\mathcal{S}_p$. This observation necessitates the following definition.

DEFINITION 17. Given a set $\mathcal{S}$ of tiles, a set $\mathcal{R}$ is called an admissible set of placement rules for $\mathcal{S}$ if the isometries of $\mathcal{R}$ can be used to tile in the plane using all the tiles of $\mathcal{S}$.

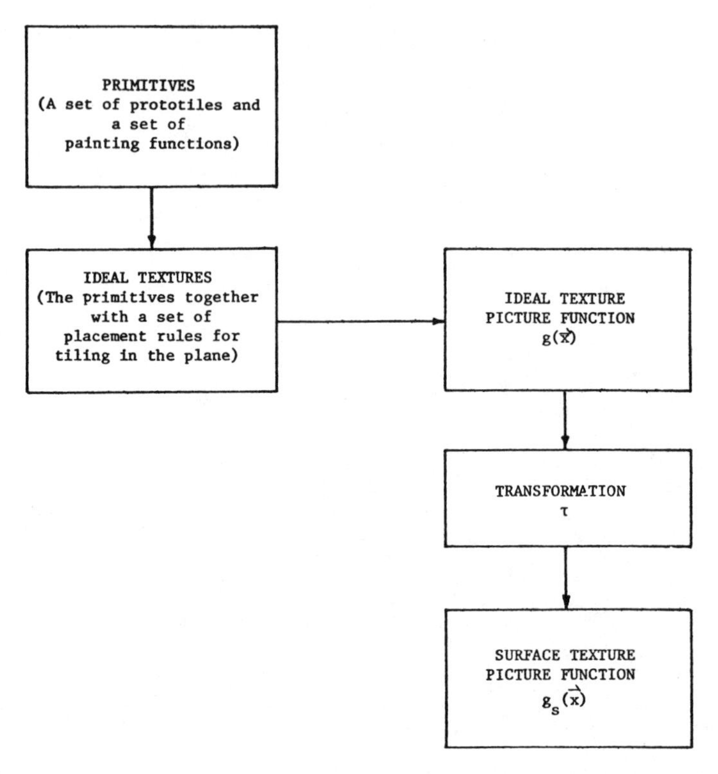

FIG. 6. Major components of the texture model and their interrelationships.

It should be noted that a placement rule of $\mathcal{R}$ must be indexed as to the tile or tiles of $\mathcal{S}$ to which it may be applied.

DEFINITION 18. An ideal texture is a set $\mathcal{I}_T = \{\mathcal{S}_p, \mathcal{R}_p\}$, where $\mathcal{S}_p$ is an admissible set of primitives and $\mathcal{R}_p$ is an admissible set of placement rules.

An ideal texture defines a picture function $g(\bar{x})$ where $\bar{x} = (x_1, x_2)$. For purposes of this presentation we assume that a texture $g(\bar{x})$ covers the plane.

3. *Transforms and Textures*

And ideal texture, or rather the picture function $g(\bar{x})$ that it represents is, transformed into a real-world observable surface texture by a transform τ.

For purposes of this discussion information about the particular forms of τ is of little use. Consequently all that will be stated is that τ embodies such transforms as

(1) projective transforms,
(2) degradations such as the point-spread distortion of an imaging system,
(3) stretching and twisting of the underlying fabric,
(4) noise.

4. *Observations about the Tiling Model*

Two observations are immediately apparent regarding the tiling model for texture. The first observation, which is only of peripheral interest with regard to

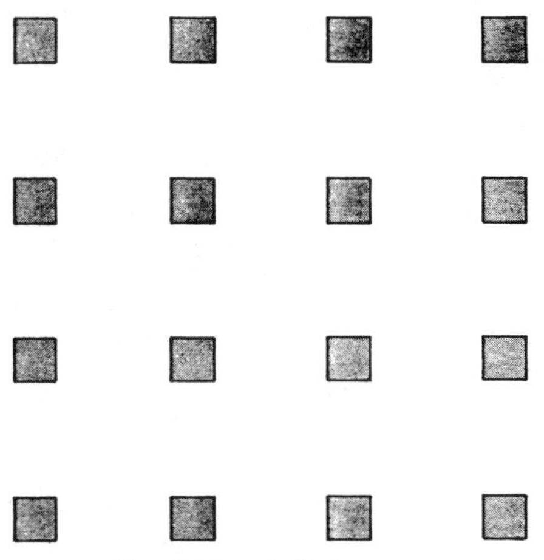

FIG. 7. A periodic texture.

the argument that is to follow, is that given an admissible set of primitives there may be several textures which can be generated from these primitives. This point emphasizes the importance of the placement rules and makes one doubt the power of structural analyzers which do not characterize the placement rules. Figure 2 illustrates this observation by showing two tilings which are obtainable from the same set of prototiles. The extension to textures generated from unit patterns is straightforward.

The second and most important observation is that given a particular surface texture $g_s(\bar{x})$ there are many possible admissible sets of primitives which can be used to generate it. That is, given $g_s(\bar{x})$ there is no unique ideal texture representation for $g_s(\bar{x})$. Consider, for example, the texture shown in Fig. 7. Figure 8 shows a number of unit patterns which can be used to generate the texture of Fig. 7.

C. Structural Analysis in Terms of the Tiling Model

The structural analysis of texture in terms of the tiling model is a procedure where one starts with a real-world observable texture, $g_s(\bar{x})$, and attempts to find a set of unit patterns and placement rules which can be used to generate this texture. That is, one is looking for an ideal texture representation for $g_s(\bar{x})$. Once the ideal texture representation is found, features are extracted to characterize the unit patterns and the placement rules.

As an aside it should be noted that the tiling model suggests that two types of features should be used to characterize a unit pattern $\{T, f\}$. That is, there should be features which measure qualities of the tile T, e.g., size, shape, and orientation; and there should be features which measure qualities of the painting function, f. Ehrich and Foith [33] use both feature types in their analyzer. They use the relative peak height and absolute peak height as features characterizing f, and peak area as a feature characterizing T. The structural analyzer, of Zucker, *et al.* [29] and the analyzer of Lu and Fu [34, 35], on the other hand, do not use both of these feature types. However, in one instance the form of f

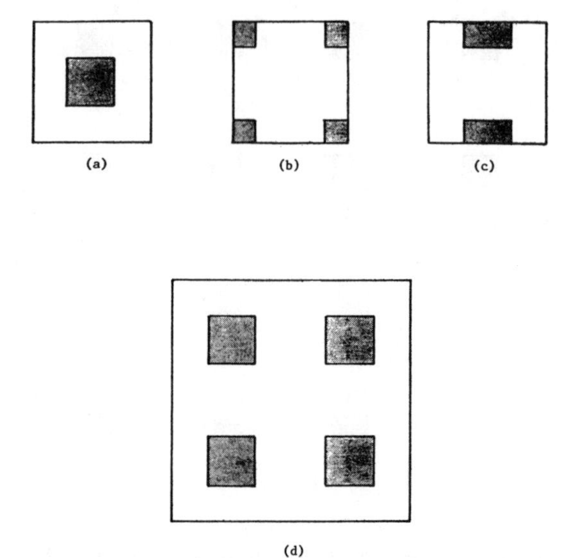

FIG. 8. Possible unit patterns which can be used to generate the texture of Fig. 7.

and the shape of T were assumed (Zucker *et al.*) and in the other shape and size of the tiles were assumed (Lu and Fu).

In the next few sections, the problem of characterizing the painting functions will be temporarily ignored. Rather, the problem of selecting and characterizing the tiles as well as the placement rules will be considered.

With this in mind, recall that given a surface texture $g_s(\bar{x})$ there is no unique ideal texture representation for $g_s(\bar{x})$. Consequently, one has some flexibility in chosing a family of unit patterns to use in a structural analyzer. In particular, one has flexibility in choosing the shape of the tiles to be used. Because there is this flexibility, it seems reasonable to choose the unit patterns in such a way that the analysis task is simplified, that is, the dimensionality of the problem is reduced. Note that

1. all else being equal one would like a structural analyzer which produces the fewest and simplest shaped unit patterns;
2. all else being equal, one would like a structural analyzer which produces the fewest and simplest form of placement rules.

It will be shown in the following sections that for repetitive textures the statistical structural analyzer defines an ideal texture representation whose set $\mathcal{S}_p$ contains only one unit pattern (called the period parallelogram unit pattern) and whose set of placement rules $\mathcal{R}_p$ is completely defined by the shape of the unit pattern. More precisely, the unit pattern shape, size, and orientation together with all the placement rules are completely defined by two vectors. In this way the very simplest ideal texture representation is chosen by this analyzer.

4. FORMALIZING THE CONCEPT OF PERIOD PARALLELOGRAM UNIT PATTERNS

We formalize the concept of a periodic texture in two definitions.

Definition 19. An isometry σ is a *symmetry* of the texture $g(\bar{x})$ if $g(\bar{x}) = g(\sigma(\bar{x}))$ for all $\bar{x} \in E^2$. Let $\mathcal{S}(g)$ denote the group of symmetries of the texture $g(\bar{x})$.

Definition 20. A texture $g(\bar{x})$ is said to be periodic if $\mathcal{S}(g)$ contains at least two translation isometries in nonparallel directions.

This leads us to the following theorem. This theorem is the adaptation of Theorem 2 to periodic textures.

Theorem 3. *Given any periodic texture $g(\bar{x})$ there exists a parallelogram tiling $\mathcal{T}_{\mathcal{P}}$ and a function f defined on the prototile $\mathcal{P}$ of $\mathcal{T}_{\mathcal{P}}$ such that for any $T_i \in \mathcal{T}_{\mathcal{P}}$ and $\bar{x} \in T_i$, $g(\bar{x}) = f(\sigma^{-1}(\bar{x}))$, where σ is the isometry which maps $\mathcal{P}$ onto T_i.*

This theorem is important for two reasons. First, it states that for periodic textures one need only consider parallelogram-shaped unit patterns. Secondly, it points out that if one knows $\{\mathcal{P}, f\}$ one can generate the original texture $g(\bar{x})$. That is, $\{\mathcal{P}, f\}$ mathematically represents a pattern tile which can be used to tile in the plane and create the texture $g(\bar{x})$.

Because of the nice properties of $\{\mathcal{P}, f\}$, we give it a special name.

Definition 21. Given a periodic texture $g(\bar{x})$, a unit pattern $\{\mathcal{P}, f\}$ of $g(\bar{x})$ satisfying the conditions of Theorem 3 is called a period parallelogram unit pattern.

The benefit gained from considering a period parallelogram unit pattern as the unit pattern of a periodic texture $g(\bar{x})$ stems from the fact that the shape, size, and orientation of $\mathcal{P}$ together with the placement rules which define $\mathcal{T}_{\mathcal{P}}$ in terms of $\mathcal{P}$ are completely defined by two nonparallel vectors $\bar{a}$ and $\bar{b}$, This fact is illustrated in Fig. 9. The vectors $\bar{a}$ and $\bar{b}$ are such that if $\sigma_a(\bar{x}) = \bar{x} + \bar{a}$ and $\sigma_b(\bar{x}) = \bar{x} + \bar{b}$ then $\sigma_a \in \mathcal{S}(g)$ and $\sigma_b \in \mathcal{S}(g)$ where $\mathcal{S}(g)$ is the symmetry group of the periodic texture $g(\bar{x})$.

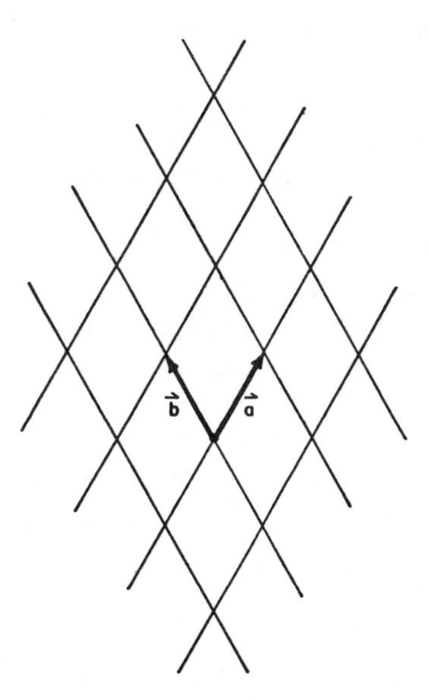

Fig. 9. Two vectors $\bar{a}$ and $\bar{b}$ completely define the tile $\mathcal{P}$ and the placement rules of a period parallelogram unit pattern.

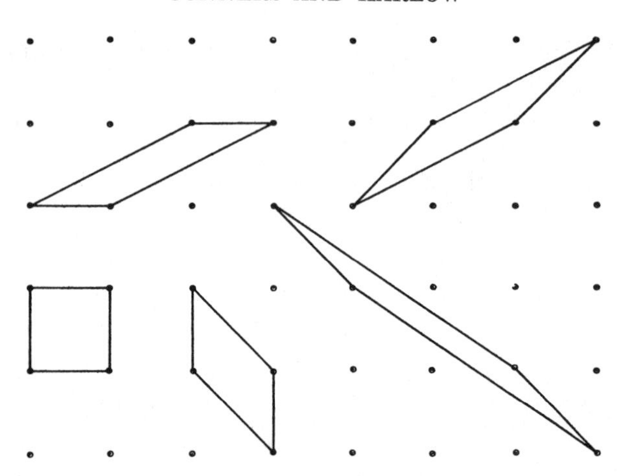

FIG. 10. These parallelograms represent a few of the possible candidates which can be used to create a period parallelogram unit pattern for a periodic texture $g(\bar{x})$ which has a periodic structure as indicated by the lattice points shown.

However, there is one minor difficulty in using period parallelogram unit patterns. Given a periodic texture $g(\bar{x})$ there are many possible parallelogram prototiles which can be used to generate a period parallelogram unit pattern. This fact is illustrated in Fig. 10. The black dots in this figure represent lattice points of the periodic texture $g(\bar{x})$. These lattice points are generated the same way lattice points of periodic tilings are generated. (See the discussion immediately following Definition 11.) The parallelograms shown in the figure are a few of the possible candidates which can be used to create a period parallelogram unit pattern for $g(\bar{x})$.

Consequently, for computational purposes one must establish a criterion for selecting the parallelogram prototile which will be used to generate the period parallelogram unit pattern so that uniqueness is guaranteed in the selection process. Further, the criterion used should satisfy a consistency requirement and it should also minimize the computations required to find the two defining vectors $\bar{a}$ and $\bar{b}$.

The consistency requirement which seems essential is that the same textural pattern regardless of its orientation should give the same size and shape parallelogram prototile $\mathcal{P}$. Obviously the orientation of the prototile should vary with the orientation of the textural pattern, but the size and shape should always remain the same.

Before stating the algorithm which will be used for selecting the vectors $\bar{a}$ and $\bar{b}$ that define the parallelogram prototile $\mathcal{P}$ which can, in turn, be used to generate the period parallelogram unit pattern $\{\mathcal{P}, f\}$, a lemma must be presented.

LEMMA 1. *If $g(\bar{x})$ is any texture and if σ is a translation symmetry of g then $-\sigma$ is also a translation symmetry of g.*

The proof of this lemma follows directly from the fact that $\mathcal{S}(g)$ is a group.

For purposes of starting the algorithm it is convenient to express vectors not in their Cartesian form but rather in polar form in terms of ρ and θ. Also, since one is dealing with discrete data in digital images it is convenient to let $\rho = \max\{x_1, x_2\}$ rather than the usual Euclidean distance when converting the Cartesian vector $\bar{x} = (x_1, x_2)$ to polar form. However, the conversion for θ remains, as is usual, $\theta = \text{arc tan } x_2/x_1$. Throughout the rest of this paper whenever we refer to polar coordinates or polar vectors we will be referring to the polar coordinate system just defined.

The algorithm for selecting the vectors $\bar{a}$ and $\bar{b}$ for a given texture $g(\bar{x})$ can now be given. The heart of the procedure is based on determining whether for a given value of ρ there is any θ such that the vector $\bar{c} = (\rho, \theta)$ defines a translation symmetry of a texture $g(\bar{x})$. In other words, the procedure is based on determining whether for a given value of ρ the texture $g(\bar{x})$ is periodic in any direction θ with period ρ. The values of θ considered are $0 < \theta < 180°$. One need only consider values of θ in this range because of Lemma 1. To use the procedure one starts with $\rho = 1$ and examines all values of θ where $0 < \theta < 180°$. One then sets $\rho = 2$ and follows the same procedure again examining all values of θ, $0 \leq \theta < 180°$. One continues iterating the procedure until two nonparallel vectors $\bar{a}$ and $\bar{b}$ have been found which define translation symmetries of $g(\bar{x})$. The desire is to find the two vectors $\bar{a}$ and $\bar{b}$ which have the smallest Euclidean magnitudes of all other vectors which can be used to define translation symmetries of $g(\bar{x})$. Further, for labeling purposes it is assumed $\|\bar{a}\|_2 \leq \|\bar{b}\|_2$. That is, it is desired that $\|\bar{a}\|_2 \leq \|\bar{b}\|_2 < \|\bar{c}\|$ where $\bar{c} \neq \bar{a}$, $\bar{c} \neq \bar{b}$, and $\bar{c}$ defines a translation symmetry of $g(\bar{x})$.

Figure 11 illustrates this algorithm. The black dots indicate lattice points of g. The arrows indicate a few of the possible vectors which can be used to define translation symmetries of g. The concept is to choose the two vectors which have the smallest Euclidean magnitude. For this illustration these are the vertical vector ($\theta = 90°$) and the horizontal vector ($\theta = 0°$).

Note that for a given value of ρ it is possible that one can find many vectors which define translation symmetries of $g(\bar{x})$. However, the algorithm is based on finding the two vectors which have the smallest Euclidean magnitude. Consequently a check of the Euclidean magnitude of each vector is required. Also, even though two candidate vectors have been found other iterations may be required to assure that these two vectors have the smallest Euclidean magnitudes. This determination is very simple and is based on comparing the magnitudes of these vectors to determine whether there are possibly other vectors with smaller Euclidean magnitudes which have longer ρ values.

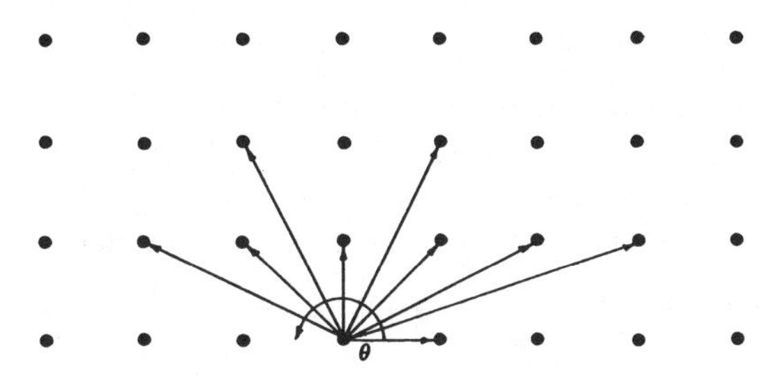

FIG. 11. A few of the many vectors which can be used to define translation symmetries for the texture $g(\bar{x})$ with lattice points shown above.

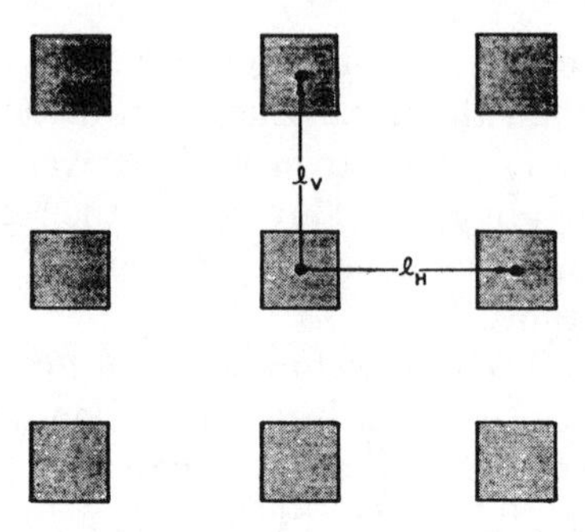

FIG. 12. A simple periodic texture composed of squares which a:e regularly spaced on a white background.

Unfortunately, even with the restrictions described above the algorithm presented does not always guarantee uniqueness in the selection process. The problem is meeting the uniqueness requirement and at the same time meeting the consistency requirement. More research is needed in this area.

The algorithm given above does, however, minimize the computations needed for finding the vectors $\bar{a}$ and $\bar{b}$.

To use this procedure, indeed to even be able to consider using period parallelogram unit patterns, one must be able to detect the periodicity of a pattern in a given direction θ. An algorithm for performing this task will be given in the next section.

5. FINDING THE SIZE, SHAPE, AND ORIENTATION OF A PERIOD PARALLELOGAM UNIT PATTERN OF A PERIODIC TEXTURE

In this section a procedure will be presented which allows one to detect the periodicity of a textural pattern in any given direction θ. To explain this procedure, a simple periodic textural pattern will be used for illustration. This textural pattern is shown in Fig. 12. As one will observe it is made up of small black squares regularly spaced on a white background. From the previous section it should be clear that the two nonparallel vectors one would like to use in the formation of a period parallelogram unit pattern are $(l_H, 0°)$ and $(l_V, 90°)$ where these vectors are given in the polar coordinate system described in the last section.

Consider, for the moment, only the $\theta = 0°$ (horizontal) direction. It can be shown that the following observations about the horizontal spatial gray level dependence matrices, $S_0(d)$, $d = 1, 2, \ldots$, are true.

1. $S_0(l_H)$ is a diagonal matrix. It has no nonzero off-diagonal elements.
2. $S_0(d)$, $d = 1, 2, \ldots, l_H - 1$ have nonzero off-diagonal elements.
3. $S_0(l_H + n) = S_0(ml_H + n)$, $m, n = 1, 2, \ldots$, that is, the horizontal spatial gray level dependence matrices are periodic with period l_H.

These observations indicate that any feature of the form

$$\sum_{i=0}^{N_G-1} \sum_{j=0}^{N_G-1} |i - j|^n s_0(i, j|d), \tag{6}$$

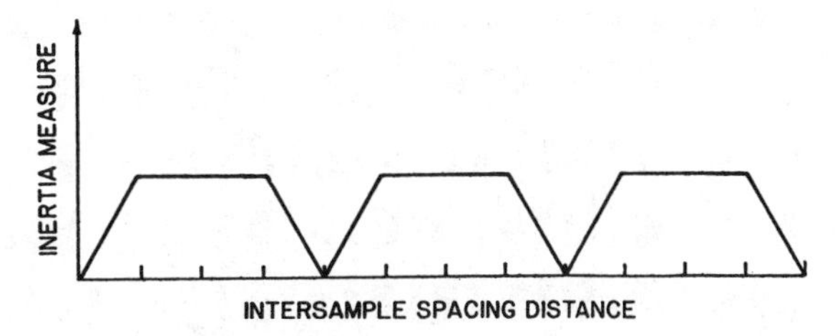

FIG. 13. A plot of the horizontal interia measure extracted from the texture given in Fig. 12.

where n is a natural number and $S_0(d) = [s_0(i, j|d)]$ can be used to determine the horizontal periodicity of the textural pattern given in Fig. 12, in that any feature of this form can be used to detect the presence or absence of nonzero off-diagonal elements. Consequently, an examination of feature values considered as a function of intersample spacing distance can be used to detect the period of the pattern.

In particular, let us consider the inertia measure defined in Eq. (5). Note that the inertia measure is of the form given by Eq. (6). Hence the above observations tell us that this feature can be used for periodicity detection. The reason for selecting the inertia measure as the one feature to be used to do periodicity detection will be explained later. For now, let us just accept this choice and consider the behavior of the inertia measure for the simple texture being considered.

Figure 13 shows a plot of the horizontal inertia feature for the texture of Fig. 12 as a function of the intersample spacing distance d. Two points should be noted about this plot. The points where the inertia measure is zero correspond to intersample spacing distance values which are integer multiples of l_H. These zero points represent local minimum values of the inertia measure and they result because the translations $\sigma_n(\bar{x})$ defined by the Cartesian form of the vectors $(nl_H, 0°)$ are such that $g(\bar{x}) = g(\sigma_n(\bar{x}))$. Secondly, note the periodic structure of the inertia measure. This reflects the periodic structure of the matrices $S_0(d)$, $d = 1, 2, \ldots$.

It should be clear that when $\theta = 90°$, all the above is still true for the texture of Fig. 12. In particular, the three observations made above hold, except that the $S_0(d)$ matrices become the $S_{90}(d)$ matrices and l_H is replaced by l_V. Similarly, the statements about the inertia measure made above are also true when $\theta = 90°$. These statements follow directly from the three observations about the properties of the spatial gray level dependence matrices.

The comments concerning the texture of Fig. 12 generalize to any periodic testure $g(\bar{x})$. If (ρ, θ) is a vector which defines a translation symmetry of $g(\bar{x})$ then it can easily be shown that

1. $S_\theta(\rho)$ is a diagonal matrix.
2. $S_\theta(d)$, $d = 1, 2, \ldots, \rho - 1$ have nonzero off-diagonal elements.
3. $S_\theta(\rho + n) = S_\theta(m\rho + n)$, $m, n = 1, 2, \ldots$, that is, $S_\theta(d)$ is periodic with period ρ.

From these observations, all the important properties of the inertia measure follow.

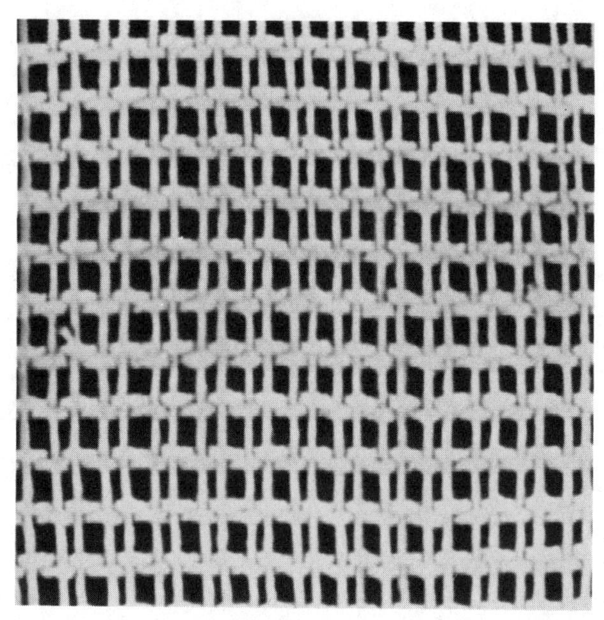

FIG. 14. French canvas.

Most important, the above concepts generalize to repetitive textures which have had their underlying fabric stretched and twisted, which have scanning noise present, or which have other types of distortions. In such cases, the inertia measure is never equal to zero. Rather, local minima represent candidate points from which the periodicity can be computed. However, a point being a local minimum is not a sufficient condition for it to be some integer multiple of the period. To ensure that a local minimum is a multiple of the period, one must also examine the periodic structure of the inertia measure.

Consider, for example, the French canvas shown in Fig. 14. This image was scanned from Brodatz book *Textures* [37] at a resolution 256 × 256 with 256 gray levels. Before the computation of the inertia values the equal probability quantizer (EPQ) [38] was applied to the image to reduce the number of gray levels from 256 to 16. The inertia values were computed from the reduced image. Figure 15 shows the horizontal and vertical inertia values plotted against the intersample spacing distance. The arrows in the figure point out the minimum inertia values, and the numbers above the arrows indicate the intersample spacing distance associated with that minimum inertia value. Note the reasonably periodic structure of the inertia values with only some minor distortion caused by the stretching and twisting of the texture fabric. The plots shown in Fig. 15 indicate that the size of the period parallelogram unit pattern for the French canvas is 20 × 18 pixels where 20 is the vertical dimension and 18 is the horizontal dimension. Finally, observe Fig. 16. This figure shows the French canvas texture with the appropriate period parallelogram unit patterns overlaid on the image. Note that each unit cell contains approximately the same pattern and that the dimensions of the rectangles agree well with a perceptually obtained unit pattern size. Inertia values computed along other angular orientations showed either no periodicity or demonstrated longer periods than the vertical and horizontal inertia measures. It should be noted that the periods referred to here are computed using the Euclidean metric rather than using ρ in the polar vector (ρ, θ).

For another example, consider the raffia shown in Fig. 17. This image was also scanned from Brodatz's book at a resolution of 256 × 256 with 256 gray levels. Again, before extraction of the inertia values the EPQ was applied to the image to reduce the number of gray levels to 16. Figure 18 shows the plots of the horizontal and vertical inertia values. The arrows indicate the minimum values. Again, observe the nearly periodic structure in these values. The minimum inertia values being significantly different from zero indicates that the replication process and/or the unit patterns are not always the same. These plots indicate that the

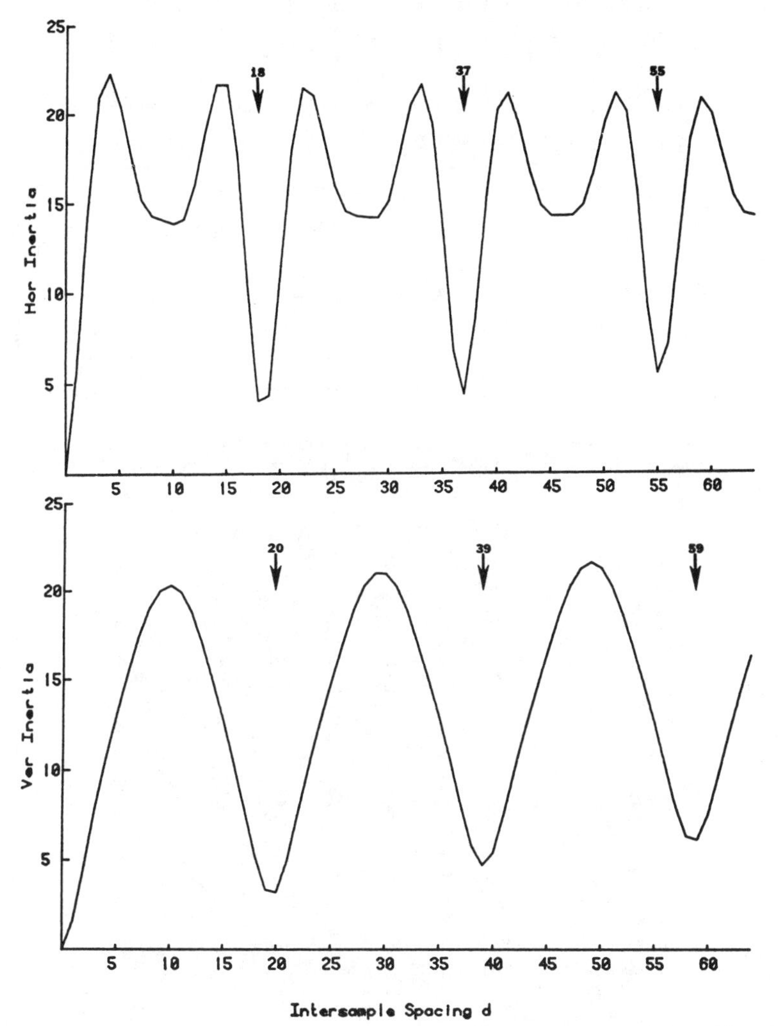

FIG. 15. Plots of the horizontal and vertical inertia measures computed from French canvas given as a function of intersample spacing distance d.

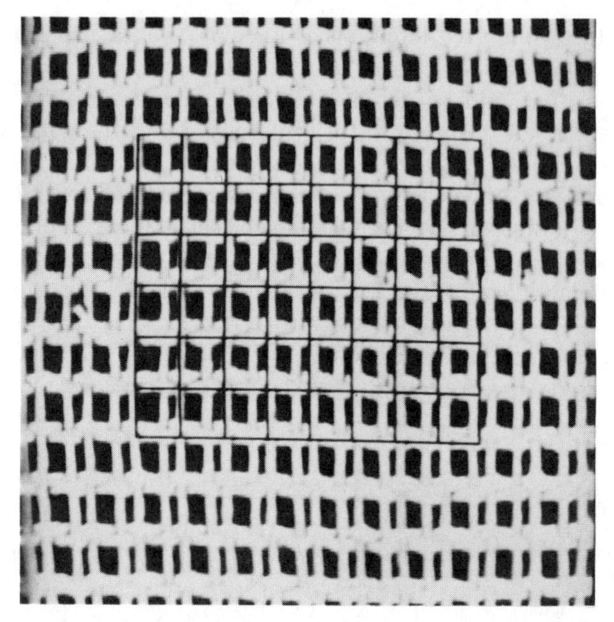

Fig. 16. French canvas with the computed unit cell rectangles overlaid.

period parallelogram unit pattern size of raffia is 9×12 pixels. Finally, Fig. 19 shows the raffia texture with the unit cell rectangles overlaid. Note that while each rectangle does not always contain exactly the same pattern they do seem to be of the right dimensions with good agreement locally.

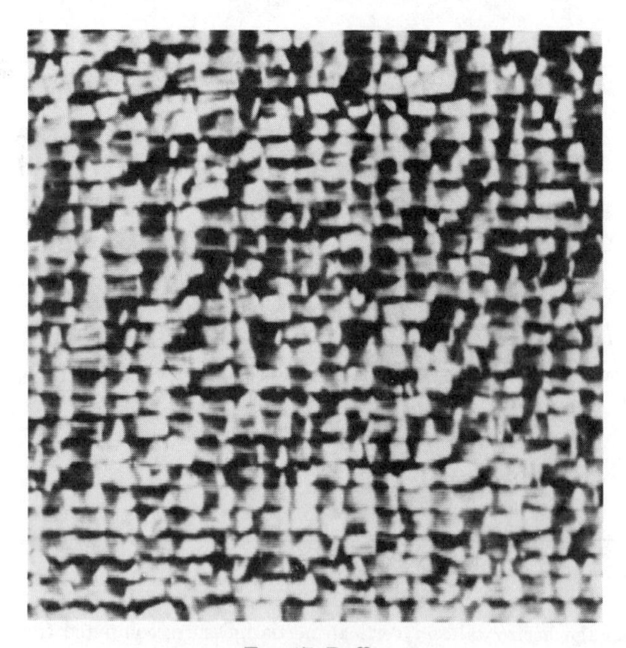

Fig. 17. Raffia.

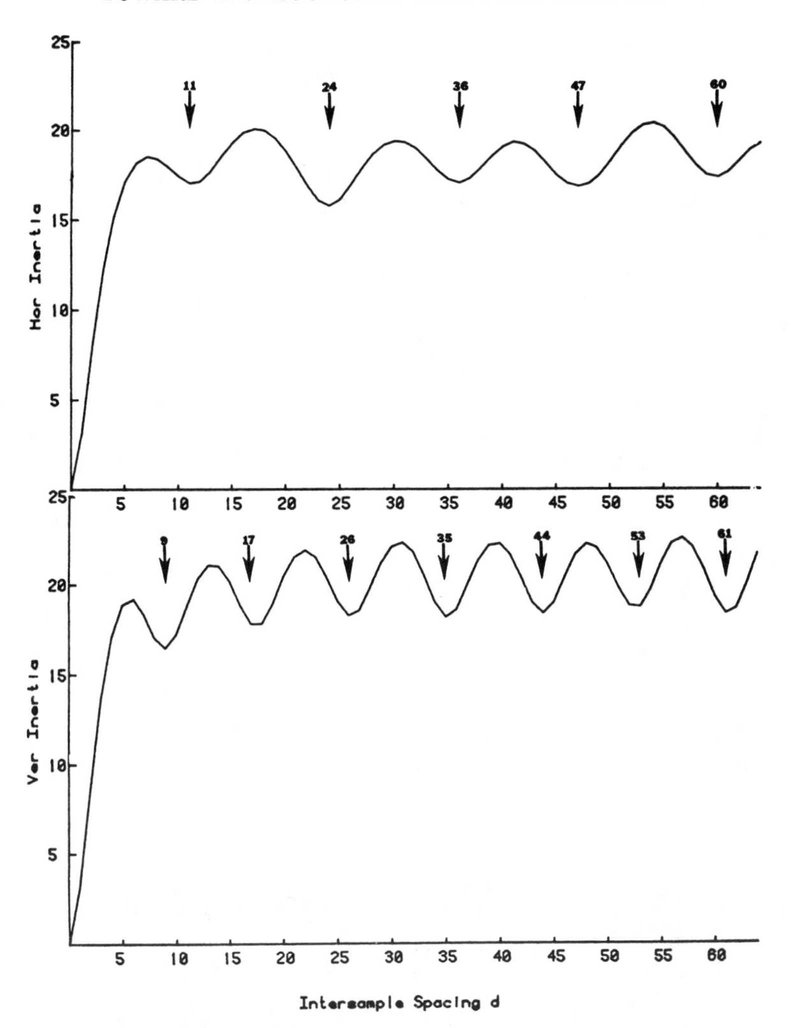

FIG. 18. Plots of the horizontal and vertical inertia measures computed from raffia given as a function of intersample spacing distance d.

The two examples which have been presented hopefully demonstrate how the inertia measure can be used in conjunction with the procedure described in the last section to compute the size, shape, and orientation of a period parallelogram tile which can be used to create a period parallelogram unit pattern. Other textures from Brodatz's book have also been analyzed. The textures examined include cane, woven wire, handmade paper, oriental straw cloth, and loose burlap. Most of these patterns and others have been processed several times at different magnifications. In each instance, a period parallelogram was easily obtained and the resulting period parallelogram unit pattern corresponded very well with the visually obtained unit pattern.

Now that the method for determining period parallelogram dimensions has been demonstrated, the reasons for selecting the inertia measure as the feature to detect periodicity should be given. First and foremost, the inertia measure applied in the manner described above can be shown to detect periodicity in the mean square sense. That is, for any given d and θ the inertia measure value is equal to $E\{(g(\bar{x}) - g(\bar{x} + \Delta\bar{x}))^2\}$ where $\Delta\bar{x}$ is the Cartesian form of the polar vector (d, θ). For images which have noise or some type of distortion, mean square periodicity would seem a logical quantity to measure. However, it should be noted that as of this writing it cannot be ensured that the periodicity detected by humans is periodicity in the mean square sense.

Another reason for selecting the inertia measure over other features of the form given by Eq. (6) is that the inertia measure provides the type of information which has classically been used in statistical communication theory [37] and in the statistical analysis of random mosaic patterns [1–6]. For example, it can be shown that for a translation-stationary random field of order 2, $X(n, m)$, $I(S_\theta(d)) = 2(\sigma^2 - R(\Delta\bar{x}))$, where $\Delta\bar{x}$ is the Cartesian form of the polar vector (d, θ), σ^2 is the variance of $X(n, m)$, and R is the autocorrelation function. Further, since the autocorrelation (or as the case may be, the autocovariance) function and the power spectrum are Fourier transform pairs, it can be argued that the inertia measure embodies all the information contained in the power spectrum. Finally, the inertia measure considered as a function of intersample spacing is mathematically identical to what is referred to as the "variogram" by researchers working in the area of random mosaics [1, 3, 6, 5].

What has been definitely shown in this section is the importance of the projection

$$p_D(k \mid d, \theta) = \sum_{\substack{i=0}}^{N_G-1} \sum_{\substack{j=0 \\ j-i=k}}^{N_G-1} s_\theta(i, j \mid d) \tag{7}$$

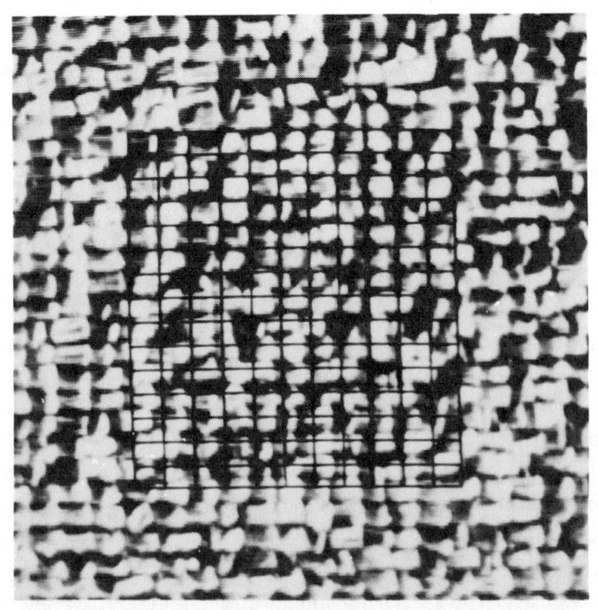

FIG. 19. Raffia with computed unit cell rectangles overlaid.

in determining the periodicity of textural patterns. While there may be some doubt that the inertia measure, or indeed any single measure of the form given by Eq. (6), can detect periodicity as well as the eye can, there can be no doubt that this projection $p_D(k|d, \theta)$ contains much of the information present in the spatial gray level dependence matrix $S_\theta(d)$ relevant to periodicity detection.

6. FURTHER PROPERTIES OF THE INERTIA MEASURE

Thus far it has been demonstrated that the inertia measure can be used to detect textural periodicity. However, the inertia measure has some other interesting qualities which make it an important primitive feature in the statistical structural analyzer. In this section, a number of these qualities will be described. Finally, it will be shown that the inertia measure does not contain all the important texture-context information contained in the SGLDM. Consequently, other features must be added to the feature set if this algorithm is to match human performance.

To begin this discussion it should be remembered that a unit pattern is composed of two entities. These are a tile T and a painting function f. For periodic textures it has been shown that one need consider only parallelogram shaped tiles. Further it has been shown that the inertia measure can be used to define the size, shape, and orientation of these parallelogram tiles. Consequently, for periodic textures the only point that remains is to somehow measure qualities of the painting function defined on this parallelogram tile.

An important point about the inertia measure is that while one is using it to compute textural periodicity it is simultaneously measuring qualities of the painting function. To see this, consider the plot given in Fig. 20. The plot shows a sketch of the values of the horizontal inertia measure computed from French canvas and woven wire. Note that the detected periodicity of both patterns is the same but that shapes of the curves are markedly different. This difference results from the marked differences in the painting functions of the two textures.

Another advantage of the inertia measure and, indeed, of any feature based on the SGLDM method is that they have a very desirable insensitivity to the arbitrary selection of unit patterns. Consider, for example, the texture shown in

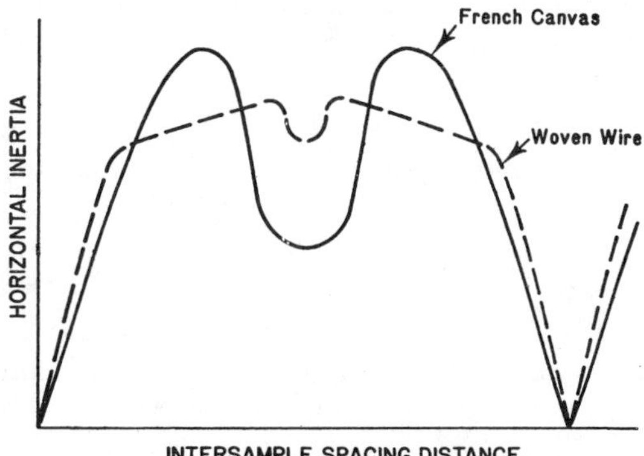

FIG. 20. The fact that the inertia measure contains information on the painting functions of textures is illustrated by this plot.

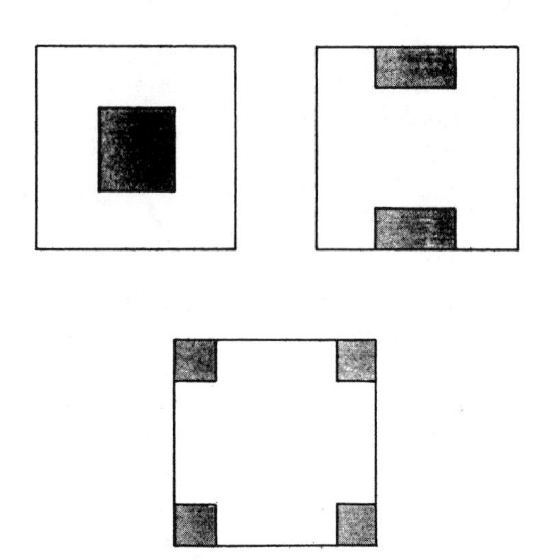

FIG. 21. A few of the possible unit patterns of the texture given in Fig. 12.

Fig. 12. Based on the tiling model for texture there are many possible unit patterns $\{T_i, f_i\}$, even given that each T_i has the same size, shape, and orientation. Figure 21 shows a few of the many possible unit patterns for the texture given in Fig. 12.

Note that if one were to apply a structural analyzer such as the one of Lu and Fu [34, 35] to these unit patterns the resulting measurements used to characterize these patterns would seem very different. That is, the grammars generated from these patterns are very different. This seems a most undesirable trait since all these unit patterns generate the same texture.

On the other hand, if one computes the value of the inertia measure or other features based on the SGLDM from an area containing a few repetitions of any of the unit patterns shown in Fig. 21 the values of these features will be identical. They are insensitive to particular unit pattern selection.

In actuality, features based on the SGLDM measure what could be considered global qualities of a painting function f of the unit pattern $\{T, f\}$, e.g., how the values of the painting function along the right-hand boundary of a unit pattern will match up with the values on the left-hand boundary when the plane is tiled with $\{T, f\}$.

This fact corresponds well with what one desires from a texture analysis algorithm. A texture is defined by the global interaction of its parts. For texture, the whole would seem greater than the sum of its parts. The texture analyzer should measure quantities indicative of global characteristics. Further, its role should be to localize each fundamental building block of the whole. Then form perception can be used to analyze each part in detail if that is desired. This concept was illustrated by Julesz [10]. Figure 22 shows an example similar to the one used by Julesz. At first glance, it appears that the figure contains a field of uniform texture. But, after careful examination, it can be determined that the left field is composed of seven-letter English words, while the right field contains nonsense words of the same length (actually they are the same English words

```
FURTHER FORMULA MIXTURE RADICAL TEXTURE ERUTXET LACIDAR ERUTXIM ALUMROF REHTRUF
FURTHER FORMULA MIXTURE RADICAL TEXTURE ERUTXET LACIDAR ERUTXIM ALUMROF REHTRUF
FURTHER FORMULA MIXTURE RADICAL TEXTURE ERUTXET LACIDAR ERUTXIM ALUMROF REHTRUF
FURTHER FORMULA MIXTURE RADICAL TEXTURE ERUTXET LACIDAR ERUTXIM ALUMROF REHTRUF
FURTHER FORMULA MIXTURE RADICAL TEXTURE ERUTXET LACIDAR ERUTXIM ALUMROF REHTRUF
FURTHER FORMULA MIXTURE RADICAL TEXTURE ERUTXET LACIDAR ERUTXIM ALUMROF REHTRUF
FURTHER FORMULA MIXTURE RADICAL TEXTURE ERUTXET LACIDAR ERUTXIM ALUMROF REHTRUF
FURTHER FORMULA MIXTURE RADICAL TEXTURE ERUTXET LACIDAR ERUTXIM ALUMROF REHTRUF
FURTHER FORMULA MIXTURE RADICAL TEXTURE ERUTXET LACIDAR ERUTXIM ALUMROF REHTRUF
FURTHER FORMULA MIXTURE RADICAL TEXTURE ERUTXET LACIDAR ERUTXIM ALUMROF REHTRUF
FURTHER FORMULA MIXTURE RADICAL TEXTURE ERUTXET LACIDAR ERUTXIM ALUMROF REHTRUF
FURTHER FORMULA MIXTURE RADICAL TEXTURE ERUTXET LACIDAR ERUTXIM ALUMROF REHTRUF
```

FIG. 22. An example which requires what Julesz referred to as the deliberate level of discrimination.

written backwards). Note the need for form perception and cognitive processing in this example.

Thus, there seems to be at least two levels to the analysis process, the global analysis of the pattern and then, if needed, a localized analysis based on form perception. Note that the area considered in the localized analysis is determined from the global processing. Julesz presents approximately the same argument in his latest articles [13–15]. In these articles Caelli and Julesz give a number of counterexamples to the Julesz conjecture. To explain these counterexamples they develop a theory based on Type A and Type B texture features. The Type A features are the type of second-order probabilities measured by the spatial gray level dependence matrices. The Type B features appear to be based largely on form perception.

However, there appear to be more than just the global and form levels to texture analysis. Global processing, within itself, appears to have a number of what could be termed sublevels. To see this, consider the plot shown in Fig. 23. This figure shows the values of the inertia measure computed from two textural patterns, Pattern 1 and Pattern 2. If one's only desire is to descriminate these two textural patterns then really only one value of intersample spacing distance,

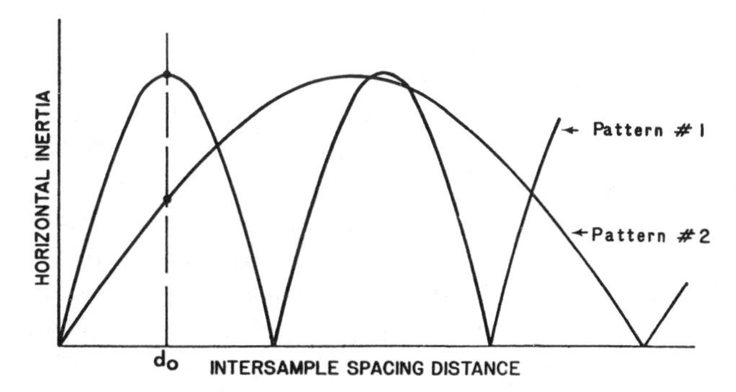

FIG. 23. An illustration of the fact that there are various levels of analysis which can be done with the context of global processing.

namely, d_0, need be used. It should be added that while the horizontal inertia measure computed for d_0 allows the discrimination of the textures, the difference in the computed values tells one nothing about why the two patterns are different visually. To explain the difference in visual qualities of all the textures to be discriminated one must consider several intersample spacing distances, as was done in creating the figure. An examination of the plot shown immediately indicates that Pattern 1 has a shorter period in the horizontal direction than does Pattern 2. This points out that the explanation of the discrimination requires more information than the mere discrimination of the patterns, a fact which agrees well with studies done on human perception [40]. It also points out that there is a hierarchial structure to the global analysis of texture. This structure which is based on the consideration of several intersample spacing distances should allow more sophisticated analysis of textural patterns in the future.

An interesting aside to the statements made thus far concerns the resolution at which a textural pattern is scanned. In a 1975 article [41], the SGLDM was being used in conjunction with the five features given in Eqs. (1)–(5) to do discrimination of two textural patterns. In this study only two intersample spacing distances were being considered, namely, $d = 1$ and 2. As a part of this study a parametric examination was conducted to determine the optimal resolution at which the film images should be scanned. To do the examination the film data base was scanned at various resolutions. The optimum resolution was defined to be the one which gives the best overall percentage of correct classification. The results indicated that as the resolution increased the classification results got better. This is as one would expect. However, after a point, the classification accuracies began to fall.

From a theoretical point of view such a decrease should never occur. As the resolution is increased one would expect the digital image to become a better representation of its film counterpart, at least up to the point where the digital image and film image have matching modulation transfer function, MTFs.

Within the context of what has been presented above it should be clear that this seeming disparity between the performance of the SGLDM and what one knows from the sampling theorem is really not a disparity at all. That is, based on what has been said about considering multiple intersample spacing distances, it should be clear that if the intersample spacing distance $d = l$ allows discrimination of two textures when a resolution of $N \times N$ is used over a film area of $s \times s$ then an intersample spacing distance of $d = 2l$ should be used when the $s \times s$ area is scanned at a $2N \times 2N$ resolution. A so-called optimum resolution was found in [41] only because the d values of 1 and 2 were arbitrarily chosen. The optimum resolution was the one which gave the spatial gray level dependence matrices computed for $d = 1, 2$ the most discriminatory power.

Thus far we have discussed periodic and almost periodic textures. The usefulness of the inertia measure in the analysis of these textural types has been demonstrated. A related question will now be considered. This question concerns the determination of the texture gradient of nonrepetitive natural textures such as grass and sand. It should be clear that determining the texture gradient of nonrepetitive textures is a generalization of the problem of determining the period of periodic textures. Consequently, it seems logical that the same basic methodologies should be used in both cases. For purposes of the discussion here, this observation implies that the inertia measure should play a key role in determining the texture gradient.

There is historical evidence which indicates that the inertia measure can be used to detect the texture gradient. In 1970 Rosenfeld [42] showed that the inertia measure could be used to determine the texture gradient. He considered only one intersample spacing distance $d = 1$. Also Bajcsy [43] used the power spectrum to determine the texture gradient of natural scenes. If one remembers that for translation-stationary random fields of order 2 the inertia measure and the power spectrum are informationally equivalent, then the Bajcsy study provides further evidence of the utility of the inertia measure in detecting the texture gradient.

The authors' investigations on this subject are just beginning. However, since the investigations in this area which are complete are essentially an elaboration of the Rosenfeld experiments it seems worth while to present them.

Consider three pictures of sand, 1, 1.2, and 1.4× magnifications. The results presented previously in this section indicate that if $I_\theta^{(1)}(d)$ is the inertia measure in the θ direction for the 1× magnification of sand, then

$$I_\theta^{(1.2)}(d) = I_\theta^{(1)}(d/1.2) \tag{7}$$

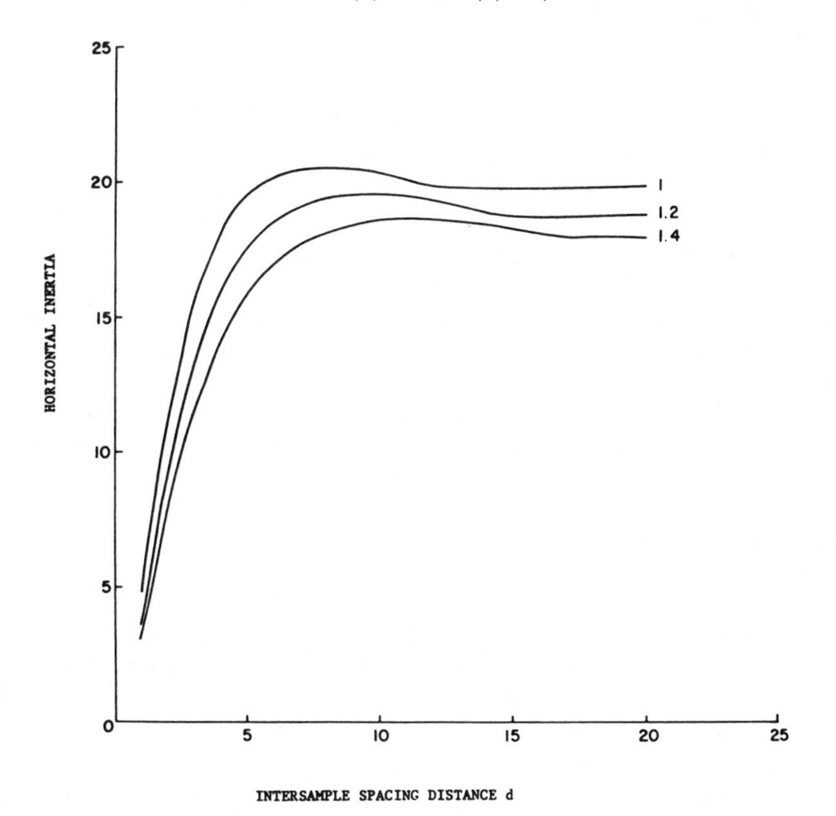

Fig. 24. Plots of the horizontal inertia measure computed from a 1, 1.2, and 1.4× magnification of sand.

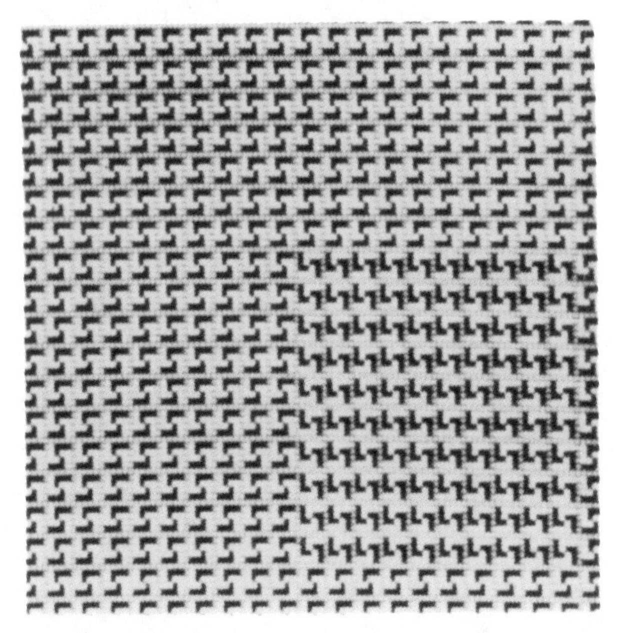

FIG. 25. A visually distinct texture pair which cannot be discriminated by the inertia measure or indeed any measure of the form given by Eq. (6). These textures were first created by Julesz [10] in his 1962 article on visual texture discrimination.

and

$$I_\theta^{(1.4)}(d) = I_\theta^{(1)}(d/1.4), \tag{8}$$

where $I_\theta^{(1.2)}(d)$ and $I_\theta^{(1.4)}(d)$ represent the inertia measure computed from the 1.2 and 1.4× magnifications.

What these equations indicate is that all three magnifications of sand have the same basic shape inertia measure curve but there is a stretching of the d axis which is proportional to the magnification.

Figure 24 shows the plots of the horizontal inertia measure computed from three magnifications. The images used were digitized at a resolution of 256 × 256 with 256 gray levels. The pictures were processed so that exactly the same area of the original picture was used in each instance. Consequently as the magnification increased the subregion used for processing got larger. The EPQ was used to reduce the number of gray levels from 256 to 16. The EPQ was applied to only the subregion chosen for processing.

Note that the stretching indicated by Eqs. (7) and (8) is evident. However, also note that there are some noticeable differences in the magnitudes of the inertia values computed from the different magnifications. It is believed that this difference results from resolution-dependent effects which are basically unrelated to the texture gradient problem. It is believed that if these same images are scanned at a 512 × 512 resolution the experimental curves will more closely match the results predicted in Eqs. (7) and (8). This is currently being investigated.

In any event, at this time it is not clear whether one should consider one intersample spacing distance, $d = 1$, as suggested by Rosenfeld [42], or many as suggested here.

There is one final point to be addressed in this section. Up until now only the usefulness of the inertia measure has been discussed. A question of concern must be whether this one feature is all that is needed to perform texture discrimination. The answer to this question is given in Figs. 25 and 26. These figures give two visually distinct texture pairs which cannot be discriminated by the inertia measure. That is, the texture pair given in Fig. 25 represents two textures which for any value of d and θ have the same values for their inertia measures. It is of interest to note that the inertia measures are the same for any d and θ, but $S_{\theta}^{(1)}(d) \neq S_{\theta}^{(2)}(d)$. The superscripts used in this inequality indicate that one matrix was computed from one texture comprising the pair. A similar statement is true for the texture pair given in Fig. 26. One last example of the inadequacy of the inertia measure is given in Fig. 1.

Furthermore, it can easily be shown that the texture pairs given in Figs. 1 and 25 cannot be discriminated by "any" feature defined from projections of the form given by Eq. (7). Thus not only is the inertia measure inadequate to discriminate all visually distinct texture pairs, but also given $S_{\theta}(d)$ the projection given in Eq. (7) does not contain all the important texture-context information contained in $S_{\theta}(d)$.

In a companion paper we will pursue other features which can be defined from the spatial gray level dependence matrices which can augment the inertia measure and add to the discriminatory power of the SGLDM algorithm.

7. CONCLUSIONS

In this paper we have shown that the inertia measure can be used to gauge certain visually meaningful qualities of textural patterns. In particular it has been shown that this measure can be used to characterize the unit pattern and

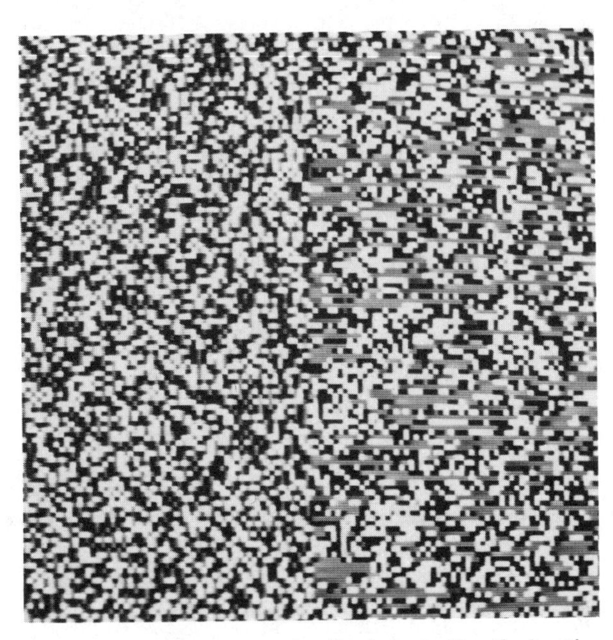

FIG. 26. Two textures which cannot be discriminated by the inertia measure.

placement rules of almost periodic textures. Hence, a structural approach to texture can be formulated based on the statistical SGLDM.

It has also been shown that the inertia measure does not gauge all of the important texture-context information contained in the spatial gray level dependence matrices. Consequently, other features are required. In a companion paper currently under preparation other features measuring visually perceivable qualities of textural patterns will be defined. These additional features can be used to further characterize the global properties of the painting function. As such they represent additional information which can be used by the structural statistical analyzer to discriminate textures.

REFERENCES

1. B. J. Schachter, A. Rosenfeld, and L. S. Davis, Random mosaic models for textures, *IEEE Trans. Systems, Man Cybernet.* **SMC-8**, Sept. 1978, 694–702.
2. B. J. Schachter and N. Ahuja, Random pattern generation processes, *Computer Graphics Image Processing* **10**, 1979, 95–114.
3. R. Miles, On the homogenous planar Poisson point-process, *Math. Biosci.* **6**, 1970, 85–127.
4. E. Gilbert, Random subdivisions of space into crystals, *Ann. Math. Statist.* **33**, 1962, 958–972.
5. R. Miles, Random polygons determined by random lines in the plane, *Proc. Nat. Acad. Sci. USA* **52**, 901–907, 1157–1160, 1969.
6. P. Switzer, Reconstructing patterns from sample data, *Ann. Math. Statist.* **38**, 1967, 138–154.
7. E. M. Darling and R. D. Joseph, Pattern recognition from satellite altitudes, *IEEE Trans. Systems, Man Cybernet.* **SMC-4**, March 1968, 38–47.
8. R. M. Haralick and K. Shanmugam, Computer classification of reservoir sandstones, *IEEE Trans. Geosci. Electron.* **GE-11**, Oct. 1973, 171–177.
9. R. M. Haralick and K. Shanmugam, Textural features for image classification, *IEEE Trans. Systems, Man Cybernet.* **SMC-3**, Nov. 1973, 610–621.
10. B. Julesz, Visual pattern discrimination, *IRE Trans. Inform. Theory* **8**, Feb. 1962, 84–92.
11. B. Julesz, E. N. Gilbert, L. A. Shepp, and H. L. Frisch, Inability of humans to discriminate betweeu visual textures that agree in second-order statistics—revisited, *Perception* **2**, 1973, 391–405.
12. W. K. Pratt, O. D. Faugeras, and A. Gagalowicz, Visual discrimination of stochastic texture fields, *IEEE Trans. Systems, Man Cybernet.* **SMC-8**, Nov. 1978, 796–804.
13. B. Julesz, E. N. Gilbert, and J. D. Victor, Visual discrimination of textures with identical third-order statistics, *Biol. Cybernet.* **31**, 1978, 137–140.
14. T. Caelli and B. Julesz, On perceptual analyzers underlying visual texture discrimination, I, *Biol. Cybernet.* **28**, 1978, 167–175.
15. T. Caelli and B. Julesz, On percetpual analyzers underlying visual texture discrimination, II, *Biol. Cybernet.* **29**, 1978, 201–214.
16. A. Gagalowicz, Stochastic texture synthesis from a priori given second-order statistics, in *Proceedings IEEE Computer Society Conference on Pattern Recognition and Image Processing, Chicago, Ill.*, Aug. 1979, 376–381.
17. R. J. Tully, R. W. Conners, C. A. Harlow, and G. S. Lodwick, Towards computer analysis of pulmonary infiltration, *Invest. Radiol.* **13**, Aug. 1978, 298–305.
18. R. P. Kruger, W. B. Thompson, and F. A. Turner, Computer diagnosis of pneumoconiosis, *IEEE Trans. Systems Man Cybernet.* **SMC-4**, Jan. 1974, 40–49.
19. E. L. Hall, R. P. Kruger, and F. A. Turner, An optical–digital system for automatic processing of chest X-rays, *Optical Engr.* **13**, May/June 1974, 250–257.
20. Y. P. Chien and K. S. Fu, Recognition of X-ray picture patterns, *IEEE Trans. Systems Man Cybernet.* **SMC-4**, March 1974, 145–156.
21. J. S. Weszka, C. R. Dyer, and A. Rosenfeld, A comparative study of texture measures for terrain classification, *IEEE Trans. Systems Man Cybernet.* **SMC-6**, April 1976, 269–285.

22. R. W. Conners and C. A. Harlow, A theoretical comparison of four texture analysis algorithms, *IEEE Trans. Pattern Anal. Machine Intelligence*, in press.
23. G. Lendaris and G. Stanley, Diffraction pattern sampling for automatic pattern recognition, *Proc. IEEE* **58**, Feb. 1970, 198–216.
24. M. Galloway, Texture analysis using gray level run lengths, *Computer Graphics Image Processing* **4**, 1974, 172–199.
25. S. W. Zucker, Toward a model of texture, *Computer Graphics Image Processing* **5**, 1976, 190–202.
26. E. Mach, *Analysis of Sensation*, Dover, New York, 1959.
27. J. J. Gibson, *The Perception of the Visual World*, Houghton–Mifflin, Boston, 1950.
28. M. Wertheimer, Principles of perceptual organization, in *Readings in Perception* (D. C. Beardslee and M. Wertheimer, Eds.), pp. 115–134, Van Nostrand, New York, 1958.
29. H. Tamura, S. Mori, and Y. Yamawaki, Textural features corresponding to visual perception, *IEEE Trans. Systems Man Cybernet.* **SMC-8**, June 1978, 460–472.
30. S. Tsuji and F. Tomita, A structural analyzer for a class of textures, *Computer Graphics Image Processing* **2**, 1973, 216–231.
31. S. W. Zucker, A. Rosenfeld, and L. Davis, Picture segmentation by texture discrimination, *IEEE Trans. Computers* **C-24**, Dec. 1975, 1228–1233.
32. R. Ehrich and J. P. Foith, Representation of random waveforms by relational trees, *IEEE Trans. Computers* **C-25**, July 1976, 725–736.
33. R. Ehrich and J. P. Foith, A view of texture topology and texture description, *Computer Graphics Image Processing*, **8**, 1978, 174–202.
34. S. Y. Lu and K. S. Fu, A syntactic approach to texture analysis, *Computer Graphics Image Processing* **7**, 1978, 303–330.
35. S. Y. Lu and K. S. Fu, Stochastic tree grammar inference for texture synthesis and discrimination, *Computer Graphics Image Processing* **9**, 1979, 234–245.
36. B. Grünbaum and G. C. Shephard, *Tilings and Patterns*, to be published.
37. P. Brodatz, *Textures*, Dover, New York, 1966.
38. R. W. Conners and C. A. Harlow, Equal probability quantizing and texture analysis of radiographic images, *Computer Graphics Image Processing* **8**, 1978, 447–463.
39. A. Papoulis, *Probability, Random Variables, and Stochastic Processes*, McGraw–Hill, New York, 1965.
40. J. D. Gould and A. B. Dill, Eye-movement parameters and pattern discrimination, *Perception Psychology* **6**, 1969, 311–320.
41. E. L. Hall, W. O. Crawford, Jr., and F. E. Roberts, Computer classification of pneumoconiosis from radiographs of coal workers, *IEEE Trans. Biomed. Eng.* **BME-22**, Nov. 1975, 518–527.
42. A. Rosenfeld and E. B. Troy, *Visual Texture Analysis*, Computer Science Center, University of Maryland, College Park, Technical Report 116, June 1970.
43. R. Bajcsy and L. Lieberman, Texture gradient as a depth cue, *Computer Graphics Image Processing* **5**, 1976, 52–67.

Stochastic Boundary Estimation and Object Recognition

D. B. COOPER,* H. ELLIOTT,† F. COHEN,* L. REISS,** AND P. SYMOSEK*

** Division of Engineering, ** Department of Computer Science, Brown University, Providence, Rhode Island 02912; and † Department of Electrical Engineering, Colorado State University, Fort Collins, Colorado 80523*

1. INTRODUCTION

A few years ago, Cooper formulated the problem of object boundary finding in noisy images as a maximum likelihood estimation problem. The motivation was the elegant formulation of Martelli [1] (based on Montanari's work [2]) in which boundary finding was treated as deterministic function minimization and solved using a sequential search procedure with backtracking known in the artificial intelligence literature as the A* algorithm or as *branch and bound.* Following the statistical formulation, the authors working individually or in subgroups have developed aspects of this approach. At the present time, what has emerged is a body of results which is taking the form of a comprehensive probabilistic/statistical theory for object recognition and object boundary estimation. There are obvious gaps and refinements requiring attention which we are individually or collectively working on, but we do feel that the kernel for a theory has been established. The results of this work include: analysis of limiting boundary estimation accuracy; tools for partially analyzing algorithm accuracy, *thus providing for the comparison of algorithms;* insights into algorithm design based on experimentation and on use of these tools; and the development of a *Sequential Boundary Finder* and a region growing/shrinking algorithm, the *Ripple Filter.* These algorithms are just two ways of carrying out maximum likelihood estimation. Other ways are possible, and either of these algorithms can be run in parallel in many regions throughout an image field in order to realize a fast and robust parallel boundary estimation algorithm. Finally, results are presented on object recognition. The paper briefly summarizes much of this work [3–6, 13], and presents some new ideas which help in tying the individual pieces together.

Our approach is to first carefully model the data and then treat the boundary finding or object recognition as a problem in maximum likelihood estimation or hypothesis testing. The images treated are simple, consisting of an underlying object of constant gray level and highly varying boundary, surrounded by a constant gray level background. It is assumed that a white Gaussian noise field is superimposed on this in order to create the resulting *picture function,* i.e., the

ISBN 0-12-597320-9

image data. The basic approach of careful data modeling followed by information extraction or recognition within a statistical estimation or decision theoretic framework is not limited to these relatively simple images, but is extendable to complex indoor or outdoor scenes. Here, texture, line or curve structure, etc. are all modeled probabilistically as random fields and parameterized geometric structures, and estimation and decision theoretic techniques applied. We are presently studying this approach.

The value of the data modeling and statistical estimation approach is that performance analysis is possible and good algorithms can be designed for the data anticipated. This is necessary, since if the images are not noisy and computation cost is not a consideration, then any of many proposed algorithms appear to yield good boundary estimates or reliable object classification. It is when the images are highly variable and noisy that theory and insight become important, and it is toward this end that the work of our group has been directed.

The interesting work of Nahi [7] in which a maximum likelihood boundary estimator is used in a line-scanning mode bears some similarity to our approach. A brief comparison of the two approaches is given in the concluding section.

1.1. The Picture Function

The image x and y axes are quantized into intervals, and the image is therefore quantized into square picture elements, pixels. Numerical values are assigned to image gray levels. The resulting *picture function* at the (j, k)th pixel (jth row, kth column) is denoted g_{jk}. The picture function is modeled as a family of independent Gaussian random variables of constant variance σ^2. The function b_{jk}, the expectation of the picture function, is taken to be the constant r_{in} within the object and r_{out} in the background, with $\Delta \equiv r_{\text{in}} - r_{\text{out}} > 0$. Then $g_{jk} \equiv b_{jk} + n_{jk}$, where n_{jk} is a white Gaussian noise field and the image can be thought of as an underlying object of constant gray level, an underlying background of constant but different gray level, and the addition of a white Gaussian noise field (see Fig. 4).

1.2. A Few Examples of Boundary Estimation

Figures 7c and 2 are examples of artificially generated images of interest to us. The use of artificially generated data has the advantage of enabling us to experimentally determine the accuracy of our estimators by designing test patterns to test various properties of our algorithms.[1] We feel that the use of such data is crucial for developing algorithms for image analysis and might be viewed as analogous to the use of basic signal inputs such as impulses, steps, ramps, pure sinusoids and stochastic processes in testing control systems. In Fig. 7c, the object consists of a quantized circle with a rectangular protrusion. The signal-to-noise ratio, Δ/σ, is 3.2. Δ is the difference in average picture function values inside the object and the background, and σ is the standard deviation of the additive Gaussian noise. The boundary quantization occurs because the picture function is constant over pixels. In Fig. 2, the object is a quantized perturbed ellipse-like

[1] This also appears to be the philosophy of Riseman and Hanson in their very interesting and pattern recognition oriented work on scene segmentation, e.g. [8].

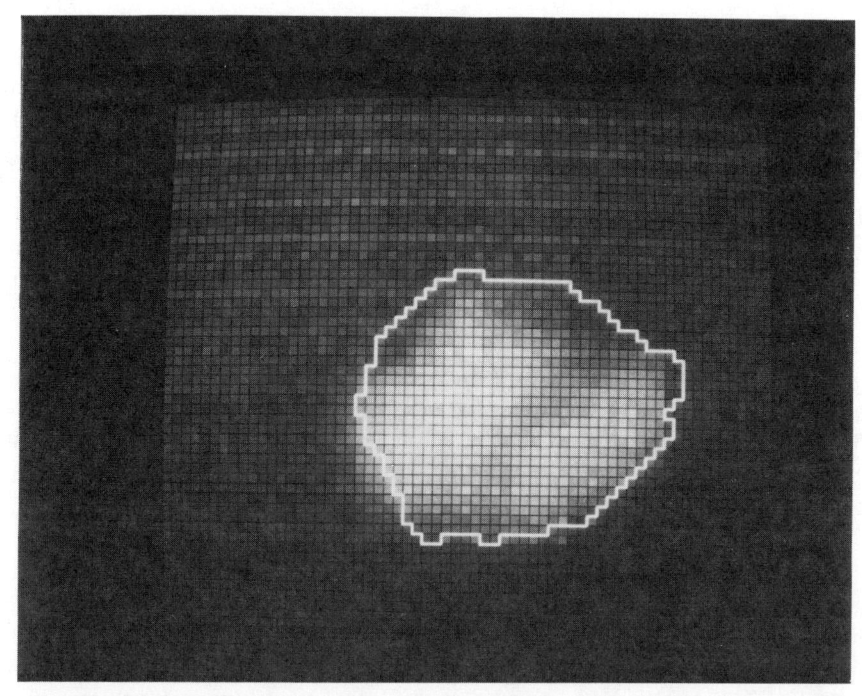

FIG. 1a. Infrared outdoor view of a tank, and an initially hypothesized boundary in white line.

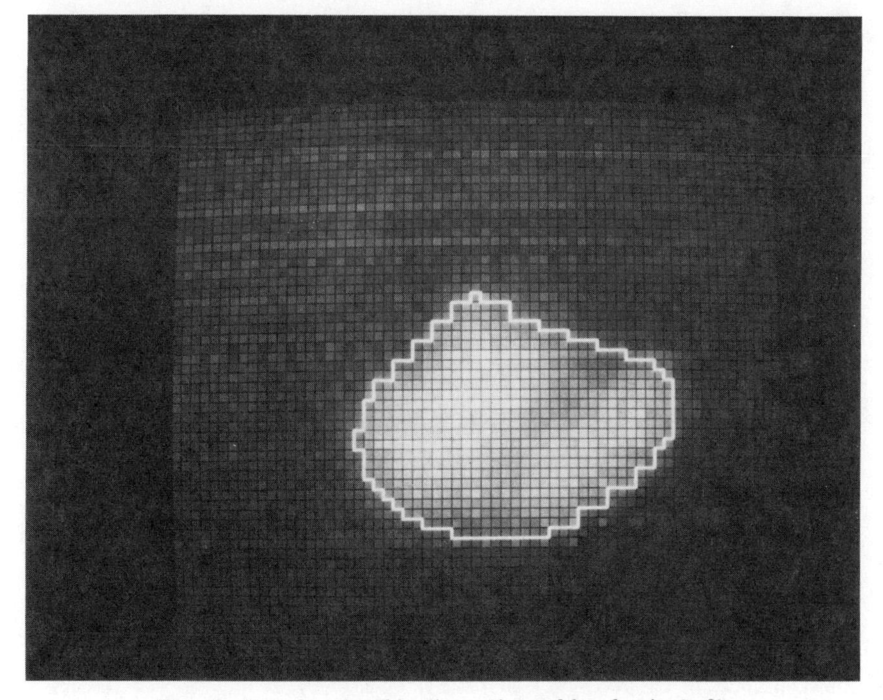

FIG. 1b. Boundary in white line estimated by the ripple filter.

figure. Δ/σ here is 1. This latter boundary was generated by driving an appropriate second-order difference equation with white noise and then quantizing the resulting boundary. Figure 7e shows in white line the boundary found by the ripple filter, and Fig. 3 shows the boundary found by the sequential boundary finder. The object pixel layer adjacent to the true boundary is marked by 0's, and the layer adjacent to the estimated boundary is marked by asterisks (*). Figures 1a and b show a magnified outdoor infrared image of a tank. The white-line boundaries in Figs. 1a and b are an initial guess at the object boundary and the ripple filter boundary estimate, respectively. Though the expectation of the picture function is not discontinuous in passing from object to background, the estimator works well. We estimate Δ/σ to be between 3 and 4 for this image. It should be emphasized that the ultimate accuracy with which highly variable random boundaries can be estimated is not as good as that with which smooth or deterministic boundaries can be estimated. Our sequential estimator, though suboptimal, appears to be highly accurate even at signal-to-noise ratios as low as 1.

2. BOUNDARY ESTIMATION AS LIKELIHOOD MAXIMIZATION

A data representation model is posed in this section, and boundary estimation is then formulated as a problem in likelihood maximization. Hence, the object here is to attack boundary finding by first carefully modeling the structure of the data and then directing the boundary search by the maximization of a performance functional. We consider this concept to be a general approach which can be extended to treating scenes of arbitrary complexity.

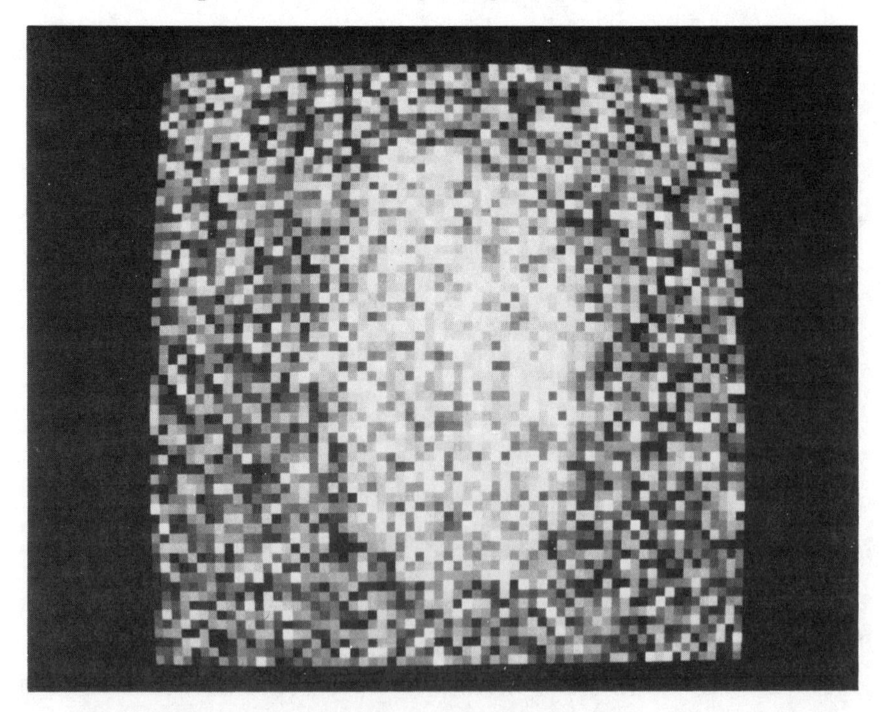

FIG. 2. Quantized perturbed ellipse with additive Gaussian noise. $\Delta/\sigma = 1$.

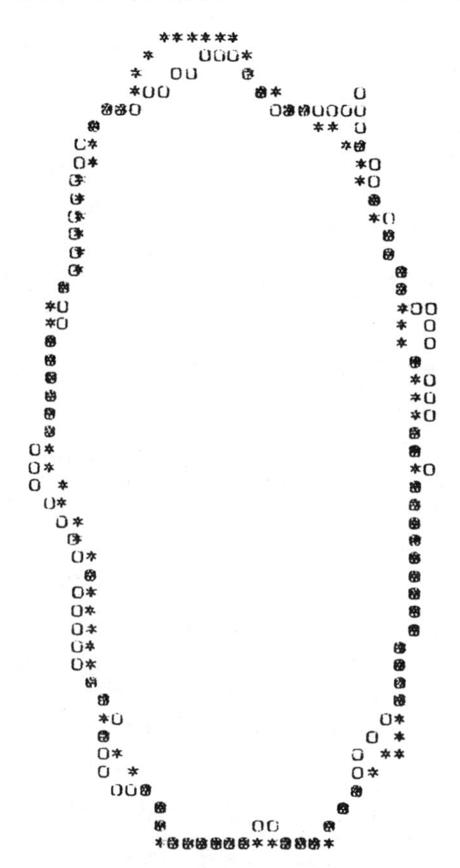

FIG. 3. Perturbed ellipse boundary marked by 0's and boundary estimated by Kalman filter sequential boundary finder marked by asterisks. $\Delta/\sigma = 1$.

With no loss of generality, assume $(r_{\text{in}} + r_{\text{out}})/2$ is subtracted from the picture function. Then the picture function g_{jk} used in the paper will have expected value $\Delta/2$ in the object and $-\Delta/2$ in the background. Let $P_G(g)$ denote the noise distribution. Let g_{jk} take a finite number of values, perhaps the 256 integer values 0, 1, 2, ..., 255. The $P_G(g)$ used in this paper is $(2\pi\sigma^2)^{-\frac{1}{2}} \exp[-(\frac{1}{2}/\sigma^2)g^2]$, a quantization of a Gaussian probability density function.

An edge element is a boundary between two adjacent pixels. An object boundary is a closed directed sequence of such edge elements which does not intersect itself. We arbitrarily label the edge elements in order, clockwise around the object, as $t_1, t_2, \ldots, t_N$ for a boundary of length N (see Fig. 4). We model the boundary $t_1, t_2, \ldots, t_N$ as a Kth-order Markov process. Given $t_1, \ldots, t_i$, t_{i+1} can be any of three edge elements. We treat the probabilities for these three possible choices as dependent on the preceding K edge elements. Thus, a state of the Markov process is a sequence of K edge elements.

$$P_B(t_i, t_{i-1}, \ldots, t_{i-K+1} | t_{i-1}, t_{i-2}, \ldots, t_{i-K})$$

denotes the transition probability to state $t_i, \ldots, t_{i-K+1}$ from state $t_{i-1}, \ldots, t_{i-K}$. Figure 5 illustrates the essence of one useful model for the state transition probabilities. Let $K = 7$ and consider the eight edge elements $t_{i-7}, t_{i-6}, \ldots, t_i$. Use angle

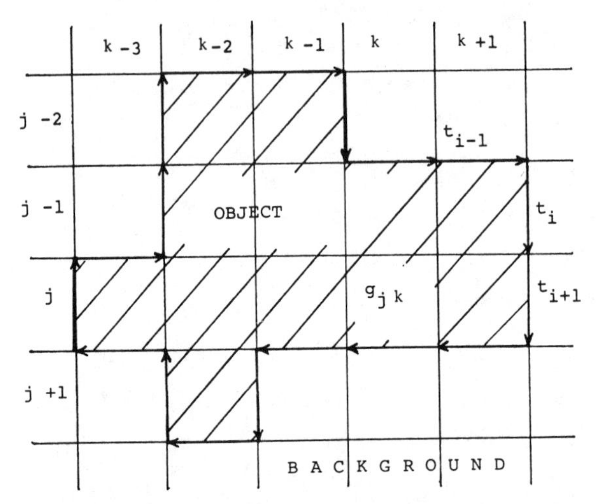

FIG. 4. Image data and boundary model.

θ_i as a measure of the curvature of the sequence $t_{i-7}, \ldots, t_i$. Now design the state transition probabilities for the three possible choices of t_i to be a decreasing function of $|\theta_i|$. Such a boundary generation mechanism favors smooth straight boundary segments.

The joint likelihood of a hypothesized boundary and the entire picture function can be written as

$$P\text{ (hypothesized boundary) } P\text{ (image date} \mid \text{hypothesized boundary)} \tag{1}$$

which in terms of the image model is

$$P(N)P(\text{starting point})P_{\mathrm{B}}(t_i, \ldots, t_K) \prod_{i=K+1}^{N} P_{\mathrm{B}}(t_i, \ldots, t_{i-K+1} \mid t_{i-1}, \ldots, t_{i-K})$$
$$\times \prod_{\substack{j,k \\ \text{within} \\ \text{object}}} P_{\mathrm{G}}(g_{jk} - \Delta/2) \times \prod_{\substack{j,k \\ \text{outside} \\ \text{object}}} P_{\mathrm{G}}(g_{jk} + \Delta/2). \tag{2}$$

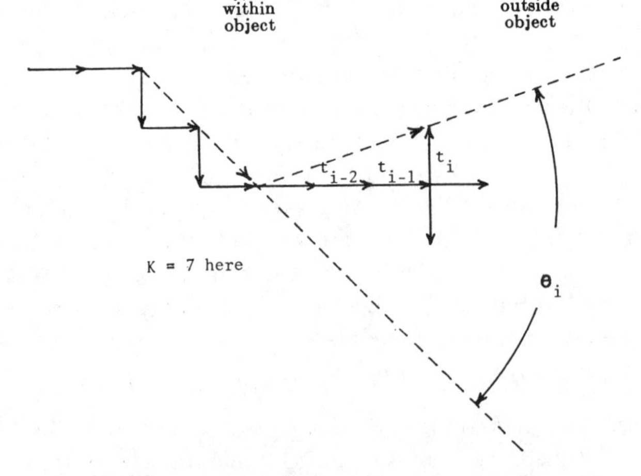

FIG. 5. A curvature boundary-process model.

The first line of (2) is the likelihood of a hypothesized boundary, with $P(N)$ denoting the a priori probability of boundary length N, $P_B(t_i, \ldots, t_K)$ denoting the likelihood of the initial state, and the product specifying the likelihood of the sequence of transitions in the boundary. The second line of (2) is the likelihood of the data given a hypothesized boundary. Upon taking the log of (2), expression (3) is easily derived:

$$\begin{aligned}
\ln \{P(\text{hypothesized boundary})\, P(\text{data} \mid \text{hypothesized boundary})\} \\
= \text{constant} + \ln P\{\text{starting point}\} + \ln P_B(t_1, \ldots, t_K) \\
+ \ln P(N) + \sum_{i=K+1}^{N} \ln P_B(\cdot \mid \cdot) \\
+ \sum_{\substack{\text{within}\\ \text{object}}} (\Delta/2\sigma^2) g_{jk} - \sum_{\substack{\text{outside}\\ \text{object}}} (\Delta/2\sigma^2) g_{jk}. \qquad (3)
\end{aligned}$$

By an *optimal boundary* we refer to one which maximizes (3) over all possible N, starting points, and $t_1, \ldots, t_N$. Note that though (3) is multimodal, the true boundary is likely to be on the most prominent hill, and it is for this reason that reasonably efficient hill climbing algorithms such as those to be discussed are possible, and computationally prohibitive exhaustive search is usually not required.

A few remarks on (3) are in order. The constant in the first line is a function of the data but is not a function of N nor $t_1, \ldots, t_N$. $\ln P_B(t_1, \ldots, t_K)$ contributes little to (3), except perhaps in the early part of sequential search. $P\{\text{starting point}\}$ may or may not be important. If this distribution function is very broad, we can ignore the first line and concentrate on the next two. If $P\{\text{starting point}\}$ is very peaked, its use is essentially in limiting the location of a segment of the boundary. $\ln P(N)$ is important, and we discuss it later.

$$\sum_{i=K+1}^{N} \ln P_B(\cdot \mid \cdot)$$

contains one summand for each edge element, t_i, in the hypothesized boundary, but the influence of each t_i is felt in more than one summand. The data enter in a very simple way in the last line of (3). If the picture function is multiplied by $(\Delta/2\sigma^2)$ at the outset, the data contribution is simply the sum of the weighted picture function values inside and the sum of the negative of the weighted picture function values outside the hypothesized object boundary, respectively.

2.1. *Image-Dependent Noise*

A slight but significant generalization of the data model is to let $\sigma = \sigma_{\text{in}}$ within the object and $\sigma = \sigma_{\text{out}}$ in the background. This situation will pertain if the noise is due largely to the detector. A model for the picture function which is often appropriate then is a set of independent Poisson random variables, in which case the mean and the variance of the picture function at a pixel are proportional to one another. Since the Gaussian approximation is usually appropriate, the Gaussian random field model is valid for the picture function, but the variances inside and outside the object will differ. The joint loglikelihood of a hypothesized boundary and the data will be a modification of (3), for this new data model, with

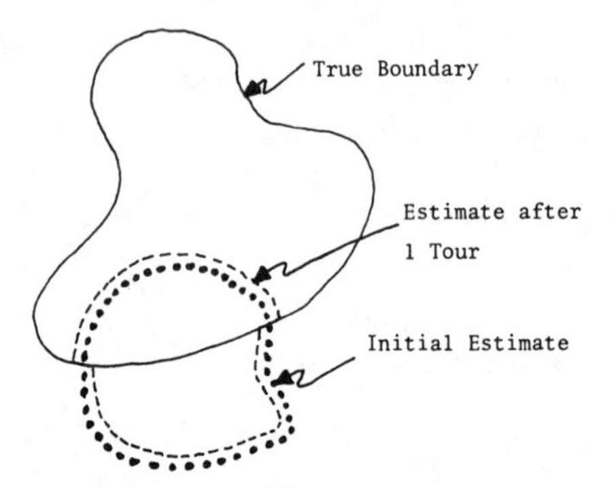

FIG. 6. Ripple filter boundary estimate after one tour.

the bottom data-dependent line now containing weighted sums of the g_{jk} and weighted sums of the g_{jk}^2 as well. Though the required loglikelihood maximization for boundary estimation will then be computationally more complex, the problem is still computationally manageable and conceptually no more difficult.

2.2. *Comments on Boundary Models*

Section 5.1 contains a discussion of Markov process boundary models. In this section, we discuss a few points concerning boundary models generally. A one-sided Markov process may seem a contrived boundary model. However, it is easily seen from the first line of (2) that, for $K < i < N - K$,

$$P_B(t_i | t_1, t_2, \ldots, t_{i-1}, t_{i+1}, \ldots, t_N) = P_B(t_i | t_{i-K}, \ldots, t_{i-1}, t_{i+1}, \ldots, t_{i+K})$$

so that t_i has a two-sided dependence on the K edge elements to either side. Furthermore, for many cases of interest this dependence will be symmetric in the forward and backward directions. Such is the case in most of the examples treated in this paper. (Also, see Section 6.3.) Hence, the one-sided Markov process can be designed to have the two-sided dependence that is intuitively desirable. (A brief pertinent discussion of two-sided Markov process representations is given in [25].)

Another point to be made here is that the boundaries generated by a process favoring straight lines will obviously not be closed self-nonintersecting boundaries. This lack of global structure is often unimportant, since the processes will have the local and semiglobal structure that is usually important for highly variable object boundaries. However, it is possible to have global structure as well, as we discuss in Section 5.1.

An edge element sequence modeled as a Markov process obviously is not the only useful boundary representation. Boundaries can be represented in polar coordinates, or as a two-dimensional vector process of some index parameter (e.g., time), or in terms of more exotic curvilinear coordinate systems which may be hierarchical and may be partially characterized by parametric curves such as polygons, rectangles, circles, etc., as discussed in [9, 10] and more recently in [11]. The models can be largely local, or they can also capture global structure as in the case of the hierarchical systems or through the use of Fourier series. Through use of Fourier series it is possible to specify bona fide probability measures for closed boundaries, but Fourier series appear to be computationally less attractive for estimating highly variable boundaries than the Markov process models we have used. The point is that there are many ways for modeling boundaries of interest and computing the likelihood of a hypothesized boundary. The most effective algorithm for maximizing the joint loglikelihood of the data and a hypothesized boundary will depend on the boundary model used.

3. THE RIPPLE FILTER: A REGION GROWING/SHRINKING BOUNDARY ESTIMATOR

The concept of the ripple filter introduced in [3] is the following. Assume an initial boundary estimate $\bar{t}_1, \bar{t}_2, \ldots$ has been obtained in some way. Then each pixel having an edge element of the initial boundary estimate as a side is examined in order to determine whether to maintain its classification or to change its classification from "object" to "background" or vice versa depending on whether it is inside or outside the initially estimated boundary. The decision is based on whether the change increases or decreases the loglikelihood of the hypothesized boundary and the image data. There are many ways of implementing such an approach, perhaps the simplest being to *ripple* around the initial boundary estimate, sequentially testing pixels adjacent to the initial boundary estimate and reclassifying each pixel upon testing it. The boundary is therefore reestimated as pixels adjacent to the boundary are tested. In the absence of noise, the change in the estimated boundary at the completion of one tour around the boundary is illustrated in Fig. 6. After such a tour, the process is repeated until no further boundary changes take place. A stable equilibrium will occur at a local maximum of the loglikelihood function. Our first generation algorithm works well in the presence of a modest amount of noise, $\Delta/\sigma = 3$. (The implementation is discussed in [6].) If $\Delta/\sigma \ll 3$, the system will not reach the true boundary. If the boundary model is stiff, one or more noisy pixels will act as a barrier, and the estimate will not proceed further, Fig. 7a. If the boundary model is highly flexible the estimate will bend around the noisy pixels, leaving long tentacle-like structures, Fig. 7b. Figures 7c, d, and e illustrate the behavior of this algorithm for $\Delta/\sigma = 3.2$. Since $\Delta/\sigma = 3.2$ is fairly large, the true boundary of the object, a quantized circle with a projection, is readily apparent. Figure 7c shows the image data and the initial boundary guess. Figure 7d shows the data and the estimated boundary after nine tours. Figure 7e shows the data and the stable boundary estimate reached after 28 tours. Note that this approach is one in which the estimate at any stage of the procedure is a closed boundary with simply connected interior (and exterior), thus utilizing the object-background topology of the image structure throughout the estimation computation. The algorithm is simple; there is no backtracking. In these respects, the algorithm shares the simplicity of the relaxation approach introduced by Rosenfeld and colleagues [12]. In operation, the ripple filter oozes along, growing into the object and shrinking within the background. Unfortunately the simple algorithm sketched is slow. It is slow because each edge element in a hypothesized boundary influences K summands in (3). A change in pixel classification during operation of the algorithm requires removing more than K summands from (3) and replacing them by more than K others. However, the basic ripple filter concept is *attractive* and *sound*, and improvements for eliminating its drawbacks are discussed in the next section.

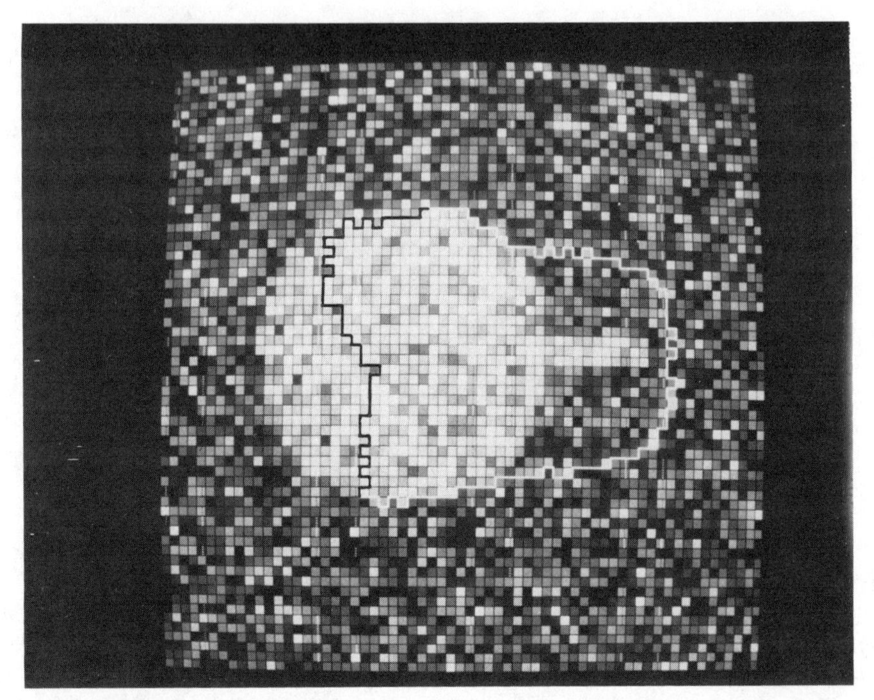

FIG. 7a. Ripple filter final boundary estimate using stiff boundary model. Estimate shown in white and black line. $\Delta/\sigma = 2$.

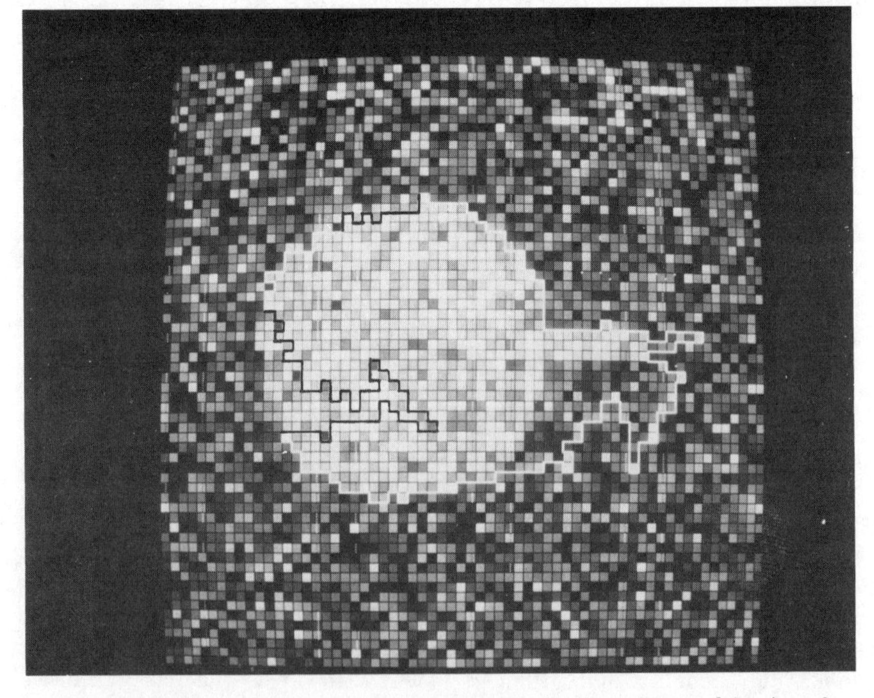

FIG. 7b. Ripple filter final boundary estimate using flexible boundary model. $\Delta/\sigma = 2$.

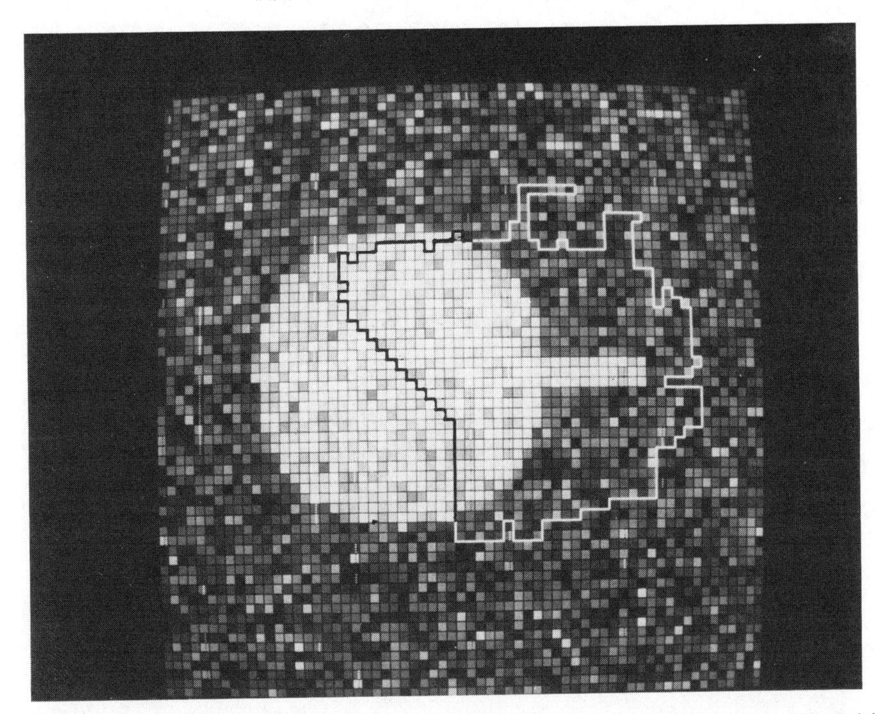

FIG. 7c. Quantized circle with projection, and additive Gaussian noise. $\Delta/\sigma = 3.2$. Initially hypothesized boundary shown in white and black line.

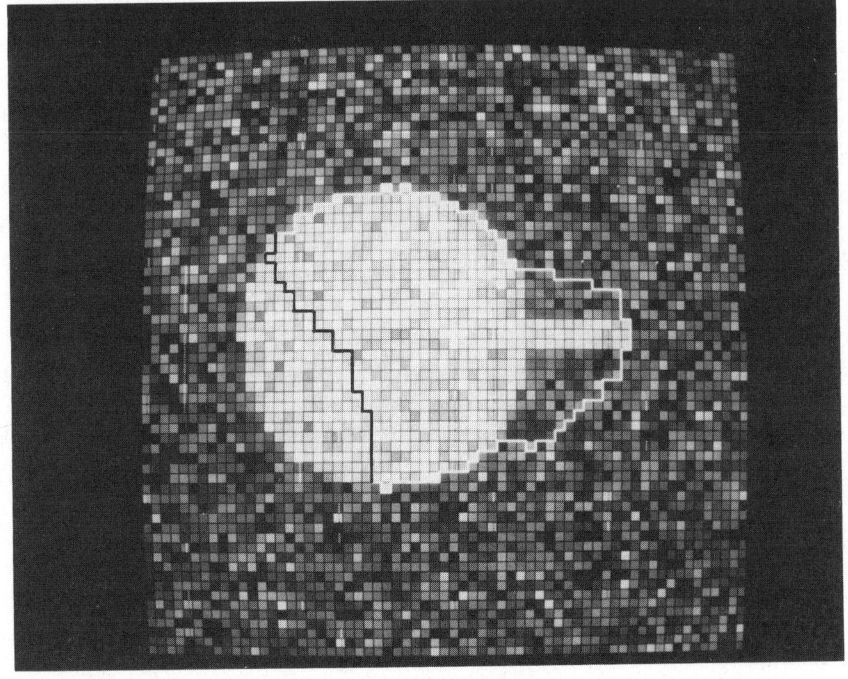

FIG. 7d. Ripple filter boundary estimate after nine tours. $\Delta/\sigma = 3.2$.

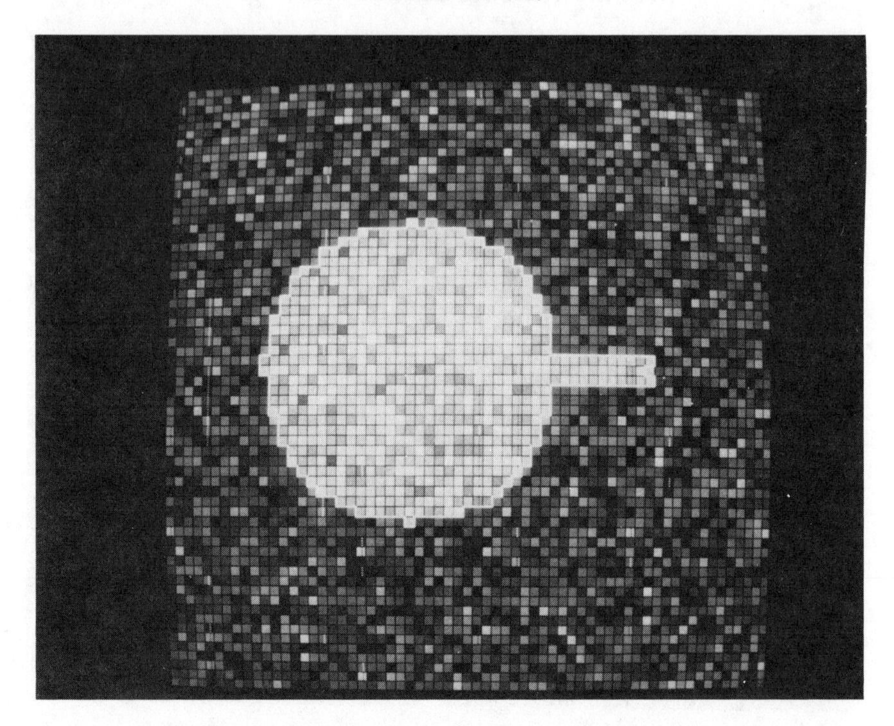

FIG. 7e. Ripple filter boundary estimate after 28 tours. $\Delta/\sigma = 3.2$.

The basic perturbation algorithm is illustrated in Figs. 8a and b. Figure 8a illustrates the simplest case. The perturbation is to be across the edge element t_i which is followed by an edge element in the same direction. The ripple filter conducts a three-way test here, namely, keep the boundary as it was, reclassify pixel A as *object*, or relcassify pixel B as *background*. Figure 8b illustrates a slightly more complicated case. Here, the edge element to be perturbed, t_i, occurs at a corner. A four-way test is now used, namely, keep the boundary as it is, reclassify pixel A as *object*, reclassify pixel B as *background*, or reclassify pixel C as *object*. Why use a three-way test in one case and a four-way in another rather than simply use two-way tests? Suppose for the case of Fig. 8a, a two-way test is used for deciding between the boundary as it is or reclassifying pixel B as *background*. Suppose the latter choice has a larger associated loglikelihood but that reclassifying pixel A as *object* has the largest associated loglikelihood. Then the two-way test results in a suboptimum solution which will have to be corrected during subsequent tours. In general, the larger the number of simultaneously considered choices, the smaller is the chance of an incorrect decision. But of course, the amount of computation involved in a test is an increasing function of the number of choices. The three- and four-way tests discussed are simple and appear to be reasonable compromises for the associated geometries.

3.1. Extensions and Comments

We comment on two generalizations of the preceding model. First, if Δ and σ are a priori partially or completely unknown, they can be estimated. If, for

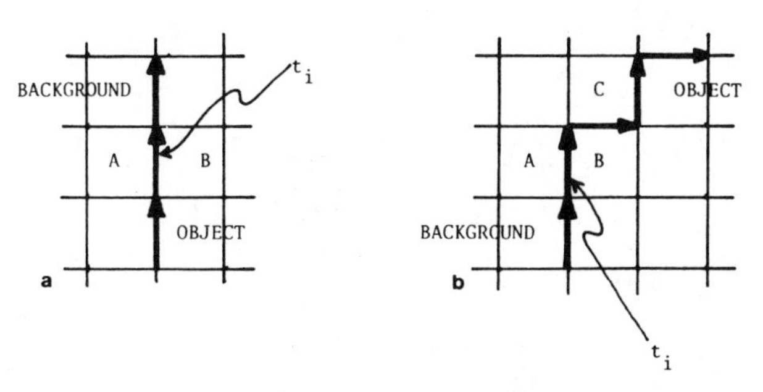

FIG. 8. (a) Ripple filter three-way test. (b) Ripple filter four-way test.

example, Δ and σ are known to be constant, they can be estimated from a histogram of all the data prior to the boundary search. Alternatively, if Δ and σ are a priori partially known, updated estimates can be computed during the boundary finding process. A number of obvious ad hoc schemes can be concocted for guessing which patches are appropriate for estimating r_{in} and which are appropriate for estimating r_{out}. The second comment is about the behavior of the algorithm when two or three objects are present in the field. Well-separated objects should lead to ripple filter behavior as illustrated in Fig. 9a. The initial boundary estimate is assumed to intersect the three objects, and we also assume that the same r_{in} and boundary model are appropriate for all three objects. Then the ripple filter estimate should flow into the objects and also shrink down about the objects, finally stabilizing with a simply connected region such as that shown. The tubes exist because the ripple filter as presently implemented contains a test which precludes splitting the estimated object into two. When waists are detected in the estimated boundary, simple tests can be used, based on local measurements and global information, for deciding whether or not to divide the estimated object into two or more parts. In Fig. 9b, the two objects are close together, and whether the final estimated boundary is as shown or is more like the convex hull of the union of the two objects will depend on how stiff a boundary model is used (i.e., whether for most states one of the three state transition probabilities is much larger than the other two).

Many variations of the algorithm are possible, and whereas the true data and boundary model should be used in the vicinity of the true boundary, *modifications of the algorithm may result in faster and more robust behavior during the earlier stages of the estimation process.*

3.2. Speed-Up of the Ripple Filter

In this section we introduce three modifications designed to reduce the number of iterations required for convergence, reduce the required computation per iteration, and increase the estimation accuracy in a noisy environment.

3.2.1. Look-up table. In experiments with our first sequential boundary finder, it was observed that the curvature measure used (that of Fig. 5) had significant drawbacks, one being a tendency toward oscillatory boundary estimates.

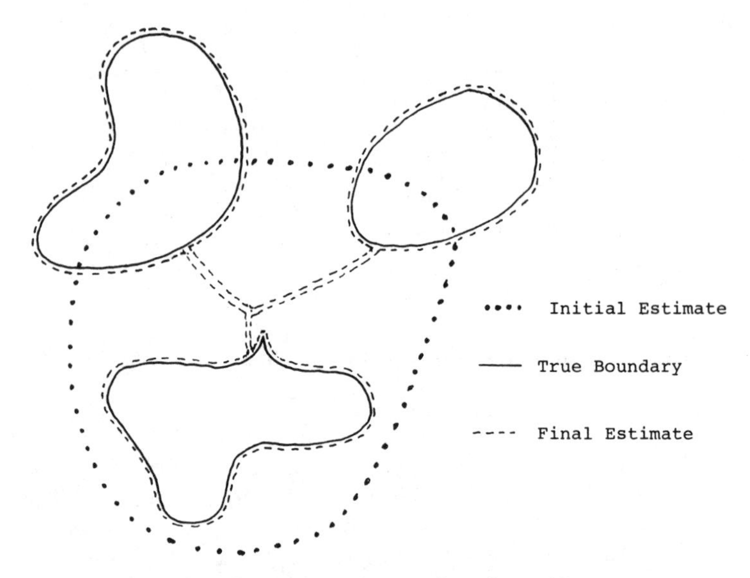

FIG. 9a. Ripple filter boundary estimate when three objects are present.

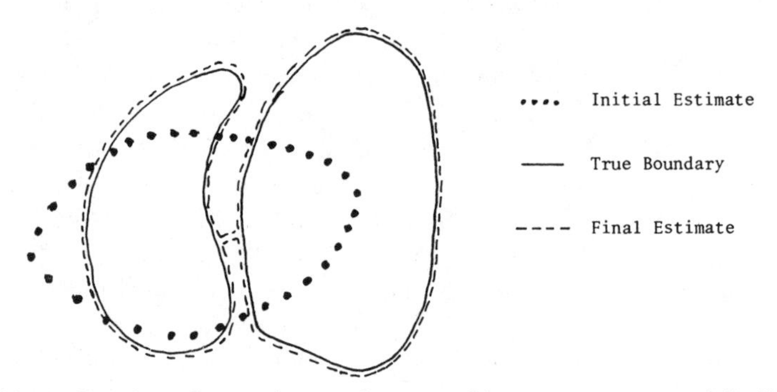

FIG. 9b. Ripple filter boundary estimate when two objects are present and flexible boundary model is used.

In addition to these shortcomings, the amount of computation required in its use in order to compute three state transition probabilities is considerable. Our solution [13, 14] was to introduce a look-up table of state transition probabilities. The entries in the table are state transition weights for all 4×3^7 edge sequences of length 8, the number used in the curvature measure in Fig. 5. State transition probabilities are directly proportional to state transition weights. The probabilities of the transitions from a specific state are obtained by summing the state transition weights and dividing each weight by the sum, to obtain probabilities. In practice, sequences of 8 edge elements are grouped into physically meaningful classes and a state transition weight is assigned to each class. Hence, to use the table the system merely identifies the state transition class and then looks up the associated transition weight. The weight must then be normalized through dividing by the sum of the weight and two other appropriate weights obtained from the table. This provides the required transition probability. The nicest feature of the table is that the designer can design the table to favor any boundary type of interest. This design procedure can be as simple as to assign high probabilities to those edge sequences that look correct to the eye. Furthermore, if simple properties of the edge sequence can be measured in order to group the sequences into meaningful classes, then the state transition probability for an edge sequence can be determined with a small amount of computation. The allowable edge sequences have a simple graph representation, Fig. 10. The graph represents all possible sequences of four edge elements where the tail of the first edge element is the center node of the graph and the head of the fourth edge element is one of the other points on the graph. Assume that the transition probability depends only on the angle, as in Fig. 5, between two vectors each representing a sequence of four edge elements. Then if the first such vector is in the upward vertical direction and its head enters the graph of Fig. 10 at the center node, all possible angles of interest are those between the radial lines in the graph and the positive vertical axis. Note that there is an inner layer and an outer layer of nodes in the graph. The inner layer corresponds to wiggly boundary segments, and the outer layer to straighter segments. Hence, to favor boundary segments which are straightish lines, if two nodes in the Fig. 10 graph lie along the same radial line, the outer node should have a higher associated probability than the inner node.

There are seven angle magnitudes between a vector which has its head at the center node and a vector which runs from the center node to one of the other nodes in Fig. 10. A useful generalization is to index these angles in accordance with increasing magnitude and assign transition weights based on index value rather than specific angle magnitude. Hence, the transition probability associated

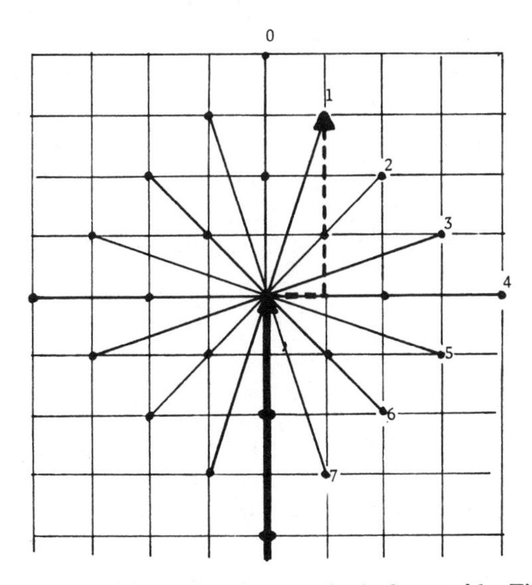

Fig. 10. All sequences of second four edge elements for look-up table. First four edge elements are assumed to be vertical vector.

with the fourth largest angle, e.g., will be the same irrespective of the direction of the first vector, i.e. the one having its head at the center node.

3.2.2. Alternative state structures. For historical reasons, the state model we have used a great deal is that previously described, namely, a sequence of seven edge elements. In boundary generation it has been assumed that edge elements are added one at a time, each concatenation depending on the immediately preceding seven edge elements. That means that each edge element in a hypothesized boundary occurs in eight states and influences eight summands in (3). Suppose that boundary curvature is still measured in terms of a sequence of eight edge elements as in Fig. 5, but that the Markov process used is such that four edge elements are added each time and a state consists of four edge elements. Then an edge element in a hypothesized boundary occurs in only *one state* and influences only *two summands* in the modified sum which would be the appropriate replacement for (3). Hence, much less computation is then required in recomputing the joint loglikelihood each time the estimated boundary is perturbed during operation of the ripple filter. The nature of the boundary generated by this model is somewhat different from that generated when one edge element is added at a time. The new model is less influenced by the boundary generated up to the point at which the extension is made. A number of practical problems arise in implementing the new model and compromises are necessary, but the problems are all manageable. Perhaps the seemingly most difficult problem is the computation of state transition probabilities when four edge elements are added at each iteration. The stumbling block here is that if something like the curvature measure of Fig. 5 is used for determining state transition probabilities, there are 3^4 possible transitions from a state to a following state, and a state transition associated weight must be divided by the sum of the state transition weights in order to determine the state transition probability. Hence, determining the boundary likelihood associated with a perturbation during one perturbation cycle of the Ripple Filter requires a prohibitive amount of computation simply in order to obtain a transition probability from a transition curvature measure. The look-up table comes to the rescue here, since the weights for all 3^4 possible transitions can be summed off line, and look-up-table entries can be the state transition probabilities rather than just the state transition weights.

3.2.3. Varying resolutions. Suppose the image is very noisy and a segment of estimated boundary is smooth and nearly straight, i.e., the situation represented in Fig. 8a. In addition, suppose the true object extends further to the left than the estimate shown. Then if the boundary is perturbed correctly in order that the estimated object include pixel A, the new value of (3) may not be higher than the value for the boundary prior to perturbation. This condition can occur because the new boundary will be more wiggly and its loglikelihood will therefore be lower than prior to perturbation. Furthermore, since the image is noisy, including pixel A in the new estimated object may not increase the data loglikelihood by an amount larger than the decrease in the boundary loglikelihood. There are a number of ways around this dilemma, one being to operate the ripple filter at resolutions of various coarseness. For example, if an image is 512×512 pixels, the ripple filter can first be applied to coarse pixels, where each such pixel is an

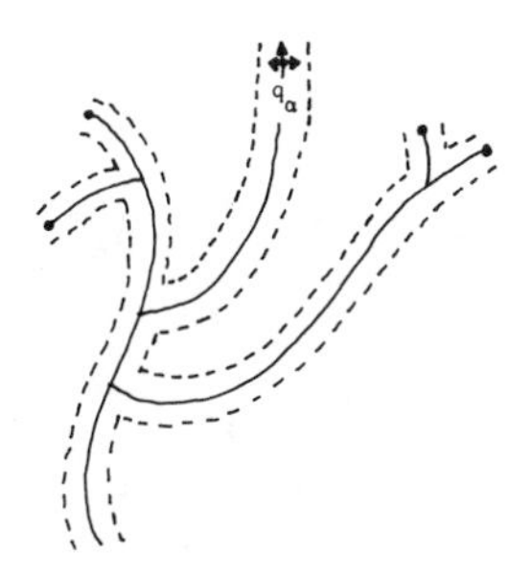

FIG. 11. Boundary segments found by sequential boundary finder at an intermediate stage.

average over a 16 × 16 or smaller block of the original pixels. Then reapply the ripple filter to 8 × 8 or 4 × 4 blocks, etc. These coarser pixels are less noisy. Hence, this approach should provide the required boundary motion plus increased speed of convergence, since if an initially estimated boundary must move over by roughly q pixels, this can now be done in fewer than q tours. (Use of varying resolution in scene analysis of course has a prior history [15, 16].) The final comment is that movement of the ripple filter in the vicinity of the true boundary may still be hampered by the additive noise. For example, the ripple filter may not be able to grow an estimate out into a projection. It is much easier for the filter to shrink an estimate down around a projection. There are a number of possible solutions to this problem, perhaps the most straightforward being to use the basic algorithm until convergence is achieved (hopefully in the vicinity of the true boundary) and then a systematic exhaustive perturbation through all local boundary shapes of possible interest. We are presently exploring this matter.

3.3. Significance of the Ripple Filter

The significance of the Ripple Filter can now be appreciated. If the boundary model and picture function distribution are correct, the joint likelihood of a hypothesized boundary and all the data completely characterizes the useful information required for an *optimal* boundary estimate. Maximization of this joint likelihood provides a good estimate which has asymptotically minimum estimation error properties as discussed in Sections 6.0–6.3. The Ripple Filter directly maximizes the joint likelihood function, and hence can be said to make *full use* of the data and a priori model information.

4. SEQUENTIAL BOUNDARY FINDER

This section is devoted to a very brief discussion of the *Sequential Boundary Finder*, an alternative to the ripple filter for maximizing (3). The SBF is a stochastic generalization of a deterministic algorithm appearing earlier in [1]. Many of the issues involved in understanding the SFB are subtle and require considerable development for exposition. Consequently, much of this section is a summary of a few portions of [5]; extensive detail will appear in [13, 14, 17]. The concept of the SBF is to start off with an edge element reliably determined to be on the true boundary, and to then sequentially estimate the true boundary. Edge elements

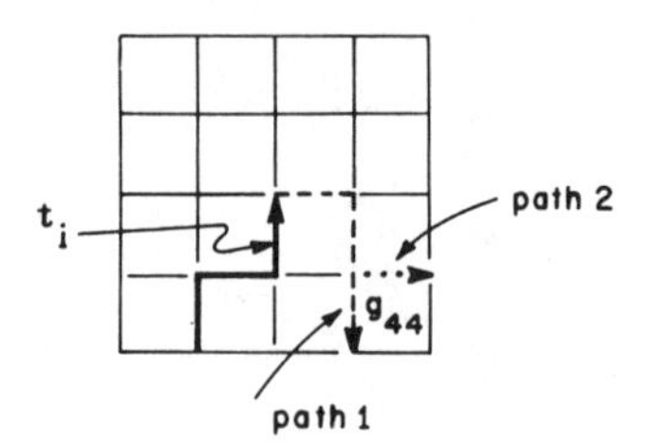

FIG. 12. Two paths through a window.

are concatenated one at a time. If the system decides it has deviated from the true boundary, it backtracks and explores a new path. Suppose the search is at an estimated boundary point i edge elements from the start. Denote the state at that point by q_α. The question of interest here is, How is the system to assign a joint loglikelihood to this boundary segment and the entire picture function in order to maximize (3)? Since the accuracy of the boundary estimator depends on data only in the vicinity of the true boundary, one solution is to use a subset of the data in the image, specifically, a narrow swath about the boundary segment under exploration. In the remainder of this section we comment on the swath and associated likelihood concept, locally optimal decisions, an *analytical* tool for *comparing the relative accuracies of algorithms*, and touch briefly on alternative boundary models. We consider the analysis technique extensively developed in [4, 13, 14] to be an extremely important design tool.

The essence of our approach is the following. At each stage of the boundary estimation process, associated with the *end state* in any path explored is a loglikelihood or *pseudo-loglikelihood* based on the data in the path swath. (See Fig. 11). These states with their loglikelihoods are stored on an *open* list. The path with the greatest loglikelihood is the one which is next extended by one edge element. Suppose q_α is the state stored on the open list for the path. Following the extension, q_α is removed from the open list, and three new state nodes are entered, corresponding to the three possible extensions by one edge element. Conceptually, states can be viewed as nodes in a graph, and arcs between nodes will have associated loglikelihoods, the loglikelihood of a path between the two nodes. An optimal boundary is one which returns to the start node and has an associated maximal loglikelihood. The A* algorithm (branch and bound) from the artificial intelligence literature is the search algorithm we use. The search procedure, as we use it, is not guaranteed to produce an optimal boundary because the increment to the path pseudo-loglikelihood upon adding a new edge element is not always negative. However, in the algorithm presently in use the expectation of this increment is less than or equal to 0, and the algorithm works well.

4.1. Locally Optimum Data Usage (Data Windows and Locally Recursive Template Matching)

The algorithm used is *locally optimum* in the sense that it does use *common data* in making a true maximum likelihood decision among the three successors of edge element t_i. The mechanism for this is to put a window around t_i as shown in Fig. 12. Though any size may be used, a 4×4 pixel window has been used in our experiments. Since the three paths under consideration are identical up to node q_α, the data swaths used in computing their associated loglikelihoods are identical, and the ways the data contribute to the three loglikelihoods differ only for the data within the window. Hence, the window is to be thought of as a local universe about the hypothesized boundary in the vicinity of t_i, and the intention is for the algorithm to make optimal use of this window data. Toward this end, a subset of all paths which pass through t_i and reach the window boundary is considered. These represent an approximation to all reasonable paths which include t_i and pass through the window, and subgroups of these paths are associated with each of the three possible choices for t_{i+1}. Two of the paths extending q_α to the right are shown in Fig. 12. The loglikelihoods for all paths considered are computed, and the choice t_{i+1} to extend q_α is that extension for which one of the associated paths has a maximal loglikelihood. Though the computation of the data contribution to the loglikelihood for each of the paths we use appears to be considerable, it is in fact not because the contributions can be computed recursively. For example, the pixel (4, 4) is the only one in the window which is classified differently for paths 1 and 2. Hence, the data contribution to the loglikelihood associated with path 2 will differ from that associated with path 1 by the amount $(\Delta/\sigma^2)g_{44}$. We consider the window concept for locally optimum data usage for path choice, and recursive template matching for many highly variable paths through a window to be two very important conceptual contributions of [5]. This example (discussed in detail in [5]) appears to introduce a new and important class of problems. The problem is, given a window and a family of objects with highly variable boundaries occuring in the window, design a useful class of templates and a matching algorithm so that the template matches can be made with an amount of computation small compared with the total amount of computation required by matching the templates to the data individually. Of course, the boundary finding problem is more complicated because boundary likelihood must be computed in addition to data likelihood. One very effective approach to this problem has already been formulated [19], and others are presently under investigation.

Section 6.2 contains material which provides additional insight into the benefit of usage of the local window.

4.2. Data Swaths and Loglikelihood Functions

The data swath we associate with a path is the swath swept out by the sequence of windows used on the path. If the obvious joint loglikelihood for the path and data are used, peculiarities in algorithm performance occur which are undesirable. Good estimation accuracy and computational efficiency require the use of *pseudo-joint-loglikelihood functions.* These conclusions are based not only on experimentation, but more importantly on theoretical analysis. We have introduced the analysis technique of computing the probability that the estimator under study

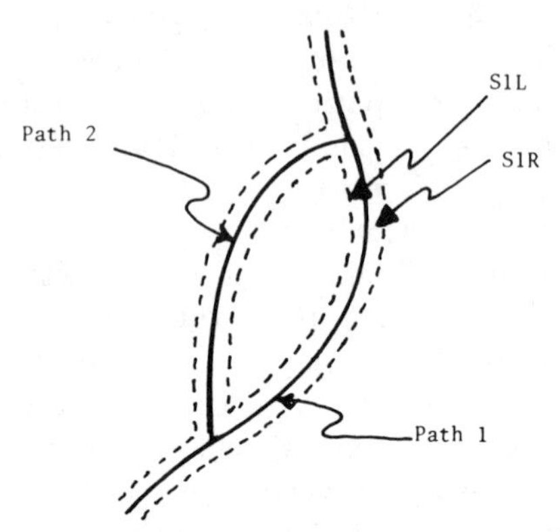

FIG. 13. Two hypothesized boundary segments and associated data swaths for sequential boundary finder.

will choose a specific incorrect path instead of a specific correct path. For example, paths 1 and 2 in Fig. 13 might be two such paths. This computation is carried out for all reasonably likely pairs of correct and incorrect paths, and the results tabulated for two or more different estimators provides a measure of comparison of the relative accuracy of the estimators. The method is especially meaningful for paths which differ only over short intervals and are the same otherwise. The method is developed and extensively applied in [3, 5, 13]. It turns out that the choice of pseudo-loglikelihood function used affects both estimation accuracy and the amount of backtracking (and hence computation) incurred.

The pseudo-loglikelihood for a hypothesized path used in obtaining the results shown in Fig. 3 is

$$\ln P(N) + \ln P\{\text{starting point}\} + \ln P(t_1, \ldots, t_K) + \sum_{\text{path}} \ln P_{\text{B}}(\cdot \mid \cdot) + \sum_{\text{SR}} \Delta(g_{jk} - \Delta) - \sum_{\text{SL}} \Delta(g_{jk} + \Delta),$$

where SR and SL denote the swath pixel-set to the right and to the left, respectively, of the hypothesized boundary. $\ln P\{\text{starting point}\}$ is not used in Fig. 3 because it is assumed that a starting point on or next to the boundary is found through ad hoc means. $P(N)$ is important for object recognition purposes, but its role in sequential estimation is not completely understood. Some insight into this matter is provided in Section 5.2 and in [13]. Note that each pixel in a swath contributes to the above likelihood only once. This and the fact that the algorithm makes local decisions in an optimal way give our algorithm a greater accuracy for curvy boundaries than is possible through the standard approach of using large overlapping templates throughout the image for local gradient measure and then some linking algorithm for linking together those edge elements which have large associated gradients. (See [13] for one comparison.)

Reference [13] contains extensive discussion of the best use of windows and of the relative merits of various pseudo-loglikelihood functions. It is pointed out there that it may be advantageous to estimate a complete path through a window and then go on to the next adjacent window rather than use a moving window as is done in the algorithm just described.

5. MORE ON BOUNDARY MODELS

5.1. Boundary Models

We have explored two types of boundary models for the sequential finder, either of which can have global or local structure, or both. The first is the curvature model previously described. This representation should be thought of as change in angle as a function of arc length as a parameter. The other representation is x, y position as a function of some parameter (the parameter can be thought of as time), through use of dynamical systems models. Among the desired properties of a good model are global structure which might favor high curvatures over certain sections of boundary, and low curvatures over others, or perhaps specific shapes over certain portions of the boundary; and local structure which might mean the wiggliness (harmonic content) of the boundary locally. Such a model is useful for boundary estimation but can also be useful for object recognition. It is well known that discrete-time linear dynamical systems driven by sequences of independent identically distributed random variables generate Markov processes. These systems provide good models for many object boundaries of interest. For example, the model

$$\begin{pmatrix} x(m+1) \\ y(m+1) \end{pmatrix} = A\begin{pmatrix} x(m) \\ y(m) \end{pmatrix} + b + \xi(m), \tag{4}$$

where $[x(m), y(m)]^T$ represents the mth *unquantized boundary point* in the plane, can generate a perturbed ellipse. If the A matrix is chosen properly, the unforced solution ($\xi(m) = 0$) will be an ellipse, and the forced solution will be a perturbed ellipse. The boundary marked by 0's in Fig. 3 is one generated by such a system. Since (4) is a lossless system, its response might be viewed as a generalization of the integral of white noise and the boundaries generated by the model have a probability less than $\frac{1}{2}$ of closing. This closure will not be exactly at the starting point. The points generated are also not equispaced in the plane. Nevertheless, if boundaries generated by (4) are quantized, they do have local and global properties which are characteristic of many boundaries of interest to us, and boundary finders built around this model perform very well. By going to a fourth-order model (rather than the second-order model in (4)), one can control global behavior but also local behavior more interestingly. Specifically, one can generate an *expected* trajectory given by a second-order model such as that in (4) but operating in the autonomous mode, and add to this a stationary contribution due to driving a lossy second-order system by white noise. The resulting fourth-order system will generate *almost closed* ellipse-like boundaries, but the boundaries will be perturbed by a low-pass stationary random process with bandwidth determined by a pair of the system's poles.

Computation of the loglikelihood of a path generated by (4) is simple, and is based on use of a one-step-ahead predictor to compute the conditional mean

$$E\{[x(m+1), y(m+1)]^T \,|\, [x(m), y(m)]^T, \ldots, [x(0), y(0)]^T\} \tag{5}$$

FIG. 14. A wiggly boundary segment.

(see [4]). However, the pixel quantization of permissible boundary paths destroys the Markov process boundary model. We handle this by treating the quantization effect as additive noise and replace (5) by a conditional expectation computed by a Kalman filter [13, 23]. Theoretical justification for this is questionable because the boundary quantization amounts almost to hard-limiting; nevertheless the algorithm performs very well in practice. See Fig. 3.

Whereas a sequential boundary estimator based on the boundary model of Fig. 5 works well down to $\Delta/\sigma = 1$, its accuracy is not quite as good as anticipated. A number of aspects of this problem are under study.

A number of other important matters are covered in [13], e.g., the important role of the hypothesized boundary likelihood in reducing the amount of search required in sequential estimation and in the size of estimation errors.

5.2. *The Role of* $P(N)$

The a priori length distribution $P(N)$ serves two purposes. First it prevents the finding of a boundary of length radically different from the range anticipated. For example, it prevents an estimated boundary of length 20 edge elements when a length an order of magnitude larger is anticipated. Second, it has an influence on the occurrence of small estimation errors. For example, suppose that the three state transition probabilities for adding the next edge element at any stage of boundary estimation are each $\frac{1}{3}$. Then all $3^{N'-1}$ boundary segments beginning at a common point and of length N' edge elements are equally likely. Consider the estimation of a boundary segment between points a and b, as shown in Fig. 14. The likelihood of the true boundary segment of length M' is $3^{-M'}$, whereas the loglikelihood of an erroneous segment of length M'' with $M'' < M'$ is $3^{-M''}$. Hence, the contribution to (4) of the hypothesized boundary likelihood tends to favor the shortest boundaries even though they may be greatly in error. More generally, if the boundary generating Markov process is stationary, long generatable boundary sequences will fall into a high probability and a low probability

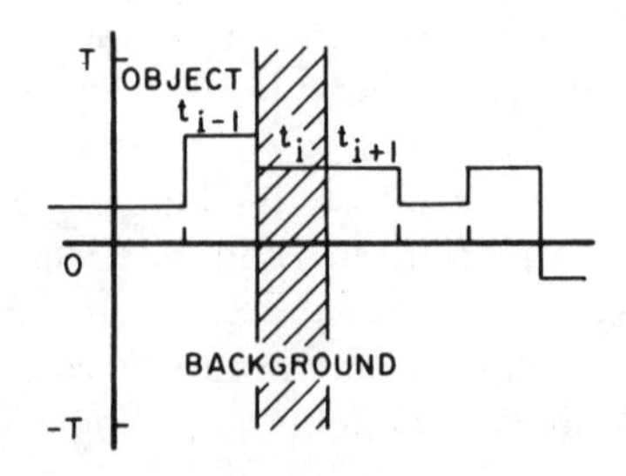

FIG. 15. A boundary segment and a data strip.

class. The former contains *roughly* e^{HN} equally likely sequences of length N where H is the entropy in nats of the state transition probability distribution [18]. The loglikelihood of each such sequence is roughly $-HN$. Hence if it is desired to treat all highly likely paths with lengths between N_1 and N_2 as equally likely, choose $P(N) = ce^{HN}$.

Then

$$\ln P(N) + \sum_{i=K+1}^{N} \ln P_{\mathrm{B}}(\cdot \mid \cdot) = \ln c + KH + \sum_{i=K+1}^{N} [H + \ln P_{\mathrm{B}}(\cdot \mid \cdot)].$$

The summation has expectation equal to zero and hence neither discourages nor favors long boundaries. The effect of adding such constants is discussed in [13, 14]. A second point of interest here is that the branching factor for this boundary process, i.e., the number of edge elements in the set from which the process can extend a boundary segment by 1 is 3. However, we see that there is an effective branching factor which is e^H. When the three state transition probabilities are each $\frac{1}{3}$, the effective branching factor is 3. However, it is constrained to lie between 1 and 3 and will usually be closer to 1.

6. BOUNDARY ERROR ESTIMATION

Boundary error estimation is readily studied [3, 4] if one is willing to treat boundary finding in a line-by-line scan mode. An approach is discussed for determining the variance of the distance between true and estimated boundaries when the boundaries are highly variable Markov processes.

The following simple model is used. Assume a portion of the boundary can be specified by a sequence of horizontal edge elements $\ldots, t_{i-1}, t_i, t_{i+1}, \ldots$ as shown in Fig. 15. The notation used here is somewhat different from that for the preceding portion of the paper. t_i is now the distance of the ith horizontal edge element from the x axis. The horizontal edge element length can be taken to be 1, and the vertical edge element length is dy. The output of an optical sensor is usually proportional to its area, so let $r_{\mathrm{in}} - r_{\mathrm{out}} = \Delta dy$, and let the variance of g_{jk} be $\sigma^2 dy$. The picture function signal-to-noise ratio, formerly Δ/σ, is now $(\Delta/\sigma)(dy)^{\frac{1}{2}}$. Without affecting the results sought, we can use $r_{\mathrm{in}} = \Delta dy$, $r_{\mathrm{out}} = 0$. $t_1, t_2, \ldots$ is a stochastic process to be specified shortly. This one-dimensional type of description may appear somewhat contrived, but axis models are commonly used in scene analysis, and the model is useful for study if the length of the boundary segment used is larger than a few correlation lengths of the boundary process.

6.1. *Isolated Strip Estimation Error*

To begin, treat the t_i as nonrandom a priori unknown constants to be estimated. Since the g_{jk} are independent random variables, the estimate $\hat{t}_i$ of t_i depends on the data in the ith strip only (see Fig. 15), and the $\hat{t}_i$'s are independent random variables. We assume $(\Delta/\sigma)(dy)^{\frac{1}{2}} \ll 1$ so that the estimation error will usually be at least a few multiples of dy. Then a good approximation to the result sought can

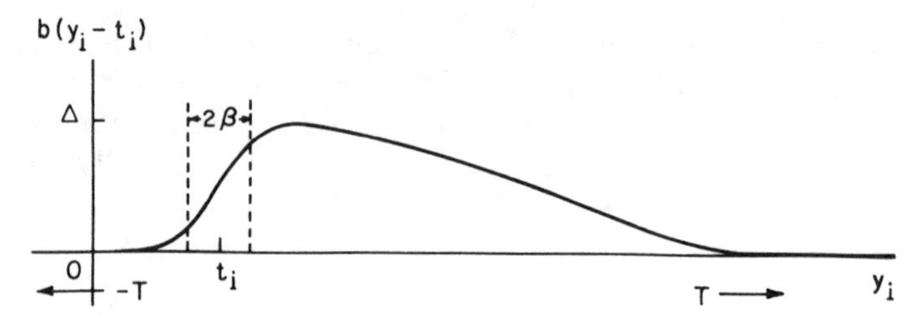

FIG. 16. Expectation of continuous-edge picture function.

be had through a greatly simplified analysis by letting $dy \to 0$. The likelihood of the data in the ith strip given t_i is proportional to

$$\exp\{-(1/2\sigma^2 dy)[\sum_{j=-T}^{y_i} g_{ji}^2 + \sum_{j=y_i+1}^{T} (g_{ji} - \Delta dy)^2]\}. \tag{6}$$

Upon dividing (6) by

$$\exp\{-(1/2\sigma^2 dy) \sum_{j=-T}^{T} g_{ji}^2\}$$

and letting $dy \to 0$, a simple integral results and the value of y_i for which it is a maximum is $\hat{t}_i$, the *unbiased* maximum likelihood estimator sought. Known bounding techniques can be applied to derive good lower bounds for the variances of *unbiased* estimates. However, we can use already derived results by treating the estimation of t_i as the estimation of the time of arrival of a radar pulse. Since the error $|t_i - \hat{t}_i|$ depends on the data in the vicinity of t_i only, the analysis is not compromised if we let b_{ji} go to 0 slowly with increasing j for $j > t_i$. Assume b_{ji} reaches 0 at some j less than T. Then the preceding model and likelihood (6) is that for a finite duration and energy radar pulse of height Δ and leading edge center t_i, received in white Gaussian additive noise of power spectral density σ^2. For discontinuous edges, a good lower bound on $\mathrm{Var}(\hat{t}_i)$ is derived from the Barankin bound [3, 4], and is $2(\sigma^2/\Delta^2)^2$. Hence, the standard deviation of the estimation error is approximately $2^{\frac{1}{2}}[\sigma^2 dy/(\Delta dy)^2]$ pixels, where the quantity in brackets is the square of the picture function noise-to-signal ratio. A more accurate edge model is a continuous one. Consider that of Fig. 16 having a leading edge which is the integral of a Gaussian pulse. The particular edge shape is unimportant here. The edge width is the important parameter. Hence, the signal pulse used is

$$b(y_i - t_i) = \int_{-T}^{y_i} \Delta(2\pi\beta^2)^{-\frac{1}{2}} \exp[-(1/2\beta^2)(\alpha - t_i)^2] d\alpha.$$

Application of the Cramér–Rao lower bound on the variance of unbiased estimates leads to [3, 4]

$$E[|\hat{t}_i - t_i|^2] \geq 2\pi^{\frac{1}{2}}(\sigma^2/\Delta^2)\beta. \tag{7}$$

This can be written

$$\text{std } \hat{t}_i \geq \pi^{\frac{1}{4}}\{[\sigma^2 dy/(\Delta dy)^2][2\beta/dy]\}^{\frac{1}{2}} \text{ pixels.}$$

The square root is the picture function noise-to-signal ratio multiplied by the square root of the number of pixels in the continuous edge. As an example, if $[\sigma^2 dy/(\Delta dy)^2]^{\frac{1}{2}} = 1$ and $2\beta dy = 6$, then std $\hat{t}_i \geq 3.3dy$. Under reasonable smoothness conditions, as $\sigma/\Delta \to 0$, $\hat{t}_i$ tends to become Gaussian with variance converging to (7).

Hence, the $\hat{t}_i$ are independent random variables with means t_i and variance lower bounded by (7). In a noisy image, the variance of $\hat{t}_i$ may well be much greater than the bound. However, if the dependence among the t_i is taken into account, the resulting variance of $\hat{t}_i$ can realistically be much smaller than (7), *in which case treating* $\hat{t}_i$ *as a Gaussian random variable with mean* t_i *and variance given by a suitable modification of* (7) *is a good approximation.* As an example, suppose a sequence of n edge centers constitute a straight line with t_c as the center element. Then there are n strips of independent data available for estimating t_c, and the variance of $\hat{t}_c$ is approximately $2\pi^{\frac{1}{2}}(\sigma^2/\Delta^2)\beta/n$ [3, 4].

6.2. Varying σ^2

As discussed in Section 2.1, σ^2 may not be constant, but rather $\sigma^2 = \sigma_{\text{in}}^2$ within the object and $\sigma^2 = \sigma_{\text{out}}^2$ within the background. Again, results from signal detection theory are applicable here. Specifically, assume the ith strip data is a Poisson process with intensity function $\mu(y_i)$, where $\mu(y_i) = r_{\text{out}} + b(y_i - t_i)$. Then $\mu(y_i)$ runs from r_{out} to a peak of $r_{\text{in}} = r_{\text{out}} + \Delta$ and back down to r_{out}. Hence, upon using the Cramér–Rao bound for Poisson processes [20] and the fact that σ^2 is a function of y_i, with $\sigma^2(y_i) = \mu(y_i) = r_{\text{out}} + b(y_i - t_i)$, we have

$$E[|\hat{t}_i - t_i|^2] \geq \left[\int_{-T}^{T} \frac{\dot{b}^2(\alpha)}{r_{\text{out}} + b(\alpha)}\, d\alpha\right]^{-1}. \tag{8}$$

For $r_{\text{out}} \gg \Delta$, the right side of (8) is approximately (7) with $\sigma^2 = r_{\text{out}}$.

6.3. Joint Estimation of the t_i

In view of the preceding, we make the following observation. If the theoretically minimum achievable boundary estimation error is such that the error in estimating an edge center is small compared with an edge width, then the error analysis problem can be viewed as the estimation of a stochastic signal in the presence of additive white Gaussian noise. The stochastic signal here is the sequence of edge centers $t_1, t_2, \ldots,$ and the additive white noise is the error between the edge centers t_i and the *isolated strip* maximum likelihood estimates of the t_i. Hence, we have the very attractive decomposition of the *error analysis problem* into isolated strip error analysis followed by error analysis where the stochastic dependence among edge centers is taken into account. (If the error in the isolated strip estimate of an edge center is much larger than the Cramér–Rao lower bound for the standard deviation of the error, then the decomposed model cannot be used for the purpose of the *implementation* of a minimal error boundary finder. Minimal error estimation then involves estimation of many edge centers jointly.) Hence, for theoretically minimal achievable edge center error analysis, we can use the rich collection of results available for signal estimation theory.

The model we will use for $t_1, t_2, \ldots$ is that of a known mean value function plus a 0-mean stationary stochastic process of the kind which can be generated by driving a constant coefficient linear difference equation with white Gaussian noise. If desired, the model can be extended to replacing the known mean value function by a large class of functions with partially unknown parameters, but the first model posed is adequate for our purposes. Since the mean value function is assumed known, it does not enter into the error analysis, and we can therefore treat it as being everywhere 0. The minimal achievable error is that for which all the data are used in estimating each edge element. Hence, if we use the model

$$x_{i+1} = Ax_i + w_i, \qquad w_i \sim \mathfrak{N}(0, R),$$
$$t_i = Cx_i + v_i, \qquad v_i \sim \mathfrak{N}(0, R_t)$$

where $\{w_i\}$ and $\{v_i\}$ are independent Gaussian white noise sequences, the steady-state infinite lag estimation error can be readily obtained using the Kalman smoothing filter approach [21], or equivalently the standard Wiener filtering approach. As an example, let A be a scalar and $C = 1$. Then the boundary process is generated by inputting white Gaussian noise to a discrete low-pass filter (specifically, an RC-type filter). The minimal achievable estimation error variance (using an infinite number of strips to either side of strip c) is given by [22]

$$E[|\hat{t}_c - t_c|^2] \geq R\gamma_0/A(1 - \gamma_0^2), \tag{9}$$

where γ_0 is the root

$$\gamma^2 - \gamma[R_t(1 + A^2) + R]/R_tA + 1 = 0, \qquad \gamma = \gamma_0, \gamma_0^{-1}, \text{ with } |\gamma_0| < 1.$$

Note that R_t is (7). To the extent that this one-dimensional model captures the essence of the two-dimensional problem, it provides a good lower bound on the limiting accuracy of our ripple filter (region growing/shrinking algorithm).

Even for the simple boundary model used here, (9) must be evaluated numerically. However, there are a few limiting cases for which the effects of the boundary and data model parameters on the estimation error variance can be seen in simple form.

Case 1. Let $A \to 0$. Then the edge centers become independent and all the useful data for estimating t_c are in the cth strip. It is easily seen that $\gamma_0 \to AR_t/(R_t + R)$ and (9) converges to $R_tR/(R_t + R)$. For $R_t/R \gg 1$, Var $\hat{t}_c \approx R$. The interpretation here is that the only use of the boundary model when $A = 0$ is that we know t_c to lie within roughly a distance of $R^{\frac{1}{2}}$ of the x axis. Hence, as R_t becomes very large, the data are not of much use and the estimate of t_c is close to 0. For $R_t/R \ll 1$, (9) is approximately R_t and the boundary model is of little use. Var $\hat{t}_c$ is described by (7).

Case 2. Let $A \to 1$. Then γ is a solution of $\gamma^2 - \gamma[2 + R/R_t] + 1 = 0$. If we now let $R/R_t \to 0$, it follows that $\gamma_0 \to 1 - (R/R_t)^{\frac{1}{2}}$ and Var $\hat{t}_c \to \frac{1}{2}(RR_t)^{\frac{1}{2}}$. This result is more interesting. In case 1, for R near 0 it follows that t_c must be near 0. However, for case 2 with $A \approx 1$ and R small, t_c can be very far from the x axis. Thus, t_c can be essentially arbitrary and yet Var $\hat{t}_c \to 0$ as $R \to 0$. The interpretation here is that for $A \approx 1$ and $R \approx 0$, the boundary becomes a very long straight line, there are many data strips to estimate its location, and its location can therefore be determined with negligible error. Note that for Var $\hat{t}_c \to 0$, $\frac{1}{2}(RR_t)^{\frac{1}{2}}$ with $R_t = 2\pi^{\frac{1}{2}}(\sigma^2/\Delta^2)\beta$, from (7), becomes an exact expression for Var $\hat{t}_c$ rather than just a lower bound. The result for the case $A \approx 1$ is in agreement with that in [3].

In order to improve estimation accuracy and reduce the amount of backtracking in our *sequential* boundary finding, we use lookahead within a window as discussed in Section 4.1. In terms of our one-dimensional signal estimation interpretation for the error analysis, this lookahead can be viewed as fixed-lag smoothing for the estimation of a stochastic signal in the presence of additive Gaussian noise. A lookahead of L edge elements can be viewed as fixed interval signal smoothing of L time units. For *continuous time* smoothing, it is well known that using all the data (L essentially ∞) results in an estimation error variance which is one-half that incurred in optimal filtering (i.e., $L = 0$). For discrete time smoothing, the use of all the data results in an estimation variance which is slightly greater than half that incurred by using the data to the estimation time only. Hence, in theory the use of $L > 0$ buys at most a factor of 2 improvement in estimation accuracy. If the small error approximation used here is not valid because the estimation error is comparable to or larger than the edge width, *the benefit of the lookahead can be much greater than a factor of 2 reduction in estimation error variance.*

A final comment here is that in general, an optimal estimator based on a two-dimensional model should be more accurate than one based on the one-dimensional model of this section. The reason for this, upon referring to Fig. 4, is that the latter makes use of the change in data across horizontal edge elements only, whereas the former makes use of the change in data across the vertical edge elements as well.

7. OBJECT RECOGNITION

7.1. *Object Recognition Functions*

Suppose the image contains an object in background, as before, but the object belongs to one of c classes, and each class may have a different r_{in}, distribution $P(L)$, boundary model, and a priori probability of occurrence. The problem of interest is to design a set of reasonable discriminant functions (i.e., statistics), one for each class, which can be used for deciding the class membership of the object.

We define the following symbols and functions:

ω_j, $j = 1, \ldots, c$	jth object class in a set of c classes
$P(\omega_j)$	a priori probability of occurrence of object class ω_j
t	a hypothesized boundary
$P(t \mid \omega_j)$	likelihood of t given that the boundary model used is that for class ω_j
g	the entire set of image data

$P(g \mid t, \omega_j)$ likelihood of the image data given boundary t and that the object belongs to class ω_j

A desirable goal is to design a Bayes classifier for the object, i.e., a minimum probability of error classifier. This reduces to computation of the conditional likelihoods $P(\omega_j \mid g)$, $j = 1, \ldots, c$, and choosing class ω^* for which

$$P(\omega^* \mid g) \geq P(\omega_j \mid g), \qquad j = 1, \ldots, c,$$

equivalently, for which

$$P(\omega^*, g) \geq P(\omega_j, g), \qquad j = 1, \ldots, c.$$

This requires the computations

$$P(\omega_j, g) = \int P(g \mid t, \omega_j) P(t \mid \omega_j) P(\omega_j) dt. \tag{10}$$

If the boundary is described by a few parameters, e.g., a few coefficients in a Fourier series, t is the finite-dimensional vector of these coefficients and (10) is well defined. If t denotes the sequence of many edge elements in a hypothesized boundary, more delicacy is required in defining and handling (10), but the results obtained by our informal treatment appear to be valid. The problem, of course, is that for the latter case (10) cannot be computed. The only recourse seems to be approximation based on extreme cases. Fortunately, these approximations are often appropriate.

Case 1. The data influence is sufficiently strong that roughly the same boundary is estimated irrespective of which of the c boundary models are used. Note, as discussed in [13], we found that data and boundary model parameters can be in error by at least 20% and the images can be noisy, but roughly the same boundary estimate is computed. This must represent a case where $P(g \mid t, \omega_j)$ and $P(t \mid \omega_j)$ are relatively impulsive and broad functions of t, respectively, for all j. Then (10) is *approximately*

$$P(\hat{t}^j \mid \omega_j) P(\omega_j), \tag{11}$$

where $\hat{t}^j$ is the maximum likelihood boundary estimate using the boundary model for class ω_j. Note that even though the $\hat{t}^j$, $j = 1, \ldots, c$, will all be *roughly* the same and approximately the true boundary, there will be appreciable differences in the $P(\hat{t}^j \mid \omega_j)$. If Δ/σ is large so that almost any estimator works, the computationally simplest can be used, and functions (11) can be computed using the same boundary estimate for all ω_j.

Case 2. Boundary estimates are highly influenced by boundary class model, and the boundary estimates associated with different ω_j will be distinctly different. This is more of a case of $P(g \mid t, \omega_j)$ and $P(t \mid \omega_j)$ being relative broad and impulsive, respectively. Then (10) is roughly

$$P(g \mid \hat{t}^j, \omega_j), \tag{12}$$

where $\hat{t}^j$ is in the vicinity of the peak of $P(t \mid \omega_j)$. This is the case discussed under Conclusions in [3]. It would seem that for boundaries described by a small number of parameters, and hence t a vector in a low-dimensional space, (12) would have to be computed for all $j = 1, \ldots, c$. On the other hand, for t representing a very large number of edge elements and for case 2, it would seem that, in practice, a meaningful estimate $\hat{t}^j$ would be obtained using the correct object class model and meaningful boundary estimates would not be obtained using the other class boundary models. In other words, the boundary estimator for a class would recognize whether it was an incorrect model before completing a boundary estimate.

In summary, an approximation to the Bayes object classifier can be computed for cases 1 and 2 above, and these should cover *most situations of practical interest.*

Case 3. The last case we mention is a fairly general case where the integrand in (10) may have a few modes. Equation (10) can be viewed as an average of $P(g|t, \omega_j)$ with respect to the measure $dP(t|\omega_j)$. A sufficiently good approximation to (10) should be obtainable by very simple numerical summation, e.g., by considering only values of t for which $P(g|t, \omega_j)P(t|\omega_j)$ has its few large peaks and replacing $dP(t|\omega_j)$ by a *roughly equivalent* measure which has its mass concentrated at these values of t. If the boundary models used in this case are simple, e.g., a few harmonics in a Fourier series, or a few parameters for a simple spline approximation, or approximation by one or a few quadratic arcs, or approximation by simple geometric figures such as rectangles, the evaluation of (10) as suggested above should require only modest computation.

7.2. *Recognition Error Analysis*

By returning to the axis boundary model of Sections 6.0–6.3, recognition error analysis becomes possible. We consider case 1 of Section 7.1. Here, *roughly* the same estimate is obtained irrespective of which of the c boundary models is used. The question of interest then is the probability of correctly recognizing the class association of the object. The problem here is that object differences may be subtle. For example, one boundary model may be that generated by driving a second-order system with white noise, another may be generated by driving a third-order system with white noise, etc., or all boundary models may be of the same order but have different bandwidths, etc. Assume Δ/σ is the same for all c object classes. Then upon using the model of Sections 6.0–6.3, we see the problem to be that in which a stationary stochastic process is observed. The process consists of one of c stationary Gaussian processes in the presence of additive white Gaussian noise. A decision must be made as to the class association of the process. This is a standard signal detection problem in communication theory, and the form of the detector and bounds on the probability of correct detection are well known [24]. The probability of correct detection will be an increasing function of boundary length and will be close to 1 if the boundary is long enough. Of course if Δ/σ and $P(L)$ are different for the different object classes, the probability of correct detection can be close to 1 even if the object boundary is short.

8. COMMENTS

8.1. *Comparison with Other Approaches*

Nahi's model is an axis model, i.e., a one-dimensional model such as that treated in Section 6. His experimental results are interesting, but there is no assessment of the accuracy of the algorithm. The significant advantage of this algorithm is that the required computation is small, and the algorithm runs in a line-by-line scan mode, thus using the data as they are output by many optical sensing systems. The relative disadvantages of the model compared with the two-dimensional models we have worked with are: (i) in many cases, describing a boundary with respect to an axis is more awkward and more sensitive to parameter specification error than is a two-dimensional model; (ii) the model is restricted to objects for which the intersection of the object with a horizontal line is a single connected interval; (iii) our sequential algorithm ought to be more accurate because it uses information across horizontal as well as vertical edges, and this improved accuracy should be most apparent for almost horizontal boundary segments; (iv) item (iii) assumes greater importance when edges are continuous and broad rather than discontinuous. Of course, drawbacks (iii) and (iv) can be somewhat mitigated by using a horizontal axis in portions of the image and a vertical axis in other portions. A computational price is paid for our lookahead, backtracking, or the use of the ripple filter, but these things provide greater resistance to large errors and improved accuracy in suppressing small errors.

Other contributions of our papers include our error analysis techniques, both for minimum achievable error and algorithm specific error.

8.2. Comments

Our experience has been that for high signal-to-noise ratios (i.e., high Δ/σ) almost any obvious procedure is adequate for boundary finding. However, our first primitive Ripple Filter worked well for $\Delta/\sigma > 3$. Our Sequential Boundary Finders work beautifully for $\Delta/\sigma > 2$, with many boundary models, and for $\Delta/\sigma = 1$ with well-designed boundary models. One reason the SBF is robust down to $\Delta/\sigma = 2$ is the use of the local 4×4 window. For $\Delta/\sigma < 2$, subteties such as pixel quantization noise and properties of boundary models are felt. *This is especially true if objects of interest comprise only a small number of pixels.*

On the surface, it appears that the present Sequential Boundary Finder can be viewed as an algorithm which makes optimal use of the boundary model and suboptimal use of the data; whereas the standard Ripple Filter makes optimal use of the data and suboptimal use of the boundary model—except in the vicinity of the true boundary where the Ripple Filter seems to make optimal use of the model and the data. We believe that with work in progress, both the SBF and the Ripple Filter can be designed to use *almost all* the useful boundary and data information available and still be computationally attractive. However, additional study is needed to sort out the relative merits of the Sequential Boundary Finder, the Ripple Filter, and variants or new algorithms. The questions of interest here pertain somewhat to estimation accuracy, but more so to computational requirements and the handling of more complex situations than the simple image models posed.

ACKNOWLEDGMENTS

This work was partially supported by the U.S. Army Research Office under Grant DAAG29-78-G-0124, by the National Science Foundation under Grant ENG77-26788, and by the Office of Naval Research under Grant N00014-75-C-0518. The authors are appreciative of this support and especially of the interest and technical suggestions of Dr. Robert Launer of ARO. The considerable good humor, effort, and skill of Ms. Ruth Santos in typing a number of drafts of this paper are gratefully acknowledged. Francis Sung made important software contributions and contributed to the experimentation with the Ripple Filter.

REFERENCES

1. A. Martelli, An application of heuristic search methods to edge and contour detection, *Comm. ACM* **19**, 1976, 73–83.
2. U. Montanari, On the optimal detection of curves in noisy pictures, *Comm. ACM* **14**, 1971, 335–345.
3. D. B. Cooper, Maximum likelihood estimation of Markov process blob boundaries in noisy images, *IEEE Trans. Pattern Recognition Machine Intelligence* **PAMI-1**, 1979, 372–384.
4. D. B. Cooper and H. Elliott, A maximum likelihood framework for boundary estimation in noisy images, in *Proc. IEEE Comput. Soc. Conf. Pattern Recognition and Image Processing, Chicago, May 31–June 2, 1978*, pp. 25–31.
5. H. Elliott, D. B. Cooper, and P. Symosek, Implementation, interpretation, and analysis of a suboptimal boundary finding algorithm, in *Proc. IEEE Comput. Soc. Conf. Pattern Recognition and Image Processing, Chicago, Aug. 6–8, 1979*, pp. 122–129.
6. L. Reiss and D. B. Cooper, The ripple filter: an algorithm for region growing in scene analysis, in *Proc. IEEE Comput. Soc. Conf. Computer Software and Applications, Chicago, Nov. 6–8, 1979*, pp. 849–853.
7. N. E. Nahi and M. H. Jahanshahi, Image boundary estimation, *IEEE Trans. Computers* **C-26**, 1977, 772–781.
8. P. Nagin, R. Kohler, A. Hanson, and E. Rieseman, Segmentation, evaluation, and natural scenes, in *Proc. IEEE Comput. Soc. Conf. on Pattern Recognition and Image Processing, Chicago, Aug. 6–8, 1979*, pp. 515–522.
9. D. B. Cooper, Feature selection and super data compression for pictures occurring in remote conference and classroom communications, in *Proc. 2nd Int. Joint Conf. Pattern Recognition, Copenhagen, Aug. 13–15, 1974*, pp. 416–422.
10. D. B. Cooper, Super high compression of line drawing data, in *Proc. 3rd Int. Joint Conf. Pattern Recognition, Coronado, Calif. Nov. 8–11, 1976*, pp. 638–642.
11. T. Pavlidis and D. J. Sakrison, Applications of a simple statistical model for curves, in *Proc. IEEE Comput. Soc. Conf. Pattern Recognition and Image Processing, Chicago, Aug. 6–8, 1979*, pp. 599–603.
12. S. W. Zucker, R. A. Hummel, and A. Rosenfeld, An application of relaxation labeling to line and curve enhancement, *IEEE Trans. Computers* **C-26**, 1977, 394–403.
13. H. Elliott, D. B. Cooper, F. Cohen, and P. Symosek, Implementation, interpretation and analysis of a suboptimal boundary finding algorithm, submitted.
14. P. Symosek, *Implementation and Analysis of a Sequential Boundary Finder for Stochastic Boundaries*, Brown University, Division of Engineering, M.Sc. Thesis, in preparation.
15. S. L. Tanimoto and T. Pavlidis, A hierarchial data structure for picture processing, *Computer Graphics Image Processing* **3**, 1975, 104–119.
16. D. H. Ballard and J. Sklansky, A ladder-structured decision tree for recognizing tumors in chest radiographs, *IEEE Trans. Computers* **C-25**, 1976, 503–513.

17. F. Cohen, *Sequential Boundary Estimation Using Dynamical System Boundary Models and Kalman Filtering*, Brown University, Division of Engineering, M.Sc. Thesis, in preparation.
18. R. Ash, *Information Theory*, pp. 14–16, Interscience, New York, 1966.
19. L. L. Scharf and H. Elliott, *Aspects of Dynamic Programming in Signal and Image Processing*, Colorado State Univ. Tech. Report, November 1979.
20. D. L. Snyder, *Random Point Processes*, pp. 85–86, Wiley, New York, 1975.
21. A. Gelb (Ed.), *Applied Optimal Estimation*, MIT Press, Cambridge, Mass., 1974.
22. D. R. Cox and H. D. Miller, *The Theory of Stochastic Processes*, p. 330, New York, Wiley, 1965.
23. F. Cohen, D. B. Cooper, H. Elliott, and P. Symosek, Two-dimensional image boundary estimation by use of likelihood maximization and Kalman filtering, in *Proceedings, 1980 IEEE International Conference on Acoustics, Speech and Signal Processing, Denver, April 9–11, 1980*, to appear.
24. H. L. Van Trees, *Detection, Estimation, and Modulation Theory*, Part II, Wiley, New York, 1971.
25. J. R. Woods, Two-dimensional discrete Markovian fields, *IEEE Trans. Information Theory* **IT-18**, 1972, 233–240.

Edge Detection in Textures

LARRY S. DAVIS AND AMAR MITICHE

Computer Sciences Department, The University of Texas at Austin, Austin, Texas 78712

1. INTRODUCTION

Detecting edges is an important first step in the solution of many image analysis tasks. Edges are used primarily to aid in the segmentation of an image into meaningful regions, but are also extensively used to compute relatively local measures of textural variation (which, of course, can subsequently be used for segmentation purposes). Although there has been a considerable amount of research concerning quantitative models for edge detection (e.g., Nahi [1], Modestino and Fries [2], Shanmugam *et al.* [3], Cooper and Elliot [4]), very little work has been devoted to developing such models for images described by texture models. This paper addresses the problem of detecting edges in what are called *macro-textures*, i.e., cellular textures where the cells, or texture elements, are relatively large (at least several pixels in diameter).

Once edges are detected in textured regions, they can be used to define texture descriptors in a variety of ways. For example, one can compute "edge per unit area" (Rosenfeld [5]). More generally, one can compute first-order statistics of edge properties [6, 7], such as orientation, contrast, fuzziness, etc., or higher-order statistics which can measure the spatial arrangement of edges in the texture. Such statistics can be computed from generalized cooccurrence matrices (Davis *et al.* [8, 9]) which count the number of times that specific pairs of edges occur in specific relative spatial positions. Clearly, the utility of such tools depends on the reliability with which edges can be detected in textures.

This paper is organized as follows: Section 2 contains a description of the image texture models which will be considered. These models are one dimensional, since the edge detection procedures, described in Section 3, are one dimensional. Section 4 contains derivations of the expected value and variance of the edge operator described in Section 3 and describes optimal edge detection procedures based on that analysis. Finally, Section 5 contains conclusions and a summary.

* This research was supported in part by funds derived from the Air Force Office of Scientific Research under Contract F49620-79-C-0043.

ISBN 0-12-597320-9

2. TEXTURE MODELS

There are a large number of formal image texture models which have been proposed and studied during the past few years. These can be broadly classified as *pixel-based* and *region-based* models (Ahuja [10]). All of these models treat textures as two-dimensional phenomena, which is appropriate for many applications (e.g., some medical applications, geographical applications). However, for other applications, regarding textures in this way is inappropriate; one should, instead, model the texture as a surface in space with certain reflectance properties. An image of such texture is then determined by the spatial disposition of the surface and the viewer, the frequency response of the viewer, and the positions of all light sources. Horn [11] should be consulted for an introduction to this branch of image science. Such models will not be considered in this paper.

Pixel-based models are ordinarily time-series models or random field models. Time-series models have been investigated by McCormick and Jayaramamurthy [12] and by Tou *et al.* [13]. Random field models are discussed in Wong [14] and in Pratt *et al.* [15]. For further references, see [16, 17].

This paper will be concerned with region-based texture models. In particular, we will consider one-dimensional models which are related to two-dimensional cell structure models. Cell structure models describe textures as mosaics, and can be generated by the following two-step process:

(1) A planar region is tessellated into cells, ordinarily convex.
(2) Each cell is independently assigned one of m colors, $c_1, \ldots, c_m$, using a fixed set of probabilities, $p_1, \ldots, p_m$.

This process partitions the original region into subregions, which are the unions of cells of constant color. If A is the original region, than $A_1, \ldots, A_m$ are the subregions. Note that the simple colors can be replaced by more complex coloring processes, e.g., the gray levels in a cell can be chosen according to a given distribution, d, which is itself chosen from a set of distributions, D, according to the given probability vector, P. Ahuja [10] contains an extensive survey of such models.

We will consider a similar class of one-dimensional models. A texture model is an ordered pair $\langle P, C\rangle$ where

(1) P is a cell width model, which successively drops intervals along a line, and
(2) C is a coloring model, consisting of coloring processes, $c_1, \ldots, c_m$, and probabilities, $p_1, \ldots, p_m$. As P produces cells, C colors the cells.

If we let w be the random variable corresponding to cell width, then the following are examples of cell width models:

(1) Constant cell width model

$$P_c(w) = 1, \quad w = b, \qquad = 0, \quad w \neq b.$$

(2) Uniform cell width model

$$P_u(w) = 1/b, \quad 0 \leq w \leq b, \qquad = 0, \quad w > b.$$

(3) Exponential cell width model

$$P_e(w) = \lambda \exp[-\lambda w].$$

To simplify the analysis, we assume that there are only two coloring processes, c_1 and c_2. Therefore, there is only one relevant probability for choosing cell colors, which will be denoted by p. Each coloring process colors a cell by choosing the intensity of each point in that cell independently from a normal distribution of intensities. The distributions are denoted $N(m_i, v_i)$, $i = 1, 2$, with mean m_i and variance v_i.

Notice that given a one-dimensional cell structure model, $\langle P, C\rangle$, one can derive a one-dimensional *component* structure model, where a component is a contiguous set of identically colored cells. For example, for the cell model $\langle P_c, C\rangle$ the corresponding component model has components whose lengths are distributed geometrically. Of course, in a component model, the various colors alternate, since by definition two adjacent components must have different colors. The component model is required to determine the prior probabilities of various types of pixels (see Section 4).

3. A ONE-DIMENSIONAL EDGE DETECTOR

In this section we will describe a simple, one-dimensional edge detection procedure. There are a variety of reasons for considering one-dimensional edge detectors:

(1) Computational efficiency on conventional, sequential computers. Since we are interested in detecting edges for the purpose of describing texture, it is important that the edge detection process be made as efficient as possible.

(2) Suitability for implementation on special-purpose image processing hardware. There has been a significant amount of research and development of image processing hardware over the past few years (CLIP [18], PICAP [19]). One-dimensional edge detectors can be easily implemented on, for example, series-parallel machines, where a row of processors "scan" and process the image one row at a time.

(3) Mathematical tractability. Direct analysis of two-dimensional edge detectors is complicated by many factors, including the mathematical complexity of most two-dimensional cellular texture models and the greater variety of edge-like features in two dimensions.

The class of edge operators which we have considered is based on differences of averages between adjacent, symmetric one-dimensional image neighborhoods. Specifically, let f be a one-dimensional image function. Then the edge operator is

$$e_k(i) = (1/k) \sum_{j=1}^{k} (f(i-j) - f(i+j))$$

$$= (1/k)(LS(i) - RS(i)),$$

where

$$LS(i) = \sum_{j=1}^{k} f(i - j)$$

and

$$RS(i) = \sum_{j=1}^{k} f(i + j).$$

By noting that

$$e_k(i + 1) = (1/k)[LS(i) - f(i - k) + f(i) - RS(i) + f(i + 1) - f(i + k + 1)]$$

we see that e_k can be computed in a constant number of operations per picture point, independent of k, on a conventional sequential computer.

The operator e_k is used to detect edges by the following three step process:

(1) Compute $e_k(i)$ for all points i.

(2) Discard all i with $|e_k(i)| < t$. This *thresholding* step is intended to discriminate between points which are edges of texture elements and points which are in the interior of texture elements, but far from edges.

(3) Discard all i with $|e_k(i)| < |e_k(i + j)|$, $|j| < d$. This *non-maxima suppression* step is intended to discriminate between points which are edges and points which are interior to texture elements, but are close to edges.

Step 3 is crucial since e_k gives high response not just at edges, but also near edges, so that thresholding alone would result in a cluster of detections about each true edge point. The above procedure involves three classes of texture pixels:

(1) *edge pixels,* which are pixels located directly at the edges between texture elements;

(2) *near-edge pixels,* which are located within distance d of an edge pixel and are discarded by the non-maxima suppression step; and

(3) *interior pixels,* which are located at distances greater than d from the nearest edge pixel, and are ordinarily eliminated by the thresholding step (but may be eliminated due to proximity to above-threshold, near-edge pixels).

Optimizing the above edge detection procedure involves choosing k, t and d in order to minimize the probability of error—i.e., minimizing the frequency of discarding edge pixels, and not discarding near-edge and interior pixels. This paper considers the edge–interior discrimination problem only. Therefore, we will be concerned with choosing values for k and t only. The complete edge detector, including the non-maxima suppression step, is discussed in Davis and Mitiche [20].

4. ANALYSIS OF e_k

In this section we will derive the expected value and variance of e_k at edges and at interior points. We will regard an interior point as a point whose distance from the nearest edge is greater than k. Expressions for the prior probabilities of edge and interior points are developed. Finally, by assuming that e_k is normally distributed at edges and interiors, a minimum error thresholding procedure for distinguishing between edge points and interior points is developed.

4.1. The Expected Value of e_k, $E[e_k]$

The definition of e_k was originally given for a discrete function f. If f is continuous, then we can redefine e_k as

$$e_k(i) = 1/k\left[\int_{-k}^{0} f(i+j)dj - \int_{0}^{k} f(i+j)dj\right].$$

Then the expected value of e_k is

$$E[e_k(i)] = 1/k\left[\int_{-k}^{0} E[f(i+j)]dj - \int_{0}^{k} E[f(i+j)]dj\right].$$

If i is an interior point, then all points $f(i+j)$, $-k \leq j \leq k$ are colored by the same process. Therefore, the expected values are all the same, and thus $E[e_k(i) \mid i$ is an interior point$] = 0$.

Now suppose that i is an edge point, and assume, without loss of generality, that the cell to the left of i, C_l, is colored by process c_1, and that the cell to the right of i, C_r, is colored by process c_2, and that $m_1 > m_2$. Let w_l be the width of C_l and w_r be the width of C_r (see Fig. 1). We will also make the simplifying assumption that the points to the left of C_l (or the right of C_r) are individually colored by processes c_1 and c_2 with probabilities p and $(1-p)$. For k much greater than w_l or w_r, this assumption is not unreasonable. As w_l or w_r approaches k, it is more likely that only one cell will be found to the left of C_l or the right of C_r. However, large cells are ordinarily less likely than small cells. Letting $a = pm_1 + (1-p)m_2$, we can then write

$$E[e_k(i)] = 1/k\left[\left(\int_0^k \left(m_1w_l + a(k-w_l)P(w_l)dw_l + \int_k^\infty m_1kP(w_l)dw_l\right)\right)\right.$$
$$\left. -\left(\int_0^k \left(m_2w_r + a(k-w_r)P(w_r)dw_r + \int_k^\infty m_2kP(w_r)dw_r\right)\right)\right].$$

Since w_l and w_r are drawn from the same distribution, terms can be grouped to obtain

$$\begin{aligned} E[e_k(i)] &= 1/k\left[\int_0^k (m_1-m_2)wP(w)dw + \int_k^\infty (m_1-m_2)kP(w)dw\right] \\ &= 1/k\left[(m_1-m_2)w_0 - \int_k^\infty (m_1-m_2)wP(w)dw + \int_k^\infty (m_1-m_2)kP(w)dw\right] \\ &= \frac{(m_1-m_2)}{k}\left[w_0 - \int_k^\infty (w-k)P(w)dw\right], \end{aligned}$$

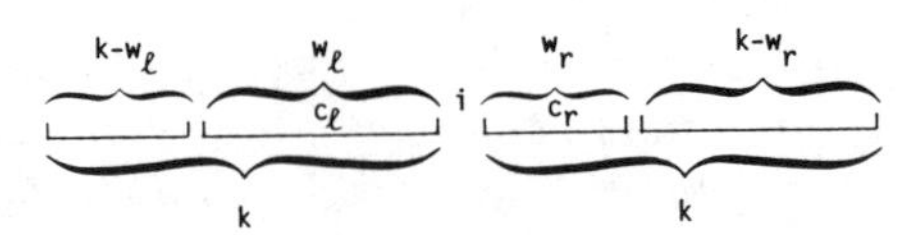

FIG. 1. Neighborhood of an edge point, i.

where

$$w_0 = \int_0^\infty wP(w)dw.$$

4.2. The Variance of e_k

From the definition of e_k, we have

$$\text{Var}[e_k(i)] = \text{Var}\left[1/k \int_0^k f(i-j) - f(i+j)dj\right].$$

Suppose i is an interior point and is in a cell colored by c_1. Then

$$\begin{aligned}\text{Var}[e_k(i)] &= (1/k^2)k2v_1 \\ &= 2v_1/k.\end{aligned}$$

If i is an interior point and is in a cell colored by c_2, then $\text{Var}[e_k(i)] = 2v_2/k$.

Next we will consider the case where i is an edge point. Let

$$LS(i) = \int_0^k f(i-j)dj,$$

$$RS(i) = \int_0^k f(i+j)dj.$$

Then

$$e_k(i) = 1/k(LS(i) - RS(i)).$$

We will derive expressions for $\text{Var}[LS(i)]$ and $\text{Var}[RS(i)]$. There are two cases to consider:

(1) $$w_l \quad \text{or} \quad w_r \geq k,$$

(2) $$w_l \quad \text{or} \quad w_r < k.$$

If $w_l \geq k$, then the variance of $LS(i)$ is kv_1. If $w_r \geq k$, then the variance of $RS(i)$ is kv_2.

Next, suppose $w_l < k$. We will assume that of the $k - w_l$ pixels not in C_1, l_1 pixels are colored by process c_1 and $l_2 = (k - w_l) - l_1$ pixels are colored by process c_2, where l_1 is a random variable described by a binomial distribution with parameter p and l_2 is a random variable described by a binomial distribution with parameter $(1 - p)$.

In general, if y is the sum of a random number, n, of independent experimental values of a random variable x, then

$$\text{Var}[y] = E[n]\,\text{Var}[x] + (E[x])^2\,\text{Var}[n].$$

Thus, if w_l is a fixed value less than k,

$$\mathrm{Var}[LS(i) \mid w_l < k] = w_l v_1 + E[l_1]v_1 + m_1^2 \,\mathrm{Var}[l_1] + E[l_2]v_2 + m_2^2 \,\mathrm{Var}[l_2]$$

with

$$E[l_1] = (k - w_l)p,$$
$$E[l_2] = (k - w_l)(1 - p),$$
$$\mathrm{Var}[l_1] = \mathrm{Var}[l_2] = (k - w_l)p(1 - p).$$

A similar expression can be obtained for $RS(i)$.

Combining the cases $w_l \geq k$ and $w_l < k$, we can write

$$\mathrm{Var}[LS(i) \mid i \text{ an edge}] = \int_k^\infty k v_1 P(w_l)\,dw_l + \int_0^k (w_l v_1 + (k - w_l)(pv_1 + (1 - p)v_2) + (k - w_l)p(1 - p)(m_1^2 + m_2^2))P(w_l)\,dw_l.$$

A similar expression is obtained for $RS(i)$. Then

$$\mathrm{Var}[e_k(i)] = 1/k^2[\mathrm{Var}[LS(i)] + \mathrm{Var}[RS(i)]].$$

Since w_l and w_r are drawn from the same distribution, we group terms and finally obtain, for edge points,

$$\mathrm{Var}[e_k(i) \mid i \text{ is an edge}] = 1/k^2\left[k(v_1 + v_2)\int_k^\infty P(w)\,dw + (v_1 + v_2)\int_0^k wP(w)\,dw + 2(pv_1 + (1 - p)v_2)\int_0^k (k - w)P(w)\,dw + 2p(1 - p)(m_1^2 + m_2^2)\int_0^k (k - w)P(w)\,dw\right].$$

For the exponential model with $p = 0.5$, for example, we find that

$$E[e_k(i) \mid i \text{ is an edge}] = \frac{(m_1 - m_2)(1 - e^{-\lambda k})}{\lambda k}$$

and

$$\mathrm{Var}[e_k(i) \mid i \text{ is an edge}] = \frac{(v_1 + v_2)}{k} + \frac{m_1^2 + m_2^2}{2k}\left(1 - \frac{1 - e^{-\lambda k}}{\lambda k}\right).$$

Other simple expressions can be obtained for the constant and uniform models.

In order to derive a minimum error threshold for discriminating between edge and interior points using e_k, it is necessary to:

(1) determine the prior probabilities of edge and interior points, and
(2) specify a form for the distribution of e_k at edges and at interior points.

To compute the prior probabilities of edge and interior points, we must derive a *component model* from the cell width model. A component is a set of connected, identically colored cells. Let a c_1-component (c_2-component) be a component whose cells are colored by process c_1 (c_2). Then the length of a component is defined to be the number of cells that compose it, and the width of a component is its actual measure.

In the following we will show how a component model can be derived from a cell model. As mentioned before, the cell coloring process is a Bernoulli process which selects coloring process c_1 with probability p and coloring process c_2 with probability $(1 - p)$. Therefore, the c_1-component and c_2-component lengths are random variables described by a geometric probability mass function. In the following analysis we will consider only c_1-components and simply refer to them as components. The same analysis will hold for c_2-components.

If n is the random variable that represents component length and $p_1(n)$ is the probability mass function for it, then

$$p_1(n) = (1 - p)p^{n-1}.$$

Now, if w and r are random variables that describe the cell width and the component width, respectively, then r is the sum of a random number n of independent identically distributed experimental values of w,

$$r = \sum_{i=1}^{n} w_i.$$

Expressions for the expected value and variance of r can be obtained in terms of the expected value and variance of n and w,

$$E[r] = E[n]E[w],$$
$$\mathrm{Var}[r] = E[n]\,\mathrm{Var}[w] + (E[w])^2\,\mathrm{Var}[n].$$

If w is continuous and if f and g are the probability density functions for w and r, then

$$g^{\mathrm{T}}(s) = p_1{}^{\mathrm{T}}(f^{\mathrm{T}}(s)),$$

where f^{T} and g^{T} are the exponential transforms (s-transforms) of f and g, respectively, and $p_1{}^{\mathrm{T}}$ is the z-transform (discrete transform) of p_1.

If w is discrete, then the above expression holds with f and g being the probability mass functions for w and r, and f^{T} and g^{T} being their respective z-transforms. By taking the inverse transforms one can obtain the distribution of component widths.

Examples

1. *Constant distribution of cell width.* For the constant cell width model, g is described by the geometric probability mass function

$$g(r) = (1 - p)p^{(r/b)-1} \quad \text{for} \quad r = b, 2b, 3b, \ldots$$
$$= 0 \quad \text{otherwise},$$

$$E[r] = b/(1 - p),$$
$$\mathrm{Var}[r] = b^2 p/(1 - p)^2.$$

2. *Exponential distribution of cell width.* For the exponential model with parameter λ, g is described by the probability density function

$$g(r) = (1 - p)\lambda \exp(-(1 - p)\lambda r).$$

Thus component widths are still exponentially distributed and

$$E[r] = 1/((1 - p)\lambda).$$

3. *Uniform distribution of cell width.*

$$E[r] = b/(2(1 - p)),$$

$$\mathrm{Var}[r] = \frac{b/(4(1 - p))}{(\frac{1}{3} + p/(1 - p))},$$

and

$$g(r) = e^{(p/b)r} g'(r),$$

where

$$g'(r) = (1 - p)\Bigg[\sum_{n=0}^{\infty} \frac{(-1)^n p^n e^{-np}}{b^{n+1}} \frac{(r - bn)^n}{\Gamma(n + 1)} S_{bn}(r) - \frac{1}{e^p} \sum_{n=0}^{\infty} \frac{(-1)^n p^n e^{-np}}{b^{n+1}} \frac{[r - b(n + 1)]^n}{\Gamma(n + 1)} S_{b(n+1)}(r)\Bigg]$$

and

$$S_k(t) = 0, \quad 0 < t < k,$$
$$= 1, \quad t \geq k,$$

is the unit step function.

In order to perform minimum-error thresholding on a given image texture, it is necessary to know the prior probabilities of an edge point, p_e, and of an interior point, p_i. An edge point is defined to be a point on the image line which is no more than a fixed distance Δ away from a true edge. We choose $\Delta > 0$ so that $p_i \neq 0$. An interior point, as before, is a point that is at least distance k away from a true edge.

In the following we will consider, without loss of generality, only points within c_1-components. Given that a point is picked at random, let h be the probability density function for the distance y_1 to the next edge (the forward edge or edge on the right). From general results on random incidence into a renewal process we have [21]

$$h(y_1) = (1 - \mathrm{prob}[r \leq y_1])/E[r].$$

The function h is also the probability density function for the distance y_2 to the backward edge (the nearest edge on the left of the point). Then we have

$$h_j(y_1, y_2) = h(y_1)h_c(y_2/y_1),$$

where h_c is the conditional probability function for y_2 and h_j is the joint probability function for y_1 and y_2. If y_1 and y_2 are independent, then

$$h_j(y_1, y_2) = h(y_1)h(y_2).$$

From the above we now can derive expressions for the priors:

$$p_i = \int_k^\infty \int_k^\infty h_j(y_1, y_2)dy_1dy_2,$$

$$p_e = 1 - \int_d^\infty \int_d^\infty h_j(y_1, y_2)\Delta y_1 \Delta y_2.$$

For the discrete case we have

$$p_i = \sum_{y_1=k}^{\infty} \sum_{y_2=k}^{\infty} h(y_1, y_2),$$

$$p_e = 1 - \sum_{y_1=\Delta+1}^{\infty} \sum_{y_2=\Delta+1}^{\infty} h(y_1, y_2).$$

The set of interior points and the set of edge points are not complementary in the sense that they do not form a partition of the set of all points of the image line. Thus it is necessary that we normalize p_e and p_i:

$$p_e \leftarrow p_e/(p_e + p_i),$$
$$p_i \leftarrow p_i/(p_e + p_i).$$

Examples

1. *Geometric distribution of component widths.* Assume for simplicity that $b = 1$. Then

$$h(y_1) = (1 - p) \sum_{\nu=y_1}^{\infty} (1 - p)p^{\nu-1}$$
$$= (1 - p)p.$$

Thus

$$\text{Prob}[y_1 \geq k] = \sum_{y \geq k}^{\infty} (1 - p)p^y$$
$$= p^{k-1}$$

Since the geometric distribution is memoryless, y_1 and y_2 are independent random variables, and

$$p_i = \text{Prob}[y_1 \geq k, y_2 \geq k] = p^{2(k-1)},$$
$$p_e = 1 - p^{2\Delta}.$$

Normalizing, we have

$$p_i = p^{2(k-1)}/(p^{2(k-1)} + 1 - p^{2\Delta}),$$
$$p_e = 1 - p_i.$$

2. *Exponential distribution of component widths.* Let λ be the parameter of the distribution. Then

$$h(y_1) = \frac{1 - (1 - \exp(-\lambda y_1))}{1/\lambda} = \lambda \exp(-\lambda y_1),$$
$$\text{Prob}[y_1 \geq k] = \exp(-\lambda k).$$

The exponential distribution is also memoryless so that y_1 and y_2 are independent. We then have

$$p_i = \exp(-2\lambda k),$$
$$p_e = 1 - \exp(-2\Delta\lambda).$$

Normalizing, we get

$$p_i = \exp(-2\lambda k)/(\exp(-2\lambda k) + 1 - \exp(-2\Delta\lambda)),$$
$$p_e = 1 - p_i.$$

In order to use p_e, p_i, and the expected values obtained above for minimum-error edge detection, we will make the assumption that e_k is normally distributed at edge points as well as interior points. More precisely, we assume that e_k is $N(0, 2v/k)$ at interior points and $N(E[e_k], \text{Var}[e_k])$ at edges. This assumption is certainly valid for interior points since, in this case, LS and RS are each the sum of k independent experimental values sampled from the same normal distribution. At edges, each of LS and RS is the sum of k_1 independent experimental values drawn from the normal distribution that represents one of the coloring processes and k_2 independent experimental values from the normal distribution that describes the other coloring process. The sum of k_1 and k_2 is k, but k_1 and k_2 will in general vary from point to point on the image line.

The assumption of normality of e_k at edges will hold well if the variances of k_1 and k_2 are low. This means that in the neighborhood of any edge point on the image line the number of pixels colored by either process remains almost constant. For example, this is trivially true for the constant distribution of cell widths, with $k \leq b$, b being the width of a cell. In this extreme case, $k_1 = k$ and $k_2 = 0$ or vice versa. However, if k_1 and k_2 have a high variance, then the normality assumption will not hold very well.

Thus, an important factor for the validity of the assumption is the variance of cell widths. Ideally, this variance should be small; however, another property that would tend to make the assumption hold well is that the image model be more likely to contain cells whose widths are close to k. This will keep the probability that the neighborhoods C_l and C_r extend over more than one cell quite low.

Given the normality assumption, the following two-step process can be used to compute an optimal k and t for a minimum-error edge detector.

(1) For a range of k, find the minimum-error threshold for discriminating between edges and interior points. Since both e_k at edges and e_k at interior points are modeled by normal distributions with known parameters and priors, this is straightforward. Let er(k) be the probability of error for the minimum-error threshold for e_k and let $t(k)$ be the threshold.

(2) Choose k such that er$(k) \leq$ er(k'), for all k' considered. Then $(k, t(k))$ define the minimum-error edge detector.

Fig. 2 shows plots of er(k) as a function of k for the three cell width models presented in Section 2. Notice that the value of k which minimizes total errors is the mean cell width for all three models. The reason that the curves tend to level off at high k, rather than rise to a higher error, is that as k becomes very large, the prior probability of interior points, p_1, approaches zero. For very high values of k, the near zero value of p_i causes the programs which compute minimum error thresholds to become unstable. Therefore, we arbitrarily stopped computing er(k) for $k > 20$. Figs. 3a–e shows an example of the effect of k on the performance of e_k. Fig. 3a contains a checkerboard texture with $b = 16$, $p = 0.5$, $m_1 = 30$, $m_2 = 20$, and $v_1 = v_2 = 10$. Fig. 3b shows the true edges, while Figs. 3c–e show

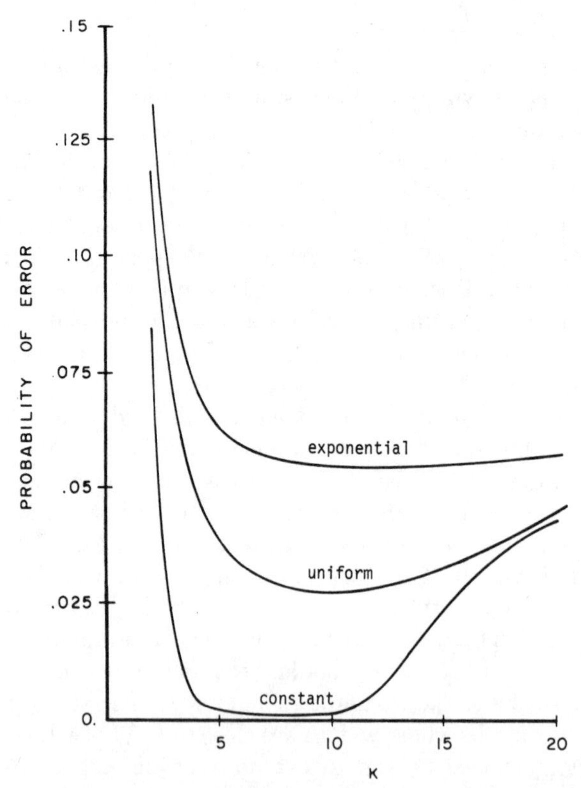

FIG. 2. Plots of er(k) as a function of k for the three cell width models. All parameters are the same for all three models: $m_1 = 10$, $m_2 = 5$, $v_1 = v_2 = 2$, $p = 0.5$ and $w_0 = 10$.

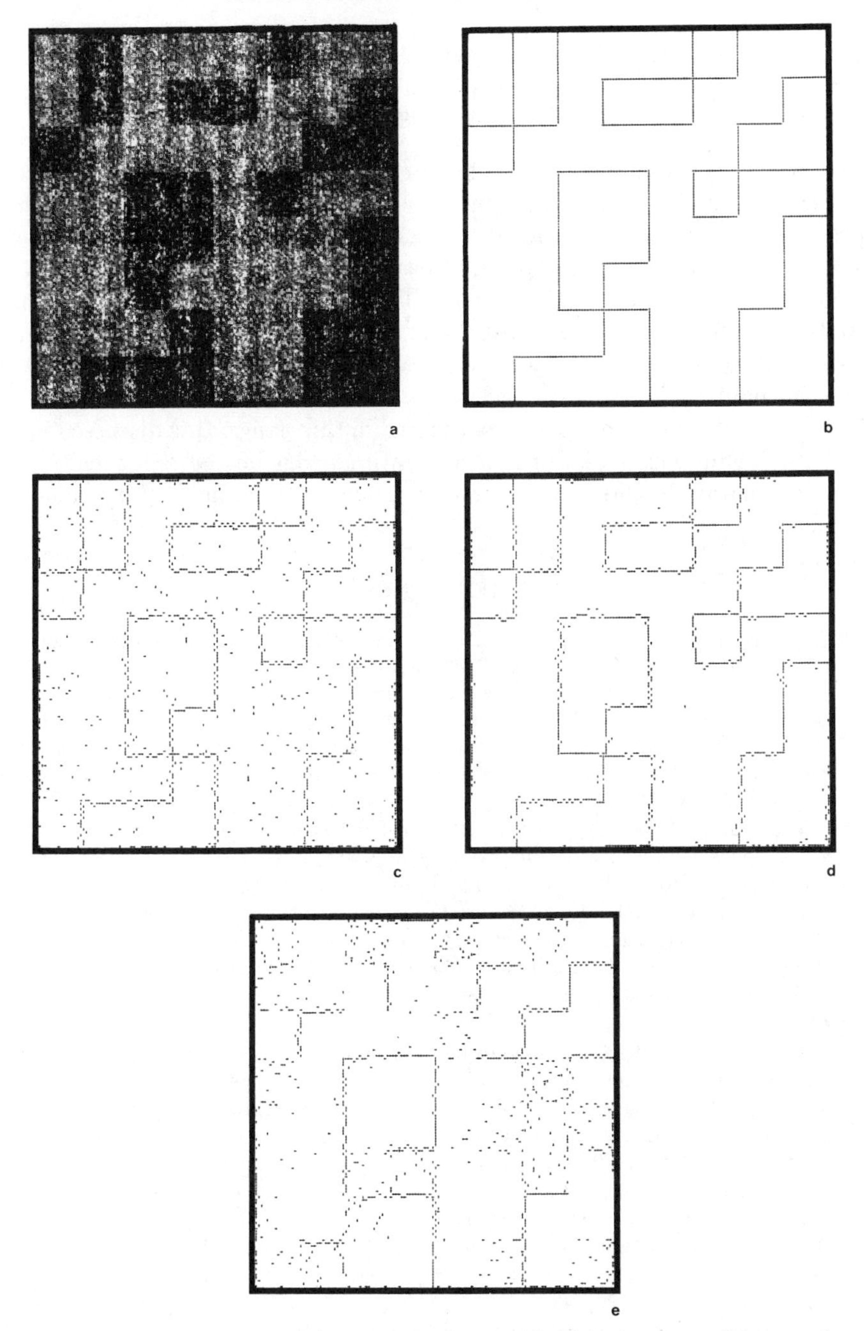

FIG. 3. Application of e_k to a checkerboard texture: (a) original texture, (b) true edges, (c) $k = 8$, (d) $k = 16$, (e) $k = 24$.

the results of applying e_k, $k = 8$, 16, and 24, thresholding at the minimum error threshold for the appropriate k, and then performing non-maxima suppression across eight pixels. Note that the results for the optimal value of k (16) are significantly better than when we choose k too small (8) or too large (24).

5. DISCUSSION

This paper has discussed the problem of detecting edges in cellular textures. A general edge detection procedure was proposed. The procedure involved applying an edge-sensitive operator to the texture, thresholding the results of the edge operator, and finally computing "peaks" from the above-threshold points. This paper concentrated on the thresholding process and developed a minimum-error thresholding procedure based on an analysis of the edge operator e_k. The thresholding procedure assumed that e_k was normally distributed at edges and at interior points.

The peak selection step was not considered in this paper. It is discussed in [20], which also includes examples of choosing optimal edge detectors for real textures, and a comparative classification study using optimal and suboptimal edge detectors.

REFERENCES

1. N. Nahi and S. Lopez-Mora, Estimation-detection of object boundaries in noisy images, *IEEE Trans. Automatic Contr.* **AC-23**, 1978, 834–846.
2. J. Modestino and R. Fries, Edge detection in noisy images using recursive digital filtering, *Computer Graphics Image Processing* **6**, 1977, 409–433.
3. K. Shanmugam, F. Dickey, and R. Dubes, An optimal frequency domain filter for edge detection in digital pictures, *IEEE Trans. Pattern Anal. Machine Intelligence* **PAMI-1**, 1979, 37–49.
4. D. Cooper and H. Elliot, A maximum likelihood framework for boundary estimation in noisy images, *Proc. IEEE Computer Society Conf. on Pattern Recognition and Image Processing*, Chicago, May 31–June 2, 1978, pp. 25–31.
5. A. Rosenfeld and A. Kak, *Digital Picture Processing*, Academic Press, New York, 1976.
6. J. Weszka, C. Dyer, and A. Rosenfeld, A comparative study of texture features for terrain classification, *IEEE Trans. Systems, Man Cybernet.* **SMC-6**, 1976, 269–285.
7. D. Marr, Early processing of visual information, *Phil. Trans. Royal Society B* **275**, 1976, 483–524.
8. L. Davis, S. Johns, and J. K. Aggarwal, Texture analysis using generalized cooccurrence matrices, *IEEE Trans. Pattern Anal. Machine Intelligence* **PAMI 1**, 1979, 251–258.
9. L. Davis, M. Clearman, and J. K. Aggarwal, A comparative texture classification study based on generalized cooccurrence matrices, in *Proc. IEEE Conf. on Decision and Control*, Miami, Dec. 12–14, 1979, to appear.
10. N. Ahuja, *Mosaic Models for Image Analysis and Synthesis*, Ph.D. dissertation, University of Maryland, Computer Science Dept., 1979.
11. B. Horn, Understanding image intensities, *Artificial Intelligence* **8**, 1977, 208–231.
12. B. McCormick and S. Jayaramamurthy, Time series models for texture synthesis, *J. Compt. Inform. Sci.* **3**, 1974, 329–343.
13. J. Tou, D. Kao, and Y. Chang, Pictorial texture analysis and synthesis, in *Proc. 3rd Int. Joint Conf. on Pattern Recognition*, Coronado, Calif., Nov. 8–11, 1976.
14. E. Wong, Two-dimensional random fields and representations of images, *SIAM J. Appl. Math.* **16**, 1968, 756–770.
15. W. Pratt and O. Faugeras, Development and evaluation of stochastic-based visual texture fields, in *Proc. 4th Int. Joint Conf. on Pattern Recognition*, Kyoto, Japan, Nov. 7–10, 1978, pp. 545–548.

16. M. Hassner and J. Sklansky, Markov random fields of digitized image texture, in *Proc. 4th Int. Joint Conf. on Pattern Recognition*, Kyoto, Japan, Nov. 7–10, 1978, pp. 538–540.
17. K. Abend, I. Harley, and L. Kanal, Classification of binary random patterns, *IEEE Trans. Inform. Theory* **IT-11**, 1965, 538–544.
18. M. Duff, CLIP-4: A large scale integrated circuit array parallel processor, in *Proc. 3rd Int. Joint Conf. on Pattern Recognition*, Coronado, Calif., Nov. 8–11, 1976, pp. 728–732.
19. B. Kruse, A parallel picture processing machine, *IEEE Trans. Computers* **C-22**, 1973, 1075–1087.
20. L. Davis and A. Mitiche, Optimal texture edge detection procedures, in preparation.
21. A. Drake, *Fundamentals of Applied Probability Theory*, McGraw–Hill, New York, 1967.

Comparative Analysis of Line-Drawing Modeling Schemes *

H. FREEMAN AND J. A. SAGHRI

Rensselaer Polytechnic Institute, Troy, New York, 12181

The computer processing of line drawings necessarily requires that the line drawings be first quantized and then encoded. Invariably, the quantization process forces the computer approximation to connect nodes lying on a lattice. The lattice normally is uniform square (though it could also be rectangular, logarithmic, or curvilinear) and is either explicitly or implicitly defined; its size is determined by the limit in the ability to resolve neighboring points in the quantization process. The coded representation of the drawing may be based on the use of straight or curved segments (approximants). The segments may be restricted either to a small set of fixed lengths or their lengths may take on any of the discrete values permissible on the lattice field. The paper examines some of the possibilities for line drawing modeling and shows that in many cases they may be regarded as special cases of the so-called generalized chain coding scheme. A procedure is developed for computing the relative probabilities of the approximating segments in a line drawing representation. The latter are necessary for any efficient (compact) line-drawing modeling scheme.

1. INTRODUCTION

To process line-drawing information with a digital computer, the line drawing information must first be quantized and encoded. The two operations—quantization and encoding—are intrinsic to any description process. Whatever the object to be described, it must first be conceptually broken down ("quantized") into small entities, sufficiently small so that no finer subdivision is of interest, and then names must be assigned to these entities (encoding) to obtain the desired linguistic model (description). In examining the modeling of line drawings (a description process), it is instructive to be explicitly aware of the roles quantization and encoding play.

Of the two operations, quantization is more subtle and is much more dependent on the specific characteristics of the object to be described. Encoding is more general; its application to line-drawing modeling differs little from other applications and it has no effect on the precision with which the information is rendered. Hence in this paper, which is concerned with line-drawing modeling, we shall

* The research described here was supported by the U.S. Air Force through the Rome Air Development Center under Contract F30602-78-C-0083. The information presented does not necessarily reflect the position or policy of the Government and no official endorsement should be inferred.

ISBN 0-12-597320-9

place emphasis on the quantization process and the effect it has on the information a line drawing is intended to convey.

Given a line drawing, there are innumerable ways in which we may quantize it. The method we select may depend on some statistical properties of an ensemble of line drawings, or it may be highly specific to the particular curve. In either case, there are three identifiable variables in the operation—the *form* of the quantization, its *size*, and the *approximant* used to represent the quanta.

In one common line-drawing quantization scheme, a uniform square lattice is overlaid on the given line drawing, the intersections between the curve and the lattice are noted, and the nodes lying closest to these intersections are connected in sequence to form a straight-line-segment approximation to the curve. In this case (the well-known chain coding scheme[1]), the form of the quantization scheme is that of a uniform square lattice, the size is the lattice spacing, and the approximant is a straight-line segment [1]. Alternatively, one could use rectangular, triangular, hexagonal, nonuniform, or even curvilinear lattices (different quantization form), one could change the basic lattice dimension (different quantization size), and one could use mathematically defined curve segments other than straight-line segments to connect the selected lattice nodes (different approximants).

Our objective in this paper is to develop increased insight into the line-drawing modeling process. We shall begin by examining the basic factors affecting line-drawing quantization. We shall show that in spite of the wide range of possibilities, practical considerations almost always lead us to the chain-coding scheme or some variant thereof. We shall describe a generalized-chain-code quantization algorithm and then develop a method for computing the relative probabilities of the approximating line segments for a particular curve or curve ensemble. These probabilities are essential to the design of any efficient line-drawing encoding scheme. However, we shall not pursue the subject of encoding itself as it would make the paper excessively long.

2. LINE-DRAWING QUANTIZATION

The problem of quantizing a line drawing is basically one of analog-to-digital conversion where the line drawing is given in some analog form and we desire a suitably encoded digital representation. The precision of our line-drawing data is thus at once limited by the dynamic range of the original analog medium (e.g., 1:500, 1:2000, etc.) and is independent of any magnification through which the drawing may be viewed. The analog-to-digital conversion process, of course, also has a dynamic range limitation. Whichever is of lower precision (coarser resolution) will determine the precision available for the digital computer representation.

Consider the simple case where we take a curve and by means of an appropriate digitizer (e.g., data tablet, flying-spot scanner) describe the curve in terms of a series of x, y pairs in a Cartesian coordinate system. The precision limit mentioned

[1] This is really a misnomer. It should be called the "chain description scheme" as it encompasses both quantization and encoding.

above will dictate the minimum differences in x and y that can be used to define the adjacent points of the curve. All points that we can use for describing the curve must thus be nodes of an implied, uniform, square lattice, oriented parallel to the coordinate axes and with spacing equal to the minimum allowed difference in the coordinate values. Since almost all two-dimensional analog-to-digital conversion devices (data tablets, scanners, etc.) utilize a Cartesian coordinate system, the uniform square lattice form of quantization is virtually forced upon us. In fact, not only is the *form* dictated to us, but the *size* and minimum *approximant* are fixed as well. The size is clearly determined by the minimum resolvable coordinate difference and since the lattice is already of minimum size, no information is available about the curve *between* two 8-adjacent lattice nodes. Hence we can link such adjacent nodes only with the most primitive approximant, namely, with a straight line segment. The result is the well-known chain code.

The foregoing provides the justification for uniform square-lattice quantization: It is clearly indicated whenever the resolution limits of the original analog medium or of the conversion mechanism are *uniform* over the plane of the line drawing, as is commonly the case. Conversely, it follows that in those (rare) cases where the resolution varies in some nonuniform manner over the image plane, a different-form lattice (rectangular, logarithmic, curvilinear, etc., as appropriate to the resolution variation) should be employed.

One important advantage of using a square lattice and straight-line approximants is that the resulting quanta are relatively simple to encode and are amenable to processing by means of fast, easily constructed algorithms. The use of nonuniform or nonsquare lattices complicates the encoding process, the processing algorithms, or both; it can be justified only when such lattices are indicated by some problem-specific considerations. The same is true for using curved approximants; though there is, of course, a tradeoff between using straight-line segments connecting adjacent or "near-adjacent" lattice nodes and using high-order curves with widely separated lattice nodes. The popularity of the chain representation for line-drawing information has been due to the fact that for the vast majority of problems it represents the simplest, most straightforward, and most universally applicable scheme [1].

There are many applications for which the resolution available from the original medium or from the conversion process is greater than needed or than known to be of significance for the particular line drawing information. In such situations we can (and should) convert to a coarser quantization. We may do this by simply forming a new, larger-size, uniform, square lattice by using, say, every third original lattice line. Once this is done, however, the available precision is governed by the size of the new lattice and all interlattice information must be presumed lost.

3. APPROXIMANTS

We have tried to show in the preceding section why the quantization of a line drawing leads (in all but unusual situations) to representation in terms of the x, y coordinates of 8-adjacent nodes on a uniform square lattice. If the 8-adjacent nodes are connected with straight-line segments, the sequence of such points leads directly to the familiar 8-direction chain code; the approximant is a straight-line segment with two allowed lengths, T and $T2^{\frac{1}{2}}$, where T is the lattice spacing.

Alternatively, we may regard the set of 8-adjacent lattice nodes merely as an interim representation and look for a final quantized representation that is based on more complex approximants. Let us first consider the so-called *polygonal approximation* scheme. In this scheme a curve is represented as a sequence of connected straight-line segments of arbitrary length, selected so as to keep some error measure within a specified bound. (The error measure may be maximum distance, average distance, mean-square distance, area, etc.). The curve will thus tend to be approximated by relatively short line segments where the curvature is high, and by relatively long segments where it is low [2, 3].

We have stated that polygonal approximation uses line segments of "arbitrary length." Actually this is not true. Since the source data is defined only in terms of nodes on a uniform square lattice of finite extent in x and y, there is, in fact, only a finite set of permissible segment lengths. Specifically, the set consists of all possible distances between lattice nodes in the field. It can be shown that for an $m \times n$ square lattice field, $m \leq n$, there will be a total of $mn - m(m-1)/2 - 1$ permissible line segment lengths (see the Appendix). The smallest segment will be of length equal to the lattice spacing T; the largest, of length equal to $T(m^2 + n^2)^{\frac{1}{2}}$.

Observe that the precision of polygonal approximation can be no higher than that of a chain representation on the same lattice field. The uncertainty in the nodes is the same for both. The precision of polygonal approximation can, however, be lower since an error tolerance greater than $T2^{\frac{1}{2}}/2$ can be specified for the separation between a segment and the curve. Polygonal approximation thus provides a convenient scheme for representing a curve to a desired precision (as long as it is no greater than that of the base lattice).

The actual coded representation for polygonal approximation would normally consist simply of the absolute x, y coordinates of successive segment end points. Alternatively we could use the changes $(\Delta x, \Delta y)$ between successive end-point coordinates. Although the magnitude of these changes may span the entire field, it will in practice usually be limited to a relatively narrow range. Knowledge of the probability distributions of the magnitudes of the segment coordinate changes can thus be used to achieve a more compact code representation, though at some expense for the added coding and decoding processes.

Algorithms for analysis and manipulation of polygonally approximated curves will be simple and fast. Since processing time tends to vary linearly with the number of approximation segments, processing for polygonally approximated curves will normally be faster than for the same curves represented in the form of chains.

Polygonal approximation is a powerful scheme for modeling line drawing data [4]. By imposing certain restrictions on it, or by permitting certain modifications, a variety of other modeling schemes can be derived. Thus if we restrict the lengths of the line segments to T and $T2^{\frac{1}{2}}$ (on a uniform square lattice), we obtain the familiar 8-direction chain representation [1]. If the permissible line segments are only those that can be drawn from a given node to the nodes lying on a set of

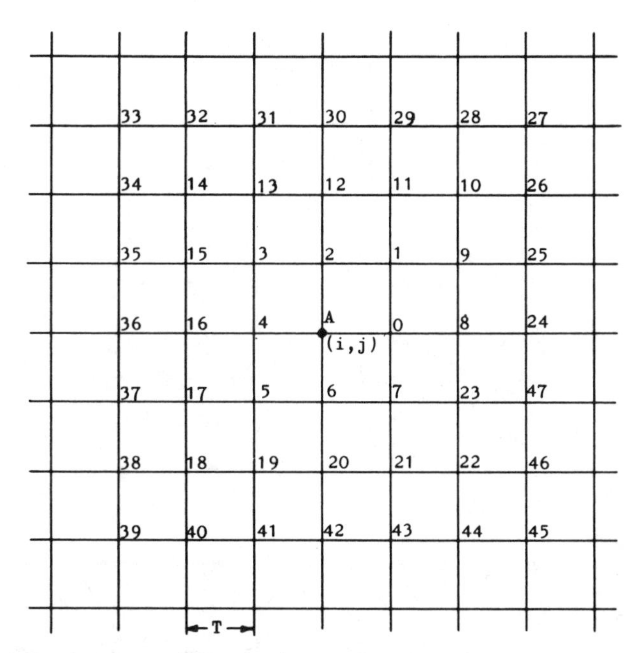

FIG. 1. The node rings surrounding a given node. Ring 1—nodes 0–7, ring 2—nodes 8–23, ring 3—nodes 24–47.

specified concentric square rings centered on this node, we obtain the so-called *generalized chain code representation* (of which the 8-direction code is a special case [5–8].

Consider the uniform square lattice shown in Fig. 1 and assume that the node marked A has already been selected as a vertex for the polygonal approximation. The node is seen to be surrounded by square rings of side $2kT$, where $k = 1, 2, \ldots$. Each square contains $8k$ nodes, evenly spaced and with nodes in the corners of the ring. We denote a generalized chain representation by the rings that are utilized. Thus in a code $-(1, 3)$ representation, a curve is represented by a polygonal approximation consisting solely of straight-line segments associated with either ring 1 (i.e., segments connecting node A in Fig. 1 to nodes 0 through 7) or ring 3 (i.e., those connecting node A to nodes 24 through 47). This is illustrated in Fig. 2. Note that code (1) is the basic chain code, and that code $(1, \ldots, p)$ where $p = \max(m, n)$, represents ordinary polygonal approximation. A section of a contour map encoded using different generalized chain codes is shown in Fig. 3.

We find that in an $n \times n$ field, a total of $n^2(n^2 - 1)$ distinct line segments may be defined. Of these, $\frac{1}{2}n(n + 1) - 1$ will be of distinct length, and $4n(n - 1)$ will be both of distinct length and distinct orientation (see the Appendix).

The generalized chain code representation can thus be regarded as covering all forms of straight-line segment approximation on a uniform square lattice. In fact, the lattice need not be square nor need it be uniform, though except for the triangular lattice, any other form would rarely be encountered.

Thus far we have discussed only straight-line segment approximation. This

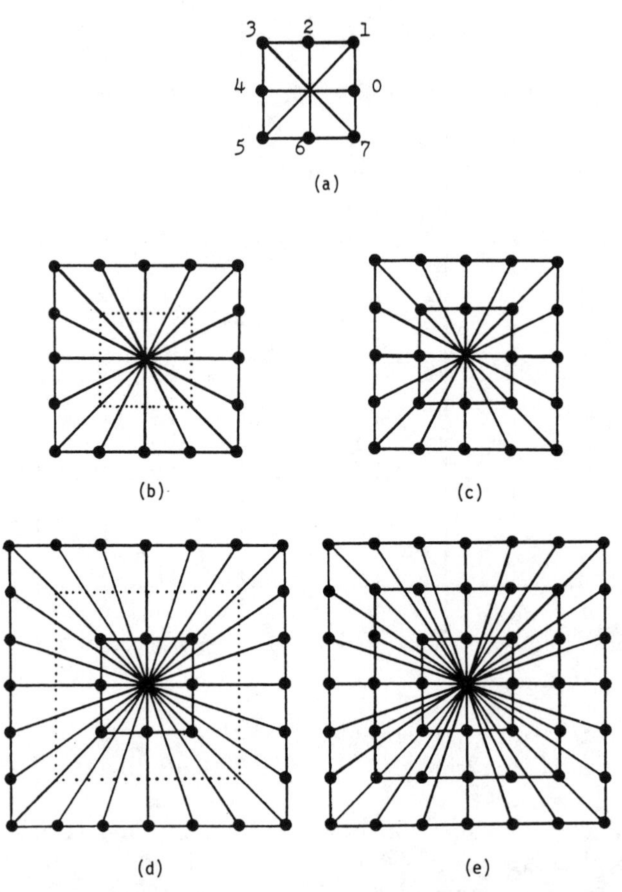

FIG. 2. Generalized chain codes. (a) (1)-code, (b) (2)-code, (c) (1, 2)-code, (d) (1, 3)-code, and, (e) (1, 2, 3)-code.

is the most common and most easily handled type. However, we can also use curved segments. Then, in addition to being able to vary the lengths of the segments, we can also vary the curvature. This is illustrated in Fig. 4. When such segments are used to represent a curve, we obtain a "polycurve," the curved-side analog of a polygon,[2] as shown in Fig. 5.

4. QUANTIZATION

To represent a line drawing in terms of the generalized chain code, we shall quantize it according to the grid-intersection scheme [1]. With this scheme, a tolerance band (link gate set) within which the curve must pass is established for each link, as follows:

1. Set $i = k$, where k is the order of the largest ring in the selected chain code.

[2] Strictly speaking, the word polygon means "many angles." Hence a "polycurve" is also a polygon. However, the term "polycurve" is descriptive and facilitates the necessary distinction.

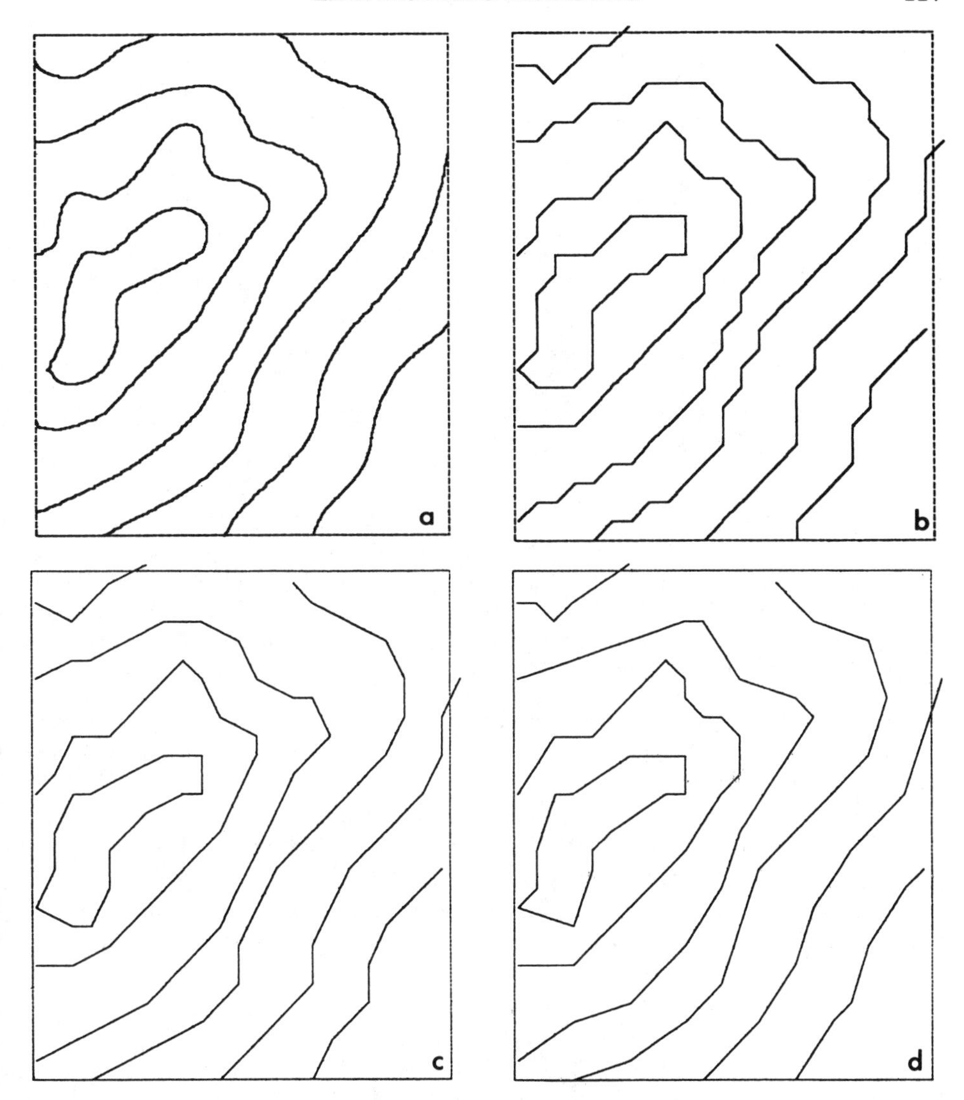

FIG. 3. A contour map section (a) encoded using (b) (1)-code, (c) (1, 2)-code, and (d) (1, 3)-code.

2. Find the midpoints of all pairs of adjacent nodes on this ring.

3. For each link of the ring, draw two lines, called midpoint lines, parallel to it from two neighboring midpoints on the sides of the link.

4. The parallel line segments cut out of each ring, 1 through k, by a pair of adjacent midpoint lines, form the *link gate set* (LGS) for the link of ring k lying between the two midpoint lines.

5. Set $i = i - 1$
 If $i = 0$, stop.
 If chain does not contain ring i, repeat step 5, else go to step 2.

The set of all link gate sets of a ring is referred to as a "template." The link gate set for the ring-3 link to node 28 is shown in Fig. 6.

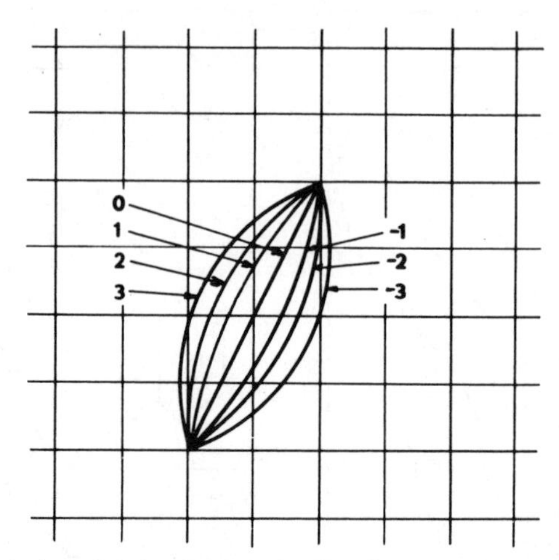

FIG. 4. A set of circular arcs for line-drawing representation.

The quantization process now consists merely of a search for the highest-order-ring link for which all link gates intersect with the curve. We begin by selecting the node closest to an end point of the curve. If the curve is closed, any node closer to the curve than one-half the lattice spacing may be selected.

1. Set $i = k$, where k is the order of the highest-order ring in the code.
2. Position the template i so that its center lies on the last-encoded node and its sides lie parallel to the grid.
3. Find the intersection points of the curve with rings $i, i - 1, i - 2, \ldots, 1$.
4. If the code does not contain any ring lower than ring i, delete from the above set all but the intersection points which lie on ring i.
5. If any LGS of ring i contains all the remaining intersection points, then the associated link is selected and we return to (1). Else set $i = i - 1$ and go to (3).

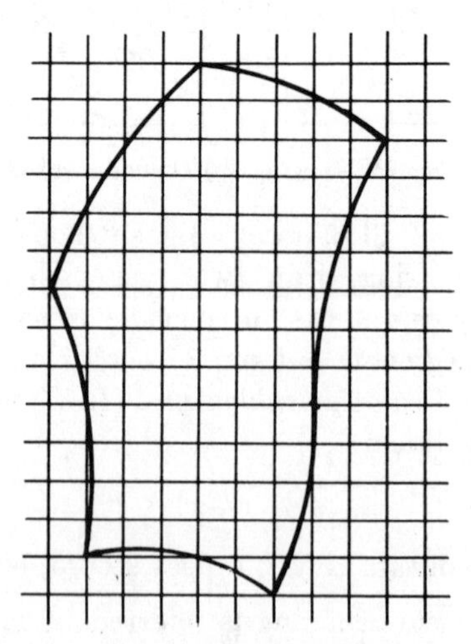

FIG. 5. A polycurve.

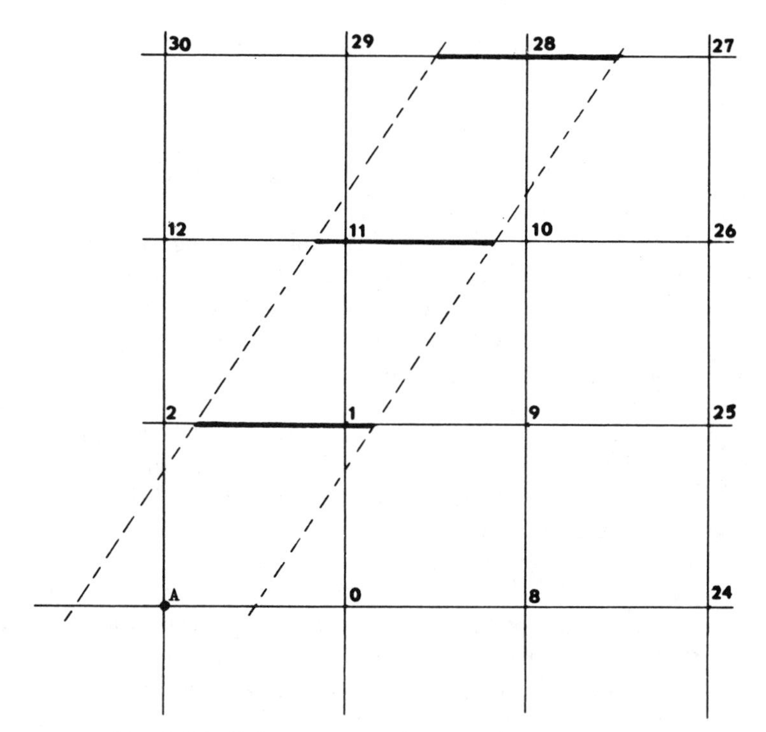

FIG. 6. The link-gate set for the ring-3 link to node 28.

The foregoing quantization algorithm assures that, provided ring 1 is included in the selected code, the resulting generalized chain approximation will never differ from the original curve by more than one-half the lattice spacing. Since the algorithm will always select the longest allowed link, subject to the tolerance requirement, the scheme is "optimum" in the sense of using the fewest number of segments to represent a given curve in the selected code.

The algorithm utilizes link gate sets formed by pairs of midpoint lines parallel to each link. One could also let the lines from the midpoints converge to the current node, thus forming a triangular rather than a parallel tolerance band within which the curve must lie to map into the associated segment [5, 7]. With the triangular bands, the link gates become progressively smaller as the current node is approached. As a result the probability distribution for the triangular bands tends to be skewed more toward lower-value rings than is the case for the parallel bands. More elaborate tolerance bands could be devised—curved, multi-segment, etc.; however, there appear to be no advantages to balance against the added complexity. In the rest of this paper it will be assumed that parallel-line tolerance bands are used exclusively.

Observe that although we have described a quantization algorithm for the generalized chain representation (in terms of straight-line segments), only a very small change is required to have the algorithm also apply to a general polycurve representation. As is readily apparent from Fig. 4, the link gate sets for curve-segment coding are more complex than for straight-line coding but the procedure is otherwise no different. A more detailed discussion of curved-segment representation will be included in a future paper.

5. LINK PROBABILITIES

Of much interest in any coding scheme are the relative frequencies of occurrence of the permissible code words. In the context of generalized chain codes, the code words are the links (or, in fact, the bit patterns used to represent the links in a computer). Knowledge of the relative link frequencies is essential for determining the optimum assignment of bit patterns to the links, for computing the average precision for a particular generalized chain code, and for establishing relative compactness among different codes.

Let us denote a link of ring n by its x and y components and write $L_{x,y}$, where, of course, $\max(x, y) = n$. Because of symmetry, we need consider only one octant; that is, we can limit consideration to the links $L_{n,a}$, where $a = 0, 1, \ldots, n$.

We wish to determine the probability of $L_{n,a}$ given that ring n of a generalized chain code is being used. We write

$$P\{L_{n,a}\} = P\{n\}P\{L_{n,a}|n\} \tag{1}$$

where $n = 1, 2, \ldots$, and $a = 0, 1, \ldots, n$. Under the assumption that the position and orientation of a curve with respect to the lattice is completely random, $P\{L_{n,a}|n\}$ will depend only on n and a. Also we shall assume that to compute these link probabilities, we need only consider straight-line segments of arbitrary position and orientation intersecting with the lattice lines.

Suppose a link is to be selected from the first octant of ring n. This requires a straight-line segment crossing $n + 1$ vertical lattice lines. The length of the segment will be $nT/\cos\theta$, where θ is the angle the segment makes with the positive x axis. Groen [9] has shown that for the universe of all curves, the product of this length and the probability density $p(\theta)$ is a constant. We write

$$p(\theta) = k\cos\theta. \tag{2}$$

Since

$$\int_0^{\pi/4} p(\theta)d\theta = \tfrac{1}{8}, \tag{3}$$

we integrate, solve for k, and obtain

$$p(\theta) = 2^{\frac{1}{2}}/8\cos\theta. \tag{4}$$

Let us consider the first-octant, ring-n configuration shown in Fig. 7. The line segment $\overline{CD}$ intersects the vertical link gate $-T/2 \leq y \leq T/2$ at node A and intersects, nT units to the right, the vertical link gate $(a - \frac{1}{2})T \leq y \leq (a + \frac{1}{2})T$ at node B. The chain link generated by $\overline{CD}$ will thus be $L_{n,a}$. The permissible angle θ which $\overline{CD}$ must make with the x axis to intersect the link gate at B must lie in the range from $\theta_{\min}$ to $\theta_{\max}$, where

$$\theta_{\min} = \tan^{-1}\frac{(a - \frac{1}{2})T - y}{nT}, \tag{5}$$

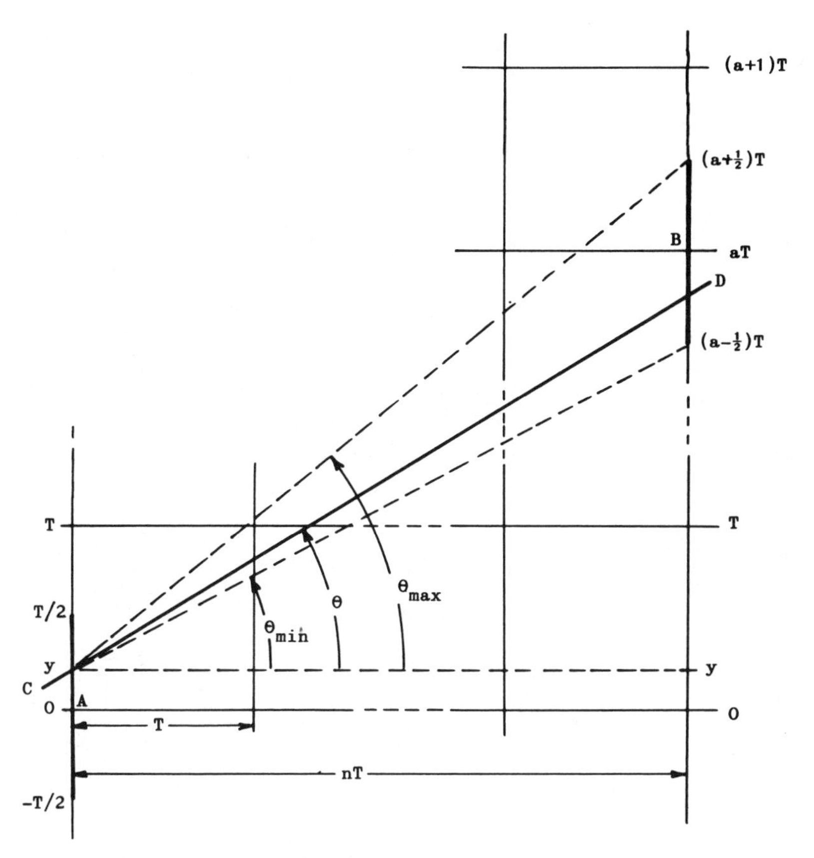

FIG. 7. First-octant, ring-n configuration for determining probability of link $L_{n,a}$, $a = 0, 1, \ldots, n-1$.

$$\theta_{\max} = \tan^{-1}\frac{(a+\frac{1}{2})T - y}{nT}. \tag{6}$$

The probability that a segment will cause the link $L_{n,a}$ to be generated is thus equal to the joint probability that the segment intersects the link gate at A and does so with an angle $\theta_{\min} \leq \theta \leq \theta_{\max}$. Since position and orientation of a segment are entirely independent, we can write

$$P\{L_{n,a}|n\} = \int_{-T/2}^{T/2} dy \int_{\theta_{\min}}^{\theta_{\max}} P(y)P(\theta)d\theta \tag{7}$$

$$= \frac{2^{\frac{1}{2}}}{8T}\int_{-T/2}^{T/2} dy \int_{\theta_{\min}}^{\theta_{\max}} d\theta \cos\theta \tag{8}$$

$$= \frac{2^{\frac{1}{2}}}{8T}\int_{-T/2}^{T/2} dy[\sin\theta_{\max} - \sin\theta_{\min}].$$

From (5) and (6),

$$P\{L_{n,a}|n\} = \frac{2^{\frac{1}{2}}}{8T}\int_{-T/2}^{T/2} dy\left[\frac{(a+\frac{1}{2})T - y}{\{(nT)^2 + [(a-\frac{1}{2})T - y]^2\}^{\frac{1}{2}}} - \frac{(a-\frac{1}{2})T - y}{\{(nT)^2 + [(a-\frac{1}{2})T - y]^2\}^{\frac{1}{2}}}\right]$$

$$= \frac{2^{\frac{1}{2}}}{8}\{[n^2 + (a+1)^2]^{\frac{1}{2}} - 2(n^2 + a^2)^{\frac{1}{2}} + [n^2 + (a-1)^2]^{\frac{1}{2}}\}. \tag{9}$$

Equation (9) gives us the probability of occurrence of $L_{n,a}$ given that ring n is being used. Our interest, however, is not in $L_{n,a}$ itself, but rather in all links of the same length. Since there is one such link in each octant, we have, for the length $l_{n,a} = (n^2 + a^2)^{\frac{1}{2}}$,

$$P\{l_{n,a}|n\} = 2^{\frac{1}{2}}\{[n^2 + (a+1)^2]^{\frac{1}{2}} - 2(n^2 + a^2)^{\frac{1}{2}} + [n^2 + (a-1)^2]^{\frac{1}{2}}\}. \tag{10}$$

Equations (9) and (10) are valid for all $a = 0, 1, \ldots, n-1$.

The case where $a = n$, shown in Fig. 8, requires separate treatment. For the

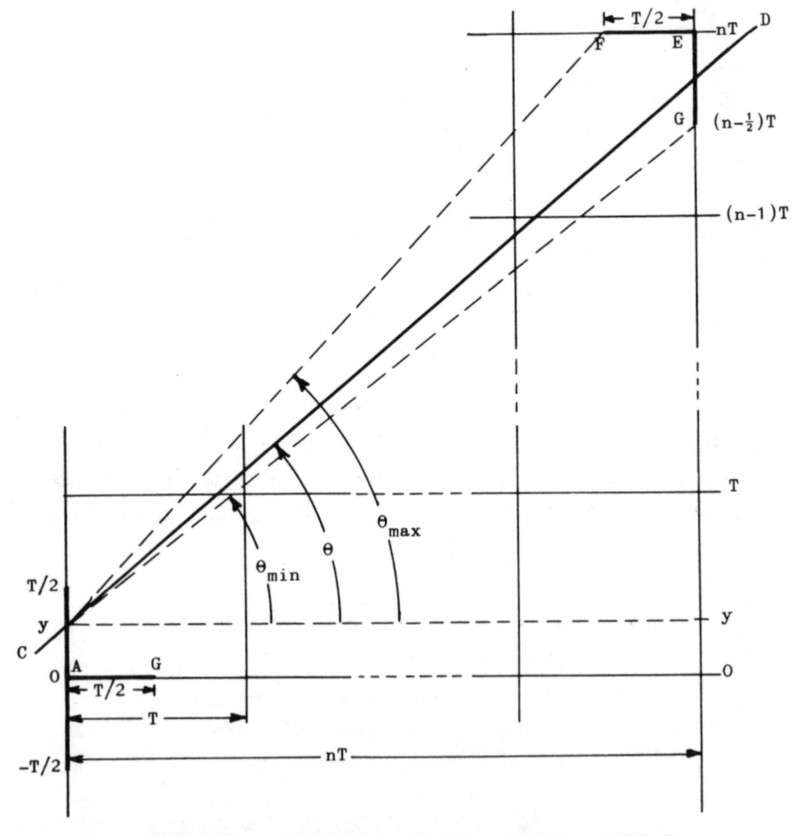

FIG. 8. Configuration for determining probability of link $L_{n,n}$.

Table 1

Conditional Chain-Link Length Probabilities, $P\{l_{n,a}|n\}$ for Rings 1 through 5

n	$a = 0$	$a = 1$	$a = 2$	$a = 3$	$a = 4$	$a = 5$
1	0.58578	0.41421				
2	0.33385	0.50387	0.16228			
3	0.22949	0.39738	0.27409	0.09902		
4	0.17409	0.31951	0.25291	0.18242	0.07106	
5	0.14003	0.26463	0.22577	0.17874	0.13544	0.05539

line segment $\overline{CD}$ to generate the link $L_{n,n}$, the segment must intersect either the vertical half-gate $[0, T/2]$ or the horizontal half-gate $\overline{AE}$ at node A, as well as either half-gates $\overline{EG}$ or $\overline{EF}$ at node E. Since the link-gate configuration is symmetrical about the line $\overline{AE}$, we shall compute the link probability only for $0 \leq y \leq T/2$ and then multiply the result by 2.

$$P\{L_{n,n}|n\} = 2\left(\frac{2^{\frac{1}{2}}}{8T}\right)\int_0^{T/2} dy \int_{\theta_{\min}}^{\theta_{\max}} d\theta \cos\theta, \tag{11}$$

where

$$\theta_{\max} = \tan^{-1}\frac{nT - y}{(n - \frac{1}{2})T}, \tag{12}$$

$$\theta_{\min} = \tan^{-1}\frac{(n - \frac{1}{2})T - y}{nT}. \tag{13}$$

We obtain

$$P\{L_{n,n}|n\} = \frac{2^{\frac{1}{2}}}{4T}\int_0^{T/2} dy\left[\frac{nT - y}{[(n - \frac{1}{2})^2T^2 + (nT - y)^2]^{\frac{1}{2}}} - \frac{(n - \frac{1}{2})T - y}{\{(nT)^2 + [(n - \frac{1}{2})T - y]^2\}^{\frac{1}{2}}}\right] \tag{14}$$

$$= \tfrac{1}{4}[1 - 2n + (4n^2 - 4n + 2)^{\frac{1}{2}}]. \tag{15}$$

There are four links of length $l_{n,a} = n2^{\frac{1}{2}}$. Hence

$$P\{l_{n,n}|n\} = 1 - 2n + (4n^2 - 4n + 2)^{\frac{1}{2}}. \tag{16}$$

Equations (10) and (16) give the probabilities of the $n + 1$ different link lengths for every ring n, $n \geq 1$. Table 1 shows these probabilities for rings 1 through 5. For ring 1, these results have also been obtained previously [1, 10, 11].

To determine the actual link-length probabilities for a class of curves, we must, of course, also know the relative probabilities of the rings used to encode the curves. Clearly these probabilities depend on the ratio of instantaneous radius of curvature to lattice width. If this ratio is large, the largest available ring in the code will be used almost exclusively. As the ratio is reduced toward unity, more and more use will be made of the smallest rings in the code. For proper quantization, the ratio of radius of curvature to lattice spacing should not be less than 2 [12].

We shall compute the ring probabilities by quantizing a circular arc of radius r at numerous orientations and positions. Each arc position will map into a link of a particular ring. The probability of each ring for that radius of curvature is then determined by counting the number of mappings into that ring.

Consider the semicircular arc of radius 1 shown in Fig. 9a. The parametric equations for this arc are

$$x = 1 - \cos t,$$
$$y = \sin t, \qquad 0 \leq t \leq \pi. \tag{17}$$

If the arc is rotated through an angle α and translated by an amount D along the diameter, we obtain (see Fig. 9b)

$$x = -\sin\alpha\sin t - \cos\alpha\cos t + \cos\alpha - D\cos\alpha,$$
$$y = -\sin\alpha\cos t + \cos\alpha\sin t + \sin\alpha - D\sin\alpha, \tag{18}$$

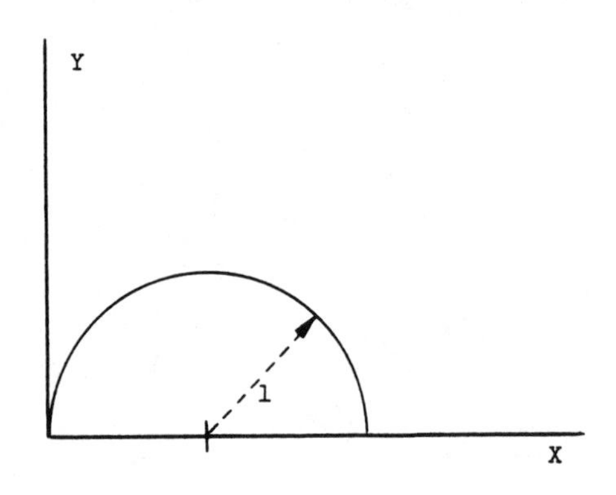

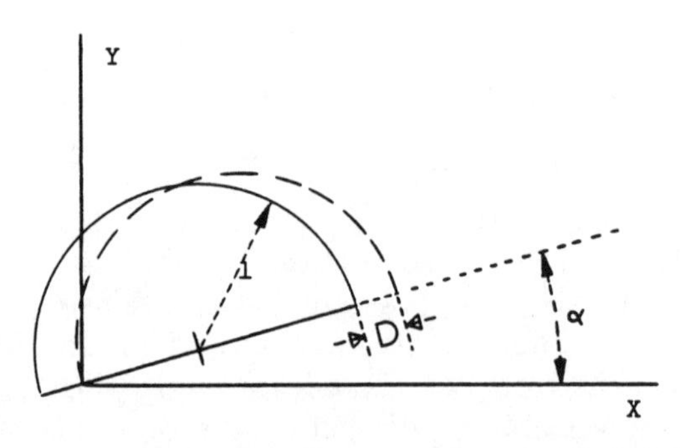

Fig. 9. Rotated and translated semicircular arc used for ring probability determination.

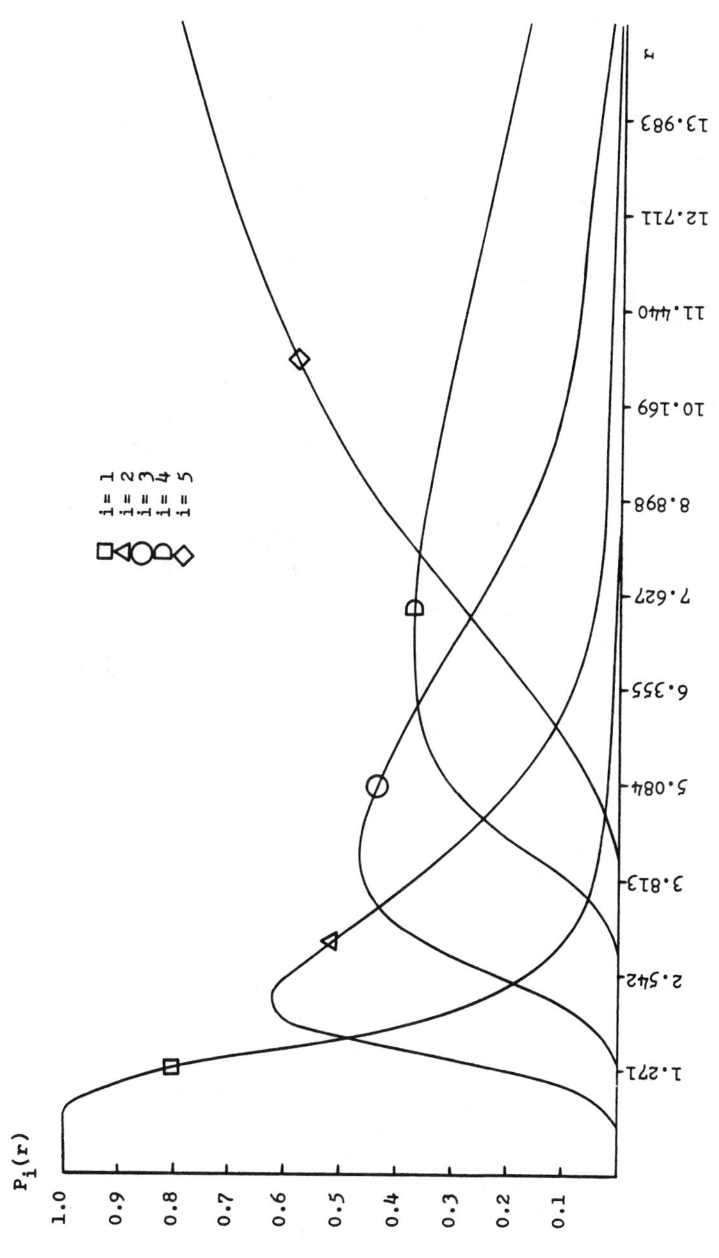

FIG. 10. Ring probabilities as a function of radius-of-curvature-to-lattice-width ratio.

or

$$t=2\tan^{-1}\left[\frac{-\sin\alpha\pm\{\sin^2\alpha-(x+D\cos\alpha)[(x+D\cos\alpha)-2\cos\alpha\}^{\frac{1}{2}}}{(x+D\cos\alpha)-2\cos\alpha}\right],$$

$$t=2\tan^{-1}\left[\frac{\cos\alpha\pm\{\cos^2\alpha-(y+D\sin\alpha)[y+D\sin\alpha)+2\sin\alpha\}^{\frac{1}{2}}}{(y-D\sin\alpha)-2\sin\alpha}\right]. \qquad (19)$$

To determine the probability of ring n for an arc of radius 4, the grid size, T, is first adjusted so that $T = r$. The template of ring n is then superimposed on the arc such that the arc starts at a node of the grid. The arc is next rotated counterclockwise in steps of $\pi/2s$ from 0 to $\pi/2$ rad, where s is the desired angular resolution. For each angle of rotation, the arc is also translated within the range $-T/2 \leq D \leq T/2$ in steps of T/s along the line perpendicular to it and passing through the center. For each position we must determine the intersection of the arc with rings 1, 2, . . ., n and then decide whether these intersection points meet the criteria for selection of a link of ring n. If they do not, then the next lower-order ring is considered. The appropriate intersection point is the one for which the parameter t is a minimum. The criterion for selecting a link based on these intersection points is the same as was described earlier.

The foregoing procedure was programmed and a computer run was made to determine the ring probabilities for rings 1 through 5. The radius of curvature was varied from 0.25 to 15 grid units in steps of 0.25. The resolution, s, was set equal to 30, yielding 900 different positions of the arc for each radius value. The results are shown plotted using a set of smoothing cubic splines in Fig. 10. The increasing use of larger rings as the radius of curvature increases is clearly evident. In a practical situation one is given a family of curves and the distribution of the radii of curvature must be determined. This can be done by selecting a representative

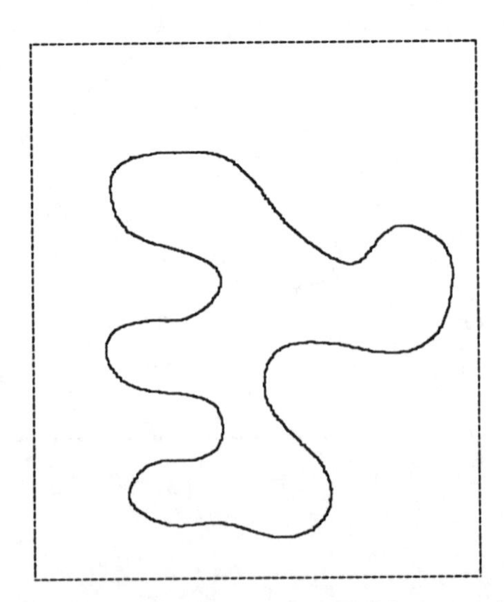

FIG. 11. A spline approximated sample curve, represented in terms of 82 cubic polynomials.

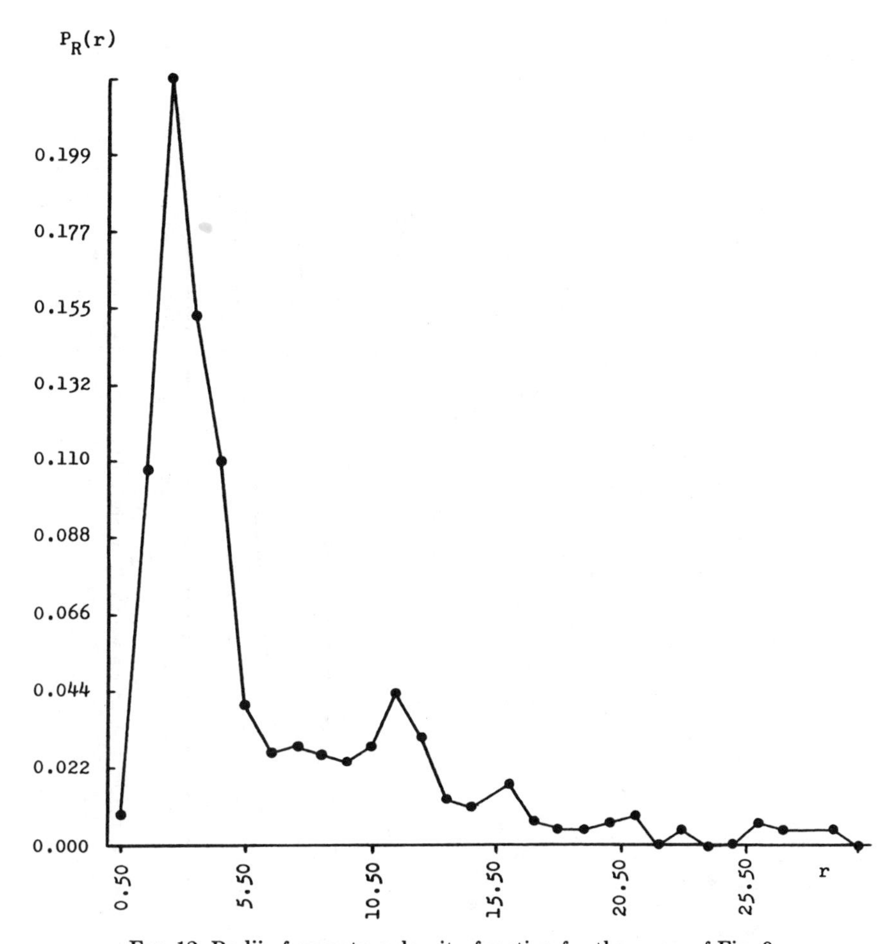

FIG. 12. Radii-of-curvature density function for the curve of Fig. 9.

set of curves from the family, finely approximating the curves with cubic splices, and then utilizing a computer to calculate the actual distribution of the radii of curvature. From this distribution and a plot such as Fig. 10, the ring probabilities $P\{n\}$ for the family and the selected code can then be determined. Once $P\{n\}$ is known, the actual chain-link probabilities can be computed with the aid of (1), (10), and (16). On the basis of this information, the optimum coding scheme for the family of curves can then be readily determined.

6. EXPERIMENTAL RESULTS

To illustrate the foregoing procedure, the sample curve shown in Fig. 11 was selected. As shown, the curve consists of 82 sections of cubic polynomials. The cubic spline representation was needed to facilitate the computer analysis. The distribution of the radii of curvature was computed; it is shown plotted in Fig. 12. This data was then combined with that of Fig. 10 to determine the actual ring probabilities for a generalized chain of type (1–5). This is shown in Fig. 13.

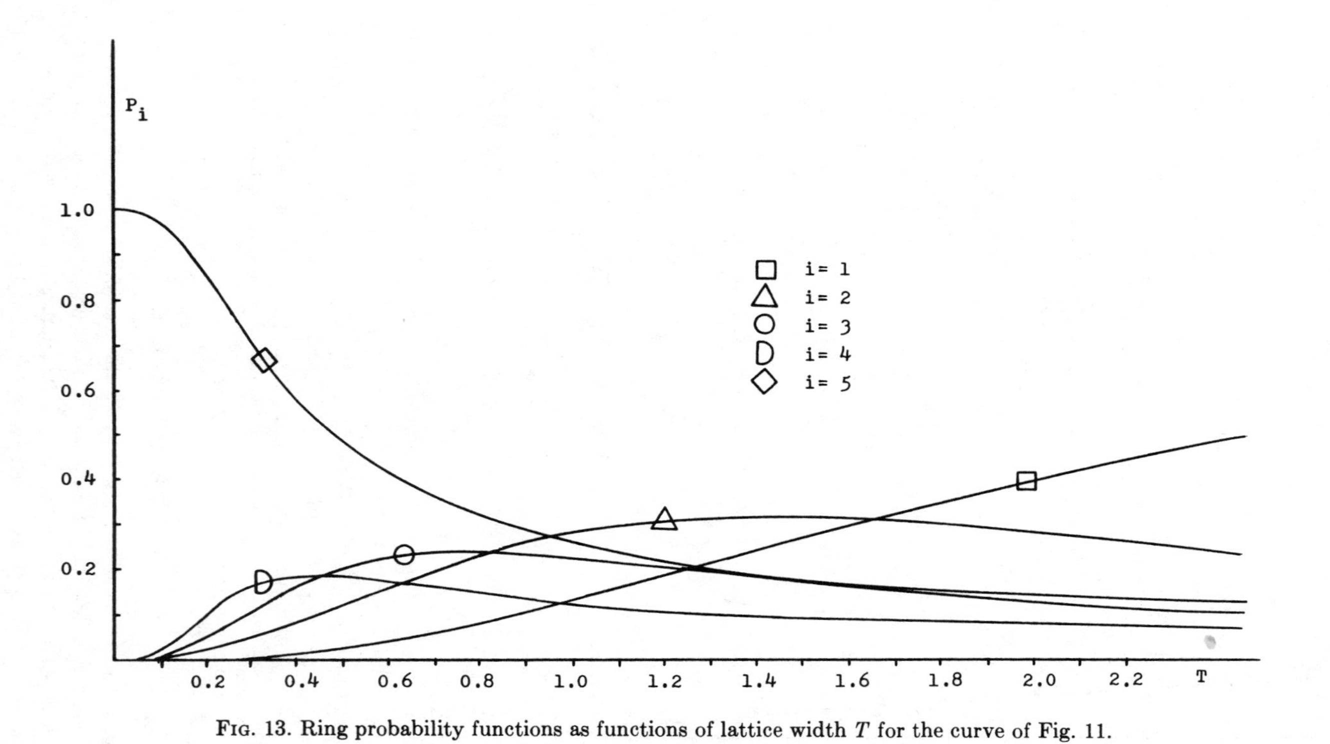

FIG. 13. Ring probability functions as functions of lattice width T for the curve of Fig. 11.

7. CONCLUSION

We have shown that insight into the process of modeling line drawings can be gained by critically examining the underlying quantization process. If a line drawing is to be approximated in terms of straight-line segments, the problems are essentially the same whether we use the tiniest possible segments (basic chain code) or the largest possible segments (polygonal approximation). All straight-line segment-modeling schemes can be viewed as being merely special cases of the generalized chain code representation. In fact, the generalized chain code representation can even be extended to include also curve-segment modeling. In any modeling scheme it is important to know the relative probabilities of the approximants used; a procedure for calculating these for straight-line-segment approximation was derived. Knowledge of these probabilities forms the basis of any compact encoding scheme that one wishes to develop for line drawing representation.

APPENDIX: *Line Segments in an $m \times n$ Lattice Field*

Consider the $n \times n$ lattice shown in Fig. A1. If we draw line segments from the origin (lower-left-corner node) to all nodes lying on or below the diagonal (nodes shown bold), we shall clearly include every permissible segment precisely once. Since the number of such segments increases by i as n increases from $i - 1$ to i, we can write for the number of lengths L permissible in an $n \times n$ lattice

$$L_{n,n} = \frac{n(n+1)}{2} - 1. \tag{A1}$$

Now for an $m \times n$ lattice field, where $m \leq n$, the only difference is that a triangular array of side $n - m$ must be subtracted from the sum for the $n \times n$

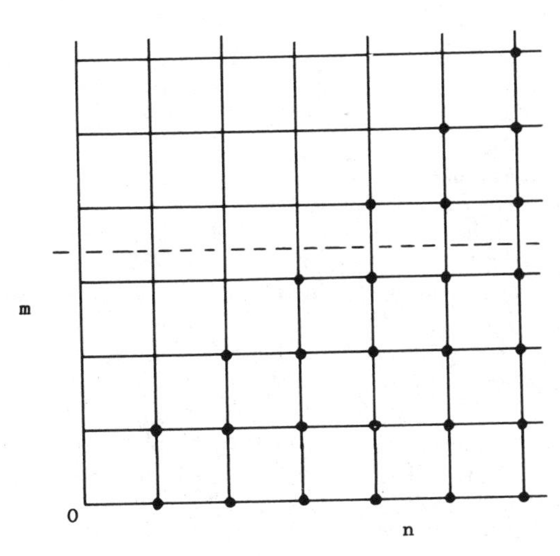

Fig. A1. Lattice, showing nodes that lead to unique-length line segments if drawn from origin.

array. Hence

$$L_{m,n} = \frac{n(n+1)}{2} - 1 - \frac{(n-m)(n-m+1)}{2}$$

$$= mn - m(m-1)/2 - 1. \tag{A2}$$

Again for an $n \times n$ field, the total number of distinct line segments is obtained by connecting each of the n^2 nodes to each other. The first node can be connected to $n^2 - 1$ nodes; the second node can be connected uniquely to only $n^2 - 2$ nodes, etc. and finally, the $(n^2 - 1)$st node can be connected uniquely to only one node. The total is then

$$\sum_{i=1}^{n^2} (n^2 - i) = \sum_{i=0}^{n^2-1} i = n^2(n^2 - 1)/2. \tag{A3}$$

To determine the number of line segments that are of *both* distinct length *and* distinct direction, we note that this corresponds exactly to the generalized chain configuration depicted in Fig. 1. When $n = 2$, there are 8 nodes, and all possible distinct-length and orientation line segments are those corresponding to the segments that can be drawn from node A to the eight ring-1 nodes (0 through 7). When $n = 3$, we have the 16 ring-2 nodes, for a total of 24 such segments, etc. Since the number of nodes increases by eight for each succeeding ring, the number of segments will be equal to

$$\sum_{i=0}^{n-1} 8i = 4n(n - 1). \tag{A4}$$

REFERENCES

1. H. Freeman, Computer processing of line drawing images, *Computing Surveys* **6** (1), March 1974, 57–97.
2. U. Montanari, A note on minimal length polygonal approximation to a digitized contour, *Comm. ACM* **13**, 1970, 41–74.
3. U. Ramer, An iterative procedure for the polygonal approximation of plane curves, *Computer Graphics Image Processing* **1**, 1972, 244–256.
4. C. M. Williams, An efficient algorithm for the piecewise linear approximation of planar curves, *Computer Graphics Image Processing* **8**, 1978, 286–293.
5. H. Freeman, Analysis of line drawings, in *Digital Image Processing and Analysis* (J. C. Simon and A. Rosenfeld, Eds.), Noordhoff, Leyden, 1977, pp. 187–209.
6. H. Freeman, Application of the generalized chain coding scheme to map data processing, in *Proc. IEEE Comp. Soc. Conf. on Pattern Recognition and Image Processing, May 31–June 2, 1978*, IEEE Computer Society Publ. 78CH1318-5C, pp. 220–226.
7. H. Freeman and J. A. Saghri, Generalized chain codes for planar curves, in *Proc. 4th International Joint Conference on Pattern Recognition, Kyoto, Japan, 7–10 November 1978*, IEEE Computer Society publ. no. 78CH1331-8C, pp. 701–703.
8. J. A. Saghri, *Efficient Encoding of Line Drawing Data with Generalized Chain Codes*, Tech. Rept. IPL-TR-79-003, Image Processing Laboratory, Rensselaer Polytechnic Institute, Troy, N. Y. August 1979.

9. F. C. A. Groen, *Analysis of DNA Based Measurement Methods Applied to Human Chromosome Classification*, Doctoral dissertation, Technical University, Delft, Dutch Efficiency Bureau, Pijnacker, The Netherlands, 1977.
10. F. C. A. Groen and P. W. Verbeek, Freeman-code probabilities of object boundary quantized contours, *Computer Graphics Image Processing* **7**, 1978, 391–402.
11. J. Koplowitz, On the performance of chain codes for quantization of line drawings, in *IEEE International Symposium on Information Theory, Grignano, Italy, June 1979.*
12. H. Freeman and J. Glass, On the quantization of line drawing data, *IEEE Trans. Systems Science Cybernet.* **SSC-5**, January 1969, 70–79.

Statistical Models for the Image Restoration Problem *

B. Roy Frieden

Optical Sciences Center, University of Arizona, Tucson, Arizona 85721

Inversion of the image formation equation for its object is an unstable or "ill-conditioned" problem. Severe error propagation tends to result. This can be reduced, however, by building a priori knowledge about the object in the form of constraints into the restoring procedure. In turn, these constraints can be accomplished by modeling the object in a suitable, statistical way. This paper is a survey of statistical models that have led to restoration methods which overcome to various degrees the ill-conditioned nature of the problem.

1. INTRODUCTION

A blurred and noisy image usually can be *processed digitally* to reduce its blur and noise level. We do not consider analog methods of processing in this paper, in the main because they are too restrictive operationally, consisting usually of a linear convolution step followed by (or preceded by) a point-to-point nonlinear mapping. These kinds of operations cannot carry through the operations required of optimum restoring approaches such as maximum likelihood, maximum entropy, etc., which require completely different operations from these.

Algebraically, the problem consists of inverting the imaging equation

$$d_m = \sum_{n=1}^{N} o_n s_{mn} + n_m, \qquad m = 1, \ldots, N, \tag{1}$$

for its unknown object scene $\{o_n\}$, given the image data $\{d_m\}$ and an estimate of the point spread function s_{mn} (the image of a point); despite the presence of an unknown, and random, noise component $\{n_m\}$ in the data.

The most naive approach to solving problem (1) for an estimate $\{\hat{o}_n\}$ of the unknown object is to ignore the noise component $\{n_m\}$ and express (1) in the matrix-product form

$$\mathbf{d} = [S]\mathbf{o}, \tag{2}$$

where $\mathbf{d}$ and $\mathbf{o}$ are obvious vector representations to sets $\{d_m\}$ and $\{o_n\}$, respectively, and $[S]$ is the spread function matrix of elements s_{mn}. Then (2) may be

* Supported by the U.S. Army Research Office.

ISBN 0-12-597320-9

simply inverted to

$$\mathbf{o} = [S]^{-1}\mathbf{d} \tag{3}$$

yielding a solution for **o**. The problem with this straightforward method is that it does not work.

An example is given in Fig. 1, where the object was the rectangle shown (dotted), the blur spread function was Gaussian with standard deviation equal to one data sampling spacing, and uniformly random noise of maximum amplitude 0.0000016 was added to the signal image. The signal image maximum was 0.16, so the maximum relative error in the data was 0.00001, a truly small error. Despite the minuscule size of this error, the restoration given by formula (3) is very erroneous (solid curve in the figure). Something goes terribly wrong when direct inversion is attempted.

The situation is little different when slightly more sophisticated estimation methods are tried, such as a least-squares solution (the $\{\hat{o}_n\}$ such that $\sum_m n_m^2$ = minimum, n_m given by Eq. (1)) or an inverse-filtered solution.

Analysis of the approach (3) shows that the problem originates in the spread function matrix $[S]$ wherein most elements are very close to zero. Then its inverse $[S]^{-1}$ has elements some of which are very large. In this case the error term due to the approximation (3), which by Eqs. (1) and (3) is $[S]^{-1}\mathbf{n}$, becomes very large. A similar problem occurs in inverse filtering, where the transfer function $\tau(\omega)$ becomes small near optical cutoff, so that the inverse filter $\tau(\omega)^{-1}$ becomes very large.

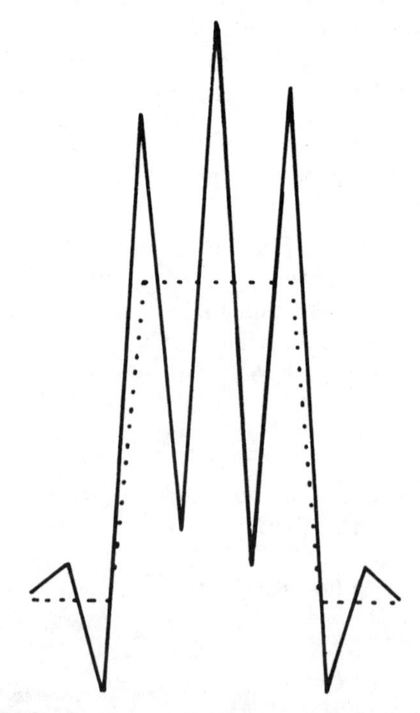

FIG. 1. Restoration (jagged curve) of the image of a rectangle object (shown dotted). Direct inversion of the imaging equation was used to form the restoration. Clearly this is not the way to go.

One may ask, at this point, why in particular the error takes the form of oscillatory spikes, and not (say) large plateau regions above the signal. This occurs because the restoration $\{\hat{o}_n\}$ in Fig. 1 must agree with the data $\{d_m\}$ when convolved with s (another way of saying Eq. (3)). This agreement can occur even when *large* errors exist in the estimate $\{o\}$, because the errors, being alternately positive and negative, cancel in the convolution operation. In other words, the imaging equation is blind to oscillatory error [1] in an estimate $\{\hat{o}\}$.

Statistical analysis shows, further, that the correlation coefficient between successive errors is very close to -1. Again, this would cause the oscillatory error shown.

Because of this notorious (sometimes called "improperly posed" [2]) effect, restoring methods have been proposed through the years that directly attempt to damp out the oscillatory errors. The best and most effective of these methods have used a priori information about the true nature of $\{o_n\}$ to effect the smoothing operation. This information often takes the form of statistical models for the objects. The subject of this paper is the nature of such statistical models.

2. AN OBJECT IS POSITIVE

The vast majority of images, used in research, industry, medicine, etc., are formed incoherently from object scenes. This means that the object to be determined is an energy distribution (or "intensity" distribution, in loose parlance). Being such, it is necessarily positive at all points, there being no such thing as negative energy. Hence, any estimate must obey the constraint

$$\hat{o}_n \geq 0. \tag{4}$$

Any number of analytic methods have been proposed through the years to effect the positive constraint. Notable are methods of Biraud [3] and Jansson [4]. The status as of about 1969 was that a few ad hoc approaches to positivity existed, and these worked quite well, except that there was no statistical rationale to prefer one over the other. Something like a minimum mean-square error (mmse) criterion had to be fashioned, and this new approach would hopefully predict a unique algorithm for the positivity problem.

Statistical methods were difficult to fashion, since mmse estimation used as a criterion seems inevitably to arrive at a Wiener [5] or Helstrom [6] filter approach, and this does not obey the constraint (4) of positivity. The textbooks on estimation are so full of this technique, and it is so convincing, that most of us found it very hard to break away from it and to enter the world of the positive constraint.

This development must now take on a personal note, since this author figured prominently in the ensuing development. We happened upon an article by workers on probability law estimation [7] which used e to a polynomial as a representation, guaranteed positive, for the unknown law. They used as justification for this procedure work of Jaynes [8] on estimating a prior probability law, i.e., a law whose functional form is to be known *prior to* seeing its data, and purely on the basis of prior information. Such information as positivity, boundedness, etc., would be used. In general these are inequality constraints.

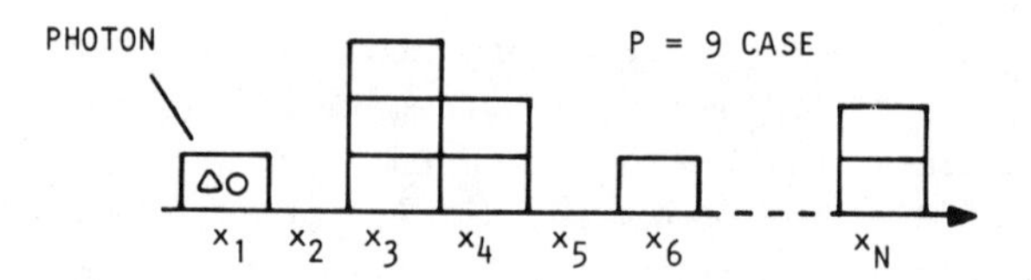

FIG. 2. Grain allocation model to accomplish a simple positivity constraint. P photons, each of intensity Δo, are dealt out among the N resolution cells of the object space with uniform probability.

Jaynes showed that there is a statistical criterion, or norm, for arriving at such a law. It is *the principle of maximum degeneracy*. By this principle, the probability law that could have existed in the maximum number of ways is presumed to be also the most likely one to have existed.

We believed it possible to adapt this principle to the object restoration problem. This was accomplished by the following model [9] for the object. See Fig. 2.

Imagine the object $\{o_n\}$ to consist of a fixed number P of photons, grains, or whatever, each of energy increment Δo. P could be known by adding up the image data values and dividing by Δo.

A discrete object space is formed by subdividing it into n cells of length Δx centered upon points $x_n = n\Delta x$. The object is imagined to be initially empty. The P photons are now allocated among N cells, one at a time, with spatially uniform probability. All photons are dealt out in this way, i.e., uniformly, since no prior knowledge exists to bias our judgment.

We now want to use Jaynes' criterion that the most degenerate object constructed in this way is also the most likely to have occurred. By most degenerate, we mean capable of being formed in the maximum number of ways through the photon allocations. We therefore have to form an expression for the number of ways W a general object $(o_1, o_2, \ldots, o_N)$ can be formed through photon allocations.

Since the photons are indistinguishable, and since any number can occupy a given cell, the answer is the well-known Boltzmann law

$$W(o_1, \ldots, o_N) = \frac{(P/\Delta o)!}{(o_1/\Delta o)! \cdots (o_N/\Delta o)!}. \tag{5}$$

Division by Δo is indicated since the law holds for particle numbers per se, not intensity values.

Proceeding with Jaynes' rationale, we set $W(o_1, \ldots, o_N) = \text{maximum}$, or equivalently, $\ln W(o_1, \ldots, o_N) = \text{maximum}$. Using Stirling's approximation $\ln m! \simeq m \ln m$, we then arrive at the criterion

$$-\sum_n (o_n/\Delta o) \ln (o_n/\Delta o) = \max.$$

This simply becomes

$$H_1 \equiv -\sum_n o_n \ln o_n = \max \tag{6}$$

after expanding out $\ln (o_n/\Delta o) = \ln o_n - \ln \Delta o$ and using $\sum o_n = P\Delta o = \text{constant}$.

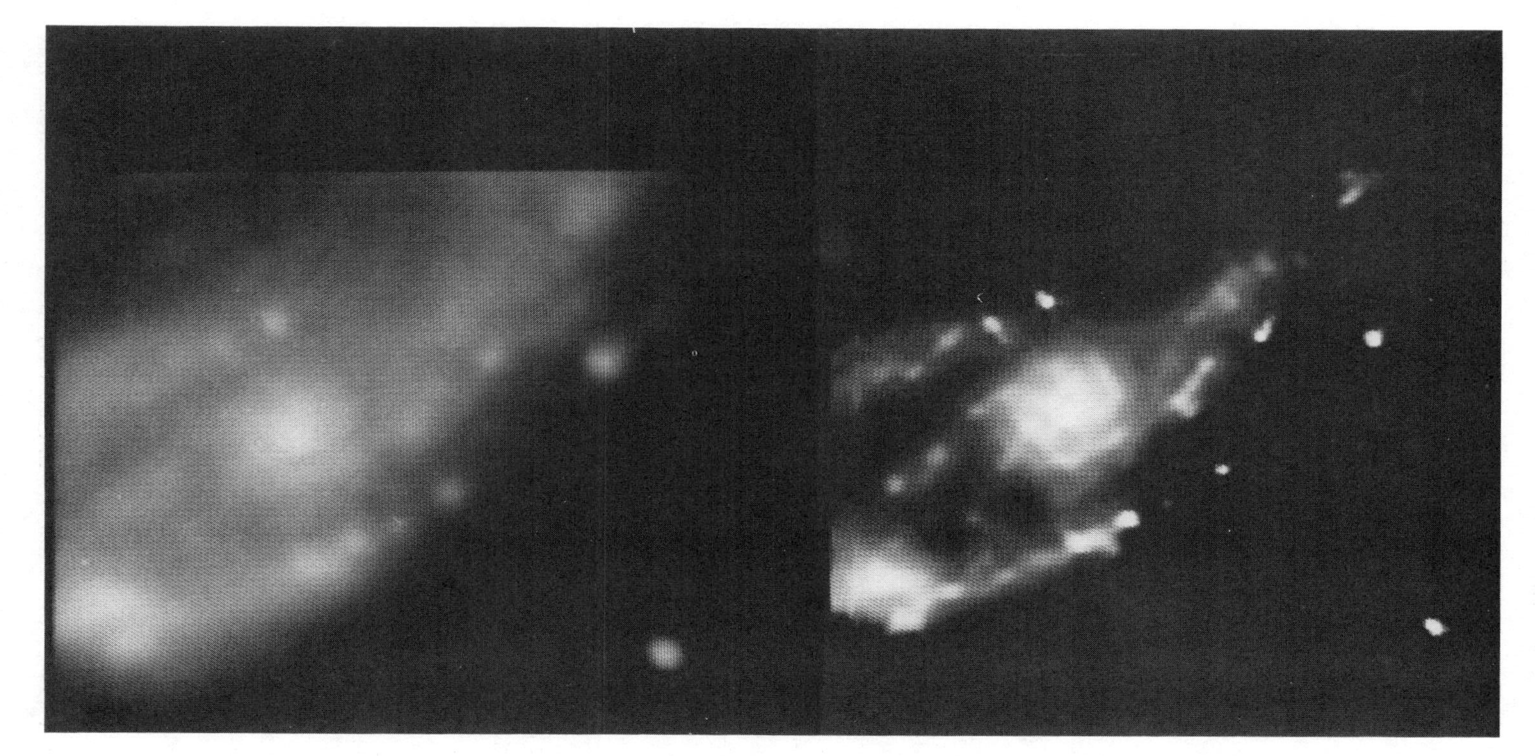

FIG. 3. Illustrative use of the maximum entropy method. On the left is a picture of a galaxy, blurred by atmospheric turbulence. On the right is its corresponding ME restoration, using an enhancement factor $\rho = 200$. Note the increase in resolution and structural detail, including a spiral set of arms emerging from the central region.

The most degenerate and likely object, therefore, obeys a principle of maximum entropy H. This was the first application of this principle to estimation of a picture. Subsequent workers [10, 11] applied maximum entropy to estimating power spectral radio images and tomographic images.

Next we will show how the use of the maximum entropy (ME) principle leads naturally to a positive-constrained object estimate.

The principle (6) must somehow be supplemented by the data inputs. These are of course $\{d_m\}$, the image values. We therefore seek the solution to (6) which also obeys the imaging equation (1). This is injected into (6) as Lagrange constraints in the usual way:

$$-\sum_n o_n \ln o_n + \sum_m \lambda_m[\sum_n o_n s_{mn} + n_m - d_m] = \max. \tag{7}$$

Seeking the extremum in $\{o_n\}$, by operating $\partial/\partial o_k = 0$, $k = 1, \ldots, N$, upon Eq. (7), we find an explicit formula for the estimate:

$$\hat{o}_n = \exp[-1 - \sum_m \lambda_m s_{mn}]. \tag{8}$$

The free parameters defining the object are $\{\lambda_m\}$, the Lagrange multipliers. We observe that any real set of $\{\lambda_m\}$ leads to an object which by representation (8) must be everywhere positive.

In this manner, a statistical model was found for arriving at a positive-constrained object estimate. The questions of how to estimate the noise $\{n_m\}$ and how to solve for the $\{\lambda_m\}$ are taken up elsewhere [9]. A restoration by this approach is shown in Fig. 3. Note the spiral galaxy emerging from out of the blur.

3. SOME OBJECTS ARE BOUNDED ABOVE AND BELOW

An absorption spectrum is a one-dimensional image which must lie between bounds of 100 and 0% absorption. Astronomical objects are usually bounded below by a finite fog level due to the nighttime airglow or other more specific causes. Many natural scenes are bounded above by the intensity from a specular reflection, which of course can be no brighter than the source. Hence, in many cases it is known a priori that

$$a \leq o_n \leq b, \tag{7}$$

with a, b known.

Can a statistical model be fashioned that will lead to a representation for $\{\hat{o}_n\}$ which obeys the *double* constraints (7)? The following model [12] leads to such an answer.

As before, subdivide the object space into cells of length Δx centered upon points $x_n = n\Delta x$. See Fig. 4. Also, as before, we imagine the object $\{o_n\}$ to be formed by the addition of discrete object intensity increments (called "grains") of size Δo to the cells. And again using Jaynes' reasoning, the most likely object to be present will be assumed to be the one which could have been formed by the grain allocations in the maximum number of ways.

The model which satisfies the upper and lower bound constraints is as follows. To satisfy the upper bound value b, imagine a given cell site to contain a total of

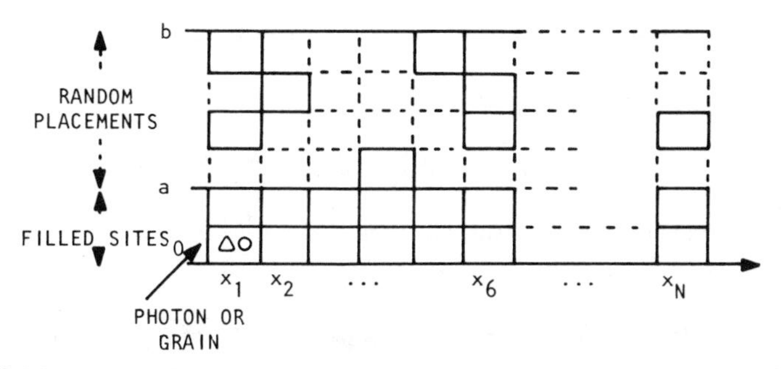

FIG. 4. Model for accomplishing upper and lower bound constraints $a \le o_n \le b$. P photons or grains, each of intensity of Δo, are dealt out randomly among grain sites above level a. Unfilled sites are dotted. All sites below level a are prefilled. Hence $o_n \ge a$. The maximum number of cell sites for each position x_n is $b/\Delta o$ so that $b \ge o_n$.

$b/\Delta o$ *subsites* for grain allocations. Each subsite can receive at most one grain, or it can remain empty. This model is in fact the "checkerboard model" [13] of O'Neill for formation of a photographic image. In statistics, it describes a Bernoulli counting process.

To satisfy the lower bound value a, imagine $a/\Delta o$ sites within the given cell to be occupied by grains, *prior to* the allocation of grains, i.e., prior to image formation with the object. In photographic work, this corresponds to a prefogged photo. In statistics, a given set of events has already occurred.

Suppose now cell n contains $o_n/\Delta o$ grains. In how many ways W_n can it have been formed? Since $a/\Delta o$ grain sites are already occupied, this leaves $b/\Delta o - a/\Delta o$ subsites. The question is then, in how many ways can $o_n/\Delta o - a/\Delta o$ grains be allocated among $b/\Delta o - a/\Delta o$ sites? The answer is analoguos to Eq. (5),

$$W_n = \frac{(b/\Delta o - a/\Delta o)!}{(o_n/\Delta o - a/\Delta o)!(b/\Delta o - o_n/\Delta o)!}. \tag{8}$$

This result is well known to describe the Bernoulli counting process we have here.

The cell sites are assumed to be filled independently, so that the number of ways W that a given object $\{o_n\}$ could be formed by grain allocation is simply

$$W = \prod_n W_n, \tag{9}$$

the product over the cells.

The most likely object $\{o_n\}$ is then presumed to have a maximum W, or ln W, so that by Eq. (8) and Stirling's approximation to the factorial we have a restoring principle

$$\ln W = -\sum_{n=1}^{N} (\hat{o}_n - a) \ln (\hat{o}_n - a) - \sum_{n=1}^{N} (b - \hat{o}_n) \ln (b - \hat{o}_n). \tag{10}$$

Note the resemblance of these sums to entropy $H \equiv$ maximum described by Eq. (6).

Adding in image constraints $\{d_m\}$ as before, the principle becomes

$$-\sum_n (\hat{o}_n - a) \ln (\hat{o}_n - a) - \sum_n (b - \hat{o}_n) \ln (b - \hat{o}_n)$$
$$- \sum_m \lambda_m (\sum o_n s_{mn} + n_m - d_m) = \max. \quad (11)$$

Solving for the maximum by operating $\partial/\partial o_m$ on the equation and setting it to zero yields an explicit solution

$$\hat{o}_n = a + b \exp[-\sum_m \lambda_m s_{nm}]/1 + \exp[-\sum_m \lambda_m s_{nm}]. \quad (12)$$

(As before, we have neglected the problem of what to do about noise n_m.) This solution (12) explicitly obeys the required constraints (7), for any set of real $\{\lambda_m\}$. Hence the model satisfies the a priori information, as required.

4. SOME OBJECTS ARE POWER SPECTRA

In the field of radio astronomy, the final estimate $\{\hat{o}_n\}$ is the Fourier transform of the data, these being two-dimensional autocorrelation values irregularly spaced over the antenna bandpass region. Since the Fourier transform of an autocorrelation function is a power spectrum, the unknown object may here be regarded as a power spectrum. In this case, the theory due to Burg [14] may be used.

Burg did not seek a maximum probability solution, as in the previous approaches of this paper, but rather a maximum entropy solution *at the outset.* Hence, maximum entropy was not derived, but assumed. The justification for this assumption was the reasoning that the autocorrelation data ought to have occurred from a maximally smooth electromagnetic field. An assumption of maximum entropy in the electromagnetic field is one way to express this. Furthermore, according to Jaynes, maximum entropy stands for a situation of maximum admitted ignorance about the process. Hence, maximum entropy represents a maximally unbiased, objective representation or a maximally conservative view of the electromagnetic field.

We note an important distinction between this reasoning, which in fact constitutes a model in itself, and that of the preceding models. In the preceding models, it was *the quantity to be estimated,* $\{o_n\}$, which was to obey maximum entropy. Here, instead it is something related to the data which is to have maximum entropy. This is perhaps the main difference between Burg's approach and Frieden's.

Now the entropy H_2 for the electromagnetic field is, assuming it to be a stationary, band-limited process,

$$H_2 = \sum_n \ln o_n + \text{const.} \quad (13)$$

Hence, the entropy relates directly to the quantity which is to be estimated, the power spectrum $\{o_n\}$.

Note the basic difference in form between this entropy expression and that of Frieden (Eq. (6)). As we have seen, their derivations are from entirely different

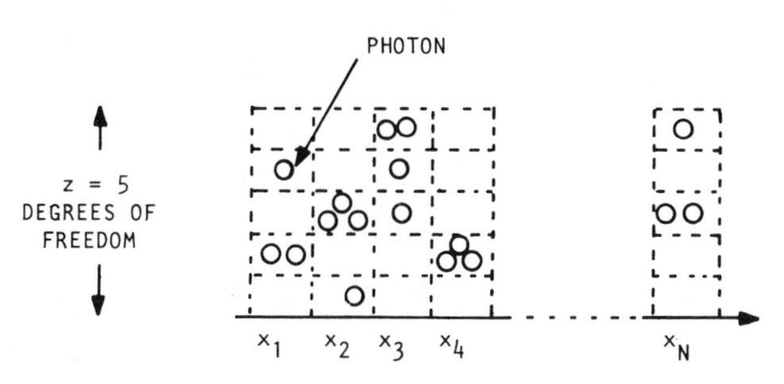

FIG. 5. Photon allocation model due to Kikuchi and Soffer. P photons, each of intensity $h\nu$, are dealt out among the N resolution cells. There are z possible subsites within each cell ready to receive photons. Any number of photons may jam into a subsite.

viewpoints of what is to be regarded as a smooth function. Since the models are so basically different, we would not expect the results to be the same.

5. THE RECONCILIATION MODEL OF KIKUCHI AND SOFFER [15]

For a while, there was great mystery surrounding the question of whether entropy H_1 of Eq. (6) or form H_2 of Eq. (13) was correct. See, for example, Ref. [11]. As we saw, the two forms of entropy arise out of two different models for the object. The question that remains is, which model is correct?

Kikuchi and Soffer answered the question by taking a quantum optics approach. They modeled the object as a random assortment of photons within an object and maximized their degeneracy on the assumption that they are Bose particles. The model is very close to being Frieden's of Eq. (5), the main difference being that Bose particles act differently from Frieden's grains. Whereas the latter were to be distributed over cell sites, Kikuchi and Soffer's photons are distributed over cell sites where *each cell has itself a finite number z of subsites* for photon allocation. See Fig. 5. These subsites are the finite number of degrees of freedom present within an image cell due to physical reasons such as cell area, bandwidth, optical aperture size and time detection interval. Also, because the photons are Bose particles, *any number* may occupy a degree-of-freedom subsite. (Note that this approach also resembles the upper-lower bound model of Eq. (8), which also postulated the existence of subsites within each object cell. However, there the particles could only *singly* occupy a cell site, since they were imagined to be photographic grains modeled by the checkerboard model.)

The authors then followed the reasoning of prior sections of this paper, solving for that object which maximizes the degeneracy W of photon placements. Hence, the number of ways W_n for forming an object value o_n out of $o_n/\Delta o$ photons was sought, assuming that the photons are indistinguishable and are to be distributed over z subsites. This follows the well-known law from Bose statistics

$$W_n = \frac{(o_n/\Delta o + z - 1)!}{(o_n/\Delta o)!(z - 1)!}. \tag{14}$$

Next, statistical independence from cell to cell was assumed, so that the total degeneracy W for placement of photons in all the cells of the object field is

$$W = \prod_n W_n, \tag{15}$$

and once more a solution

$$\log W = \text{maximum} \tag{16}$$

was sought.

In summary of Eqs. (14) through (16), we have to solve the problem

$$\ln W = \sum_n \ln \frac{(o_n/\Delta o + z - 1)!}{(o_n/\Delta o)!(z - 1)!} = \text{maximum}. \tag{17}$$

At this point it will become apparent why we have entitled the model a "reconciliation" model.

The authors examined the restoring principle (17) in two different limits, $o_n/\Delta o \ll z$ and $o_n/\Delta o \gg z$, of particle number relative to degrees of freedom. In the first limit, the top factorial quantity in Eq. (17) approximates the bottom $(z - 1)!$, so that these cancel, leaving

$$\ln W \simeq -\sum_n \ln (o_n/\Delta o)!$$

Then, using Stirling's approximation to the factorial, we get

$$\ln W \simeq -\sum_n (o_n/\Delta o) \ln (o_n/\Delta o) = \text{maximum}.$$

Expanding out the log, and using the fact that $\sum_n o_n$ is a constant, we finally arrive at the principle

$$\ln W \simeq -\sum_n o_n \ln o_n = \text{maximum}. \tag{18}$$

This was Frieden's form H_1 for the maximum entropy expression; see Eq. (6).

To evaluate (17) in the opposite limit $o_n/\Delta o \gg z$, first expand out the log and use Stirling's approximation to get

$$\ln W \simeq \sum_n (o_n/\Delta o + z - 1) \ln (o_n/\Delta o + z - 1) - \sum_n (o_n/\Delta o) \ln (o_n/\Delta o) - \sum_n (z - 1) \ln (z - 1). \tag{19}$$

In this expression

$$\ln (o_n/\Delta o + z - 1) = \ln (o_n/\Delta o) + \ln [1 + (z - 1)\Delta o/o_n] \simeq \ln (o_n/\Delta o) + (z - 1)\Delta o/o_n$$

in the limit. Substituting this result back into (19) gives

$$\ln W \simeq \sum_n (o_n/\Delta o + z - 1) \ln (o_n/\Delta o) + \sum_n (o_n/\Delta o + z - 1)(z - 1)\Delta o/o_n - \sum_n (o_n/\Delta o) \ln (o_n/\Delta o) - \sum_n (z - 1) \ln (z - 1).$$

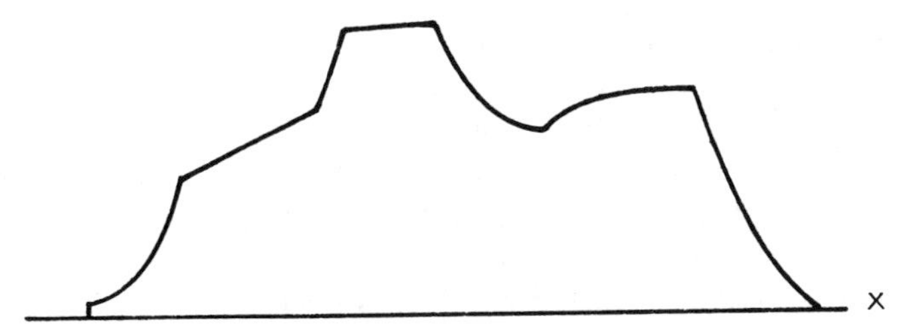

FIG. 6. A simply connected object.

Because of the limit the second sum is much smaller than the others, and so may be ignored. After cancellation of other terms, we get

$$\ln W \simeq (z-1) \sum_n \ln (o_n/\Delta o) - N(z-1) \ln (z-1) = \text{maximum}.$$

Expanding out the log, and ignoring all constant terms that do not affect the maximization, we get the restoring principle

$$\ln W \simeq \sum_n \ln o_n = \text{maximum}. \tag{20}$$

This is recognized as Burg's form H_2 in Eq. (13) for the maximum entropy expression.

It is apparent, then, that Kikuchi and Soffer have reconciled the two disparate approaches to maximum entropy restoration. These two are, in fact, limiting forms of the authors' *general* expression (17) for a maximum degeneracy restoring principle. It should be noted that the authors' form (17) is not, in fact, entropy in its conventional form, although once Stirling's approximation is used, it becomes Eq. (19), which *is* the sum of three entropy expressions. Of course Stirling's approximation is always needed, either explicitly or implicitly, to arrive at *any* entropy expression from a degeneracy expression for $\ln W$. This is true even of Jaynes' original work [8] on the subject.

The thrust of Kikuchi and Soffer's work is perhaps to reemphasize that Jaynes' principle of maximum degeneracy, and not maximum entropy, is fundamental to all. Only in certain cases does the log degeneracy approach a conventional expression for entropy.

6. SOME OBJECTS ARE SIMPLY CONNECTED

A simply connected object is, as shown in Fig. 6, one which does not contain impulses. All points on the object curve can be connected by a continuous curve, and edge gradients can even approach infinity in places, so long as two such do not exist back to back (forming an impulse).

Objects of this form are restorable by an approach which takes advantage of their monotonic behavior over limited ranges of x. This is the median window (MW) restoring algorithm [16]. The reason a window of this type is called for in the simply connected case is that it passes unaltered any monotonic region of a given curve, while smoothing out locally undulating parts of the curve. This is obviously a very nonlinear behavior. A linear filter would, by contrast, smooth out the monotonic regions (say, edge profiles) if it smoothed out local ripples. The linear price paid for noise reduction is blurring.

But the MW operation does not pay this price. Consider Fig. 7, which shows on the top row a typical line of image data representing a linear restoration of an edge. There is a low plateau region on the left, which exhibits spurious Gibbs oscillation, then the edge transition region (numbers 3 5 7), followed by a high plateau region which, again, exhibits spurious oscillation.

The second row shows the output of a median window of length seven points, as it is centered point by point across the data, the output at any emplacement being the median value of all seven points within the window. (For example, the value 2, the leftmost value, results as the median value of the first seven points 1 2 3 3 2 1 0.) Note the character of the output (second) row. The leftmost oscillatory region is replaced by essentially one value, level 2, the gradient region remains *unchanged*, and the rightmost oscillatory region is replaced by only 7's and 8's. Hence, the Gibbs errors in the original restoration have been almost entirely eliminated, while preserving the important edge gradient information as is. It is this effect which makes the MW operation a very useful weapon for restoring purposes.

The third row shows results when using a smaller window, of size three points. The output is seen to be more oscillatory than for the seven-point window, and once again the edge transition numbers 3 5 7 are preserved.

In general, it is found that a window length L will completely obliterate oscillations of length $L/2$ or less. The reader may easily verify this using a line of numbers as in the figure. Hence, if dealing with diffraction-limited imagery, where the highest signal frequency is at optical cutoff, a value of L equal to two Nyquist sampling intervals should be used. It will obliterate all oscillations at the cutoff frequency and higher; these frequencies can only arise from noise in the data, and so should be obliterated in a restoration.

The preceding has shown that application of a median window operation to linearly restored data will improve the restoration. An edge object was used to demonstrate the idea. This is a particular simply connected object. If, on the other hand, the restoration of an impulsive object (consisting of a sequence of impulses) is MW operated upon, results are not so good. The object is intrinsically not monotonic, i.e., it goes up and then down abruptly. Hence, a MW operation will tend to obliterate the object impulses along with the spurious oscillations.

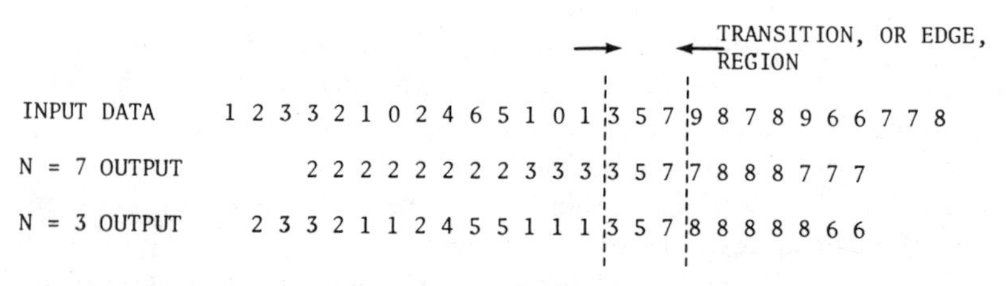

FIG. 7. Illustrative use of a median window MW operation upon a row of input numbers (top row). The second row shows the outputs for a seven-point window length, the third a three-point window length.

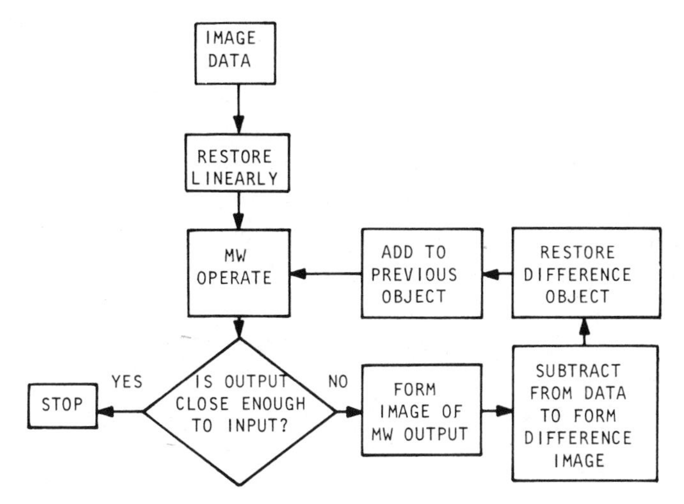

FIG. 8. Flow chart of median window restoring MWR algorithm.

Everything appears to be spurious oscillations to the median window in this case. Hence, the operation should not be used when the object scene has impulses that it is desired to exhibit in the output.

Finally, let us consider whether the edge gradient can itself be improved, i.e., be steepened in some way. One MW operation leaves the gradient unaltered. However, it does remove the adjacent spurious oscillations, as we found. This means that the output is no longer consistent with the image data, i.e., if convolved with the spread function it does not equal the image data. Suppose, then, a simple attempt was now made to again make the restoration consistent with the data. Would the gradient be steepened? The answer is yes, since consistency implies a restoration procedure, and hence, enhanced resolution.

In fact, suppose the output of the MW is convolved with the spread function to form an estimated image, and this is subtracted from the data image to yield a difference image. This is a measure of the inconsistency with the data at this step. Let this difference image now be deconvolved by any linear restoring method, to yield a difference object. This difference object is added to the prior MW output to form a new object estimate. This object is once again consistent with the image data, and hence has steepened edge gradients. But it also has Gibbs spurious oscillation, as before. Therefore, we once again pass a MW over it to remove them, preserving the existing edge gradients. Now this output is not consistent with the data once again, so once again we form the difference image and deconvolve it, further steepening the edge gradients.

Evidently, the process should be continued until the MW output is close enough to consistency with the image data to be acceptable. Closeness could be measured by the rms value of the difference image compared to the known rms error in the data, the latter known from the known noise characteristics in the imagery.

The procedure is mapped out in Fig. 8, in flow chart form. At each linear restoration, the edge gradients are steepened toward their true values, and at each MW operation the edge gradients are preserved while knocking down spurious oscillations engendered by the restoring step.

The method was tried out on edge images, as shown in Fig. 9. In Fig. 9a the noiseless image was MW restored, giving the (starred) restoration shown in the second row; the top row shows a conventional linear restoration of the same data (the latter shown by dots). In Fig. 9b a healthy amount of noise was added to the image data, to wit, uniformly random with a maximum amplitude of 20% of the signal image maximum. The bottom row shows the MW restoration, the row

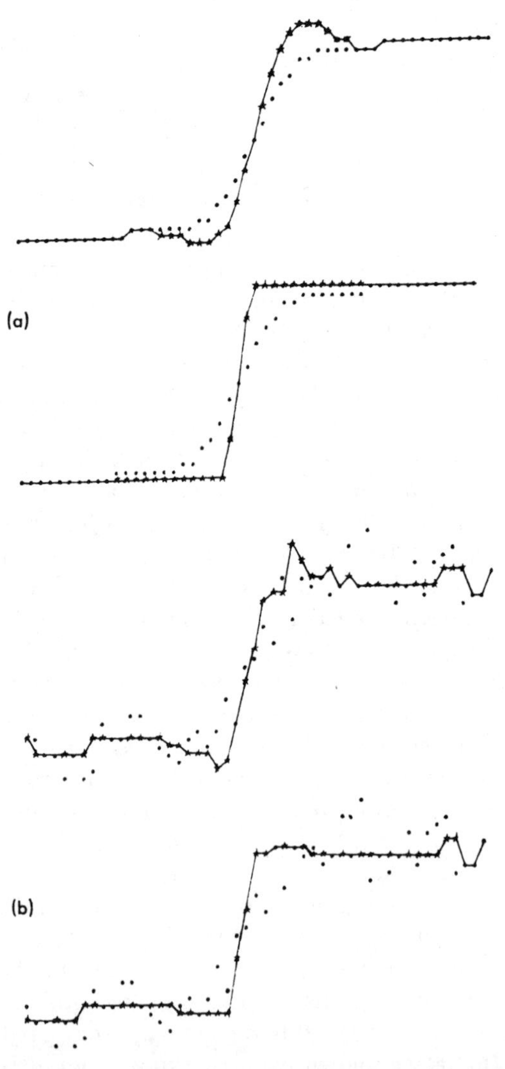

FIG. 9. (a) A step object restored linearly (top, starred curve) and by the MWR algorithm (bottom, starred curve). Five cycles of the MWR were used, with a window length of one Rayleigh resolution length (11 consecutive points). Dots show the image data, which here suffer no noise. (b) As in (a), except with 20% amplitude noise added to the image data.

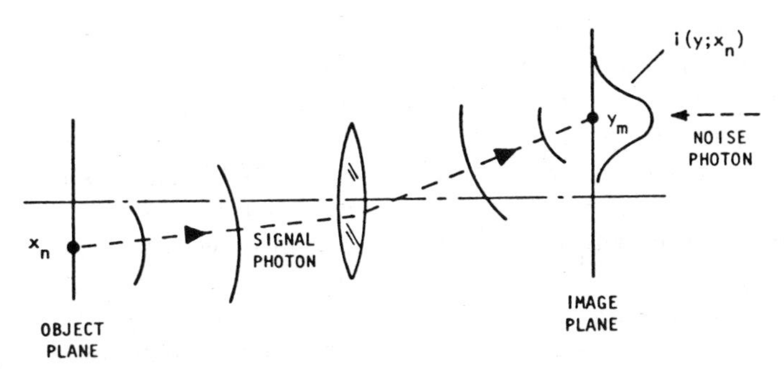

FIG. 10. The information channel. A signal photon undergoes transition from an object position x_n to image position y_m. A noise photon undergoes a transition to y_m independent of any x_n.

above it the conventional linear restoration of the same data (shown by dots). The advantages of the MW restorations over the linear ones are quite evident.

These tests were upon one-dimensional objects, whereas most objects are two dimensional. In the world of 2-D, one has to decide upon a window *shape* as well as a window size. It was found that a square shape yields a blocky looking output, although this does yield the best resolution enhancement over all window shapes. Of course, a blocky appearance is rather distracting. By contrast, a circular filled window was found to yield omnidirectional resolution and a satisfactory gain in resolution.

7. AN OBJECT IS A PROBABILITY LAW: MAXIMUM INFORMATION RESTORATION

All but one of the preceding models have assumed the object to consist of a finite number of grains or photons of energy Δo. But then if cell n contains $o_n/\Delta o$ grains, and if the total energy within the object is E, the ratio o_n/E actually represents the probability that a photon will radiate from cell n. Without loss of generality we may assume that $E = 1$, so that directly

$$o_n = p(x_n), \tag{21}$$

the probability that a photon radiates from position x_n.

By identical reasoning, an image value i_m representing the intensity at position y_m in the image plane also represents the probability $p(y_m)$ that a photon will strike the image plane at position y_m, i.e.,

$$i_m = p(y_m). \tag{22}$$

Hence, in summary, object and image intensity distributions are alternatively probability laws on position for photons.

What can be made of this correspondence? The object probability law $p(x_n)$ can also be regarded as an unknown source or input function, and the image law $p(y_m)$ is the corresponding received data or output function from a *communication channel.* The communication channel is the image forming medium, in general consisting of lenses, turbulent atmosphere, electronics, display, perhaps the human eye, etc. This modeling of image formation is illustrated in Fig. 10.

Given a communication channel, a natural question one can ask is, what is the maximum information rate of the channel? Also, what source function $p(x_n)$ gives rise to the maximum information? Moreover, this question has direct bearing on the restoration problem. The source function $p(x_n)$ giving rise to maximum information (also called "channel capacity") represents by (21) a restoration of the object which, had it existed, would maximize the throughput of information about itself into the image. This maximum information (MI) restoration would therefore be the object estimate which contains the most information about the unknown object.

Image interpreters often express the desire to "extract a maximum of information" from the image. Perhaps the above criterion will lead to an entity which satisfies this aim.

We now go about showing how to find the MI restoration. First, the quantity to be maximized is the Shannon information [17]

$$I(X, Y) \equiv \sum_m \sum_n p(x_n, y_m) \ln \frac{p(y_m|x_n)}{p(y_m)}. \tag{23}$$

Now since $p(x_n, y_m) = p(x_n)p(y_m|x_n)$, substitution into (23) shows that the only new quantity to consider in (23) is $p(y_m|x_n)$, the conditional probability of an image position y_m for a fixed object position x_n. A little thought, and a look at Fig. 10, shows that for photons that pass through the lens system $p(y_m|x_n) = s(y_m - x_n)$, the point spread function.

However, not all photons arriving at y_m pass through the image-forming medium. Some are noise contributions (see the figure), and these arrive at y_m *independent of* any source position x_n. Suppose fraction f of all photons in the image are source-dependent photons, and the remaining fraction $1 - f$ are noise photons. Then the net $p(y_m|x_n)$ obeys simply

$$\begin{aligned} p(y_m|x_n) &= fs(y_m - x_n) + (1 - f)n(y_m) \\ &\equiv fs_{mn} + \bar{f}n_m, \end{aligned} \tag{24}$$

where $n(y_m) \equiv n_m$ is the probability law for location in the image of a noise photon. Also $\bar{f} \equiv (1 - f)$.

We are now ready to express Eq. (23) for information in terms of our system model. First of all, expanding out the log of the quotient in (23) leads to

$$I(X, Y) = \sum_m \sum_n p(x_n, y_m) \ln p(y_m|x_n) - \sum_m p(y_m) \ln p(y_m), \tag{25}$$

where the sum over n has been carried through in the last sum. Now this last sum is precisely the entropy of the image data, by correspondence (22), and hence is a fixed quantity. Therefore, maximizing $I(X, Y)$ is equivalent to solving

$$\sum_m \sum_n p(x_n, y_m) \ln p(y_m|x_n) = \text{maximum}. \tag{26}$$

The negative of this quantity is usually called the "entropy of the channel," and is denoted as $H(Y|X)$. Using the identity $p(x_n, y_m) = p(x_n)p(y_m|x_n)$, correspondences (21) and (22) and identity (24), Eq. (26) becomes

$$\sum_n o_n \sum_m (fs_{mn} + \bar{f}n_m) \ln (fs_{mn} + \bar{f}n_m) = \text{maximum} \tag{27}$$

through variation of o_n.

But in (27) we also have the unknowns $\{n_m\}$ to contend with. How do these behave as $\{o_n\}$ is varied? In fact, as we show next, these are known in terms of the image data $d_m = i_m$, and in terms of the $\{o_n\}$ themselves.

From basic probability theory,

$$p(y_m) = \sum_n p(x_n)p(y_m | x_n). \tag{28}$$

Therefore, by correspondences (21) and (22), and channel model (24), Eq. (28) is

$$\begin{aligned} i_m \equiv d_m &= \sum_n o_n(fs_{mn} + \bar{f}n_m) \\ &= f \sum_n o_n s_{mn} + \bar{f}n_m. \end{aligned} \tag{29}$$

This allows us to solve for n_m in terms of the known data $\{d_m\}$ and any estimate of the object $\{o_n\}$. Hence, when $\{o_n\}$ is varied in criterion (27), n_m also varies through Eq. (29).

Equation (27) was operated upon by $\partial/\partial o_n = 0$ to find an extremum, and the resulting N equations were solved by an iterative Newton–Raphson approach using the zeros as target values. Also, the procedure involves the use of a penalty function that is constructed so as to find a feasible solution, i.e., one obeying the constraints

$$\sum_n \hat{o}_n = 1, \qquad 1 \geq \hat{o}_n \geq 0 \text{ and } \hat{n}_m \geq 0. \tag{30}$$

In the interests of brevity, we shall not linger any further on these details.

We tested out the MI restoring algorithm on computer simulations. Two basically different object types were used—edges and two-point impulses (double stars). Images were formed by convolution with a point spread function s of Gaussian profile, with standard deviation equal to the data sampling interval. Signal-independent, uniformly random noise was added to the signal image. The amount of randomness was defined by an f value of 0.95 used in all cases. This is a modest but significant amount of noise.

As a benchmark comparison, we also restored each image by a maximum entropy algorithm defined by requiring

$$H(X) = -\sum_n o_n \log o_n = \text{maximum}.$$

That is, we used Frieden's form H_1 of entropy to form an ME restoration as well.

The edge-object test results are shown in Fig. 11. The object in all cases was a broad rectangle, indicated by dots. The blurred images are almost the same in all cases, to the accuracy of these plots, although each has a different set of noise values superimposed. A typical image is indicated by the dashed curve in the upper left-hand position.

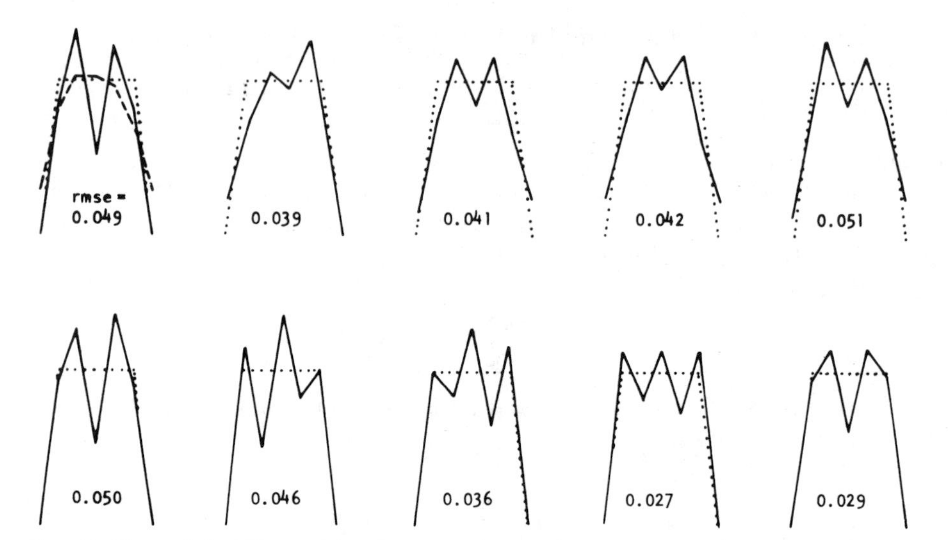

FIG. 11. Tests of the MI algorithm, in comparison with ME, upon successive images of a rectangle object.

The top row of Fig. 11 shows five successive ME restorations (the solid curves), along with the root mean square error (rmse) from the object for each. Note that although the problem of oscillation within the plateau region of the rectangle is not serious, there is some loss of gradient at the two edges.

The bottom row of Fig. 11 shows the corresponding MI restorations. Each MI restoration used the same image data as the ME restoration immediately above it. These, therefore, should be compared. The rmse value for each MI restoration is also shown within each curve. We see that (1) although the rmse values for the first two left-hand cases give the advantage to the ME approach, in fact the average rmse over the five cases (and others generated in addition) shows a smaller rmse for MI. Averaged over nine test cases, the two rmse values are 0.0395 and 0.0363 for ME and MI, respectively. Hence, MI restorations of edges are slightly more accurate than corresponding ME restorations. The gain in accuracy lies in attaining steeper edge gradients (see the figure), and in more accurate placement of the estimate edges. In fact, the figure shows that *the most accurate points within the MI restorations are those defining the two edges.* This may indicate that MI is a superior algorithm for detecting and locating object edges, in particular.

The next series of tests was done upon two-point objects, shown by dots in Fig. 12. The two impulses forming the object are separated by twice the standard deviation of the spread function, or, two sampling spaces. The blurred images do not resolve the two impulses. A typical image is shown by the dashed curve in the upper left-hand position.

As in the previous study, the top row of restorations, indicated by solid curves, is by ME, while the bottom row shows the corresponding MI restorations from the same image data. Each MI restoration should be compared with the ME restoration above it.

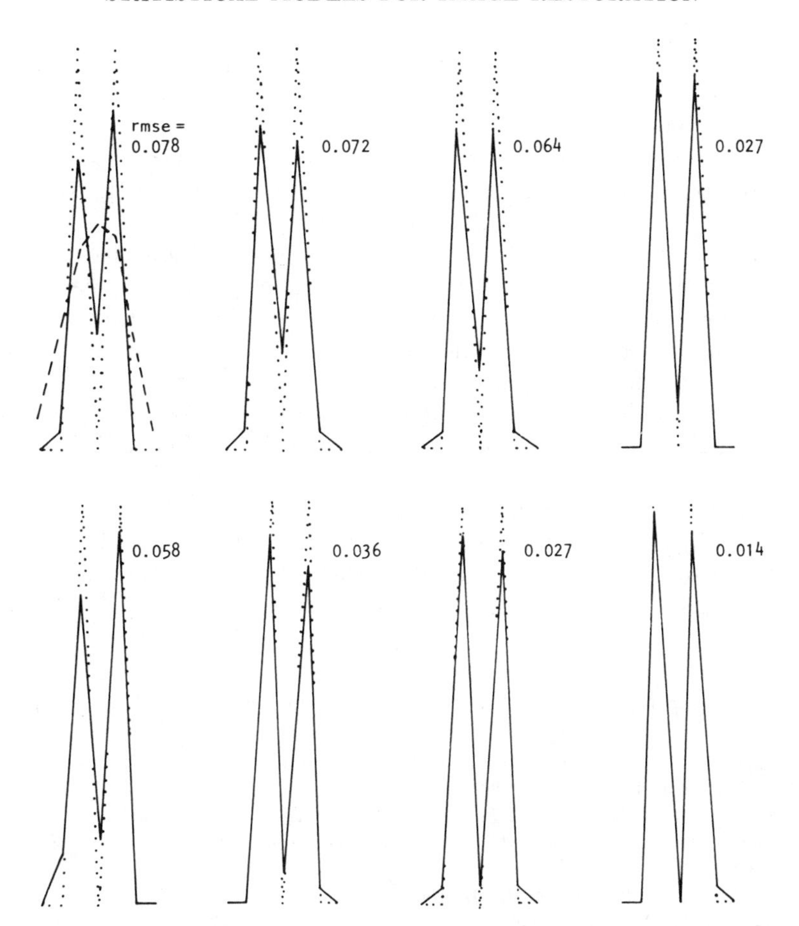

FIG. 12. Tests of the MI algorithm, in comparison with ME, upon successive images of a two-point impulse object.

The results are remarkable in that every case shows the same trend: the MI restoration has both higher peaks and a lower central dip than its corresponding ME case. Also, the rmse is lower in each case for MI, in most cases less by a factor of 2. This is an appreciable gain in accuracy. Hence, MI appears to be inherently more well suited to restoring impulses than is ME, an interesting development since, to date, ME was probably the best restoring algorithm for this purpose.

8. SUMMARY

Since image restoration is an "ill-conditioned" problem, as was demonstrated, inequality constraints have to be built into the output in order to control the severe error propagation problem. Various forms of a priori information may be input into the restoring scheme through the constraints. Depending upon the a priori information at hand, an appropriate statistical model for the object may be devised. Each model gives rise to a new restoring principle.

Simple positivity is amenable to a simple particle model and invocation of *maximum degeneracy* for the object. These lead to a condition of *maximum entropy* in the object.

Upper and lower boundedness may be attained by a "checkerboard" model, wherein each pixel site contains a multiplicity of grain sites for possible allocation of grains. This leads to an *entropy-like* expression.

A quantum statistical model and the criterion of maximum degeneracy lead to an object representation which is a *generalization of maximum entropy*. Only in two specific limits does the representation go over into entropy-like forms.

Simply connected objects are amenable to *median window* operation in conjunction with any linear restoring algorithm.

Power spectral objects are well restored if the amplitude data giving rise to them are regarded as having maximum entropy. Then *Burg's ME formalism* results.

Finally, if the object and image are regarded as probability laws on photon position, the object that maximizes *information throughput* from object to image may be regarded as the restoration. This constitutes a new norm of estimation *per se*, as well as a new norm of restoration.

REFERENCES

1. This observation seems first to have been made by D. L. Phillips, *J. Assoc. Comput. Mach.* **9**, 1962, 84.
2. R. Barakat and E. Blackman, *Opt. Comm.* **9**, 1973, 252.
3. Y. Biraud, *Astron. Astrophys.* **1**, 1969, 124.
4. P. A. Jansson, R. H. Hunt, E. K. Plyler, *J. Opt. Soc. Amer.* **60**, 1970, 596.
5. N. Wiener, *The Extrapolation, Interpolation and Smoothing of Stationary Time Series*, p. 84, Wiley, New York, 1949.
6. C. W. Helstrom, *J. Opt. Soc. Amer.* **57**, 1967, 297.
7. A. Wragg and D. C. Dowson, *IEEE Trans. Inform. Theory* **IT-16**, 1970, 226.
8. E. T. Jaynes, *IEEE Trans. Systems Sci. Cybernet.* **SSC-4**, 1968, 227.
9. B. R. Frieden, *J. Opt. Soc. Amer.* **62**, 1972, 511.
10. J. G. Ables, *Astron. Astrophys. Suppl.* **15**, 1974, 383.
11. S. J. Wernecke and L. R. D'Addario, *IEEE Trans. Computers* **C-26**, 1977, 351.
12. B. R. Frieden, *IEEE Trans. Inform. Theory* **IT-19**, 1973, 118.
13. E. L. O'Neil, *Introduction to Statistical Optics*, Addison–Wesley, Reading, Mass., 1963.
14. J. P. Burg, Maximum entropy spectral analysis, paper presented at the 37th Annual Society of Exploration Geophysicists Meeting, Oklahoma City, 1967.
15. R. Kikuchi and B. H. Soffer, *J. Opt. Soc. Amer.* **67**, 1977, 1656.
16. B. R. Frieden, *J. Opt. Soc. Amer.* **66**, 1976, 280.
17. C. E. Shannon and W. Weaver, *The Mathematical Theory of Information*, Univ. of Illinois Press, Urbana, 1949.

Syntactic Image Modeling Using Stochastic Tree Grammars *

K. S. FU

School of Electrical Engineering, Purdue University, West Lafayette, Indiana 47907

1. INTRODUCTION

Recently, formal (nonstochastic) languages and stochastic languages have been used in the modeling of image structures [1, 2]. An image is described as a composition of its components, called subimages and primitives (the simplest subimages). This approach draws an analogy between the structure of images in terms of components and relations among components and the syntax of a language in terms of grammar rules. For one-dimensional signal and line patterns, the one-dimensional string representation appears to be quite natural and efficient. However, for two-dimensional images and three-dimensional scenes, an extension from the one-dimensional string language approach to higher dimensions will often result in a more efficient representation.

One natural extension of one-dimensional string languages to high-dimensional languages is tree languages. A string could be regarded as a single-branch tree. The capability of having more than one branch often gives trees a more efficient image representation. Interesting applications of tree languages to image recognition include the classification of bubble chamber events, the recognition of fingerprint patterns, and the interpretation of LANDSAT data [3–7]. In some practical applications, image distortion and measurement noise often exist. In order to describe noisy and distorted images, the use of stochastic languages has been proposed [1]. With probabilities associated with grammar rules, a stochastic tree grammar generates trees with a probability distribution. The probability distribution of the trees representing images can be used to model the noisy situations. In this paper, we present some results on stochastic tree languages and their application to image modeling, in particular to texture modeling.

2. TREE GRAMMARS AND STOCHASTIC TREE GRAMMARS

Definitions and notations on trees that will be referred to in this paper are first summarized in this section.

* This work was supported by the ONR Contract N00014-79-C-0574.

ISBN 0-12-597320-9

(1) D is a tree domain if

(a) $0 \in D$, 0 is the root of a tree;
(b) $a \in D$, then an immediate successor of a has the form $a \cdot i$ where i is a positive integer;
(c) $a \cdot i$ and $a \cdot j \in D$, i and j are positive integers and if $a \cdot i$ is to the left of $a \cdot j$, then $i < j$;
(d) $a \cdot j \in D$ then $a \in D$ and $a \cdot 1, \ldots, a \cdot j - 1 \in D$.

(2) a is called a node of $a \in D$ for some D.
(3) Σ is a finite set of symbols.
(4) $\alpha : D \rightarrow \Sigma$, α is a tree over Σ in the domain D. $\alpha(a) \in \Sigma$ is the label of node a.
(5) $r : \Sigma \rightarrow N$, r is the rank of symbols in Σ where $r[\alpha(a)] = \max\{i \mid a \cdot i \in D\}$, that is, the rank of a label at a must be equal to the number of branches (or descendents) in the tree domain at a.
(6) T_Σ is the set of all trees over Σ.

The following are operations on trees and tree domains that remove a subtree from a tree, attach a subtree to a tree, or replace a subtree by another subtree.

(1) Suppose that $b = a \cdot c$, then $b/a = c$.
(2) $\alpha/a = \{(b, x) \mid (a \cdot b, x) \in \alpha\}$, α/a is a subtree of α rooted at a.
(3) $a \cdot \alpha = \{(b, x) \mid (b/a, x) \in \alpha\}$, this is the result of affixing the root of α to node a.
(4) $\alpha(a \leftarrow \beta) = \{(b, x) \in \alpha \mid b \not\geq a\} \cup a \cdot \beta$, this is the result of replacing the subtree α/a at a by the tree β.

Definition 1. A (regular) tree grammar G is a 4-tuple

$$G = (V, r', P, S)$$

over the ranked alphabet $\langle V_T, r \rangle$ where $\langle V, r' \rangle$ is a finite ranked alphabet such that $V_T \subseteq V$ and $r' \mid V_T = r$. $V = V_T \cup V_N$ and V_T and V_N are the set of terminals and nonterminals, respectively. P is a finite set of productions of the form $\phi \rightarrow \psi$ where $\phi, \psi \in T_V$. $S \subseteq T_V$ is a finite set of start symbols.

A generation or derivation $\alpha \rightarrow_a \beta$ is in G if and only if there is a production $\phi \rightarrow \psi$ in P such that $\alpha/a = \phi$ and $\beta = (a \leftarrow \psi)\alpha$. $\alpha \Rightarrow \beta$ is in G if and only if there exist trees $t_0, t_1, \ldots, t_m \in T_V$, $m \geq 0$, such that

$$\alpha = t_0 \rightarrow t_1 \rightarrow \cdots t_m = \beta$$

in G.

Definition 2. The language generated by $G = (V, r', P, s)$ over $\langle V_T, r \rangle$ is defined as $L(G) = \{\alpha \in T_{V_T} \mid \text{there exists } X \in S \text{ such that } X \Rightarrow \alpha \text{ is in } G\}$.

Definition 3. A tree grammar $G = (V, r', P, S)$ over $\langle V_T, r \rangle$ is expansive if and only if each production in P is of the form

$$X_0 \rightarrow \begin{array}{c} x \\ \diagup \quad \diagdown \\ X_1 \ldots X_{r(x)} \end{array} \qquad \text{or} \qquad X_0 \rightarrow x,$$

where $x \in V_T$ and $X_0, X_1, \ldots, X_{r(x)} \in V_N$.

For a given regular tree grammar one can effectively construct an equivalent expansive tree grammar. For every regular (expansive) tree grammar, one can effectively construct a deterministic tree automaton which accepts precisely the trees generated by the tree grammar [3].

DEFINITION 4. A stochastic tree grammar G_S is a 4-tuple $G_S = (V, r', P, S)$ over the ranked alphabet $\langle V_T, r \rangle$ where V, V_T, r', r, and S are the same as in Definition 1, and P is a finite set of stochastic productions of the form $\phi \to^p \Psi$ where ϕ and Ψ are trees over $\langle V, r' \rangle$ and $0 \leq p \leq 1$.

A derivation $\alpha \to_a^p \beta$ is in G_S if and only if there is a production $\phi \to^p \Psi$ in P such that $\alpha/a = \phi$ and $\beta = (a \leftarrow^p \Psi)\alpha$. We write $\alpha \to^p \beta$ in G_S if and only if there exists $a \in D_\alpha$, the domain of α, such that $\alpha \to_a^p \beta$.

DEFINITION 5. If there exists a sequence of trees $t_0, t_1, \ldots, t_m$ such that

$$\alpha = t_0,\ \beta = t_m,\ t_{i-1} \xrightarrow{p_i} t_i, \qquad i = 1, \ldots, m,$$

then we say that α derives β with probability

$$p = \prod_{i=1}^{i=m} p_i$$

and denote this derivation by $\alpha \vdash^p \beta$ or $\alpha \Rightarrow^p \beta$. The probability associated with this derivation is equal to the product of the probabilities associated with the sequence of stochastic productions used in the derivation.

DEFINITION 6. The language generated by stochastic tree grammar G_s is

$$L(G_s) = \{(t, p(t)) \mid t \in T_{V_T},\ S \overset{p_j}{\Rightarrow} t,\ j = 1, \ldots, k \text{ and } p(t) = \sum_{j=1}^{k} p_j\},$$

where k is the number of all distinctly different derivations of t from S and p_j is the probability associated with the jth distinct derivation of t from S.

DEFINITION 7. A stochastic tree grammar $G_s = (V, r', P, S)$ over $\langle V_T, r \rangle$ is simple if and only if all rules of p are of the form

$$X_0 \xrightarrow{p} \begin{array}{c} x \\ / \quad \backslash \\ X_1, \ldots, X_{r(x)} \end{array}, \qquad X_0 \xrightarrow{q} X_1, \qquad \text{or} \qquad \begin{array}{c} x \\ / \quad \backslash \\ X_1, \ldots, X_{r(x)} \end{array} \xrightarrow{r} X_0,$$

where $X_0, X_1, \ldots, X_{r(x)}$ are nonterminal symbols and $x \in V_T$ is a terminal symbol and $0 < p, q, r \leq 1$. A rule with the form

$$X_0 \to \begin{array}{c} x \\ / \quad \backslash \\ X_1, \ldots, X_{r(x)} \end{array}$$

can also be written as $X_0 \to xX_1 \cdots X_{r(x)}$.

Given a stochastic tree grammar $G_s = (V, r, P, S)$ over $\langle V_T, r\rangle$, one can effectively construct a simple stochastic tree grammar $G'_s = (V', r', P', S')$ over V_T which is equivalent to G_s [8].

DEFINITION 8. A stochastic tree grammar $G_s = (V, r', P, S)$ over $\langle V_T, r\rangle$ is expansive if and only if each rule in P is of the form

$$X_0 \xrightarrow{p} \begin{array}{c} x \\ / \quad \backslash \\ X_1 \ldots X_{r(x)} \end{array} \quad \text{or } X_0 \xrightarrow{p} x, \qquad \text{where } x \in V_T$$

and $X_0, X_1, \ldots, X_{r(x)}$ are nonterminal symbols contained in $V - V_T$.

EXAMPLE 1. The following is a stochastic expansive tree grammar. $G_s = (V, r', P, S)$ over $\langle V_T, r\rangle$ where

$$V_N = V - V_T = \{S, A, B, C\},$$
$$V_T = \{a, b, \$\},$$
$$r(a) = r(b) = \{2, 0\}, r(\$) = 2,$$

P: (1) $$S \xrightarrow{1.0} \begin{array}{c} \$ \\ / \quad \backslash \\ A \qquad B \end{array},$$

(2) $$A \xrightarrow{p} \begin{array}{c} a \\ / \quad \backslash \\ A \qquad B \end{array},$$

(3) $$A \xrightarrow{1-p} a,$$

(4) $$B \xrightarrow{q} \begin{array}{c} b \\ | \\ C \end{array},$$

(5) $$B \xrightarrow{1-q} b,$$

(6) $$C \xrightarrow{1.0} a,$$

$0 \leq p \leq 1, 0 \leq q \leq 1.$

DEFINITION 9. Define a mapping $h: T_{V_T} \to V^*_{T_0}$ as follows:

$$\text{(i)} \quad h(t) = x \qquad \text{if} \qquad t = x \in V_{T_0}.$$

Obviously, $p(t) = p(x)$

$$\text{(ii)} \ h\left(\begin{array}{c} x \\ / \quad \backslash \\ t_1 \ldots t_n \end{array}\right) = h(t_1) \ldots h(t_n) \qquad \text{if } x \in V_{T_n}, \ n > 0.$$

Obviously,

$$p\left(\begin{array}{c} x \\ / \quad \backslash \\ t_1 \ldots t_n \end{array}\right) = p(x)p(t_1) \ldots p(t_n).$$

The function h forms a string in $V^*_{T_0}$ obtained from a tree t by writing the frontier of t. Note that the frontier is obtained by writing in order the images (labels) of all end points of tree "t."

THEOREM 1. *If L_T is a stochastic tree language, then $h(L_T)$ is a stochastic context-free language with the same probability distribution on its strings as the trees of L_T. Conversely, if $L(G'_s)$ is a stochastic context-free language, then there is a stochastic tree language L_T such that $L(G'_s) = h(L_T)$ and both languages have the same probability distribution [8].*

DEFINITION 10. By a consistent stochastic representation for a language $L(G_s)$ generated by a stochastic tree grammar G_s, we mean that the following condition is satisfied:

$$\sum_{t \in L(G_s)} p(t) = 1,$$

where t is a tree generated by G_s and $p(t)$ is the probability of the generation of tree "t."

The set of consistency conditions for a stochastic tree grammar G_s is the set of conditions which the probability assignments associated with the set of stochastic tree productions in G_s must satisfy such that G_s is a consistent stochastic tree grammar. The consistency conditions of stochastic context-free grammars can be found in Fu [1]. Since nonterminals in an intermediate generating tree appear only at its frontiers, they can be considered to be causing further branching. Thus, if only the frontier of an intermediate tree is considered at levels of branching then, due to Theorem 1, the consistency conditions for stochastic tree grammars are exactly the same as those for stochastic context-free grammars and the tree generating mechanism can be modeled by a generalized branching process [8, 9].

Let $P = \Gamma_{A_1} \cup \Gamma_{A_2} \cup \cdots \cup \Gamma_{A_K}$ be the partition of P into equivalent classes such that two productions are in the same class if and only if they have the same premise (i.e., same left-hand side nonterminal). For each Γ_{A_j} define the conditional probability $\{p(t|A_j)\}$ as the probability that the production rule $A_j \rightarrow t$, where t is a tree, will be applied to the nonterminal symbol A_j where

$$\sum_{\Gamma_{A_j}} p(t|A_j) = 1.$$

Let $r_{jl}(t)$ denote the number of times the variable A_l appears in the frontier of tree "t" of the production $A_j \rightarrow t$.

DEFINITION 11. For each Γ_{A_j}, $j = 1, \ldots, K$, define the K-argument generating function $g_j(S_1, S_2, \ldots, S_K)$ as

$$g_j(S_1, S_2, \ldots, S_K) = \sum_{\Gamma_{A_j}} p(t|A_j) S_1^{r_{j,1}(t)} \ldots S_K^{r_{j,K}(t)}.$$

EXAMPLE 2. For the stochastic tree grammar G_s in Example 1:

$$\begin{aligned}
g_1(S_1, S_2, S_3, S_4) &= p\left(\begin{array}{ccc} & \$ & \\ \diagup & & \diagdown \\ A & & B \end{array}\middle| S\right) S_2S_3 \\
&= S_2S_3, \\
g_2(S_1, S_2, S_3, S_4) &= p\left(\begin{array}{ccc} & a & \\ \diagup & & \diagdown \\ A & & B \end{array}\middle| A\right) S_2S_3 + p(a \mid A) \\
&= pS_2S_3 + (1 - p), \\
g_3(S_1, S_2, S_3, S_4) &= p\left(\begin{array}{c} b \\ | \\ C \end{array}\middle| B\right) S_4 + p(b \mid B) \\
&= qS_4 + (1 - q), \\
g_4(S_1, S_2, S_3, S_4) &= p(a \mid C) \\
&= 1.0.
\end{aligned}$$

These generating functions can be used to define a generating function that describes all ith level trees.

Note that for statistical properties, two ith level trees are equivalent if they contain the same number of nonterminal symbols of each type in the frontiers.

DEFINITION 12. The ith level generating function $F_i(S_1, S_2, \ldots, S_K)$ is defined recursively as

$$\begin{aligned}
F_0(S_1, S_2, \ldots, S_K) &= S_1, \\
F_1(S_1, S_2, \ldots, S_K) &= g_1(S_1, S_2, \ldots, S_K), \\
&\vdots \\
F_i(S_1, S_2, \ldots, S_K) &= F_{i-1}[g_1(S_1, S_2, \ldots, S_K), g_2(S_1, S_2, \ldots, S_K), \ldots, \\
&\qquad g_K(S_1, S_2, \ldots, S_K)].
\end{aligned}$$

$F_i(S_1, S_2, \ldots, S_K)$ can be expressed as

$$F_i(S_1, S_2, \ldots, S_K) = G_i(S_1, S_2, \ldots, S_K) + C_i$$

where $G_i(\cdot)$ does not contain any constant term. The constant term C_i corresponds to the probability of all trees $t \in L(G_s)$ that can be derived in i or fewer levels.

THEOREM 2. *A stochastic tree grammar G_s is consistent if and only if* [8]

$$\lim_{i \to \infty} C_i = 1.$$

DEFINITION 13. The expected number of occurrences of nonterminal symbol A_j in the production set Γ_{A_i} is

$$e_{ij} = \left. \frac{\partial g_i(S_1, S_2, \ldots, S_K)}{\partial S_j} \right|_{S_1, S_2, \ldots, S_K = 1}$$

DEFINITION 14. The first moment matrix $\mathbf{E}$ is defined as

$$\mathbf{E} = [e_{ij}], \ 1 \leq i, \ j \leq K.$$

LEMMA 1. *A stochastic tree language is consistent if all the eigenvalues of* **E** *are smaller than 1. Otherwise, it is not consistent.*

EXAMPLE 3. In this example, consistency conditions for the stochastic tree grammar G_s in Example 1 (as verified in part (a)) are found, and thus the consistency criterion is verified.

(a) The set of trees generated by G_s is as follows:

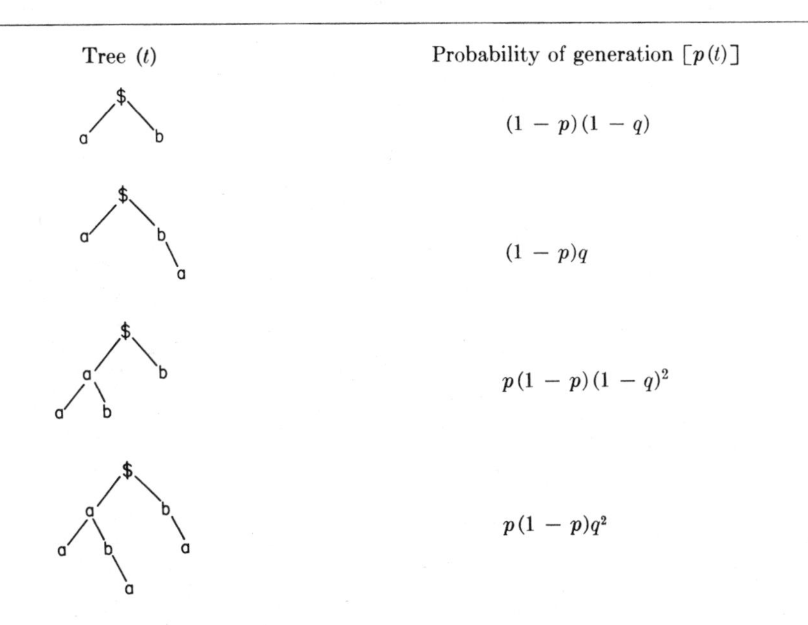

etc.

Tree (t)
Probability of generation [p(t)]
$
a
b
(1 − p)(1 − q)
(1 − p)q
p(1 − p)(1 − q)²
p(1 − p)q²

In all the above trees, production (1) is always applied. If production (2) is applied $(n - 1)$ times, there will be one A and n B's in the frontier of the obtained tree. Production (3) is then applied when no more production (2)'s are needed. In the n B's in the frontier, any one, two, three or all n B's may have production (4) applied and to the rest of B production (5) is applied. Production (6) always follows production (4).

Thus we have

$$\sum_{t\in L(G_s)} p(t) = (1-p)p^0[{}^1C_0(1-q) + {}^1C_1q]$$
$$+ (1-p)p^1[{}^2C_0(1-q)^2 + {}^2C_1q(1-q) + {}^2C_2q^2]$$
$$+ (1-p)p^2[{}^3C_0(1-q)^3 + {}^3C_1q(1-q)^2 + {}^3C_2q^2(1-q) + {}^3C_3q^3]$$
$$+ \cdots$$
$$+ (1-p)p^{n-1}[{}^nC_0(1-q)^n + {}^nC_1(1-q)^{n-1}q$$
$$+ \cdots + {}^nC_r(1-q)^{n-r}q^r + \cdots + {}^nC_nq^n] + \cdots.$$

Note that the power of p in the above terms shows the number of times

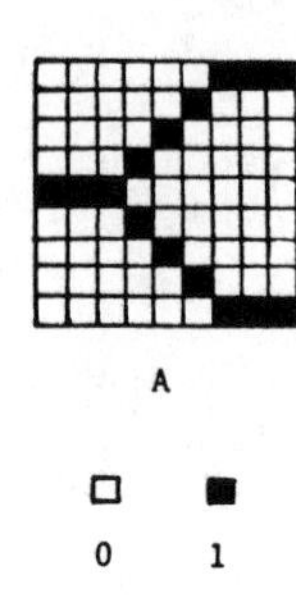

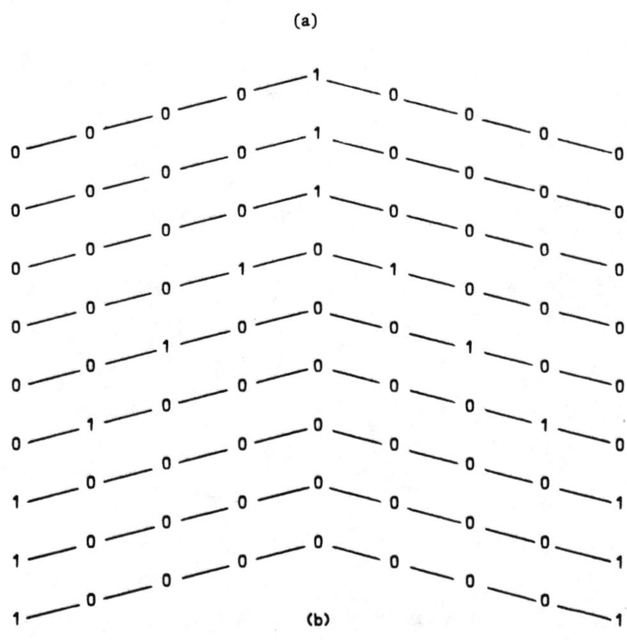

FIG. 1. An image and its tree representation.

production (2) has been applied before applying production (3). So

$$\sum_{t\in L(G_s)} p(t) = (1-p)[\overline{1-q+q}]$$
$$+ (1-p)p[(\overline{1-q}+q)^2]$$
$$+ (1-p)p^2[(\overline{1-q}+q)^3]$$
$$+ \cdots$$
$$+ (1-p)p^{n-1}[(\overline{1-q}+q)^n] + \cdots$$

or

$$\sum_{t\in L(G_s)} p(t) = (1-p) + (1-p)p + \cdots + (1-p)p^{n-1} + \cdots$$
$$= (1-p)[1 + p^1 + p^2 + \cdots + p^{n-1} + \cdots]$$

$$= (1 - p)\frac{1}{1 - p} \qquad [\text{if } p < 1]$$

$$= 1.$$

Hence, G_s is consistent for all values of p such that $0 \leq p \leq 1$.

(b) Let us find the consistency condition for the grammar G_s using Lemma 1 and verify the consistency criterion. From Example 3, we obtain

$$\mathbf{E} = \begin{bmatrix} 0 & 1 & 1 & 0 \\ 0 & p & p & 0 \\ 0 & 0 & 0 & q \\ 0 & 0 & 0 & 0 \end{bmatrix}.$$

The characteristic equation for $\mathbf{E}$ is

$$\phi(\tau) = (\tau - p)\tau^3.$$

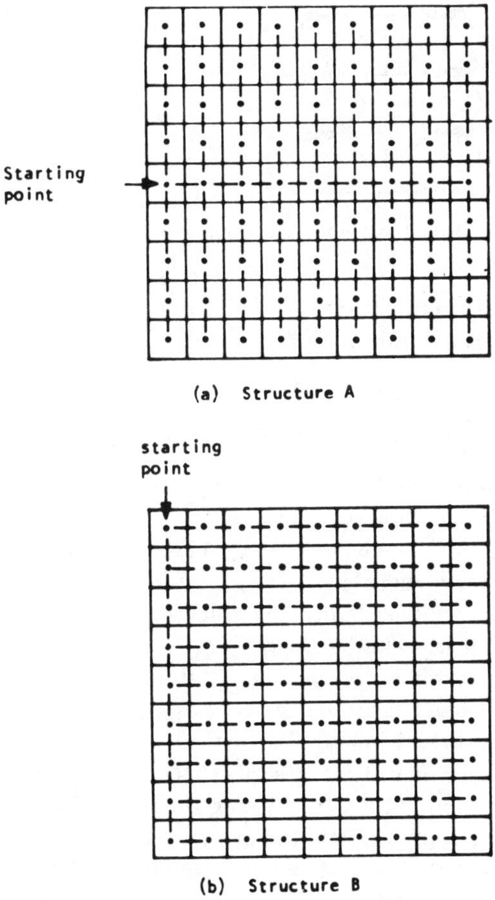

Starting point

(a) Structure A

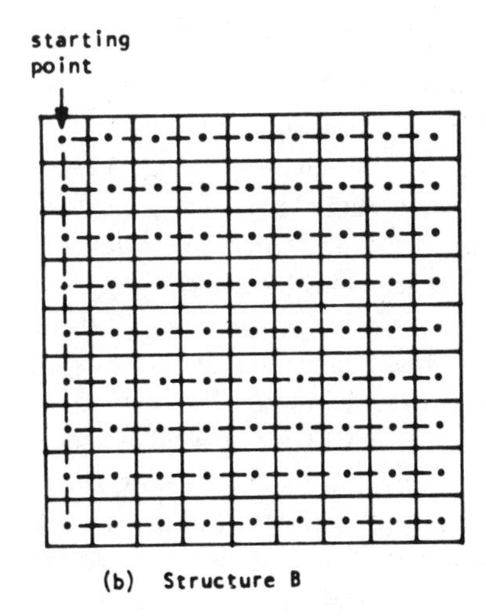

(b) Structure B

FIG. 2. Two tree structures for image modeling. (a) Structure A, (b) Structure B.

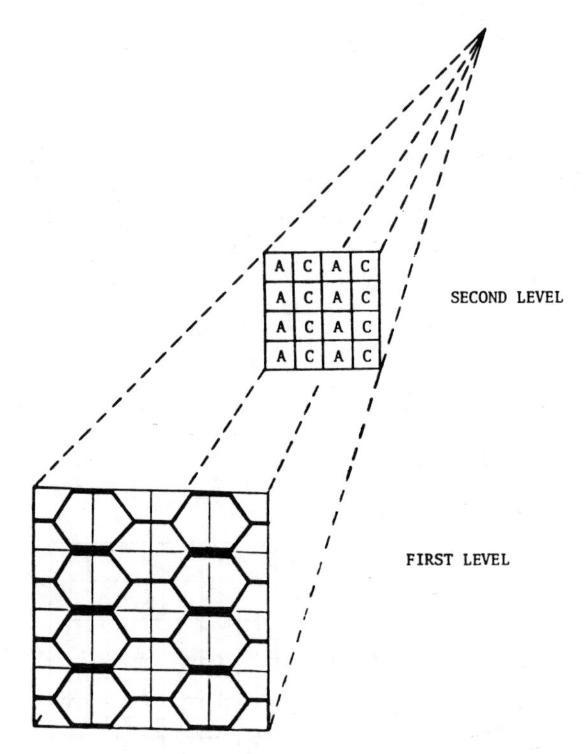

FIG. 3. A regular hexagonal tessellation image.

Thus, the probability representation will be consistent as long as $0 \leq p \leq 1$. The value of q is constrained only for the normalization of production probabilities.

Hence, G_s is consistent.

3. APPLICATION OF TREE GRAMMARS TO IMAGE MODELING

The distribution of gray level in an image can be represented by a tree. The following example illustrates such an idea.

EXAMPLE 4. The image shown in Fig. 1a can be represented by the tree in Fig. 1b using the tree structure A shown in Fig. 2a.

For complex images, based on the basic concept of the syntactic approach, multilevel tree structures can be used.

EXAMPLE 5. The image shown in Fig. 3 can be represented by a two-level tree structure. The first-level representation uses the windowed image A shown in Fig. 1a and the windowed image C shown in Fig. 4, both represented by trees. The second-level representation describes the arrangement or distribution of windowed images A and C by a tree. Using the tree structure B shown in Fig. 2b, the second-level tree representation is given in Fig. 5.

When the number of different windowed images at the first level is large, a tree grammar can be used to generate trees representing the windowed images. In such a case, the terminals or primitives of the tree grammar could be the

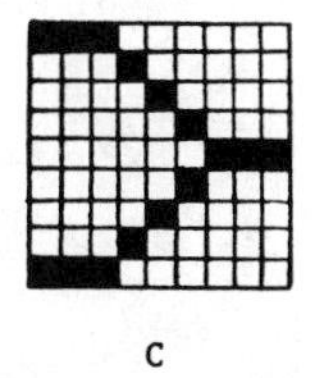

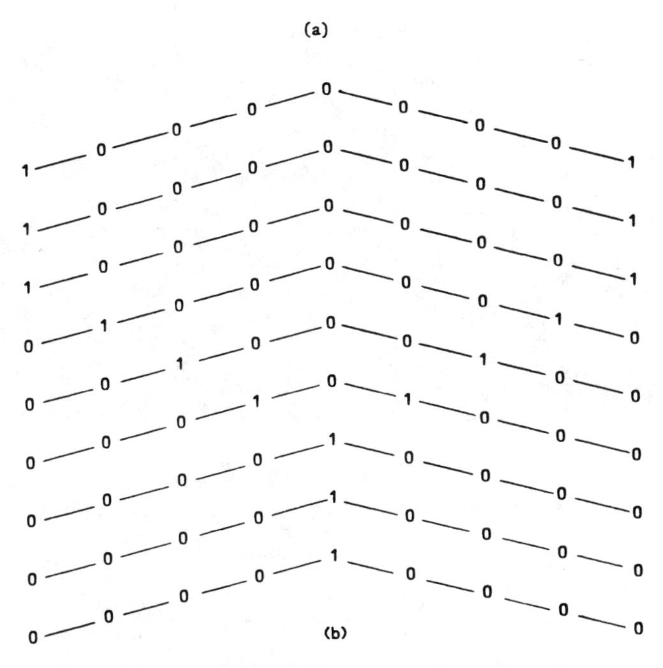

FIG. 4. Image C and its tree representation.

average gray value of a single pixel or within a small array of pixels. Similar ideas can be applied to the higher-level structure representations if necessary.

Research on texture modeling in image processing has received increasing attention in recent years [10]. Most of the previous research has concentrated on the statistical approach [11, 12]. An alternative approach is the structural approach [13]. In the structural approach, a texture is considered to be defined

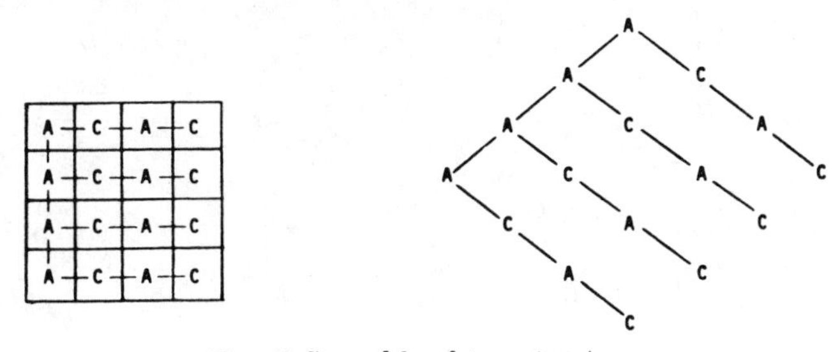

FIG. 5. Second-level tree structure.

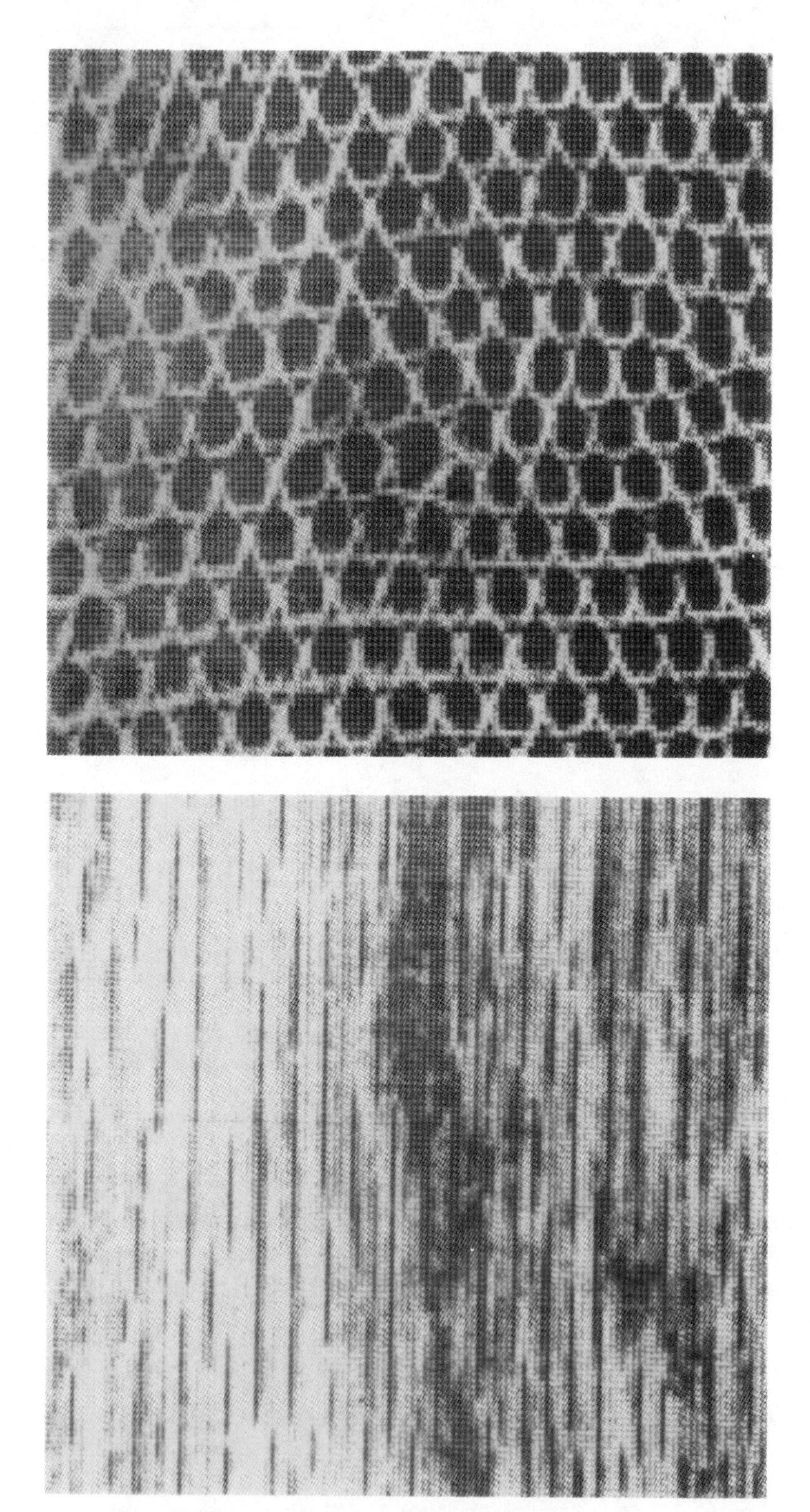

FIG. 6. Texture patterns. (Top) D22—reptile skin and (bottom) D68—wood grain.

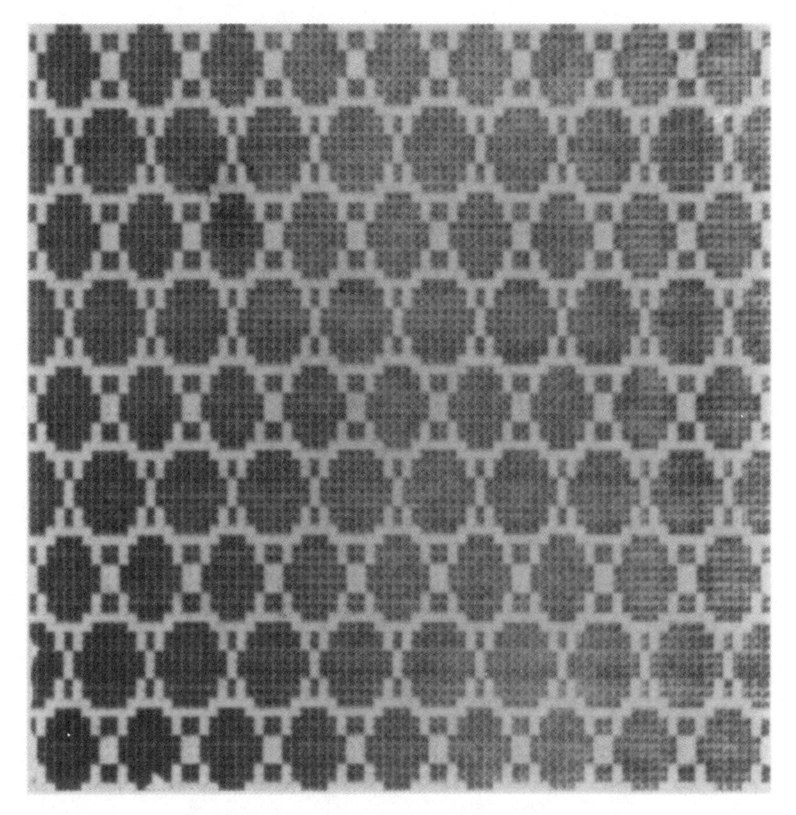

FIG. 7. The ideal texture of pattern D22.

by subpatterns which occur repeatedly according to a set of well-defined placement rules within the overall pattern. Furthermore, the subpatterns themselves are made of structural elements.

We have proposed a texture model based on the structural approach [14]. A texture pattern is divided into fixed-size windows. Repetition of subpatterns or a portion of a subpattern may appear in a window. A windowed pattern is treated as a subpattern and is represented by a tree (see Example 5). Each tree node corresponds to a single pixel or a small homogeneous area of the windowed patterns. A tree grammar is then used to characterize windowed patterns of the same class. The advantage of the proposed model is its computational simplicity. The decomposition of an image into fixed-size windows and the use of a fixed tree structure for representation make the texture analysis procedure and its imple-

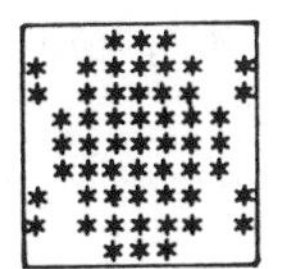

FIG. 8. Basic patterns of Fig. 7.

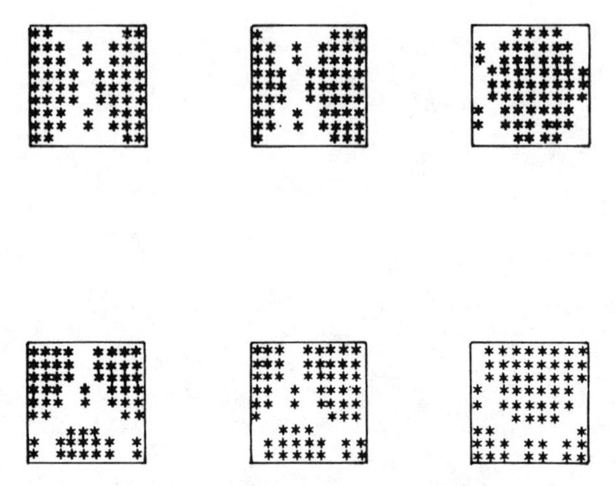

FIG. 9. Windowed pattern primitives.

mentation very easy. We will describe the use of stochastic tree grammars and high-level syntax rules to model texture with local noise and structural distortions.

Figures 6a and b are digitized images of the texture patterns D22, and D68 from Brodatz' book *Textures* [15]. For simplicity, we use only two primitives, black as primitive "1," and white as primitive "0." For pattern D22, the reptile skin, we may consider that it is the result of twisting a regular tessellation such as the pattern shown in Fig. 7. The regular tessellation image is composed of two basic subpatterns shown in Fig. 8. A distorted tessellation can result from shifting a series of basic subpatterns in one direction. Let us use the set of shifted subpatterns as the set of first-level windows. There will be 81 such windowed images.[1] Figure 9 shows several of them. A tree grammar can be constructed for the generation of the 81 windowed images [14]. Local noise and distortion of the windowed images can be taken care of by constructing a stochastic tree grammar. The procedure of inferring a stochastic tree grammar from a set of texture patterns is described in [19]. A tree grammar for the placement of the 81 windowed images can then be constructed for the twisted texture pattern. A generated texture D22 using a stochastic tree grammar is shown in Fig. 10.

The texture pattern D68, the wood grain pattern, consists of long vertical lines. It shows a higher degree of randomness than D22. No clear tessellation or subpattern exists in the pattern. Using vertical lines as subpatterns we can construct a stochastic tree grammar G_{68} to characterize the repetition of the subpatterns. The density of vertical lines depends on the probabilities associated with production rules. Figure 11 shows two patterns generated from G_{68} using different sets of production probabilities

$$G_{68} = (V, r', P, S),$$

[1] A cluster analysis procedure can be applied to determine the number of structurally different window images [16–18].

where $V = \{S, A, B, 0, 1\}$, $V_T = \{0, 1\}$, $r(0) = r(1) = \{0, 1, 2, 3\}$, and P is

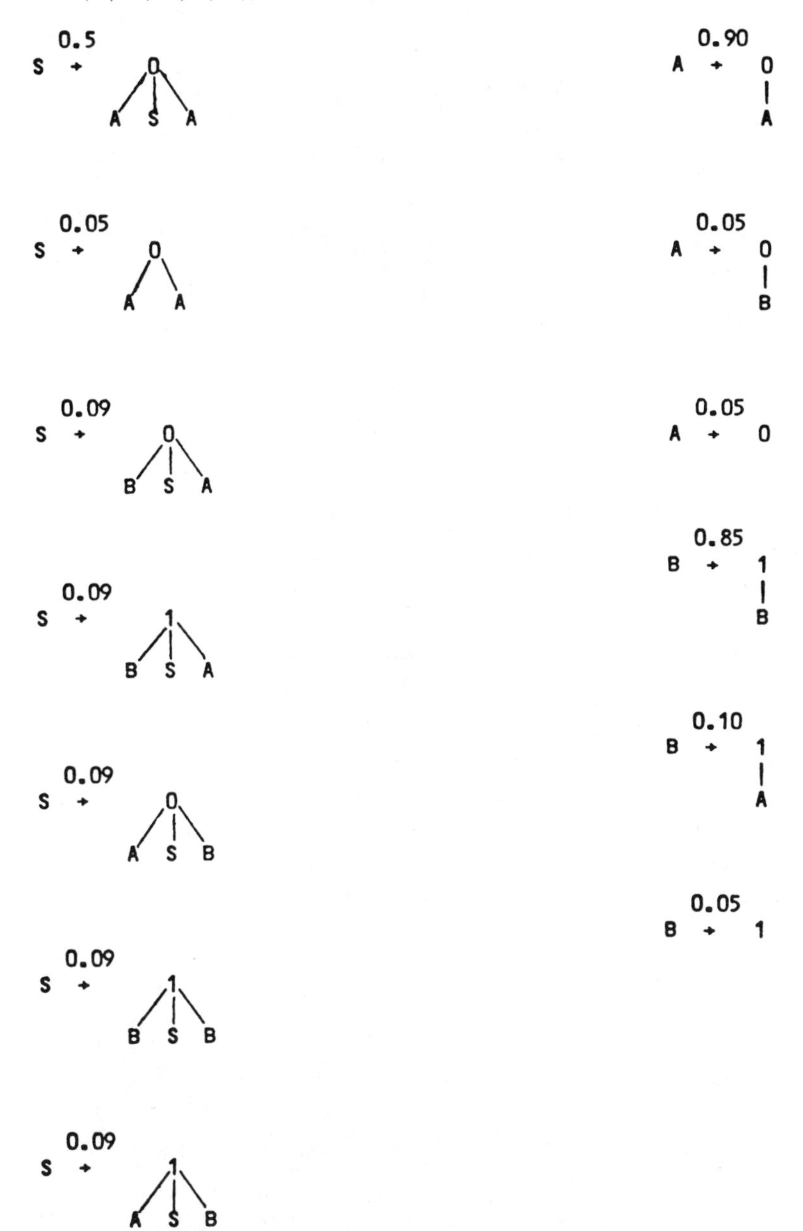

4. CONCLUDING REMARKS

In this paper, we have reviewed the use of stochastic tree grammars for image modeling. Tree grammars have been used in the description and modeling of finger-patterns, bubble chamber pictures, highway and river patterns in LANDSAT images, and texture patterns. In practical applications, noise and distortions

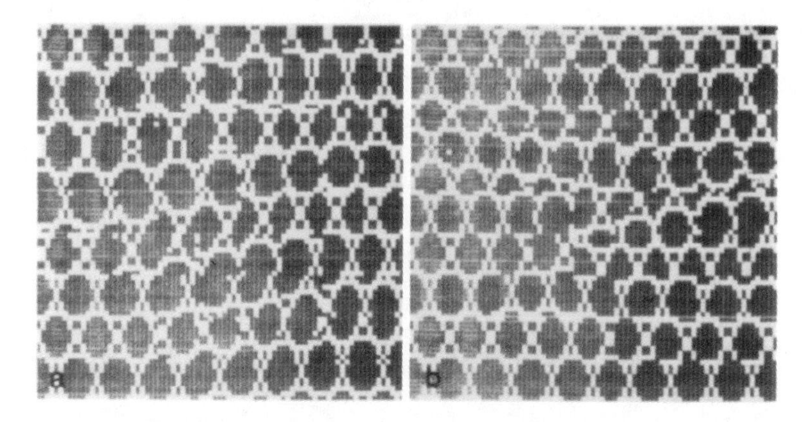

FIG. 10. Synthesis results for pattern D22.

often exist in the process under study. In order to describe and model real world patterns more realistically, stochastic tree grammars have been suggested. We have briefly presented some recent results in texture modeling using (stochastic) tree grammars. For a given stochastic (expansive) tree grammar describing a set of images, we can construct a stochastic tree automaton which will accept the set of images with their associated probabilities [8]. In the case of a multi-class image recognition problem, the maximum-likelihood or Bayes decision rule can be used to decide the class label of an input image represented by a tree [1].

In order to model the images of interest realistically, it would be nice to have the stochastic tree grammar actually inferred from the available image samples. Such an inference procedure requires the inference of both the tree grammar and its production probabilities. Unfortunately, a general inference procedure for stochastic tree grammars is still a subject of research. Only some very special cases have been treated [19, 20].

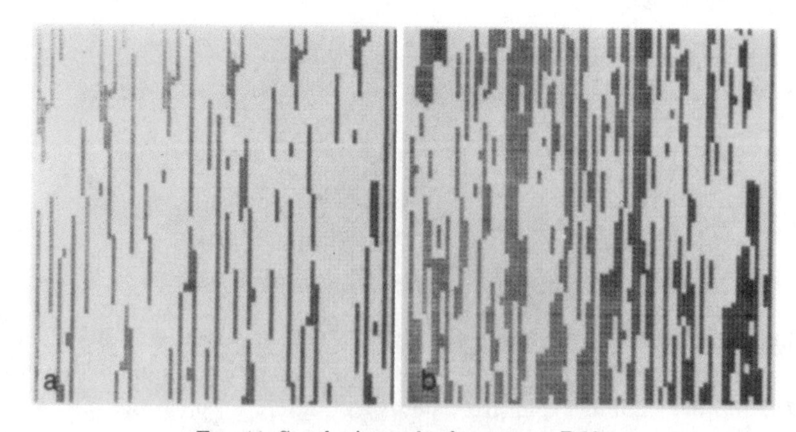

FIG. 11. Synthesis results for pattern D68.

REFERENCES

1. K. S. Fu, *Syntactic Methods in Pattern Recognition*, Academic Press, New York, 1974.
2. K. S. Fu (Ed.), *Syntactic Pattern Recognition, Applications*, Springer-Verlag, Berlin/New York, 1977.
3. K. S. Fu and B. K. Bhargava, Tree systems for syntactic pattern recognition, *IEEE Trans. Computers* **C-22**, Dec. 1973.
4. B. Moayer and K. S. Fu, A tree system approach for fingerprint pattern recognition, *IEEE Trans. Computers* **C-25**, March 1976.
5. R. Y. Li and K. S. Fu, Tree system approach to LANDSAT data interpretation, in Proc. Symposium on Machine Processing of Remotely Sensed Data, June 29–July 1, 1976, Lafayette, Ind.
6. J. Keng and K. S. Fu, A syntax-directed method for land use classification of LANDSAT images, in Proc. Symposium on Current Mathematical Problems in Image Science, Nov. 10–12, 1976, Monterey, Calif.
7. K. S. Fu, Tree languages and syntactic pattern recognition, in *Pattern Recognition and Artificial Intelligence* (C. H. Chen, Ed.), Academic Press, New York, 1976.
8. K. S. Fu, Stochastic tree languages and their applications to picture processing, in Proc. 1978 International Symposium on Multivariate Analysis (P. R. Krishnaiah, Ed.), North-Holland, Amsterdam, 1979.
9. T. E. Harris, *The Theory of Branching Processes*, Springer-Verlag, Berlin/New York, 1963.
10. S. W. Zucker, Toward a model of texture, *Computer Graphics Image Processing* **5**, 1976, 190–202.
11. R. M. Haralick, K. Shanmugam, and I. Dinstein, Texture features for image classification, *IEEE Trans. Systems Man Cybernet.* **SMC-3**, Nov. 1973.
12. J. S. Weszka, C. R. Dyer, and A. Rosenfeld, A Comparative Study of Texture Measures for Terrain Classification, *IEEE Trans. Systems Man Cybernet.* **SMC-6**, April 1976.
13. B. S. Lipkin and A. Rosenfeld (Eds.), *Picture Processing and Psychopictorics*, pp. 289–381, Academic Press, New York, 1970.
14. S. Y. Lu and K. S. Fu, A syntactic approach to texture analysis, *Computer Graphics Image Processing* **7**, 1978, 303–330.
15. P. Brodatz, *Textures*, Dover, New York, 1966.
16. K. S. Fu and S. Y. Lu, A clustering procedure for syntactic patterns, *IEEE Trans. Systems Man Cybernet.* **SMC-7**, Oct. 1977.
17. S. Y. Lu and K. S. Fu, A sentence-to-sentence clustering procedure for pattern analysis, *IEEE Trans. Systems Man Cybernet.* **SMC-8**, May 1978.
18. S. Y. Lu, Tree-to-tree distance and its application to cluster analysis, *IEEE Trans. Pattern Anal. Machine Intelligence* **PAMI-1**, April 1979.
19. S. Y. Lu and K. S. Fu, Stochastic tree grammar inference for texture synthesis and discrimination, *Computer Graphics Image Processing* **9**, 1979, 234–245.
20. J. M. Brayer and K. S. Fu, A note on the k-tail method of tree grammar influence, *IEEE Trans. Systems Man Cybernet.* **SMC-7**, April 1977.

Edge and Region Analysis for Digital Image Data

ROBERT M. HARALICK
Department of Electrical Engineering and Department of Computer Science, Virginia Polytechnic Institute and State University, Blacksburg, Virginia 24061

In this paper we provide a unified view of edge and region analysis. Our framework is based on the sloped-facet model which assumes that regions of image segments are maximal areas which are sloped planes. Edge strength between two adjacent pixels is measured by the F statistic used to test the significance of the difference between the parameters of the best-fitting sloped neighborhoods containing each of the given pixels. Edges are declared to exist at locations of local maxima in the F-statistic edge strength picture. We show that this statistically optimum procedure in essence scales the edge strength statistic of many popular edge operators by an estimate of the image noise. Such a scaling makes optimum detection possible by a fixed threshold procedure.

1. INTRODUCTION

Edge detection and region growing are two areas of image analysis which are opposite in emphasis but identical at heart. Edges obviously occur at bordering locations of two adjacent regions which are significantly different. Regions are maximal areas having similar attributes. If we could do region analysis, then edges could be declared at the borders of all regions. If we could do edge detection, regions would be the areas surrounded by the edges. Unfortunately, we tend to have trouble doing either: edge detectors are undoubtedly noisy and region growers often grow too far.

In this paper we give an even-handed treatment of both. Edges will not occur at locations of high differences. Rather, they will occur at locations of high differences between the parameters of sufficiently homogeneous areas. Regions will not be declared as just areas of similar value of gray tone. They will occur at connected areas where resolution cells yield minimal differences of region parameters, where minimal means smallest among a set of resolution cell groupings. In essence we will see that edge detection and region analysis are identical problems that can be solved with the same procedure.

Because the framework we wish to present tends to unify some of the popular techniques, the paper is organized to first give a description of our framework and then to describe related techniques discussed in the literature in terms of our framework.

ISBN 0-12-597320-9

2. THE SLOPED-FACET MODEL

The digital image g is a function from the Cartesian product of row and column index sets into the reals. The sloped-facet assumption is a restriction on the nature of the function g for the ideal image (one having no defocusing or noise). The restriction is that the domain of g can be partitioned into connected sets $\Pi = \{\Pi_1, \ldots, \Pi_M\}$ such that for each connected set $\Pi_m \in \Pi$.

(1) $(r, c) \in \Pi_m$ implies that for some K-pixel neighborhood N containing (r, c), $N \subseteq \Pi_m$,
(2) $(r, c) \in \Pi_m$ implies $g(r, c) = \alpha_m r + \beta_m c + \gamma_m$.

Condition (1) requires that the partition Π consists of connected sets each of large enough and smooth enough shape. For example, if $K = 9$ and the only neighborhoods we consider are 3×3, than each set Π_m must be no thinner in any place than 3×3. If Π_m has holes, they must be surrounded everywhere by pixels in 3×3 neighborhoods which are entirely contained in Π_m.

Condition (2) requires that the gray tone surface defined on Π_m be a sloped plane. This constraint could obviously be generalized to include higher-order polynomials. It is, of course, more general than the piecewise constant surface implicitly assumed by other techniques.

The fact that the parameters α and β determine the value of the slope in any direction is well known. For a planar surface of the form

$$g(r, c) = \alpha r + \beta c + \gamma$$

the value of the slope at an angle θ to the row axis is given by the directional derivative of g in the direction θ. Since α is the partial derivative of g with respect to r and β is the partial derivative of g with respect to c, the value of the slope at angle θ is $\alpha \cos \theta + \beta \sin \theta$. Hence, the value of the slope at any direction is an appropriate linear combination of the values for α and β. The angle θ which maximizes this value satisfies

$$\cos \theta = \frac{\alpha}{(\alpha^2 + \beta^2)^{\frac{1}{2}}} \quad \text{and} \quad \sin \theta = \frac{\beta}{(\alpha^2 + \beta^2)^{\frac{1}{2}}}$$

and the gradient which is the value of the slope in the steepest direction is $(\alpha^2 + \beta^2)^{\frac{1}{2}}$.

The sloped-facet assumption can avoid some of the problems inherent in edge detectors or region growers. Consider, for example, two piecewise linear surfaces meeting at a V-junction. A typical step edge detector applied at the V-junction would tend to find the average gray tone to the left of the junction equal to the average gray tone to the right of the junction. A region grower, for the same reason, if approaching the junction from the left is likely to grow somewhat into the part to the right of the junction before it realizes that there may be significant gray tone difference. Consider also a simple sloped surface. A typical step edge detector applied any place along this surface might declare an edge because the gray tone average to the detector's left is surely different than the average to its right.

The sloped-facet model is an appropriate one for either the flat-world or sloped-world assumption. In the flat world each ideal region is constant in gray tone. Hence, all edges are step edges. The observed image taken in an ideal flat world is a defocused version of the ideal piecewise constant image with the addition of some random noise. The defocusing changes all step edges to sloped edges. The edge detection problem is one of determining whether the observed noisy slope has a gradient significantly higher than one which could have been caused by the noise alone. Edge boundaries are declared in the middle of all significantly sloped regions.

In the sloped facet world, each ideal region has a gray tone surface which is a sloped plane. Edges are places of either discontinuity in gray tone or derivative of gray tone. The observed image is the ideal image with noise added and no defocusing. To determine if there is an edge between two pixels, we first determine the best slope fitting neighborhood for each of the pixels. Edges are declared at locations having significantly different planes on either side of them. In the sloped facet model, edges surrounding regions having significantly sloped surfaces may be the boundaries of an edge region. The determination of whether a sloped region is an edge region or not may depend on the significance and magnitude of the slope as well as the semantics of the image.

In either the noisy defocused flat world or the noisy sloped world we are faced with the problem of estimating the parameters of a sloped surface for a given neighborhood and then calculating the significance of the difference of the estimated slope from a zero slope or calculating the significance of the difference of the estimated slopes of two adjacent neighborhoods. To do this we proceed in a classical manner. We will use a least-squares procedure to estimate parameters and we will measure the strength of any difference by an appropriate F statistic.

3. SLOPED FACET PARAMETER ESTIMATION AND SIGNIFICANCE MEASURE

We employ a least-squares procedure to estimate the parameters of the slope model for a rectangular region whose row index set is R and whose column index set is C. We assume that for each $(r, c) \in R \times C$,

$$g(r, c) = \alpha r + \beta c + \gamma + \eta(r, c)$$

where η is a random variable indexed on $R \times C$ which represents noise. We will assume that η is noise having mean 0 and variance σ^2 and that the noise for any two pixels is independent.

The least-squares procedure determines an $\hat{\alpha}$, $\hat{\beta}$, and $\hat{\gamma}$ which minimize

$$\epsilon^2 = \sum_{r \in R} \sum_{c \in C} [\hat{\alpha} r + \hat{\beta} c + \hat{\gamma} - g(r, c)]^2.$$

Taking the partial derivatives of ϵ^2 and setting them to zero results in

$$\begin{bmatrix} \dfrac{\partial \epsilon^2}{\partial \hat{\alpha}} \\ \dfrac{\partial \epsilon^2}{\partial \hat{\beta}} \\ \dfrac{\partial \epsilon^2}{\partial \hat{\gamma}} \end{bmatrix} = 2 \sum_{r\in R} \sum_{c\in C} (\hat{\alpha} r + \hat{\beta} c + \hat{\gamma} - g(r, c)) \begin{bmatrix} r \\ c \\ 1 \end{bmatrix} = 0. \tag{1}$$

Without loss of generality, we choose our coordinate system $R \times C$ so that the center of the neighborhood $R \times C$ has coordinates (0, 0). When the number of rows and columns is odd, the center pixel, therefore, has coordinates (0, 0). When the number of rows and columns is even, there can be no one center pixel and the point where the four center pixels meet has coordinates (0, 0).

The symmetry in the chosen coordinate system leads to

$$\sum_{r\in R} r = \sum_{c\in C} c = 0.$$

Hence, Eq. (1) reduces to the system of three decoupled equations

$$\sum_{r\in R} \sum_{c\in C} \hat{\alpha} r^2 = \sum_{r\in R} \sum_{c\in C} r g(r, c),$$

$$\sum_{r\in R} \sum_{c\in C} \hat{\beta} c^2 = \sum_{r\in R} \sum_{c\in C} c g(r, c),$$

$$\sum_{r\in R} \sum_{c\in C} \hat{\gamma} = \sum_{r\in R} \sum_{c\in C} g(r, c).$$

Solving for $\hat{\alpha}$, $\hat{\beta}$, and $\hat{\gamma}$ we obtain

$$\begin{aligned} \hat{\alpha} &= \sum_{r\in R} \sum_{c\in C} r g(r, c) \Big/ \sum_{r\in R} \sum_{c\in C} r^2, \\ \hat{\beta} &= \sum_{r\in R} \sum_{c\in C} c g(r, c) \Big/ \sum_{r\in R} \sum_{c\in C} c^2, \\ \hat{\gamma} &= \sum_{r\in R} \sum_{c\in C} g(r, c) \Big/ \sum_{r\in R} \sum_{c\in C} 1. \end{aligned} \tag{2}$$

Replacing $g(r, c)$ by $\alpha r + \beta c + \gamma + \eta(r, c)$ and reducing the equations will allow us to explicitly see the dependence of $\hat{\alpha}$, $\hat{\beta}$, and $\hat{\gamma}$ on the noise. We obtain

$$\hat{\alpha} = \alpha + \Big(\sum_r \sum_c r\eta(r, c) \Big/ \sum_r \sum_c r^2\Big),$$

$$\hat{\beta} = \beta + \Big(\sum_r \sum_c c\eta(r, c) \Big/ \sum_r \sum_c c^2\Big),$$

$$\hat{\gamma} = \gamma + \Big(\sum_r \sum_c \eta(r, c) \Big/ \sum_r \sum_c 1\Big).$$

From this it is apparent that $\hat{\alpha}$, $\hat{\beta}$, and $\hat{\gamma}$ are unbiased estimators for α, β, and γ, respectively, and have variances

$$V[\hat{\alpha}] = \sigma^2 / \sum_{r \in R} \sum_{c \in C} r^2,$$

$$V[\hat{\beta}] = \sigma^2 / \sum_{r \in R} \sum_{c \in C} c^2,$$

$$V[\hat{\gamma}] = \sigma^2 / \sum_{r \in R} \sum_{c \in C} 1.$$

Normally distributed noise implies that $\hat{\alpha}$, $\hat{\beta}$, and $\hat{\gamma}$ are also normally distributed. The independence of the noise implies that $\hat{\alpha}$, $\hat{\beta}$, and $\hat{\gamma}$ are independent since they are normal and that

$$E[(\hat{\alpha} - \alpha)(\hat{\beta} - \beta)] = E[(\hat{\alpha} - \alpha)(\hat{\gamma} - \gamma)] = E[(\hat{\beta} - \beta)(\hat{\gamma} - \gamma)] = 0$$

as a straightforward calculation shows.

Examining the total squared error ϵ^2 we find that

$$\epsilon^2 = \sum_{r \in R} \sum_{c \in C} [(\hat{\alpha} r + \hat{\beta} c + \hat{\gamma}) - (\alpha r + \beta c + \gamma + \eta(r, c))]^2$$

$$= \sum_{r \in R} \sum_{c \in C} [(\hat{\alpha} - \alpha)^2 r^2 + (\hat{\beta} - \beta)^2 c^2 + (\hat{\gamma} - \gamma)^2 + \eta^2(r, c)$$

$$- 2(\hat{\alpha} - \alpha) r \eta(r, c) - 2(\hat{\beta} - \beta) c \eta(r, c) - 2(\hat{\gamma} - \gamma) \eta(r, c)].$$

Using the fact that

$$(\hat{\alpha} - \alpha) = \sum_{r \in R} \sum_{c \in C} r \eta(r, c) / \sum_{r} \sum_{c} r^2,$$

$$(\hat{\beta} - \beta) = \sum_{r} \sum_{c} c \eta(r, c) / \sum_{r} \sum_{c} c^2,$$

$$(\hat{\gamma} - \gamma) = \sum_{r \in R} \sum_{c} \eta(r, c) / \sum_{r} \sum_{c} 1$$

we may substitute into the last three terms for ϵ^2 and obtain after simplification

$$\epsilon^2 = \sum_{r} \sum_{c} \eta^2(r, c) - (\hat{\alpha} - \alpha)^2 \sum_{r} \sum_{c} r^2 - (\hat{\beta} - \beta)^2 \sum_{r} \sum_{c} c^2 - (\hat{\gamma} - \gamma)^2 \sum_{r} \sum_{c} 1.$$

Now notice that

$$\sum_{r} \sum_{c} \eta^2(r, c)$$

is the sum of the squares of

$$\sum_{r} \sum_{c} 1$$

independently distributed normal random variables. Hence,

$$\sum_{r} \sum_{c} \eta^2(r, c) / \sigma^2$$

is distributed as a chi-squared variate with

$$\sum_r \sum_c 1$$

degrees of freedom. Because, $\hat{\alpha}$, $\hat{\beta}$, and $\hat{\gamma}$ are independent normals,

$$((\hat{\alpha} - \alpha)^2 \sum_r \sum_c r^2 + (\hat{\beta} - \beta)^2 \sum_r \sum_c c^2 + (\hat{\gamma} - \gamma)^2 \sum_r \sum_c 1)/\sigma^2$$

is distributed as a chi-squared variate with 3 degrees of freedom. Therefore, ϵ^2/σ^2 is distributed as a chi-squared variate with

$$\sum_r \sum_c 1 - 3$$

degrees of freedom.

From this it follows that to test the hypothesis of no edge for the flat-world assumption, $\alpha = \beta = 0$, we use the ratio

$$F = ((\hat{\alpha}^2 \sum_r \sum_c r^2 + \hat{\beta}^2 \sum_r \sum_c c^2)/2)/(\epsilon^2/(\sum_r \sum_c 1 - 3)),$$

which has an F distribution with

$$2, \sum_r \sum_c 1 - 3$$

degrees of freedom and reject the hypothesis for large values of F.

Notice that F may be regarded as a significance or reliability measure associated with the existence of a nonzero sloped region in the domain $R \times C$. It is essentially proportional to the squared gradient of the region normalized by

$$\epsilon^2/(\sum_r \sum_c 1 - 3)$$

which is a random variable whose expected value is σ^2, the variance of the noise.

EXAMPLE 1. Consider the following 3×3 region:

3	5	9	
4	7	7	observed.
0	3	7	

Then $\hat{\alpha} = -1.17$, $\hat{\beta} = 2.67$, and $\hat{\gamma} = 5.00$. The estimated gray tone surface is given by $\hat{\alpha}r + \hat{\beta}c + \hat{\gamma}$ and is

3.50	6.17	8.83	
2.33	5.00	7.67	estimated.
1.17	3.83	6.5	

The difference between the estimated and the observed surfaces is the error and it is

0.50	1.17	−0.17	
−1.67	−2.00	0.67	error.
1.17	0.83	−0.50	

From this we can compute the squared error $\epsilon^2 = 11.19$. The F statistic is then

$$\frac{[(-1.17)^2 \cdot 6 + (2.67)^2 \cdot 6]/2}{11.19/6} = 13.67.$$

If we were compelled to make a hard decision about the significance of the observed slope in the given 3×3 region, we would probably call it a nonzero sloped region since the probability of a region with true zero slope giving an $F_{2,6}$ statistic of value less than 10.6 is 0.99. 13.67 is greater than 10.6 so we are assured that the probability of calling the region a nonzero sloped region when it is in fact a zero sloped region is much less than 1%. The statistically oriented reader will recognize the test as a 1% significance level test.

EXAMPLE 2. We proceed just as in Example 1 for the following 3×3 region:

1	3	11	
6	11	7	observed.
−4	1	9	

Then $\hat{\alpha} = 1.5$, $\hat{\beta} = 4.0$, and $\hat{\gamma} = 5.0$. The estimated gray tone surface is

2.5	6.5	10.5	
1	5	9	estimated
−0.5	3.5	7.5	

and the error surface is

1.5	3.5	−0.5	
−5	−6	2	error.
3.5	2.5	−1.5	

From this we compute an F statistic of 3.27 and hence we call it a zero sloped region at the 1% significance level.

For the sloped-facet world our problem is not whether the true slope of a region is zero; rather, it is determining whether two regions are part of the same sloped surface. To do this we are naturally led to examine the differences between the parameters for the estimated sloped-plane surfaces.

For simplicity, we assume that the two regions 1 and 2 are identically sized mutually exclusive rectangular regions. Let $\hat{\alpha}_1$, $\hat{\beta}_1$, and $\hat{\gamma}_1$ be the estimated parameters for region 1. Let $(\Delta r, \Delta c)$ be the coordinates of the center of region 2 relative to the center of region 1. Let $\hat{\alpha}_2$, $\hat{\beta}_2$, and $\hat{\gamma}_2$ be the estimated parameters for region 2. Then under the hypothesis that

$$\alpha_1 = \alpha_2 \qquad \text{and} \qquad \beta_1 = \beta_2,$$

$$(1/2^{\frac{1}{2}})(\hat{\alpha}_1 - \hat{\alpha}_2)\Big(\sum_r \sum_c r^2\Big)^{\frac{1}{2}} \qquad \text{and} \qquad (1/2^{\frac{1}{2}})(\hat{\beta}_1 - \hat{\beta}_2)\Big(\sum_r \sum_c c^2\Big)^{\frac{1}{2}}$$

each have a normal distribution with mean 0 and variance σ^2.

Due to the fact that the gray tone surfaces are sloped the hypothesis $\gamma_1 = \gamma_2$ is inappropriate. Instead, we must adjust the average height for each surface to account for the sloped rise or fall as we travel from the center of region 1 to the center of region 2 and test the equality of the adjusted heights. To do this we choose a place halfway between the centers of the two regions. Since each region's center had relative coordinates (0, 0), the coordinates of the halfway location from region 1 is $(\Delta r/2, \Delta c/2)$ and the coordinates of the halfway location from region 2 is $(-\Delta r/2, -\Delta c/2)$. The true height of the gray tone surface at these locations is

$$\alpha_1 \Delta r/2 + \beta_1 \Delta c/2 + \gamma_1 \qquad \text{and} \qquad \alpha_2(-\Delta r/2) + \beta_2(-\Delta c/2) + \gamma_2$$

and the hypothesis that the regions are part of the same sloped surface would imply

$$(\alpha_1 + \alpha_2)\frac{\Delta r}{2} + (\beta_1 + \beta_2)\frac{\Delta c}{2} + \gamma_1 - \gamma_2 = 0.$$

Under this hypothesis the statistic

$$(\hat{\alpha}_1 + \hat{\alpha}_2)\frac{\Delta r}{2} + (\hat{\beta}_1 + \hat{\beta}_2)\frac{\Delta c}{2} + (\hat{\gamma}_1 - \hat{\gamma}_2)$$

has a normal distribution with mean 0 and variance

$$\sigma^2\left[2\left((\Delta r/2)^2/\sum_r \sum_c r^2\right) + 2\left((\Delta c/2)^2/\sum_r \sum_c c^2\right) + \left(2/\sum_r \sum_c 1\right)\right].$$

After noting that under the same sloped-surface assumption

$$E[(\hat{\alpha}_1 + \hat{\alpha}_2)(\hat{\alpha}_1 - \hat{\alpha}_2)] = E[(\hat{\beta}_1 + \hat{\beta}_2)(\hat{\beta}_1 - \hat{\beta}_2)] = 0$$

we find that an appropriate statistic to measure the significance of the departure from the same-surface hypothesis is

$$\frac{\left\{\frac{1}{2}(\hat{\alpha}_1 - \hat{\alpha}_2)^2 \sum_r \sum_c r^2 + \frac{1}{2}(\hat{\beta}_1 - \hat{\beta}_2)^2 \sum_r \sum_c c^2 + \left(\frac{[(\hat{\alpha}_1 + \hat{\alpha}_2)(\Delta r/2) + (\hat{\beta}_1 + \hat{\beta}_2)(\Delta c/2) + \hat{\gamma}_1 - \hat{\gamma}_2]^2}{2[(\Delta r/2)^2/\sum_r \sum_c r^2 + (\Delta c/2)^2/\sum_r \sum_c c^2 + 1/\sum_r \sum_c 1]}\right)\right\} \Big/ 3}{\left[(\epsilon_1^2 + \epsilon_2^2)/(2\sum_r \sum_c 1 - 6)\right]}$$

Example 3. We proceed just as in Examples 1 and 2 for the following pair of 3×3 regions:

17	15	12				
20	16	13	18	14	12	
21	18	15	20	15	11	observed.
			19	19	16	

Here, $\hat{\alpha}_1 = \hat{\alpha}_2 = 1.67$, $\hat{\beta}_1 = \hat{\beta}_2 = -3$, $\hat{\gamma}_1 = 16.33$, and $\hat{\gamma}_2 = 16.00$. The estimated gray tone surface is

17.33	14.33	11.33				
19	16	13	17.55	14.55	11.55	
20.67	17.67	14.67	19.22	16.22	13.22	estimated
			20.88	17.88	14.88	

and the error surface is

−0.67	0.33	0.67				
−1.0	1.0	2.0	−0.44	−0.44	0.55	error.
1.67	−1.33	−1.33	2.22	0.22	−1.77	
			−2.12	0.88	0.88	

From this we compute an F statistic for the hypothesis that the two regions are part of the same sloped surface:

$$F = \left(\frac{(1.67 + (-9) + 0.33)^2}{2(0.5^2/6 + 1.5^2/6 + 1/a)} \Big/ 3 \right) \Big/ (14.64/12) = 12.68$$

At the 1% significance level we would reject this hypothesis. Note that an edge detector comparing simple differences of average gray tones would most likely not call this an edge.

To detect whether an edge exists between a pair of neighboring pixels, we can employ the following procedure. For each pixel, examine all neighborhoods containing the pixel and determine that neighborhood whose sloped-surface fitting error is least. If the best-fitting neighborhoods for the two pixels overlap, then declare no edge. If the best-fitting neighborhoods for the two pixels have no overlap, then compute the F statistic to measure the significance of the difference between the two sloped surfaces.

4. USING THE SLOPED-FACET MODEL

To find regions we must look for connected sets of resolution cells which are surely on the same sloped surface. To find edges we must look for pairs of adjacent regions having significantly different sloped surfaces. To do edge detection and region analysis we must do both. This suggests the following way to apply the sloped-facet model. Select an appropriate sized neighborhood. Run this neighborhood over the image. For each location where the neighborhood may be placed on the image, determine the α, β, γ parameters of the sloped surface fit as well as the ϵ^2 error of the fit. Use this information to create an image in which each resolution cell has the four parameters α, β, γ, ϵ^2 from its corresponding neighborhood in the given image.

Although the sloped-surface parameters are placed in a single location, they actually apply to all the pixels in the neighborhood on the original image over which the fitting was done. Hence, on the original image, each pixel has associated with it the four parameters α, β, γ, and ϵ^2 for *each* neighborhood that contains it. From all of the neighborhoods of a pixel, one will have a parameter set having the smallest ϵ^2, that is, the best fit. Now associate with each pixel on the original image the parameters associated with its best-fitting neighborhood as well as the coordinates for this best-fitting neighborhood.

To do edge detection, for each pair of adjacent pixels compute the F statistic associated with the hypothesis that the pair of best-fitting neighborhoods corresponding to these pixels have the same sloped surface. Of course, set the F

statistic to zero if the best neighborhoods overlap. In this manner, an image of these F statistics can be created for vertical and horizontal edges. Then apply a non-maximum suppression operator (Rosenfeld and Kak [6]) in a direction orthogonal to the edge direction to determine those locations having a local maximum in the F-statistic. These locations should be the edges.

To do region growth use a clustering or grouping method. Join together each pair of neighboring pixels if their best neighborhoods overlap. If their neighborhoods do not overlap join them together if the hypothesis that the pixels' best neighborhoods are part of the same sloped surface cannot be rejected.

5. LITERATURE REVIEW

The idea of fitting linear surfaces for edge detection is not new. Roberts [1] employed an operator commonly called the Roberts gradient to determine edge strength in a 2×2 window in a blocks-world scene analysis problem. The fact that the Roberts gradient arises from a linear fit over the 2×2 neighborhood may surprise some people. To see this, let

$$\begin{matrix} a & b \\ c & d \end{matrix}$$

be the four gray tone levels in the 2×2 window whose coordinates are

$$\begin{matrix} (-\frac{1}{2}, -\frac{1}{2}) & (-\frac{1}{2}, \frac{1}{2}) \\ (\frac{1}{2}, -\frac{1}{2}) & (\frac{1}{2}, \frac{1}{2}) \end{matrix}$$

The least-squares fit for α and β is

$$\hat{\alpha} = \tfrac{1}{2}[(c + d) - (a + b)] \quad \text{and} \quad \hat{\beta} = \tfrac{1}{2}[(b + d) - (a + c)].$$

The gradient, which is the slope in the steepest direction, has magnitude $(\hat{\alpha}^2 + \hat{\beta}^2)^{\frac{1}{2}}$.

$$\begin{aligned} (\hat{\alpha}^2 + \hat{\beta}^2)^{\frac{1}{2}} &= \tfrac{1}{2}([(c + d) - (a + b)]^2 + [(b + d)^2 - (a + c)]^2)^{\frac{1}{2}} \\ &= \tfrac{1}{2}(2(a - d)^2 + 2(b - c)^2)^{\frac{1}{2}} \\ &= (2^{\frac{1}{2}}/2)((a - d)^2 + (b - c)^2)^{\frac{1}{2}} \end{aligned}$$

which is exactly $2^{\frac{1}{2}}/2$ times the Roberts gradient.

Prewitt [2] used a quadratic fitting surface to estimate the parameters α and β in a 3×3 window for automatic leukocyte cell scan analysis. The resulting values for $\hat{\alpha}$ and $\hat{\beta}$ are the same for the quadratic or linear fit. O'Gorman and Clowes [8] also used a 3×3 window and linear fit. Brooks [7] and Hummel [16] discussed the general fitting idea. Meró and Vamos [10] used the fitting idea to find lines on a binary image. Hueckel [11] used the fitting idea with low-frequency polar-form Fourier basis functions on a circular disk in order to detect step edges.

Figure 1 illustrates the windows for computing $\hat{\alpha}$ and $\hat{\beta}$ for neighborhood sizes of 2, 3, 4, and 5. Note that for the larger windows, the pixels on the edges of the window have higher weights. This means that the idea of using the difference

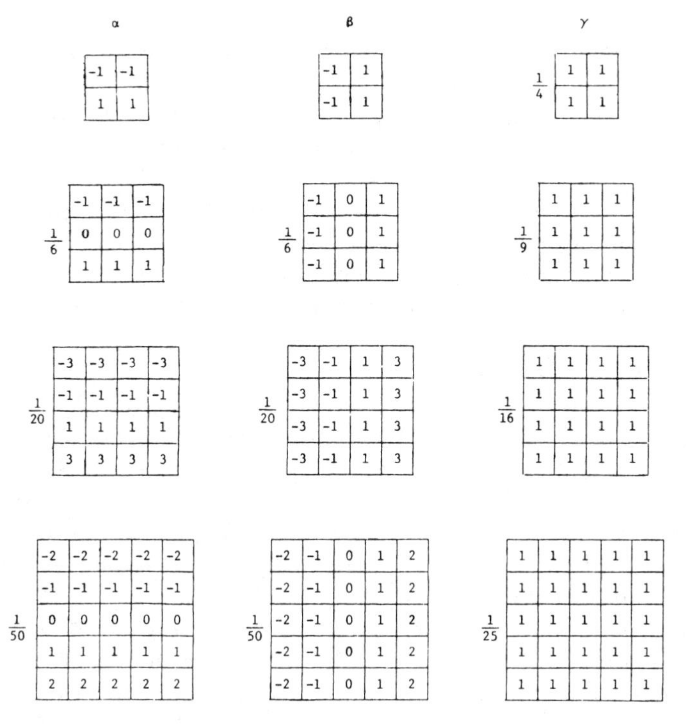

FIG. 1. The masks for estimating α, β, and γ for 2×2, 3×3, 4×4, and 5×5 neighborhoods.

of equally weighted averages to measure the slope in a given direction is incorrect. However, under the condition that $\alpha = \beta = 0$ an appropriate statistic to test whether two neighborhoods are on the same constant surface is an F statistic of the form

$$F = ([(\hat{\gamma}_1 - \hat{\gamma}_2) \sum_r \sum_c 1]/1)/((\epsilon_1^2 + \epsilon_2^2)/(2 \sum_r \sum_c 1 - 6)).$$

Under the hypothesis that $\gamma_1 = \gamma_2$, this statistic has an F distribution with 1,

$$2 \sum_r \sum_c 1 - 6$$

degrees of freedom. High values of F constitute a reason for rejecting the hypothesis that the constant surfaces are identical.

Rosenfeld *et al.* [4] suggested taking the products of $|\hat{\gamma}_1 - \hat{\gamma}_2|$ over varying size neighborhoods. Such a product will be large if each of the terms is high. High values for a set of neighborhood sizes will tend to arise in a region where the denominator of the F-statistic is low, thereby making the F statistic high for the largest neighborhood used. Of course, a better way of using the varying sized neighborhoods is to compute the F statistic itself for each of the various neighborhood sizes and then measure the strength of the step edge by the highest value F statistic computed.

There are few papers that approach edge detection statistically. Yakimovsky [5] suggests using a maximum likelihood principle for detecting step edges. His null hypothesis is that the sample values for the two adjacent neighborhoods come from the same normal distribution. The alternative hypothesis is that the distributions differ in mean and or variance. The disadvantage of Yakimovsky's test is that unless the sample is large, the distribution of the maximum likelihood test statistic is not known. Asymptotically however, -2 times the natural logarithm of the test statistic has a chi-squared distribution with 3 degrees of freedom. The F test discussed in Section 3 has the advantage of being an exact test under the assumption that the variances are identical. If the identical variance assumption is not appropriate, the maximum likelihood test could be extended to the case of the slope model.

Statistical approaches to edge detection require the use of a fitting model combined with a noise assumption to permit evaluating the significance of the edge strength measure. There seems to be few papers which do both. Besides Yakimovksy's maximum likelihood approach, Shanmugam *et al.* [20] derive an optimal edge filter whose behavior with noise is computed. Cohen and Toussaint [21] are concerned with the unequal distribution of random noise in the equal interval basis in a Hough transform space.

Nahi and Jahanshahi [15], Nahi and Assefi [12], Habibi [13], and Nahi [14] all use a statistical model of the object, background, and noise in a Bayesian and/or recursive estimation scheme to improve the image or estimate the boundary. The spatial models underlying these approaches are different from the fitting models discussed here. It would be interesting to find a method which combines both approaches.

As mentioned earlier, the fact that the slopes in two orthogonal directions determine the slope in any direction is well known in vector calculus. However, it seems not to be so well known in the image processing community. Prewitt [2], Kirsch [3] and Robinson [19] each suggested a set of masks which could be used to determine the gradient direction and magnitude for eight directions. Obviously, the masks of Prewitt, Kirsch, and Robinson not only do not give correct values for the gradient but also require two or four times more work to compute than the two masks required to determine $\hat{\alpha}$ and $\hat{\beta}$. Their application to determine the angle of step edges might seem more appropriate. However, O'Gorman [9] determined that for ideal straight-line step edges, the edge direction as determined by the direction of steepest slope using $\hat{\alpha}$ and $\hat{\beta}$ will not differ by more than $6.6°$ from the actual step edge direction. This implies that perhaps even for the problem of step edges, using many directional masks is not necessary.

6. BAYESIAN EDGE DETECTION AND REGION ANALYSIS

In Section 4 we have outlined a procedure for computing a statistic F whose distribution is known under the null hypothesis: no edge. This means that $P(F \mid \text{no edge})$ is known. An image of the computed F statistics can be histogrammed to estimate $P(F)$. Now if the prior probability of a nonedge were known, by Bayes' formula it is possible to determine the probability of no edge given the value of the statistic F:

$$P(\text{no edge} \mid F) = \frac{P(F \mid \text{no edge})\, P(\text{no edge})}{P(F)}.$$

Since there is either an edge or no edge, we have

$$P(\text{edge} \mid F) = 1 - P(\text{no edge} \mid F).$$

The image itself can tell us some about the prior probability of edge and nonedge. For example, under the assumption that all edges are of equal strength, it is clear that the histogram of the F-statistic image will be bimodal: one mode corresponding to the F statistics generated in no edge areas and the other mode corresponding to the F statistics generated in edge areas. The valley in the histogram gives us a threshold to detect edges and the probability on the side of the valley corresponding to the higher-valued F statistics gives us an estimate of the edge probability after observing the image.

We may have some prior knowledge about edge probability before observing the image and this probability can be used to help determine the edge probability after observing the image by biasing the choice of valley bottom. Also assuming a simple distributional form for $P(F \mid \text{no edge})$ and $P(F \mid \text{edge})$ it is possible to compute the priors $P(\text{no edge})$ and $P(\text{edge})$ by best fitting to the observed mixture $P(F)$ (see Haralick and Singh [18]).

There is no reason why the prior probabilities for edge and no edge cannot vary with position on the image. We can make these dynamic by making them a function of average neighborhood gray tone.

Consider, for example, a popular segmentation scheme which is appropriate when an image consists of a collection of objects which are constant in gray tone. In this case, the image histogram will have a tendency to be multimodal with the modes corresponding to the various gray tone shades of the objects. We may locate the valleys in the histogram or determine the parameters of the mixture distribution to find the valleys (Chow and Kaneko [17]) and use those points as decision boundaries in a pixel-by-pixel classification scheme. This scheme uses no spatial image structure and works poorly with moderate noise, producing a highly broken-up segmentation. But there is information in the location of the valleys. Small changes in gray tones in the valley regions should be more significant of an edge than similar-sized changes near one of the histogram modes. Knowing this, we may set the prior probability of an edge at each pixel to depend on the nearness of the average neighborhood gray tone to a valley in the histogram of the image grey tones.

From this perspective it is clear that the modal segmentation scheme corresponds to segmentation based on edge prior probabilities which are a function of gray tone. The simple F test of Section 3 corresponds to equal edge and nonedge prior probabilities.

7. CONCLUSION

We have discussed a surface fitting model by which edge detection and region delineation can be done. Our presentation has been theoretical. Future work will be empirical. We will be determining the validity of the model itself, the effectiveness of the statistical procedure for edge detection and region delineation, and possible ways of using the residual fitting error as a feature in its own right.

REFERENCES

1. L. G. Roberts, Machine perception of three-dimensional solids, in *Optical and Electroptical Information Processing* (J. T. Tippett, *et al.*, Eds.), pp. 159–197, MIT Press, Cambridge, Mass., 1965.
2. J. M. S. Prewitt, Object enhancement and extraction, in *Picture Processing and Psychopicotorics* (B. S. Lipkin and A. Rosenfeld, Eds.), pp. 75–149, Academic Press, New York, 1970.
3. R. Kirsch, Computer determination of the constituent structure of biological images, *Comput. Biomed. Res.*, **4**, 1971, 315–328.
4. A. Rosenfeld, Y. Lee, and R. Thomas, Edge and curve detection for texture discrimination, in *Picture Processing and Psychopictorics* (B. S. Lipkin and A. Rosenfeld, Eds.), pp. 381–393, Academic Press, New York, 1970.
5. Y. Yakimovsky, Boundary and object detection in real world images, *J. Assoc. Comput. Mach.* **23**, 1976, 599–618.
6. A. Rosenfeld and A. Kak, *Digital Picture Processing*, Academic Press, New York, 1976.
7. M. Brooks, Rationalizing edge detectors, *Computer Graphics Image Processing* **8**, 1978, 277–285.
8. F. O'Gorman and M. B. Clowes, Finding picture edges through collinearity of feature points, *IEEE Trans. Computers* **C-25**, April 1976, 449–456.
9. F. O'Gorman, Edge detection using Walsh functions, in *AISB Summer Conference*, 1976.
10. L. Meró and T. Vamos, Real-time edge-detection using local operators, in *3rd International Joint Conference on Pattern Recognition*, Coronado, California, November 1976.
11. M. Hueckel, An operator which locates edges in digital pictures, *J. Assoc. Comput. Mach.* **18**, 1971, 113–125.
12. N. E. Nahi and T. Assefi, Bayesian recursive image estimation, *IEEE Trans. Computers*, **C-21**, July 1972.
13. A. Habibi, Two-dimensional Bayesian estimate of images, *Proc. IEEE* **60**, July 1972.
14. N. E. Nahi, Role of recursive estimation in statistical image enhancement, *Proc. IEEE* **60**, July 1972.
15. N. E. Nahi and M. H. Jahanshahi, Image boundary estimation, *IEEE Trans. Computers* **C-26**, August 1977, 772–781.
16. R. Hummel, Feature detection using basic functions, *Computer Graphics Image Processing* **9**, 1979, 40–55.
17. C. K. Chow and T. Kaneko, Automatic boundary detection of the left ventricle from cineangiograms, *Comput. Biomed. Res.* **5**, 1972, 388–410.
18. R. Haralick and A. Singh, Boundary detection in images using a nonstationary statistical model, in *IEEE International Conference on Cybernetics and Society*, Washington, D.C., September 1977.
19. G. Robinson, Edge detection by compass gradient masks, *Computer Graphics Image Processing* **6**, 1977, 492–501.
20. K. S. Shanmugam, F. Dickey, and J. Green, An optimal frequency domain filter for edge detection in digital images, *IEEE Trans. Pattern Anal. Machine Intelligence* **PAMI-1**, January 1979, 39–47.
21. M. Cohen and G. Toussaint, On the detection of structures in noisy pictures, *Pattern Recognition* **9**, 1977, 95–98.

The Use of Markov Random Fields as Models of Texture

MARTIN HASSNER AND JACK SKLANSKY

School of Engineering, University of California, Irvine, Irvine, California 92717

We propose Markov Random Fields (MRFs) as probabilistic models of digital image texture where a textured region is viewed as a finite sample of a two-dimensional random process describable by its statistical parameters. MRFs are multidimensional generalizations of Markov chains defined in terms of conditional probabilities associated with spatial neighborhoods. We present an algorithm that generates an MRF on a finite toroidal square lattice from an independent identically distributed (i.i.d.) array of random variables and a given set of independent real-valued statistical parameters. The parametric specification of a consistent collection of MRF conditional probabilities is a general result known as the MRF–Gibbs Random Field (GRF) equivalence. The MRF statistical parameters control the size and directionality of the clusters of adjacent similar pixels which are basic to texture discrimination and thus seem to constitute an efficient model of texture. In the last part of this paper we outline an MRF parameter estimation method and goodness of fit statistical tests applicable to MRF models for a given unknown digital image texture on a finite toroidal square lattice. The estimated parameters may be used as basic features in texture classification. Alternatively these parameters may be used in conjunction with the MRF generation algorithm as a powerful data compression scheme.

1. MRFS AS STATISTICAL MODELS OF TEXTURE

The objective of this paper is to present an application of random field theory to image processing. As it has been pointed out by Dobrushin [1] the recent developments of this theory have found almost no reflection in the information-theoretic literature. To be specific mainly those properties of random fields which are a direct generalization of one-dimensional properties such as spectral representations associated with stationarity have been used in an image processing context. This has been pointed out by Wong [2] who mentioned the multi-dimensional generalization of the Markov property which is inherently dependent on the geometry and partial ordering of the parameter space as a "promising and challenging area of research" with a particular impact on image processing.

The images considered in this paper possess a finite alphabet $S = \{0, \ldots, s-1\}$ and are defined on the two-dimensional integer lattice Z^2 or on a finite subset of it. Specifically we assume M_1, M_2 nonnegative integers and consider the *toroidal* two-dimensional integer lattice $Z_{M_1} \times Z_{M_2}$ shown in Fig. 1 where Z_{M_i} is the additive group of integers modulo M_i, $Z_{M_i} = \{1, \ldots, M_i\}$, $i = 1, 2$. This imposes periodic boundary conditions on a finite rectangular picture region and allows us to ignore edge effects. We assume that the rectangular S-valued

ISBN 0-12-597320-9

data array on $Z_{M_1} \times Z_{M_2}$ is a finite sample of a *homogeneous* and furthermore *ergodic* discrete random field $\{X(i, j)\}$, $i, j \in Z_{M_1} \times Z_{M_2}$, that we define as an S-valued *digital image texture*. Arbitrary image data may be viewed as being composed of locally homogeneous and ergodic texture regions and thus the analysis of digital image texture is of general interest in digital image processing.

The generalization of the Markov property from one to two (or more) dimensions is a difficult problem mainly due to the lack of a preferred or natural direction in the parameter space Z^2 as compared with the "discrete time" interpretation of the one-dimensional integer set Z. We introduce the MRF models by making use of the *graph structure* of $Z_{M_1} \times Z_{M_2}$ where two vertices (i, j), $(i', j') \in Z_{M_1} \times Z_{M_2}$ are adjacent if $i = i'$ and $j = j' \pm 1$ or if $j = j'$ and $i = i' \pm 1$. The nearest neighbor or more consistently *first-order MRF* conditional probabilities $p\{(X(i, j) | X(i-1, j), X(i-1, j), X(i, j-1), X(i, j+1)\}$ thus correspond to the underlying graph structure of $Z_{M_1} \times Z_{M_2}$ and by generalizing the concept of neighborhood we can define a natural *hierarchy* of MRFs of increasing order (spatial "memory") as shown in Fig. 2. If we assume a unit distance between adjacent graph vertices then the first-order MRF corresponds to a neighborhood configuration of radius 1 that consists of the four nearest neighbors labeled by 1's, the second-order MRF corresponds to a neighborhood configuration of radius $2^{\frac{1}{2}}$ that further includes the diagonal neighbors labeled by 2's, and so on.

The major problem is the consistent specification of the conditional probabilities associated with the neighborhoods naturally defined by the underlying graph structure. The solution to this purely mathematical problem is surprisingly physical as it became obvious that the MRF models are identical with lattice

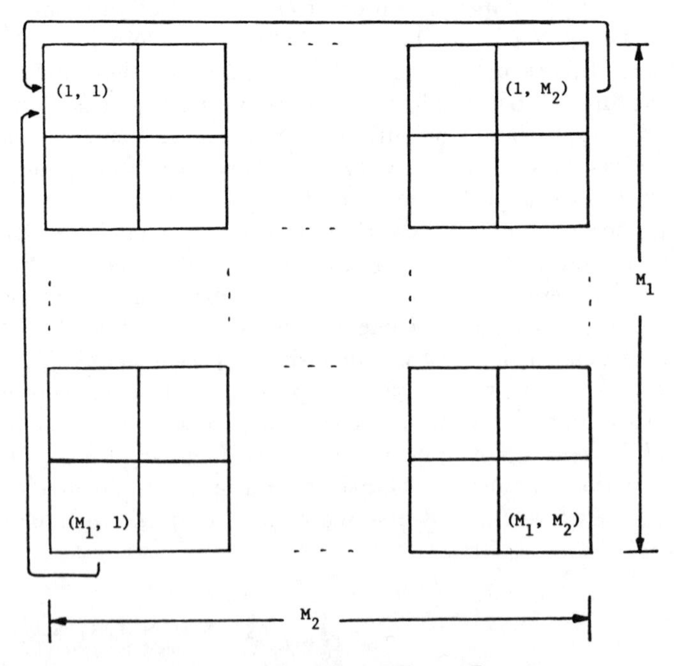

FIG. 1. Toroidal lattice $Z_{M_1} \times Z_{M_2}$.

			6			
	5	4	3	4	5	
	4	2	1	2	4	
6	3	1	0	1	3	6
	4	2	1	2	4	
	5	4	3	4	5	
			6			

FIG. 2. Natural hierarchy of MRF models determined by neighborhood configurations of increasing order.

gas models studied in statistical mechanics [3]. Specifically the classical *Ising model* is identical with a first-order binary-valued MRF. This identity allows for the consistent specification of the MRF conditional probabilities in terms of a small number of statistical parameters.

One of the main obstacles in assessing the practical significance of this result from an engineering viewpoint is the obscure language that relies heavily on

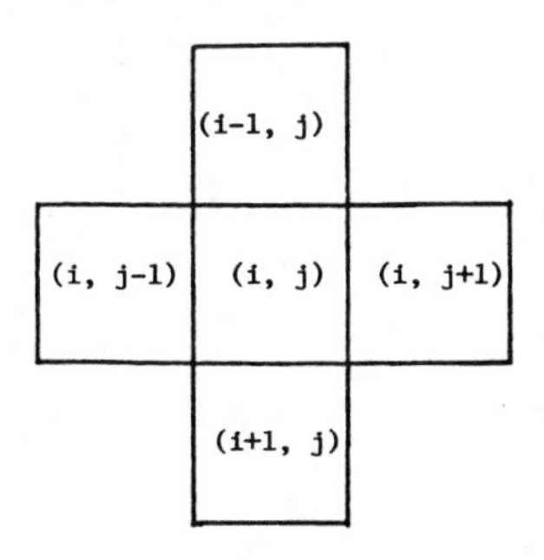

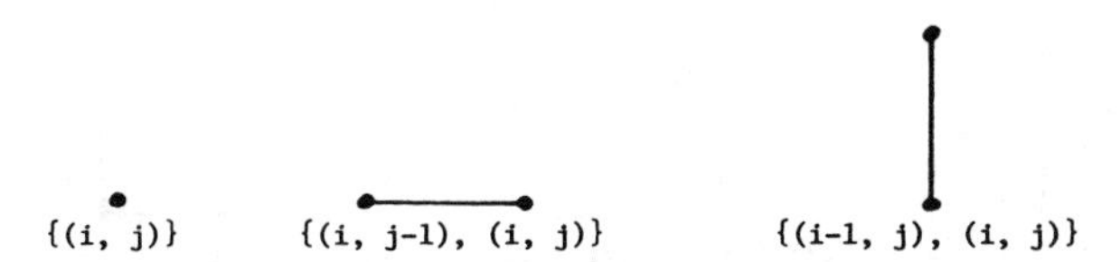

FIG. 3. First-order MRF neighborhood configuration and associated class of cliques.

statistical mechanical formalism that characterizes most of the relevant references. In this paper we will rely on an intuitive and simple description of the parametric MRF characterization as presented in Besag [4] which is based on the possible *clique* graphs that can be associated with a neighborhood configuration. A clique is a graph whose vertex set is composed of vertices such that each one is a neighbor of all the others. In Figs. 3, 4, and 5 we present the first-, second-, and third-order MRF neighborhood configurations and their associated clique classes, respectively. The collection of cliques associated with a neighborhood configuration defines a *sufficient statistic* for the corresponding MRF.

Assuming this viewpoint, the MRF–GRF equivalence essentially identifies an MRF with a *maximum entropy* probabilistic description that is compatible with a fixed expected value of the sufficient statistic. For the purpose of concreteness we consider a first-order binary-valued MRF. A sufficient statistic for this MRF is defined in terms of its cliques given in Fig. 3:

$$\begin{aligned} E\{\textstyle\sum x(i,\, j)\} &= K_o, \\ E\{\textstyle\sum x(i-1,\, j) \times (i,\, j)\} &= K_v, \qquad (i,\, j) \in Z_{M_1} \times Z_{M_2}, \\ E\{\textstyle\sum x(i,\, j-1) \times (i,\, j)\} &= K_h. \end{aligned} \tag{1}$$

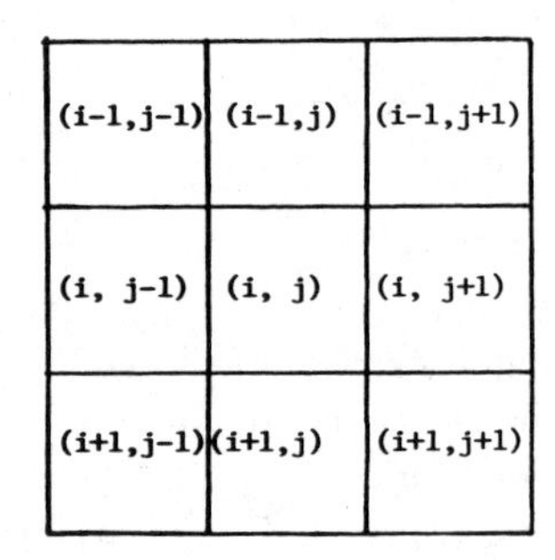

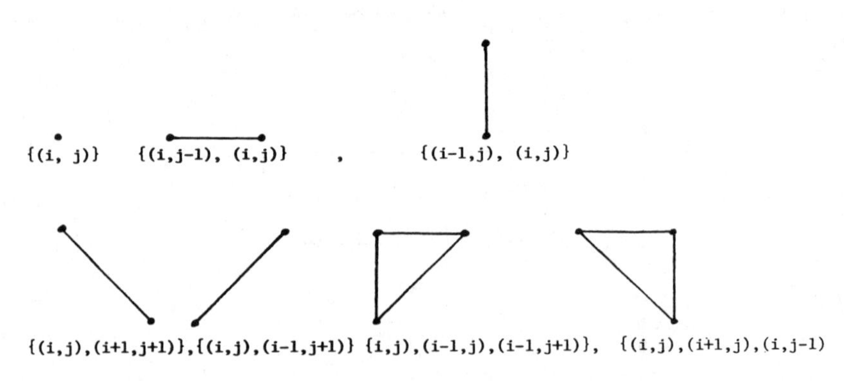

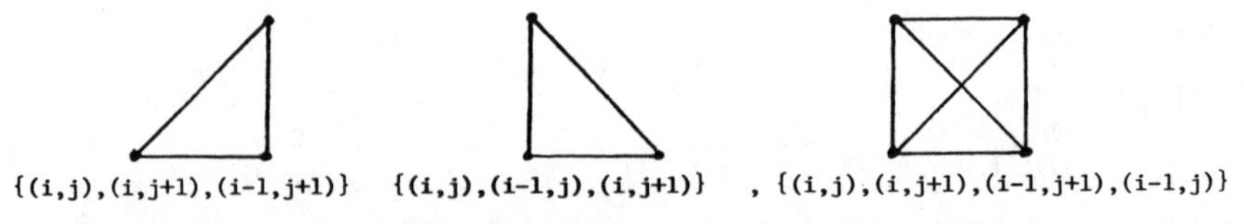

FIG. 4. Second-order MRF neighborhood configuration and associated clique class.

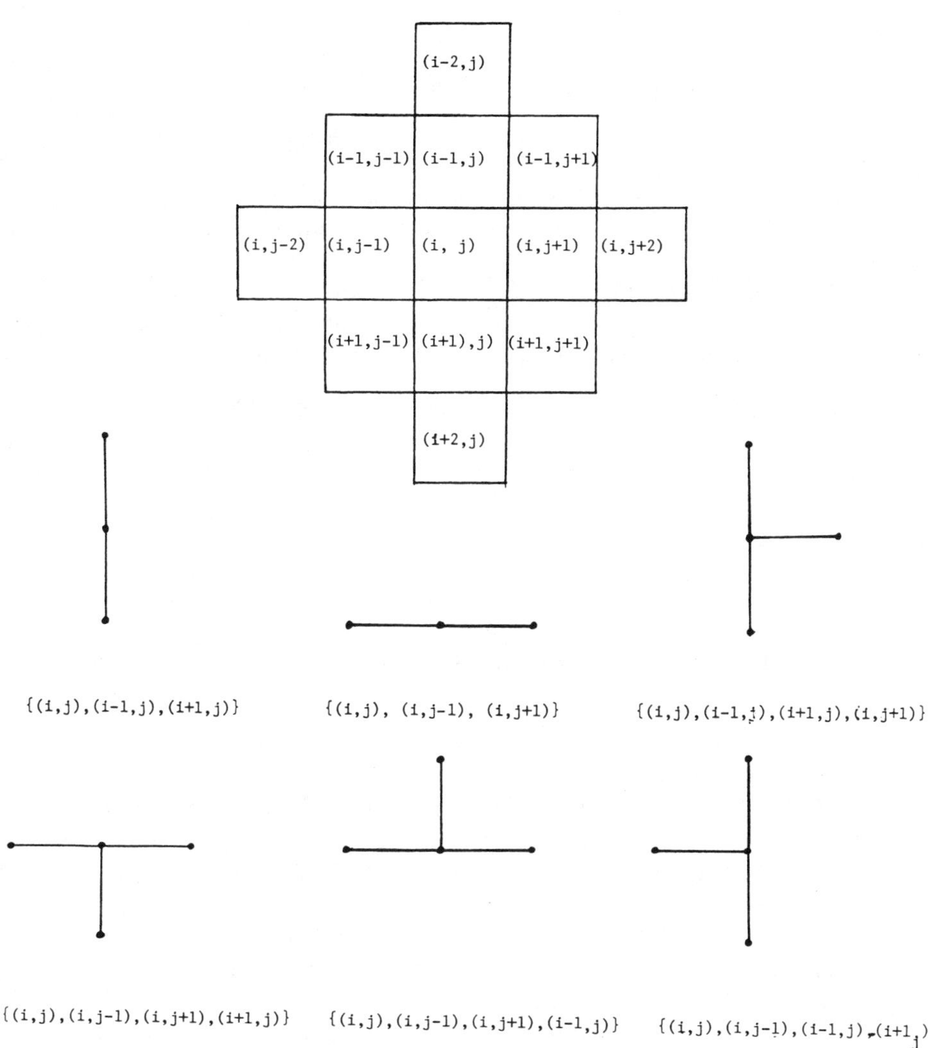

FIG. 5. Third-order MRF neighborhood configuration and associated clique classes (additional to those in Fig. 4).

To a given set of fixed expected values $\{K_o, K_v, K_h\}$ where K_o is the expected number of 1's, K_v is the expected number of vertical "dipoles" $\begin{matrix}1\\1\end{matrix}$ and K_h is the expected number of horizontal "dipoles" 1 1, there corresponds a first-order MRF and a set of statistical parameters $\{a, b_1, b_2\}$ in terms of which the MRF conditional probabilities are consistently specified. The joint probability $P(\mathbf{x})$ assigned by this MRF to a realization $\mathbf{X} = \mathbf{x}$, $\mathbf{X} = \{X(i, j)\}$, $(i, j) \in Z_{M_1} \times Z_{M_2}$, is as follows:

$$P(\mathbf{x}) = Z^{-1}(a, b_1, b_2) \exp\left[a \sum x(i, j) + b_1 \sum x(i, j-1)x(i, j) + b_2 \sum x(i-1, j)x(i, j)\right], \quad (2)$$

where

$$Z(a, b_1, b_2) = \sum_{\mathbf{x}} \exp[a \sum x(i, j) + b_1 \sum x(i, j-1)x(i, j) + b_2 \sum x(i-1, j)x(i, j)] \quad (3)$$

is the normalizing statistical sum (or the *partition function* as it is referred to in statistical mechanics) over all possible binary random configurations $\mathbf{X} = \mathbf{x}$ on $Z_{M_1} \times Z_{M_2}$.

The MRF parametric model formally described in (2) and (3) is identical with the *Gibbs ensembles* studied in statistical mechanics. These are maximum entropy probabilistic descriptions where the parameter set $\{a, b_1, b_2\}$ is chosen such that there exists

$$\sum P(\mathbf{x})L(\mathbf{x}) = K_o,$$
$$\sum P(\mathbf{x})H(\mathbf{x}) = K_h,$$
$$\sum P(\mathbf{x})V(\mathbf{x}) = K_v, \quad (4)$$

where

$$L(\mathbf{x}) = \sum x(i, j),$$
$$H(\mathbf{x}) = \sum x(i, j-1)x(i, j),$$
$$V(\mathbf{x}) = \sum x(i-1, j)x(i, j)$$

and $\mathbf{x}$ is a binary configuration on $Z_{M_1} \times Z_{M_2}$.

Conversely a set of real-valued statistical parameters associated with the cliques of a given neighborhood configuration determines a (not necessarily unique) MRF. Thus a first-order MRF is determined by three statistical parameters. If we furthermore assume that this MRF is *isotropic* then the number of parameters required for its specification is two. Generally the imposition of symmetry conditions has the effect of reducing drastically the number of parameters required for the MRF specification. (We have already made the implicit assumption that the statistical parameters associated with cliques that are translates of each other are equal).

The choice of a statistical parameter set $\{a, b_1, b_2\}$ determines a first-order MRF and thus the statistical moments or expected values specified in (4). This implies that by means of these parameters we can control the probabilities of occurrence of a 1 and of horizontal and vertical dipoles. From (2) we can immediately deduce that $b_1 > 0$ and $b_2 < 0$ determine an MRF in whose sample fields horizontal clusters of adjacent ones are more likely to occur than vertical clusters of adjacent ones. By considering higher-order MRFs we furthermore have at our disposition a hierarchy of parametric models by means of which we can control the frequencies of triangles, squares, T-shaped configurations, etc.

The preceding discussion may be viewed as providing the motivation for applying these models to the problem of texture. As stated by Julesz [5] the success of statistical considerations in texture discrimination depends on whether the clusters of adjacent similar dots which are basic to texture perception can be controlled and analyzed by a statistical model. The MRF models introduced possess the required property and thus are proposed here as an efficient model of texture. The major disadvantage associated with these models is that the *analytic* evaluation of MRF statistical moments is generally impossible. For a binary first-order MRF the moment generating function is given in Bartlett [6, p. 32]. Except for a constant factor the MRF moment generating function is the logarithm of the statistical sum function and the statistical moments are obtained by taking partial derivatives w.r.t. the corresponding parameters. Thus for a binary first order MRF the first statistical moment $E\{L(\mathbf{x})\}$ is obtained as the partial derivative of log $Z(a, b_1, b_2)$ w.r.t. to a, etc. Since the statistical sum function is generally not derivable in closed form, except for the specific case $a = 0$, $b_1 = b_2$ solved by Onsager [7], the exact evaluation of MRF statistical moments is impossible.

However there exist *numerical* methods for the approximate evaluation of the statistical sum function that corresponds to an MRF described by a set of statistical parameters. We particularly refer to the *graph*-theoretical methods as presented in Domb [8]. The statistical sum function and the statistical moments derived from it are approximately evaluated in terms of power series whose terms are the cardinalities of the possible closed random walks (polygonal contours on the square lattice) of a given order weighted by coefficients that are determined by the MRF parameters. The cardinalities of these polygon classes are *lattice constants* which are tabulated in the physical literature for a variety of lattices and interaction parameters. Computer techniques for the evaluation of these lattice constants are available and Domb [8] is a rich source of references.

These numerical methods can be used to derive a variety of textural features from a given set of MRF parameters that can be used in texture discrimination. The MRF parameters themselves may thus be viewed as *basic textural features*. The practical evaluation of these numerical methods to texture discrimination will not be pursued here, our primary concern being the MRF parameters as a means to simulate textures and furthermore the estimation of these parameters from unknown textures (which is equivalent to MRF model fitting).

2. AN MRF SIMULATION ALGORITHM

The MRF simulation algorithm presented has originally been devised for the simulation of Ising models [9]. In view of the MRF–GRF equivalence this algorithm "cleaned" of its physical context is applicable to the generation of MRF textures. The example presented is a first-order binary isotropic MRF. However, the algorithm is generalizable to higher order and multivalued MRFs. At this point we should stress that the number of MRF parameters increases rapidly with an increase in the cardinality of the MRF alphabet. Thus a first-order s-valued two-dimensional isotropic MRF requires $2(s-1)^2 + s - 1$ independent parameters for its description. Thus for large values of s even for first order MRFs the number of MRF parameters can become prohibitively large.

We have already mentioned symmetry inposed on the clique classes as a means of reducing the required number of independent MRF parameters. We can also assume "color indifference," i.e., for example, for a first-order s-valued MRF we will not differentiate between adjacent pairs of gray levels as long as

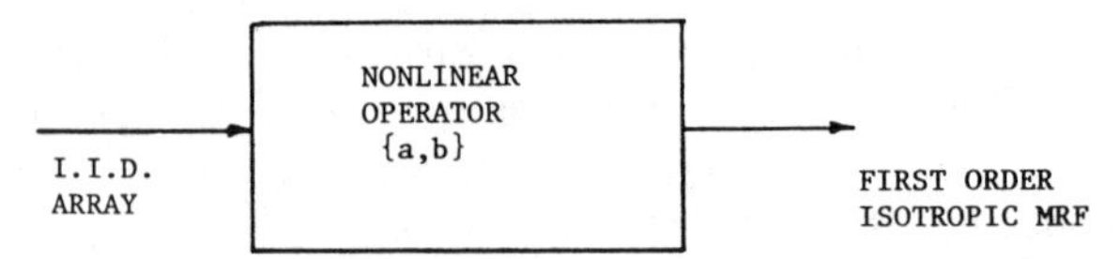

FIG. 6. MRF texture generation.

they are not identical. Another practical alternative is to consider Gaussian–MRF models [10].

We view the simulation algorithm as a *nonlinear operator* specified in terms of a collection of MRF parameters that operates on an i.i.d. binary array as shown in Fig. 6. A given set of MRF parameters $\{a, b_1, b_2\}$ specifies a consistent collection of first order MRF conditional probabilities [4]

$$p(x \mid t, t', u, u') = \frac{\exp[x\{a + b_1(t + t') + b_2(u + u')\}]}{1 + \exp\{a + b_1(t + t') + b_2(u + u')\}}, \tag{5}$$

where for simplicity we have written (x, t, t', u, u') instead of $(x(i, j), x(i - 1, j), x(i + 1, j), x(i, j - 1), x(i, j + 1))$. We furthermore assume $b_1 = b_2 = b$, i.e., we consider an isotropic first-order binary MRF. Under this assumption there are 10 possible values of (5) to which we refer as the *MRF local characteristics* all of which are consistently specified in terms of $\{a, b\}$. To each of these 10 local characteristics there corresponds a pixel class defined by its value and the values of its four adjacent neighbors. These are tabulated in Table 1. The algorithm described essentially operates on the *frequencies* of these pixel classes such that in the limit they assume the equilibrium values specified by (5) which are determined by an MRF parameter set $\{a, b\}$.

The two-dimensional i.i.d. binary random array on the toroidal lattice is converted into a pixel *class map* by sliding it through the first-order MRFneighborhood configuration that can be viewed as a *two-dimensional sliding block filter* [11]. The analogy with sliding block filters as specified in [11] is that pixel

TABLE 1

MRF Local Characteristics and Associated Pixel Classes

Pixel class	Local characteristic	Pixel value	Number of adjacent 1's
1	P_1	1	4
2	P_2	1	3
3	P_3	1	2
4	P_4	1	1
5	P_5	1	0
6	P_6	0	4
7	P_7	0	3
8	P_8	0	2
9	P_9	0	1
10	P_{10}	0	0

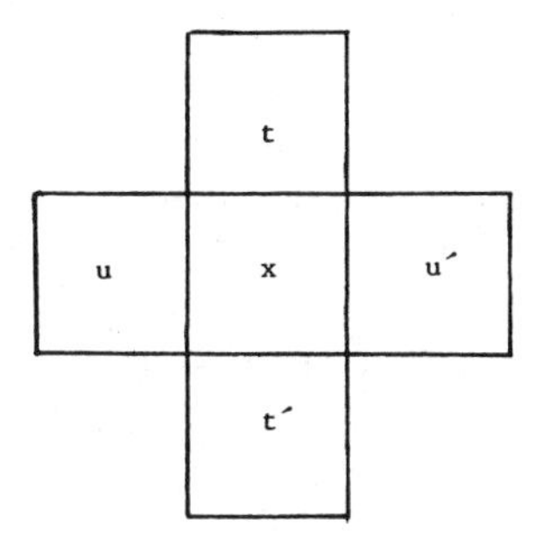

(a) First order MRF filter "window"

(b) Data array

1	1	0	0
1	1	0	1
1	0	0	0
0	1	1	0

(c) Class map

3	2	8	8
1	3	8	4
4	7	9	8
7	3	4	9

FIG. 7. Conversion of data into class map.

classes associated with zero-valued local characteristics may be viewed as a two-dimensional complementation set. The MRF parameter specification, however, not only determines the admissible nearest neighbor configurations but also their frequencies. A numerical example of the conversion into a class map is given in Fig. 7 for a $Z_4 \times Z_4$ toroidal lattice.

The pixel class array is now converted into 2 one-dimensional arrays and by means of a *Monte Carlo procedure* the pixel class rates are adjusted to assume the

MRF equilibrium values, i.e., the values of the local characteristics specified in (5). The 2 one-dimensional arrays are the LOCATION ARRAY, in which the pixels are arranged according to class membership, and the LOOKUP ARRAY in which the pixels are arranged in the order of their location on the lattice. These two arrays and their interrelationship are shown in Fig. 8 where we use the numerical example of Fig. 7.

The frequencies of occurrence of the 10 possible pixel classes are adjusted to assume their equilibrium values specified in (5) by a *flipping mechanism* specified in terms of the flip probabilities $P_i = P(i \to i + 5)$ and $P_{i+5} = P(i + 5 \to i)$, $1 \leqslant i \leqslant 5$ which are required to satisfy a *balance equation*

$$\frac{P_{i+5}}{P_i} = \frac{p_i}{p_{i+5}} \tag{6}$$

LOCATION ARRAY

1	5	1
2	2	2
3	1	
4	6	
5	14	3
6	8	
7	9	
8	15	4
9	10	
10	13	7
11	3	
12	4	
13	7	
14	12	8
15	11	
16	16	9

LOOKUP ARRAY

	CLASS	ADD
1	3	3
2	2	2
3	8	11
4	8	12
5	1	1
6	3	4
7	8	13
8	4	6
9	4	7
10	7	9
11	9	15
12	8	14
13	7	10
14	3	5
15	4	8
16	9	16

VAL (LOOK.ADD) = ADD(LOC)
VAL (LOOK.CLASS) = CLASS NUMBER OF CORRESPONDING ELEMENT

FIG. 8. One-dimensional representation of two-dimensional data.

which is subject to the constraint $p_i + p_{i+5} = 1$. The resulting flip probabilities are as follows:

$$P_i = \frac{\exp\{-a - (5 - i)b\}}{1 + \exp\{-a - (5 - i)b\}}, \qquad 1 \leqslant i \leqslant 5,$$

$$P_{i+5} = 1 - P_i. \tag{7}$$

We denote by n_i the number of pixels of class i and by m_i the number of pixels whose class number is less than i, $1 \leqslant i \leqslant 10$. We can precompute the 10 numbers

$$Q_i = \sum_{j=1}^{i} n_j P_j, \qquad i = 1, \ldots, 10$$

and perform two *random choices* $R \in [0, Q_{10})$ with $Q_{i-1} \leqslant R < Q_i$ resulting in class i and $l_i \in [1, n_i]$, resulting furthermore in the specific pixel of class i that is stored in LOC $(m_i + l_i)$. Flipping this pixel changes its class membership by ± 5. However, its nearest neighbors also change their class membership by ± 1. The function of the LOOKUP ARRAY is to serve verbally as a lookup table for these neighbors, then locate them in the LOCATION ARRAY and adjust their class membership. The flipping mechanism is stopped by using a threshold rule which compares the theoretical and empirical probabilities of the pixel classes. Examples of MRF textures generated by using this program are given in Fig. 9. By choosing positive values for b we get clustering in Fig. 9b whereas a negative value for b results in the nonclustered texture of Fig. 9c. This is exactly in agreement with what we would expect to happen from looking at the formula in (2) with $b = b_1 = b_2$.

3. MRF PARAMETER ESTIMATION

The major problem with MRF model fitting to an unknown texture is that the direct approach to statistical inference through *maximum likelihood* is intractable because of the extremely awkward nature of the normalizing statistical sum function. An alternative technique is the *coding method* introduced by Besag [4] which provides MRF parameter estimates given a *single* realization $\mathbf{x}$ on $Z_{M_1} \times Z_{M_2}$. We present the coding method for the explicit example of a first-order MRF model and binary data. However, this method is generalizable to the estimation of MRF parameters for higher-order models and also for nonbinary (including Gaussian) data.

The coding method uses only part of the data that, however, is now mutually independent. This results in a factorized conditional likelihood from which conditional maximum likelihood estimates of the MRF parameters are easily obtained. With each MRF model it is possible to associate such a coding scheme. The coding scheme for a first-order MRF model is given in Fig. 10. The random variables labelled by $\times$ *given* the random variables labeled by $\bullet$ are mutually independent and this results in the conditional likelihood

$$L = \prod p(x(i, j) | x(i-1, j), \quad x(i+1, j), \quad x(i, j-1), \quad x(i, j+1)), \tag{8}$$

FIG. 9. a,b.

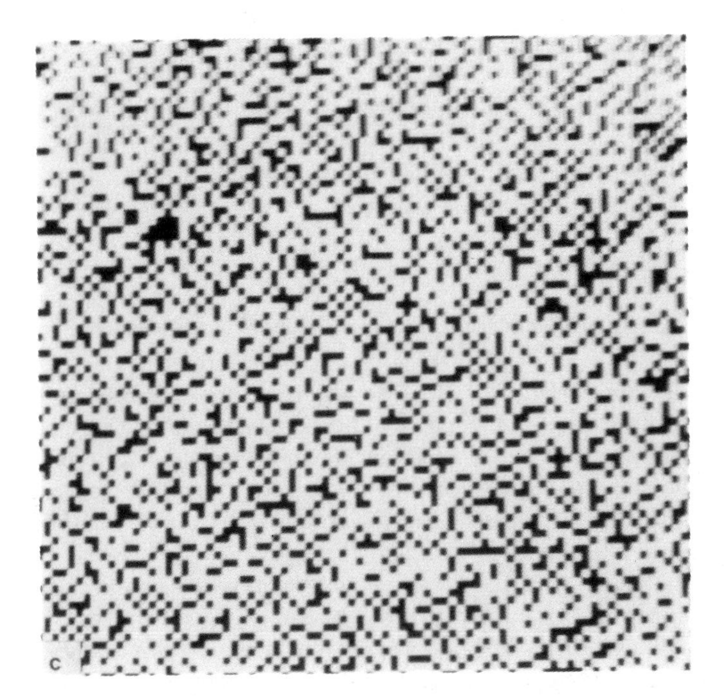

FIG. 9. Examples of binary-valued Markov random fields. (a) Initial 80×80 random field: statistically independent pixels. (b) Isotropic random field: $a = -6$, $b = 1.5$. (c) Isotropic random field: $a = 6$, $b = -3$.

where the product is taken over all the × pixels. By shifting by one unit the coding scheme in Fig. 10 we obtain a likelihood function for the • pixels now conditioned on the × pixels. These maximum likelihood estimates can be combined to yield one set of estimated MRF parameters.

The coding method results in likelihood functions as in (8) for which it is easy to construct *likelihood-ratio tests* (LRT) to examine the goodness of fit of MRF models. Given an unknown binary realization $\mathbf{x}$ on $Z_{M_1} \times Z_{M_2}$ we can represent this data in the form of a *contingency table* as described in Fig. 11; provided we want to fit to the data a first-order MRF we use the coding scheme of Fig. 10 and construct the table for the erased data. The first-order MRF parameters $\{a, b\}$ are estimated from the contingency table by maximum likelihood. By using these estimates we can compute the first-order MRF conditional probabilities as specified in (5). This model constitutes a statistical hypothesis that we test against

• × • × • ×

× • × • × •

• × • × • ×

× • × • × •

FIG. 10. First-order MRF coding scheme.

$y(i, j) = x(i-1, j) + x(i+1, j) + x(i, j-1) + x(i, j+1)$

		0	1	2	3	4	
$x(i, j)$	0	l_0	l_1	l_2	l_3	l_4	l
	1	m_0	m_1	m_2	m_3	m_4	m
		n_0	n_1	n_2	n_3	n_4	n

FIG. 11. Contingency table representation of pictorial data.

the *randomness hypothesis* ($b = 0$). The observed and expected frequencies can be used to conduct simple *chi-squared goodness-of-fit tests* for the scheme.

The major drawback of the coding method and the resulting simple contingency table analysis is the *low efficiency* that results from the fact that only a subset of the data is used. However, the MRF parameter estimation method as outlined is simple, flexible and directly testable by statistical tests. It would be interesting to compare this method to the fitting of *unilateral autoregressive models* to image data as presented in [12]. However, at this point we still have not tested these models on real textures, a task we are currently undertaking and whose results will be shortly available.

ACKNOWLEDGMENT

This research was supported by the National Institute of General Medical Sciences of the U.S. Public Health Service under Grant GM-17632.

REFERENCES

1. R. L. Dobrushin and S. A. Pirogov, Theory of random fields, in *Proc. of the 1975 IEEE–USSR Joint Workshop on Information Theory*, pp. 33–43.
2. E. Wong, Recent progress in stochastic processes—a survey, *IEEE Trans. Inform. Theory* **IT-19**, May 1973, 262–275.
3. D. Ruelle, *Statistical Mechanics*, Benjamin, New York, 1977.
4. J. Besag, Spatial interaction and the statistical analysis of lattice systems, *J. Roy. Statist. Soc. Ser. B* **36**, 192–236.
5. B. Julesz, Experiments in the visual perception of texture, *Sci. Amer.* **232**, No. 4, April 1975, 34–43.
6. M. S. Bartlett, *The Statistical Analysis of Spatial Pattern*, Chapman & Hall, London, 1976.
7. L. Onsager, Crystal lattices. I. A two dimensional model with an order–disorder transition, *Phys. Rev.* **65**, 117–149.
8. C. Domb and M. Green (Eds.), *Phase Transitions and Critical Phenomena*, Vol. 3, Academic Press, London/New York, 1974.
9. A. B. Bortz *et al.*, A new algorithm for Monte Carlo simulation of Ising spin systems," *J. Computational Physics* **17**, 1975, 10–18.
10. P. A. P. Moran, A Gaussian Markovian process on a square lattice, *J. Appl. Prob.* **10**, 54–62.
11. T. Berger and J. K. Y. Lau, On binary sliding block codes, *IEEE Trans. Inform. Theory* **IT-23**, May 1977, 343–354.
12. E. J. Delp, R. L. Kashyap, *et al.*, Image modelling with a seasonal autoregressive time series with applications to data compression, in *Proc. of Pattern Recognition and Image Processing IEEE Conference, May 1978*, pp. 100–104.

On the Noise in Images Produced by Computed Tomography

GABOR T. HERMAN*

Medical Image Processing Group, Department of Computer Science, State University of New York at Buffalo, 4226 Ridge Lea Road, Amherst, New York 14226

Radiation passing through the human body is attenuated. The nature of the structures the radiation has passed through is indicated by the total attenuation of the radiation between its source and its point of detection. Computed tomography is a recent invention which has revolutionized diagnostic radiology. Computers are used to calculate the attenuation at individual points inside the body from a collection of total attenuations along a large number of lines (this process is called reconstruction) and to display the internal structures of the body based on this information. In this paper we discuss the nature of noise in images produced by computed tomography. Noise is taken in its most general sense: any deviation from the "true" image is considered noise. The physical sources of noise in computed tomography are considered and their effects on the images produced are illustrated. The mathematical relationship between noise in the data and noise in the reconstruction is given for a particularly popular reconstruction method. Techniques of noise suppression in computed tomography images are mentioned.

1. IMAGE RECONSTRUCTION FROM PROJECTIONS

In this section we give a concise introduction to the field of image reconstruction from projections. For details, see [1].

We define a *picture* f to be a real-valued function of two polar variables r and ϕ, whose value is 0 for all r greater than some fixed positive number E.

The Radon transform $\mathcal{R}f$ of f is defined for all real number pairs (l, θ) as follows:

$$[\mathcal{R}f](l,\theta) = \int_{-\infty}^{\infty} f\left((l^2+z^2)^{\frac{1}{2}},\quad \theta + \tan^{-1}\left(\frac{z}{l}\right)\right)dz, \qquad \text{if } l \neq 0,$$

$$[\mathcal{R}f](0,\theta) = \int_{-\infty}^{\infty} f(z, \theta + \pi/2)\,dz.$$

$[\mathcal{R}f](l, \theta)$ is the line integral of f along the line segment $L_{l,\theta}$ determined by the parameters l and θ as shown in Fig. 1.

Image reconstruction from projections is the process of estimating f from approximate values (usually obtained by physical measurements) of $\mathcal{R}f$ at a finite number of points (l_i, θ_i).

An example of this process is X-ray *computed tomography* (CT). In CT, f is the distribution of the X-ray linear attenuation coefficient (at a fixed energy $\bar{e}$)

* Present address: Medical Imaging Section, Department of Radiology, Hospital of the University of Pennsylvania, 3400 Spruce Street, Philadelphia, Pennsylvania 19104.

ISBN 0-12-597320-9

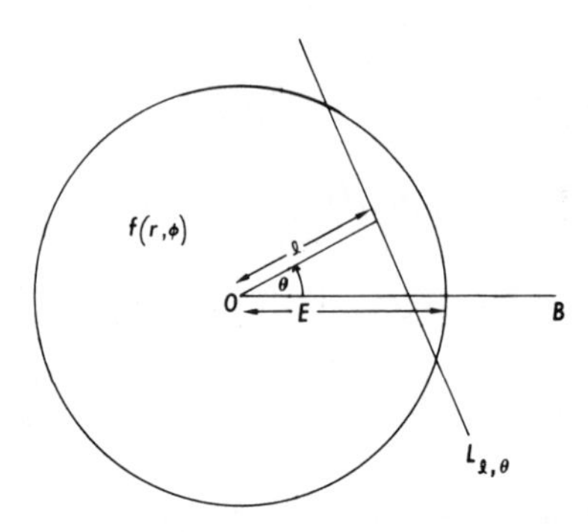

FIG. 1. A picture is defined as a function f (of two polar variables r and ϕ) whose value is zero for all r greater than E. The Radon transform $\mathcal{R}f$ of f is a function (of two variables l and θ) whose value at (l, θ) is defined as the line integral of f along the line $L_{l,\theta}$.

of the tissue in a slice of the body; see Fig. 2. It can be shown that in such a case $[\mathcal{R}f](l, \theta)$ is $-\ln \rho_{l,\theta}$, where $\rho_{l,\theta}$ is the probability that an X-ray photon of energy $\bar{e}$ moving along the line segment $L_{l,\theta}$ can get from one end to the other without being absorbed or scattered. $\rho_{l,\theta}$ is measured as the ratio of the number A of photons that get through $L_{l,\theta}$ when the body to be reconstructed is in the apparatus (A stands for *actual measurement*) to the number C of photons that get through $L_{l,\theta}$ when there is only air between the two ends of $L_{l,\theta}$ (C stands for *calibration measurement*). Due to statistical variations in photon numbers and other measurement difficulties (some to be mentioned below), A/C is only an estimate of $\rho_{l,\theta}$.

Suppose $\mathcal{R}f$ is (approximately) known for I pairs (l_1, θ_1), ..., (l_I, θ_I). For $1 \leq i \leq I$, we define $\mathcal{R}_i f$ by

$$\mathcal{R}_i f = [\mathcal{R}f](l_i, \theta_i).$$

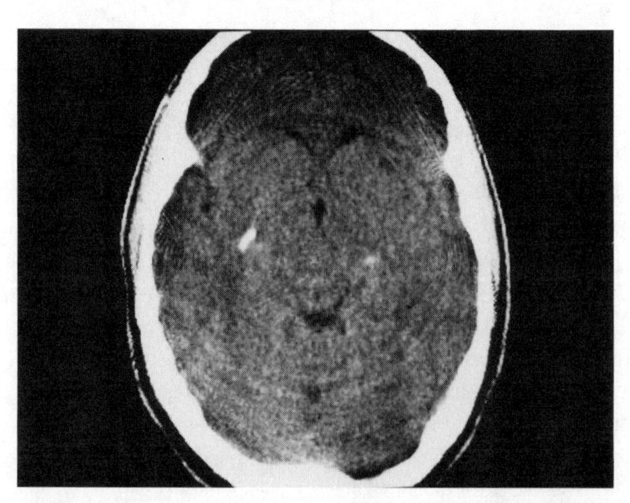

FIG. 2. Central part of a CT-produced reconstruction of a cross section of the head of a human patient. This reconstruction served as basis of the mathematically defined picture shown in Fig. 3.

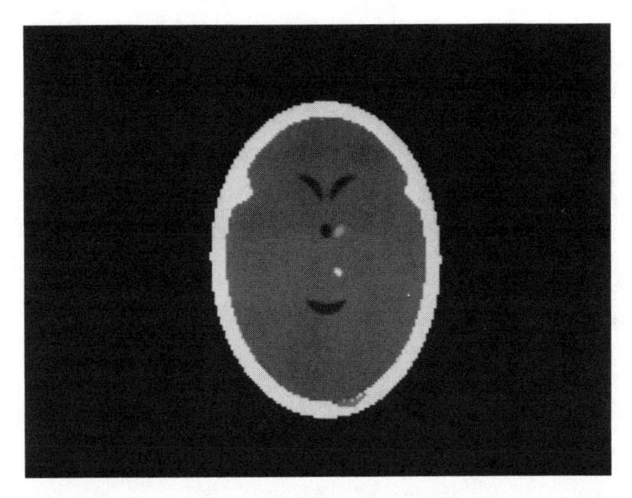

FIG. 3. A 115 × 115 digitization of the mathematically defined picture whose reconstructions from simulated measurement vectors are shown in Figs. 5–11.

(In the CT example $\mathcal{R}_i f$ is $-\ln \rho_{l_i,\theta_i}$.) We use y_i to denote the available estimate of $\mathcal{R}_i f$ and y to denote the I-dimensional vector whose ith element is y_i. (In the CT example y_i is $-\ln A/C$.) We refer to y as the *measurement vector*.

We can now restate the image reconstruction problem as:

GIVEN the data y, ESTIMATE the picture f.

An image reconstruction algorithm $\mathcal{R}^*$ produces from y a picture $\mathcal{R}^* y$, which we sometimes denote by f^* and refer to as the *image*.

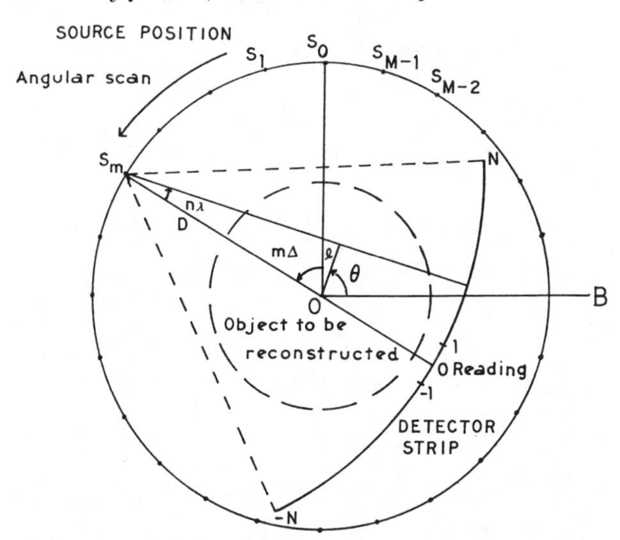

FIG. 4. Schematic of the method of data collection that was assumed in our simulated measurements. The X-ray source occupies M successive equally spaced positions $S_0, \ldots, S_{M-1}$ on a circle of radius D. The angle subtended at the center of rotation 0 by two consecutive source positions is $\Delta = 2\pi/M$. A circular strip of $2N + 1$ detectors (center of the circle at the X-ray source) moves with the source. The angle subtended at the source by two consecutive detectors in the strip is λ.

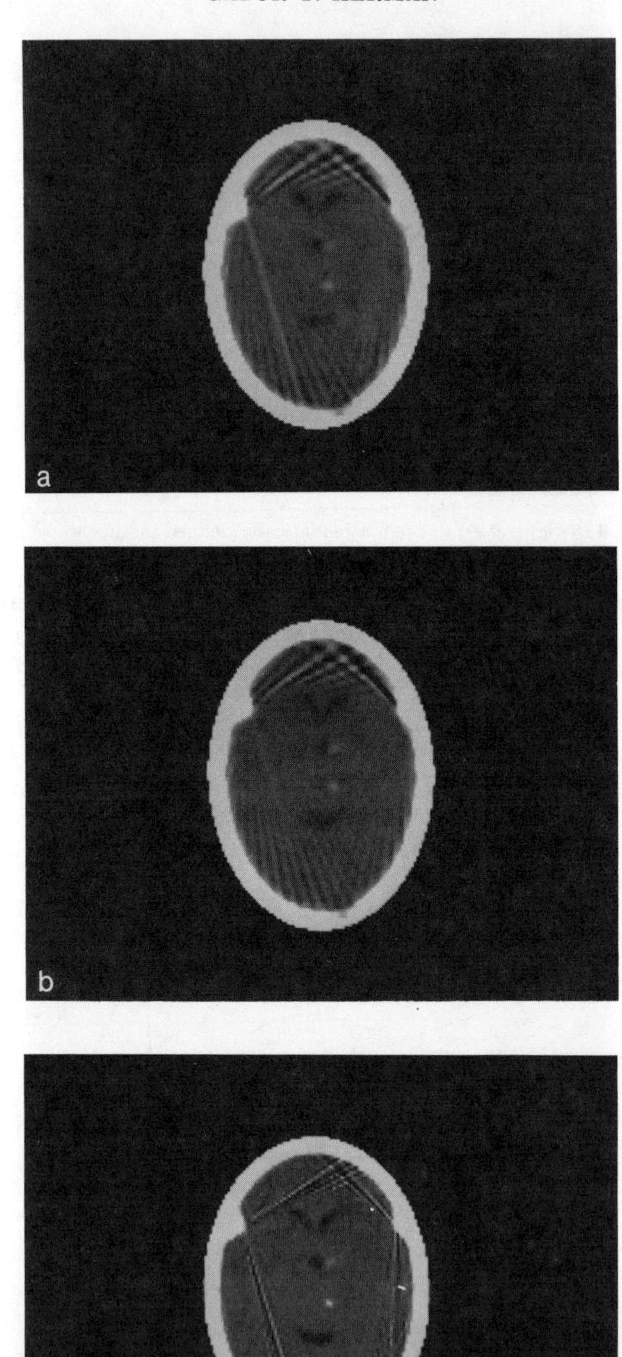
a
b
c

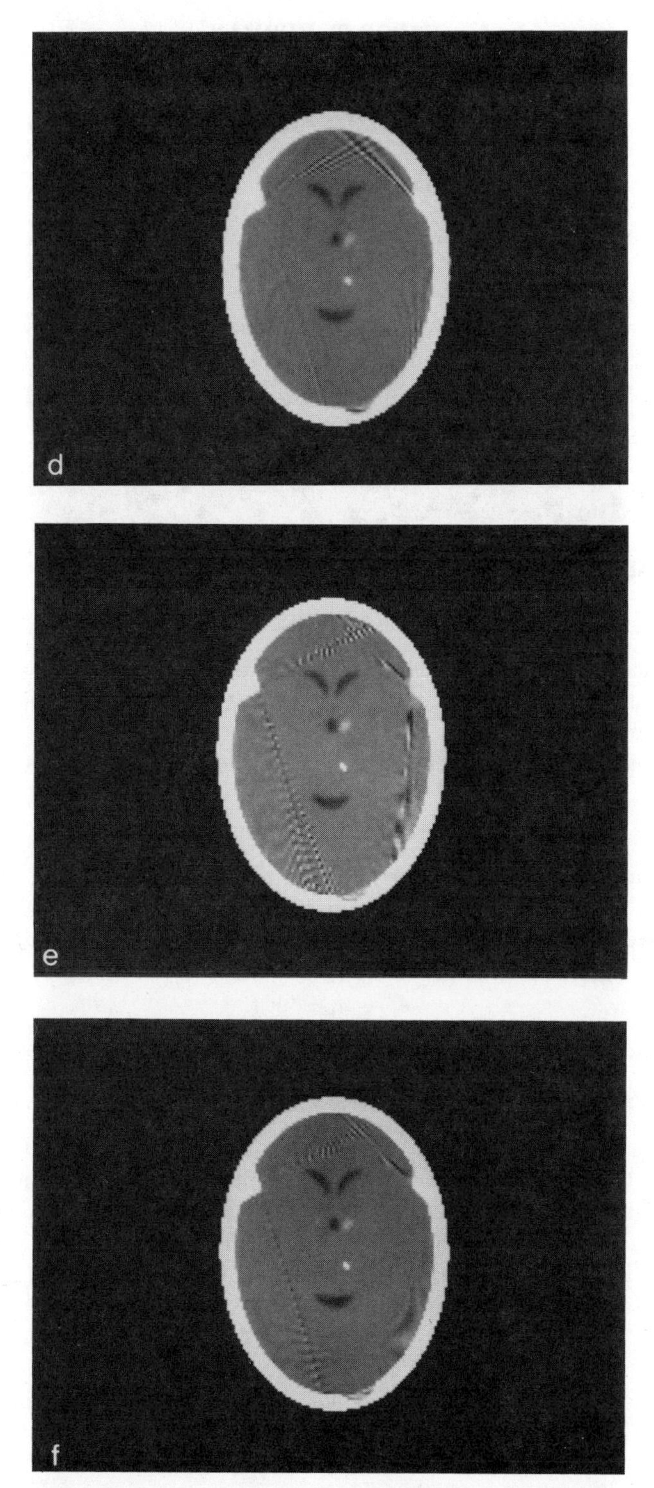

FIG. 5. Reconstruction from perfect data vectors with different values of M and N (see Fig. 4). The size of the detector strip was kept constant. (a) $M = 144$, $N = 83$. (b) $M = 288$, $N = 83$. (c) $M = 144$, $N = 165$. (d) $M = 576$, $N = 165$. (e) $M = 288$, $N = 329$. (f) $M = 576$, $N = 329$.

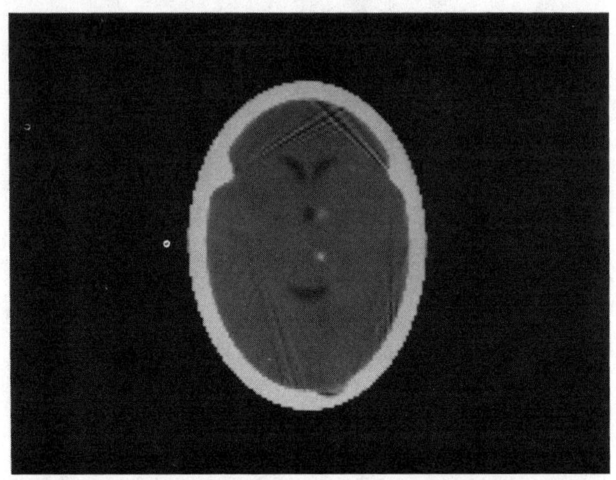

FIG. 6. Reconstruction from a perfect data vector with $M = 288$, $N = 165$. These values of M and N are used in all the reconstructions shown in Figs. 7–11.

2. NOISE IN IMAGE RECONSTRUCTION

In this section we first define noise in the all-encompassing sense that it is everything that is unwanted in the image. We then break down noise into components according to the source of the noise.

Given a picture f and its image f^*, we define the *noise in the image* to be the picture

$$n = f^* - f.$$

Thus, we call noise everything which is in the image, but not in the picture.

In order to study the noise in the image, we need to introduce the notion of *noise in the measurements*. This is defined as an I-dimensional vector e, whose ith component is

$$e_i = y_i - \mathcal{R}_i f.$$

To simplify notation we use p for the I-dimensional *perfect data vector* whose ith component is $\mathcal{R}_i f$. Then the above can be rewritten as

$$e = y - p.$$

It appears desirable to have a reconstruction algorithm $\mathcal{R}^*$ such that it reconstructs pictures from perfect projections perfectly, i.e., such that for any picture f, $\mathcal{R}^* p = f$. Alas, such reconstruction algorithms do not exist (see [1, Sect. 16.4]). Therefore, a component of the noise in the image is appropriately called *reconstruction noise* and is defined as

$$r = \mathcal{R}^* p - f.$$

The rest of the noise in the image is called *data noise* and is defined as

$$\begin{aligned} d &= n - r \\ &= f^* - \mathcal{R}^* p \\ &= \mathcal{R}^* y - \mathcal{R}^* p. \end{aligned}$$

Note that another way of defining data noise is that it is the difference picture between the reconstruction from the measurement vector and the reconstruction from the perfect data vector.

If the reconstruction algorithm is *linear*, i.e., for any measurement vectors y_1 and y_2 and for any real numbers c_1 and c_2,

$$\mathcal{R}^*(c_1 y_1 + c_2 y_2) = c_1 \mathcal{R}^* y_1 + c_2 \mathcal{R}^* y_2,$$

then

$$d = \mathcal{R}^* e,$$

i.e., the data noise is the image produced by the reconstruction algorithm from the noise in the measurements.

Clearly, the reconstruction noise is mathematically determined by the function f to be reconstructed, by the selection of the pairs $(l_1, \theta_1), \ldots, (l_I, \theta_I)$, and by the reconstruction algorithm $\mathcal{R}^*$. It does *not* depend on the mechanical and physical details of the apparatus used to collect the data. It would be nice if we could say that the other component of the noise in the image, namely data noise, depends only on the data collection apparatus and not on the function to be reconstructed. Unfortunately, this is not the case. As we shall show below, the noise in the measurements e is essentially dependent on the function to be reconstructed f. The impossibility of identification of some important components of the noise in the image which are independent of the (in practice unknown) picture to be reconstructed, makes the analysis of and correction for noise in image reconstruction very difficult.

3. ILLUSTRATIONS

All illustrations in this section are from [1], which may be consulted for details.

Based on the head cross section shown in Fig. 2, a two-dimensional distribution representing a cross section of the human head has been mathematically described as a superposition of five ellipses, eight segments of circles, and two triangles, all of different intensities. A 115×115 digitization of the resulting picture is shown in Fig. 3.

The CT X-ray data collection was simulated assuming a geometry of the kind shown in Fig. 4.

First we illustrate reconstruction noise. The perfect data vector for this was generated for different values of M and N (see Fig. 4). The so-called divergent beam convolution algorithm has been applied to these vectors (see [1, Sect. 10.1]), resulting in the images shown in Fig. 5.

Next we illustrate data noise. We used the perfect data vector with $M = 288$ and $N = 82$ and modified it in a number of ways to indicate the data noise resulting from different kinds of noise in the measurement. In each case the same reconstruction algorithm was used as previously, and so the resulting images f^* contain not only the data noise we wish to illustrate, but also the reconstruction noise that can be observed in Fig. 6, which is the reconstruction from the perfect data vector with $M = 288$ and $N = 82$.

One source of noise in the measurements is the stochastic nature of X-ray

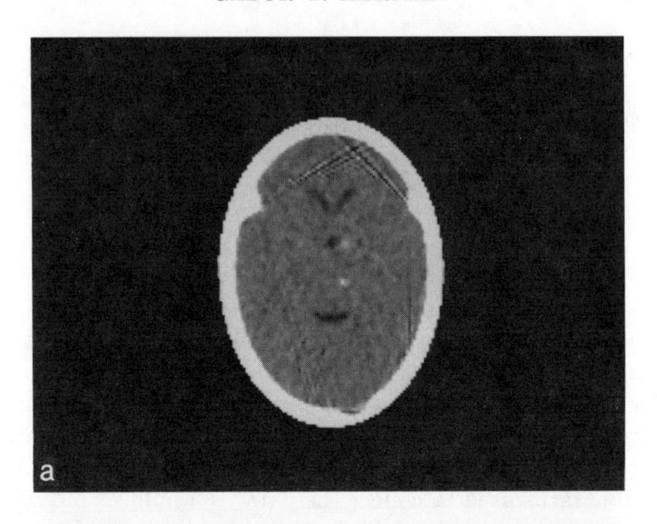
a

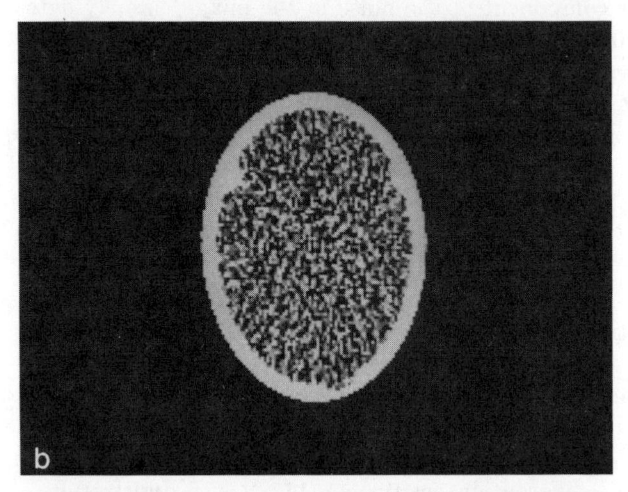
b

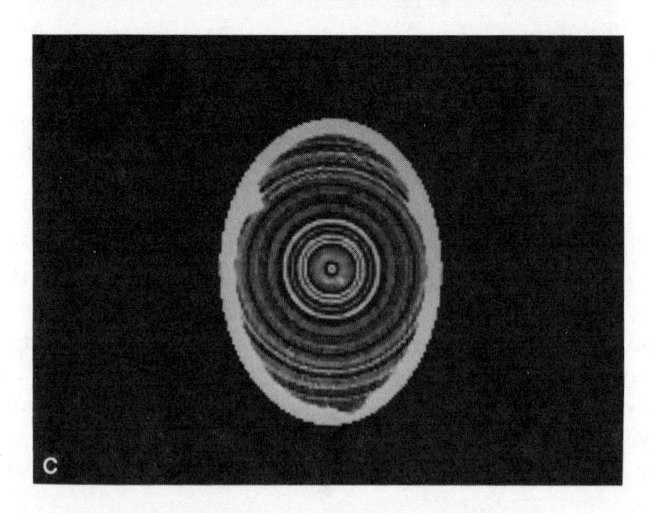
c

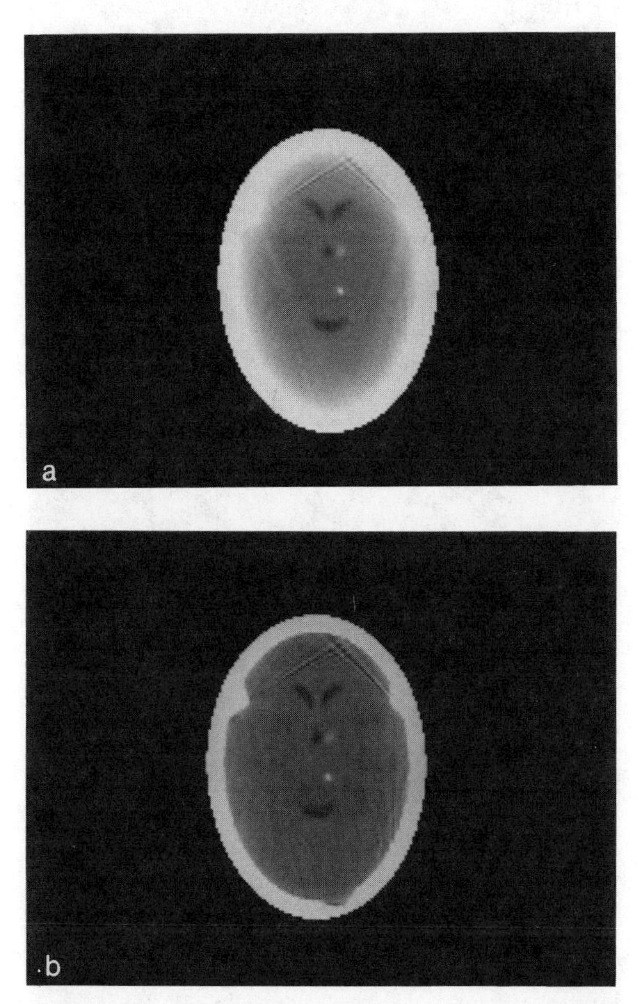

FIG. 8. (a) The effect of noise in the measurements due to the polyenergetic nature of the X-ray beam. (b) The same measurement vector has been used as in (a), but the reconstruction algorithm had an additional routine, which further processed the image shown in (a) to (partially) correct for the data noise due to the polyenergetic X-ray beam.

photon generation and interaction with matter. As a result of this, the actual measurement A and calibration measurement C are samples of Poisson random variables. If a large number of X-ray photons are used, the degradation of image quality is not great (Fig. 7a), but a substantial decrease in the number of photons makes the images essentially useless (Fig. 7b). For certain machine designs we may even get images of the kind shown in Fig. 7c; cf. Fig. 2, which shows the same type of noise, but to a lesser extent.

FIG. 7. The effects of noise in the measurements due to the stochastic nature of X-ray photon generation and interaction with matter. (a) Realistic noise levels during both actual and calibration measurements. (b) Extreme noise level during actual measurement. (c) Extreme noise level during calibration measurement.

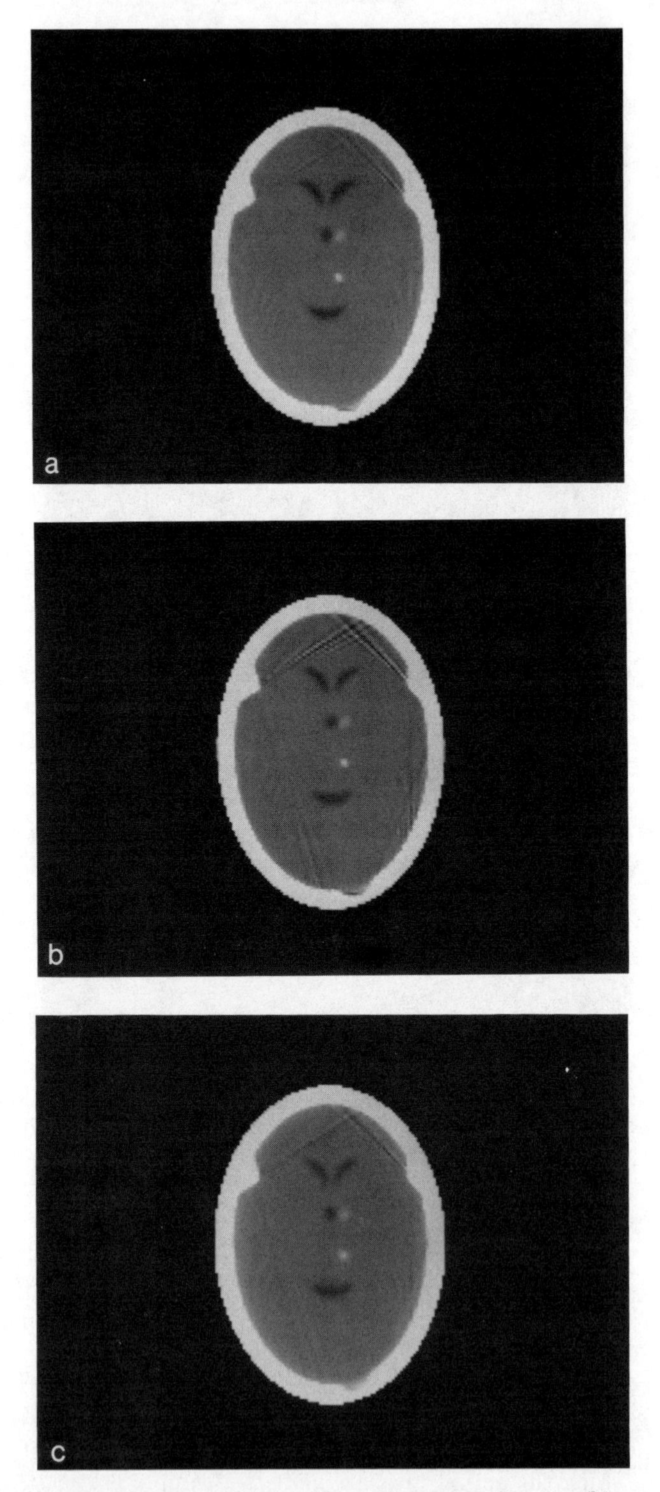

FIG. 9. The influence of detector width (a) or a small amount of scatter (b) can actually appear to be beneficial, inasmuch as some aspects of reconstruction noise are counteracted (cf. Fig. 6). A large amount of scatter (c), however, results in a blurring of edges in the image.

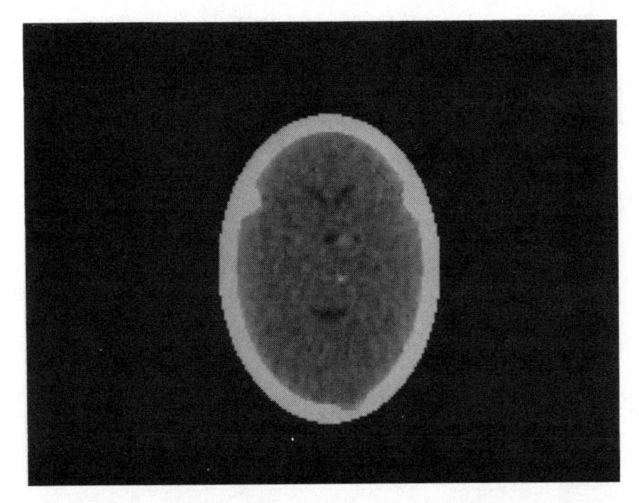

FIG. 10. Reconstruction from a measurement vector containing realistic levels of the different types of noise shown in Figs. 7–9.

The fact that X-ray sources produce photons at many different energies (rather than at a single energy) results in reconstructions of the type shown in Fig. 8a. This can be to some extent corrected for by post-processing the image, resulting in reconstructions as shown in Fig. 8b.

The fact that X-ray sources and detectors are not points, and hence the photons do not travel on a single line, influences the image only very slightly, and if anything counteracts the reconstruction noise (see Fig. 9a). The same is true for the noise in the measurements produced when detectors count photons which have scattered on their way towards other detectors (see Fig. 9b). However, if there is lots of scattering, the reconstructed skull appears noticeably wider than the skull in the original (see Fig. 9c).

Figure 10 shows a reconstruction from a measurement vector which is obtained by a realistic simulation of all the above ways that noise in the measurements is produced in CT.

Using the same measurement vector as the one used to produce Fig. 10, we have applied a number of reconstruction algorithms, to illustrate the combination of reconstruction noise and data noise. Results are shown in Fig. 11.

4. DISCUSSION

The previous section illustrated the fact that noise in the image can assume many different aspects. It is very unlikely that any simplistic analysis would lead to generally useful results. For example, noise in the measurements due to the stochastic nature of X-rays results in the components of the measurement vector y being the logarithms of the ratios of two samples from two Poisson distributions of different means. Unfortunately, the mean value of the Poisson distribution of which the actual measurement is a sample is itself dependent on the function f to be reconstructed: the denser f is, the fewer photons get through on the average. Hence, this statistical component of the noise in the measure-

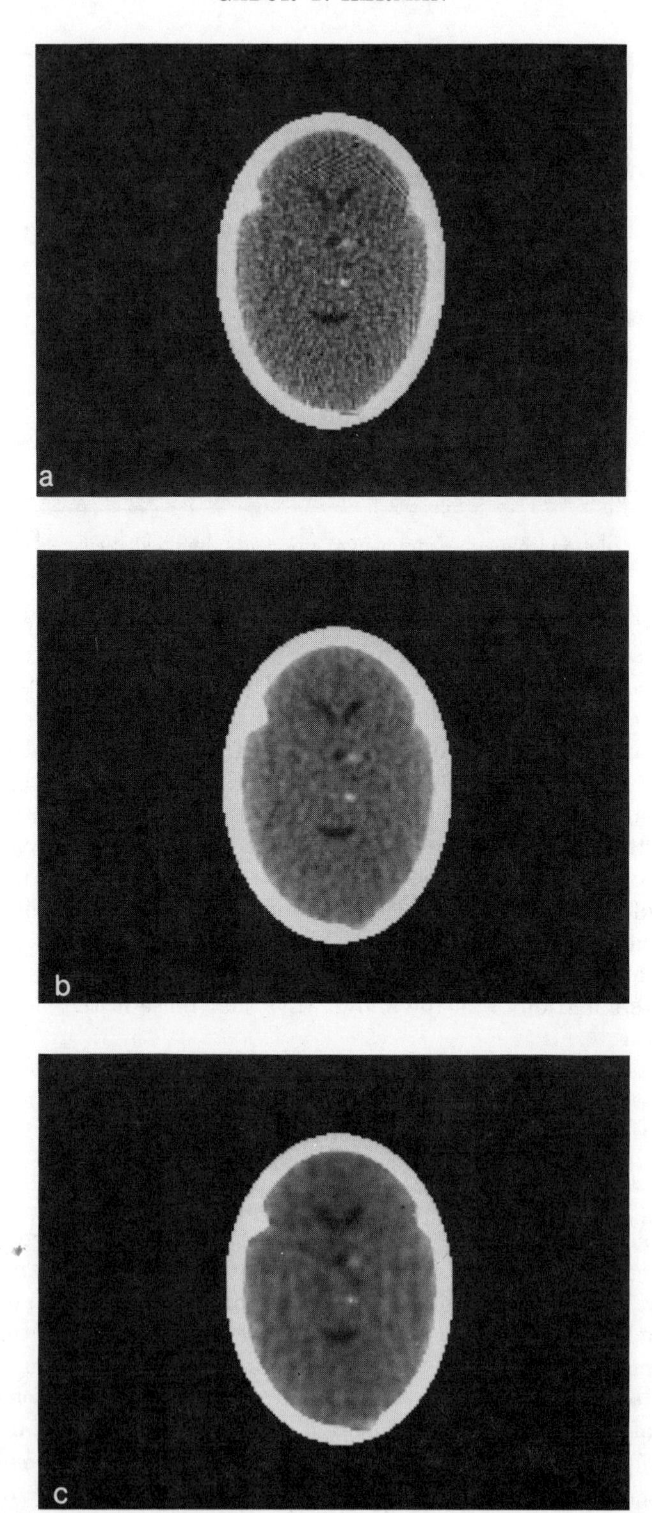
a
b
c

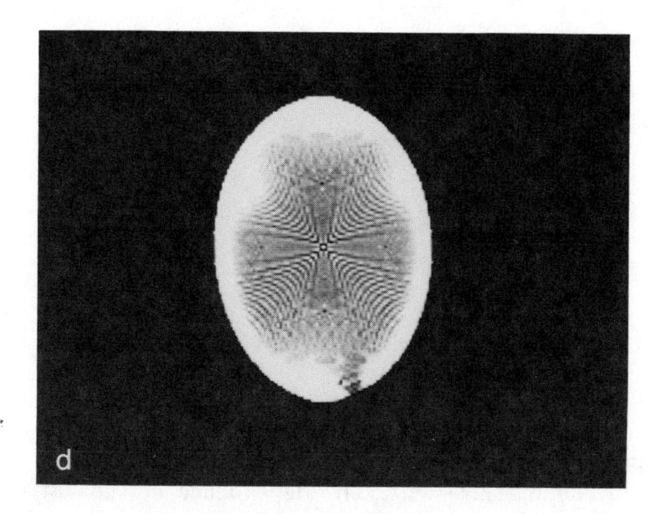

FIG. 11. Reconstructions using different algorithms on the measurement vector which was used to produce Fig. 11. (a) Divergent beam convolution algorithm with a sharp (nonsmoothing) convolving function. (b) Divergent beam convolution algorithm with a smoothing convolving function. (c) An iterative algorithm aimed at producing an image is "optimal" according to a Bayesian criterion. (d) An iterative algorithm aimed at producing an image which is "optimal" according to a criterion which is inappropriate for this application.

ments is dependent on the in practice unknown distribution f. The analysis is further complicated by the fact that the stochastic behavior of the reconstruction $f^*(r, \phi)$ at the point (r, ϕ) (in terms of the stochastic behavior of the measurement vector) is different for different values of (r, ϕ), as shown below. Here we have discussed only one of the many sources of noise, and yet a complete analysis without unrealistic simplifying assumptions appears to be very difficult.

Much work has been done regarding special cases of noise in images reconstructed from projections.

For example, the relationship between the noise e in the measurement and data noise d has been studied for specific reconstruction algorithms under certain simplifying assumptions. As has been shown above, if the reconstruction algorithm $\mathcal{R}^*$ is linear, this relationship is

$$d = \mathcal{R}^* e.$$

Suppose that each component of e is an independent sample of a zero-mean random variable whose variance is V. (This is typical of assumptions which are made for the sake of simplifying the mathematics and are not really justifiable physically.) Then the reconstructed value at a point (r, ϕ) is also a random variable, which we denote by $F^*(r, \phi)$. For the divergent beam convolution algorithm it is easy to show that the mean of $F^*(r, \phi)$ is also zero. The variance of $F^*(r, \phi)$ depends on a number of things: the variance in the data V, the number of views M, the sampling angle λ, the location of the point (r, ϕ) and the choice of two functions used in the reconstruction algorithm, the so-called convolving function q (which maps pairs of integers (n', n), $-N \leq n, n' \leq N$, into real numbers) and interpolating function ψ (which maps real numbers into real numbers). The precise formula for the variance is

$$\frac{V\lambda^2 D^2}{4M^2\pi^2} \sum_{M=0}^{M-1} \frac{1}{W^2} \sum_{n=-N}^{N} \sum_{n'=-N}^{N} [\psi(\sigma' - n'\lambda)q(n', n)]^2,$$

where

$$\sigma' = \tan^{-1} \frac{r \cos (m\Delta - \phi)}{D + r \sin (m\Delta - \phi)}, \qquad -\frac{\pi}{2} < \sigma' < \frac{\pi}{2},$$

and

$$W = \{[r \cos (m\Delta - \phi)]^2 + [D + r \sin (m\Delta - \phi)]^2\}^{\frac{1}{2}}, \qquad W > 0,$$

(see [1, Chap. 10]).

This is a complex formula, even though we have chosen a reconstruction algorithm which is quite tractable mathematically and a very simple model for the noise in the measurement. The dependence of the variance on (r, ϕ) complicates matters. They are further complicated by the fact that even if we assume that components of e are uncorrelated, the $F^*(r, \phi)$ at different points (r, ϕ) are not uncorrelated, since the reconstruction algorithm introduces some correlation between reconstructed values at different points.

What can be done to reduce noise in the image? With methods such as the divergent beam convolution algorithm one has the freedom to choose the convolving function q and interpolating function ψ. Figures 11a and 11b show reconstructions using two different convolving functions. While it is clear that data noise can be arbitrarily suppressed (extreme case: set q to be identically zero), this is usually at the expense of increasing reconstruction noise. Reconstructions can be made to conform to some optimization criterion, which contains an approximate mix of matching the measurements but suppressing the noise. Figure 11c shows an image produced according to a Bayesian optimization criterion. Such an approach is not always successful, as can be seen from Fig. 11d, which is the result produced by an algorithm designed to satisfy an alternative optimization criterion.

As can be seen from these considerations, noise in reconstructed images poses a difficult problem. Because of the practical significance of the problem, for example, in CT, there has been a large literature in recent years. We direct the attention of interested readers to [2–15], which are articles in recent issues of four publications, and to the articles referenced in them.

Many of these papers provide valid analyses of special problems; some are, on the other hand, more misleading than useful from the practical point of view. The overall conclusion of this author regarding the state of the art is the following.

In trying to find an optimal (or even good) reconstruction algorithm for a particular device to be applied to particular types of pictures, it may be useful to separate data noise and reconstruction noise for a greater insight into the problem, but in the final analysis the noise in the image has to be treated as a single entity. Since the noise in the image is highly dependent on the picture to be reconstructed, theoretical analysis is unlikely to give as useful guidance in specific situations as can be obtained by appropriate simulation studies.

ACKNOWLEDGMENTS

This work was supported by NIH grants HL4664, HL18968, and RR-7; also by NCI contract CB84235.

REFERENCES

1. G. T. Herman, *Image Reconstruction from Projections: The Foundations of Computerized Tomography*, Academic Press, New York, 1980.
2. R. A. Brooks, G. H. Glover, A. J. Talbert, R. L. Eisner, and F. A. DiBianca, Aliasing: A source of streaks in computed tomograms, *J. Comput. Assisted Tomography* **3**, No. 4, 1979, 511–518.
3. A. J. Duerinckx and A. Macovski, Nonlinear polychromatic and noise artifacts in x-ray computed tomography images, *J. Comput. Assisted Tomography* **3**, No. 4, 1979, 519–526.
4. S. C. Prasad, Effects of focal spot intensity distribution and collimator width in reconstructive x-ray tomography. *Medical Phys.* **6**, No. 3, 1979, 229–232.
5. E. T. Tsui and T. F. Budinger, A stochastic filter for transverse section reconstruction, *IEEE Trans. Nucl. Sci.* **NS-26**, No. 2, 1979, 2687–2690.
6. M. B. Katz, Rigorous error bounds in computerized tomography, *IEEE Trans. Nucl. Sci.* **NS-26**, No. 2, 1979, 2691–2692.
7. R. Alvarez and E. Seppi, A comparison of noise and dose in conventional and energy selective computed tomography, *IEEE Trans. Nucl. Sci.* **NS-26**, No. 2, 1979, 2853–2856.
8. G. T. Herman, S. W. Rowland, and M.-M. Yau, A comparative study of the use of linear and modified cubic spline interpolation for image reconstruction, *IEEE Trans. Nucl. Sci.* **NS-26**, No. 2, 1979, 2879–2894.
9. I. S. Reed, W. V. Glenn, C. M. Chang, T. K. Truong, and Y. S. Kwoh, Dose reduction in x-ray computed tomography using a generalized filter, *IEEE Trans. Nucl. Sci.* **NS-26**, No. 2, 1979, 2904–2909.
10. G. Cohen, L. K. Wagner, S. R. Amtey, F. A. DiBianca, S. F. Handel, C. Katragadda, S. Fogel, and R. G. Lester, Contrast-detail-dose evaluation of computed radiography: Comparison with computed tomography (CT) and conventional radiography, in *Proceedings of the Society of Photo-Optical Instrumentation Engineers, Application of Optical Instrumentation in Medicine VII, March 25–27, 1979, Toronto, Ontario, Canada,* Vol. 173, pp. 41–47, 1979.
11. I. J. Kalet, Uncertainties generated by computed tomography (CT) beam-hardening corrections, in *Proceedings of the Society of Photo-Optical Instrumentation Engineers, Application of Optical Instrumentation in Medicine VII, March 25–27, 1979, Toronto, Ontario, Canada,* Vol. 173, pp. 258–263, 1979.
12. G. T. Herman and R. G. Simmons, Illustration of a beam-hardening correction method in computerized tomography, in *Proceedings of the Society of Photo-Optical Instrumentation Engineers, Application of Optical Instrumentation in Medicine VII, March 25–27, 1979, Toronto, Ontario, Canada,* Vol. 173, pp. 264–270, 1979.
13. R. M. Lewitt, Aspects of the convolution method for image reconstruction from projections. *Proceedings of the Society of Photo-Optical Instrumentation Engineers, Application of Optical Instrumentation in Medicine VII, March 25–27, 1979, Toronto, Ontario, Canada,* Vol. 173, pp. 271–278, 1979.
14. K. M. Hanson, The detective quantum efficiency of computed tomographic (CT) reconstruction: The detection of small objects, in *Proceedings of the Society of Photo-Optical Instrumentation Engineers, Application of Optical Instrumentation in Medicine VII, March 25–27, 1979, Toronto, Ontario, Canada,* Vol. 173, pp. 291–298, 1979.
15. A. Fenster, D. Drost, and B. Rutt, Correction of spectral artifacts and determination of electron density and atomic number from computed tomographic (CT) scans, in *Proceedings of the Society of Photo-Optical Instrumentation Engineers, Application of Optical Instrumentation in Medicine VII, March 25–27, 1979, Toronto, Ontario, Canada,* Vol. 173, pp. 333–341, 1979.

Mathematical Models of Graphics

THOMAS S. HUANG*

School of Electrical Engineering, Purdue University, West Lafayette, Indiana 47907

1. INTRODUCTION

We shall review briefly some mathematical models for graphics as data sources, and the interplay between these models and coding schemes. The word "graphics" is used to mean images which are either strictly or nominally binary. Thus, graphics include the following three classes of images:

(1) Images which are nominally two tone. Examples are business documents (especially typewritten letters), weather maps, engineering drawings, newspaper and magazine pages, fingerprint cards.

(2) Binary images derived from continuous-tone images.

(a) To represent a continuous-tone image in digital form requires 5–8 bits per picture element (pel). We can transmit or store this multibit image in terms of its bit planes [1]. Each bit plane is a strictly binary image.

(b) Sometimes we cannot afford to transmit or store a multibit image. Then we might want to get a binary approximation of the continuous-tone image and work with that. For example, one possibility is to use the dynamic thresholding technique of Morrin [2] to reduce a continuous-tone image to a binary one.

(3) Binary images created in some continuous-tone image coding schemes.

(a) In some transform coding schemes [3], one needs to transmit or store the locations of selected transform coefficients in each block. The locations can be represented by a binary image.

(b) In contour coding schemes [4, 5], the contour information forms a binary image.

In this paper, we shall concentrate on the first image class, especially business documents and weather maps. The transmission and storage of business documents is of particular concern because of its potential commercial market.

2. MATHEMATICAL FRAMEWORK

Most past work on graphics coding [6] takes the following approach:

(1) The sampled image is quantized to 1 bit per picture element.

* Present address: Coordinated Science Laboratory, University of Illinois, Urbana, Illinois 61801

ISBN 0-12-597320-9

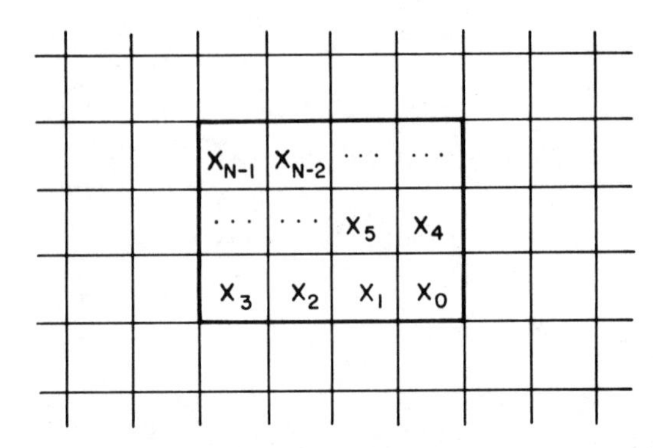

FIG. 1. A block of N pels in an image (pertaining to the definition of entropy).

(2) The source coding method preserves the binary image.
(3) Channel effects and channel coding are considered as a separate issue.

The key element of most past work in graphics coding is item (2), for which the elementary part of Shannon's information theory [7] serves as an ideal mathematical framework. What one needs here is simply the idea of entropy for a discrete source and optimum ways of encoding (Shannon–Fano code, Huffman code [8]) the output from such sources. No channel coding or rate-distortion theory is needed.

In Sections 3–6, we describe four categories of image models and their associated coding schemes.

3. JOINT PROBABILITY MODEL

We consider the digitized images as sample functions of a two-dimensional stochastic process characterized by joint probability distributions of all orders. For simplicity, we assume the images to be infinite in extent and the stochastic process to be ergodic. Then the minimum achievable bit rate is equal to the entropy of the stochastic source as defined below.

Consider a block of pels $x_0, x_1, \ldots, x_{N-1}$ as shown in Fig. 1, where each square cell represents a pel. We shall use x_i also to denote the gray level of the pel x_i, which can be either 0 (white) or 1 (black). Let the Nth-order joint probability of these N pels be $p(x_0, x_1, \ldots, x_{N-1})$. Define the Nth-order joint entropy as

$$H_N = -\sum_{x_0,\ldots,x_{N-1}} p(x_0, \ldots, x_{N-1}) \log_2 p(x_0, \ldots, x_{N-1}). \tag{1}$$

It can be readily shown that $(1/N)H_N$ is a monotone nonincreasing function of N, and that the limit of $(1/N)H_N$ as $N \to \infty$ exists. This limit is by definition the entropy of the stochastic source (in bits per pel)

$$h \triangleq \lim_{N\to\infty} \frac{1}{N} \cdot H_N. \tag{2}$$

For any finite N, $(1/N)H_N$ is an upper bound on the entropy h.

The main results of Shannon's theory are:

(1) No matter what exact coding methods are used, the bit rate can never be smaller than the entropy h.
(2) Exact coding methods do exist whose bit rates approach h.

According to the second result of Shannon's theory, exact coding methods exist whose bit rates approach the entropy of the source. In fact, one well-known way of constructing such codes is that of Huffman.

Based on the joint probability distributions, we can code in the following straightforward manner. The image is divided into blocks of N pels each. The 2^N different patterns of the block are considered as our messages and coded according to the procedure of Huffman. Then the bit rate per block, B, satisfies the inequality

$$H_N \leq B \leq H_N + 1 \tag{3}$$

and the bit rate per pel, b, satisfies

$$\frac{H_N}{N} \leq b \leq \frac{H_N}{N} + \frac{1}{N}. \tag{4}$$

It is thus obvious that we can make b as close to the source entropy h as we wish by making N large. However, for large N the implementation of the encoder and decoder becomes complicated.

To simplify the code structure, we can use suboptimum codes. One example is the white-block skipping (WBS) scheme [9]. An image is divided into $M \times N$ pel blocks. A block containing all white pels is represented by the 1-bit codeword "0." A block containing at least one black pel is represented by a $(MN + 1)$-bit codeword, the first bit being "1," and the remaining MN bits being the binary pattern of the block. The measured bit rates for six images are listed in Table 1.

The six images are labeled as follows:

A1 Typewritten letter
A2 Typewritten text (double-spaced)
A3 Circuit diagram
A4 Handwritten text
A5 Weather map
A6 Weather map

TABLE 1

Bit Rates (Bits per Pel) of WBS and Run-length Coding

Coding method	Image					
	A1 Typewritten	A2 Typewritten	A3 Circuit diagram	A4 Handwritten	A5 Weather map	A6 Weather map
WBS coding (4 × 4 blocks)	0.13	0.26	0.15	0.17	0.30	0.27
Run-length coding						
Entropy	0.06	0.15	0.11	0.09	0.23	0.21
H_1-code	0.09	0.20	0.16	0.12	0.31	0.26

These are shown in Fig. 2. The original images are all 8.2 × 7.4 in. in size. Each was sampled in 1024 × 1024 pels with a corresponding resolution of about 130 pels/in.

4. CONDITIONAL PROBABILITY MODEL

Referring again to Fig. 1, we define the Nth-order conditional entropy as

$$h_N = - \sum_{x_0, \ldots, x_{N-1}} p(x_0, x_1, \ldots, x_{N-1}) \log_2 p(x_0 | x_1, \ldots, x_{N-1}), \tag{5}$$

where $p(x_0 | x_1, \ldots, x_{N-1})$ is the conditional probability of x_0 given $x_1, \ldots, x_{N-1}$.

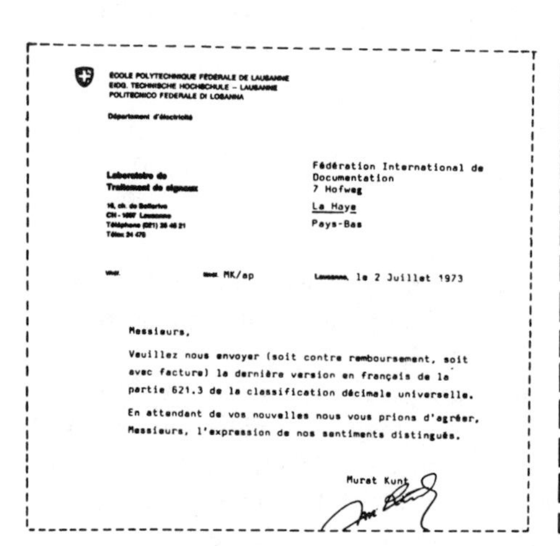

ECOLE POLYTECHNIQUE FEDERALE DE LAUSANNE
EIDG. TECHNISCHE HOCHSCHULE – LAUSANNE
POLITECNICO FEDERALE DI LOSANNA

Département d'électricité

Laboratoire de
Traitement de signaux

Fédération International de
Documentation
7 Hofweg
La Haye
Pays-Bas

MK/ap

le 2 Juillet 1973

Messieurs,

Veuillez nous envoyer (soit contre remboursement, soit avec facture) la dernière version en français de la partie 621.3 de la classification décimale universelle.

En attendant de vos nouvelles nous vous prions d'agréer, Messieurs, l'expression de nos sentiments distingués.

Murat Kunt

A1

Deux éclatements de taille se sont produits en 1968, à Paris en mai, à Prague en août, l'un pour le socialisme dans la liberté, l'autre pour la liberté dans le socialisme. Une fois dépouillés de quelques apparences et oripeaux, les deux objectifs socialisme et liberté apparaissent bien ceux de la grande majorité de l'humanité évoluée. En dehors de l'Amérique du Nord, peu nombreux sont ceux qui osent les répudier ouvertement. Du moins personne ne se prononce-t-il contre la justice sociale, ni pour la mise en condition ou en tutelle des individus, ni même pour la société de classes.

Ceux qui ont peur du socialisme ne sont pas tous des propriétaires endurcis de grandes usines ou de centaines d'hectares, mais d'accablants précédents leur font craindre pour la plus précieuse des propriétés, celle de disposer de soi-même. Et ceux que n'enthousiasme pas l'expression "monde libre" ont bien présentes à l'esprit les exactions que recouvre ce beau drapeau.

Après deux siècles de recherches, de révolutions, de théories, d'expériences en tous sens, aucun point n'apparait sur la planète, aucun îlot, où les deux objectifs socialisme et liberté soient conciliés de façon satisfaisante.

Pendant un siècle ou presque, la démocratie, appelée dans la suite démocratie bourgeoise ou démocratie occidentale, selon le degré de sympathie qui lui est porté, a vécu sous la bannière de la liberté, mais avec le remords croissant de l'injustice sociale.

Depuis un demi-siècle, a été institué un nouveau régime dont l'objectif social a relégué la liberté des individus au second plan. Qui ne refuse franchement à ces deux régimes le droit de s'appeler dé-

A2

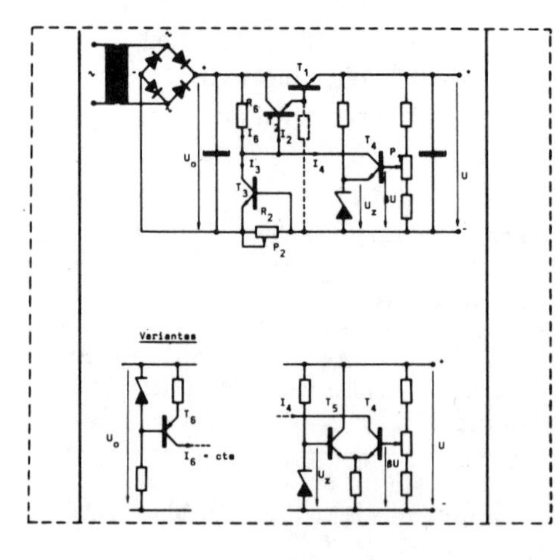

A3

GENERATEUR DE PROCESSUS DE POISSON

Fonctions générales.

A4

FIG. 2. The six images A1–A6 used for Table 1.

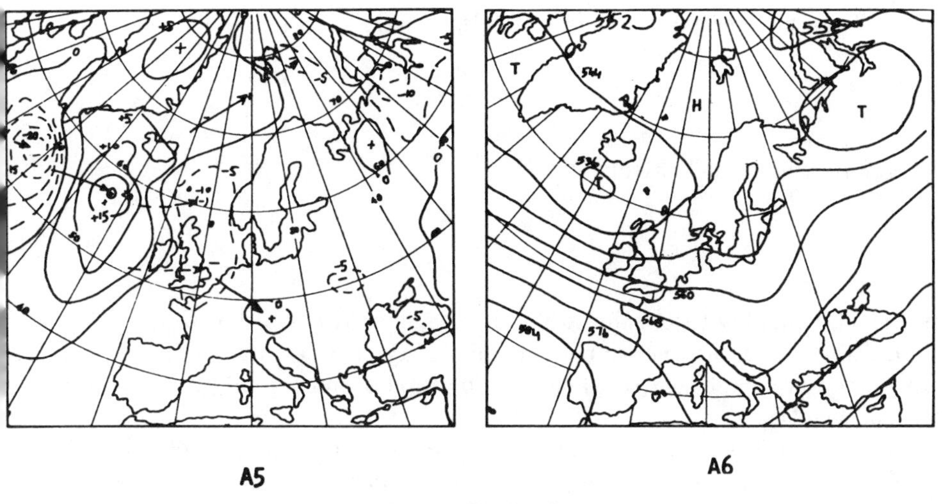

FIG. 2—*Continued.*

It can be readily shown that h_N is a monotone nonincreasing function of N, and that

$$\lim_{N \to \infty} h_N = h, \tag{6}$$

where h is the entropy of the source.

The familiar run-length coding can be considered as inspired by this conditional probability model.

Assume that a scan line contains 3 white pels followed by 2 black pels followed by 10 white pels. Then, in run-length coding, the messages we transmit are the "run lengths" 3, 2, and 10. Since white and black runs always alternate, the color of each run need not be transmitted. At both the transmitting and the receiving end, we need to store one scan line.

The message set for run-length coding contains the run lengths 1, 2, 3, . . ., N, where N is the number of pels in a scan line. To realize the advantage of run-length coding, an efficient statistical code is used for the run lengths. If we use a Huffman code based on the measured probabilities of the run lengths, $(p_1, p_2, \ldots, p_N)$, then the bit rate B in bits per run satisfies

$$H \leq B \leq H + 1, \tag{7}$$

where H is the run-length entropy

$$H = -\sum_{i=1}^{N} p_i \log_2 p_i. \tag{8}$$

The bit rate b in bits per pel satisfies

$$\frac{H}{V} \leq b \leq \frac{H}{V} + \frac{1}{V}, \tag{9}$$

where V is the average run length in pels

$$V = \sum_{i=1}^{N} i p_i. \tag{10}$$

Therefore, the per pel entropy

$$h = \frac{H}{V} \tag{11}$$

is a good estimate of the achievable bit rate using run-length coding.

In Table 1, we list the per pel entropies of the run lengths for images A1–A6. Also listed are the bit rates when a simple suboptimum code [10] was used for the run lengths.

The relation between run-length coding and the conditional probability model is as follows. Assume the image is a first-order Markov source, i.e.,

$$p(x_0 | x_1, x_2, \ldots) = p(x_0 | x_1). \tag{12}$$

Then it can be shown that the run lengths are independent [11]. Therefore, one needs only to code the run lengths independently and not worry about the correlation between them.

An extension of run-length coding using an nth-order Markov model for the image source was proposed and studied by Preuss [12]. In this model, the present pel x_0 is assumed to be statistically dependent on N previously transmitted pels. For example, for a third-order Markov image source, we have

$$p(x_0 | \text{ all previously transmitted pels}) = p(x_0 | x_1, x_2, x_3). \tag{13}$$

A particular black-and-white pattern for (x_1, x_2, x_3) is called a state of the Markov source. Thus, a third-order Markov source has eight states. Preuss' extended run-length coding scheme takes advantage of all the statistical constraints of an Nth-order Markov source by coding the runs of each state separately using codes matched to the individual run-length distributions of the states.

5. CONTOUR MODELS

It is perhaps fair to say that a majority of the current efficient coding schemes for two-tone images are based directly or indirectly on the concept of transmitting contours or boundary points.

As mentioned earlier, we assume that two-level quantization is used. Then the information in an image is completely contained in the boundary points between black and white. For convenience, let us consider the image as containing black objects on a white background. Then we have to transmit only the boundary points of the black areas. At the receiving end we reconstruct the boundaries and fill the insides with black.

An efficient way of transmitting the boundary points is to follow the boundary, i.e., to do contour tracing. Since the image is sampled, each pel has only eight neighbors. Therefore, it is sufficient to use 3 bits to indicate where the next boundary point is. Of course, to get on each boundary, we need to transmit the position of an initial point. The compression factor for contour following is approximately

$$CF = \frac{\text{Total Number of Pels in Image}}{3 \cdot (\text{Number of Boundary Points})}. \tag{14}$$

In (14), we have neglected the additional bits required to determine the initial points on the boundaries.

To do contour tracing, one needs to store the entire image at the transmitting and the receiving end, which is impractical. Schemes have been proposed which do essentially contour tracing but require much less storage. One example is PDQ [13], which requires the storage of only two scan lines.

6. PATTERN RECOGNITION MODELS

Perhaps the ultimate in efficient coding is to use the idea of pattern recognition. This is particularly appealing in the case of typewritten and printed matter, since the problem of character recognition has been studied extensively. If the transmitter can recognize the characters in the image, then it needs basically to transmit only a short codeword for each character. If we allow 100 different characters, then a codeword of length 7 bits is sufficient for each character. Additional overhead bits are necessary. For example, we need to indicate the locations of the characters. However, in most typewritten or printed matter, the positions of the characters are highly regular. The transmission of the locations of the first characters of each line should suffice. We also need to indicate the font.

We give a numerical example. Consider elite type characters. Each character occupies $\frac{1}{12} \times \frac{1}{6}$ in. At a resolution of 100 pels/in. this space contains approximately $8 \times 17 = 136$ pels. On a single character basis, the compression factor is approximately $136/7 \approx 19$. (We have neglected the overhead bits.) Note that the 7 bits per character figure is independent of resolution.

Although the scheme described above is very efficient, its application is limited. The class of images it can deal with are restricted to typewritten and printed matter with a small number of fonts and formats which the transmitter and the

TABLE 2

Dependence of Compression Factor on Resolution
(L = Linear Resolution)

Coding method	Total number of bits per image is proportional to	Compression factor is proportional to	Code book size is proportional to
PCM	L^2	1	1
Block coding	L	L	L^2
Run-length coding	$L \log_2 L$	$L/\log_2 L$	L
Contour tracing	L	L	1
Pattern recognition	1	L^2	1

receiver agree upon beforehand. Also, the implementation is expensive, since the transmitter includes a character recognition and the receiver a character generation system.

A more flexible scheme based on the idea of pattern recognition is described by Ascher and Nagy [14].

7. DEPENDENCE ON RESOLUTION

An important comparison one can make among the various coding schemes is how their compression factors depend on image resolution or sampling density. Making simplifying assumptions, we arrive at the results listed in Table 2. Also listed in the table are the codebook sizes.

8. PROSPECTS

The following areas need further investigation.

(1) The rate-distortion theory approach to graphics coding.
(2) Exploration of the pattern recognition idea.
(3) Combining source and channel coding.

In terms of applications, the two most important areas are perhaps

(a) Office document handling—at present mostly transmission, in the future also storage, retrieval, and processing.
(b) Digital handling of images in the printing industry.

ACKNOWLEDGMENT

This work was supported by the Defense Advanced Research Projects Agency under Contract F30602-75-C-0150.

REFERENCES

1. D. Spencer and T. S. Huang, Bit-plane encoding of images, in *Computer Processing in Communications*, Polytechnic Institute of Brooklyn Press, New York, 1969.
2. T. Morrin, A black-and-white representation of a gray-scale picture, *IEEE Trans. on Computers*, Feb. 1974.
3. P. A. Wintz, Transform image coding, *Proc. IEEE*, July 1972.
4. W. F. Schreiber, T. S. Huang, and O. J. Tretiak, Contour coding of images, in *Picture Bandwidth Compression* (T. S. Huang and O. J. Tretiak, Eds.), Gordon & Breach, New York, 1972.
5. J. Gupta and P. A. Wintz, A boundary finding algorithm and its applications, *IEEE Trans. Circuits Systems*, April 1975.
6. T. S. Huang, Coding of two-tone Images, *IEEE Trans. Communications*, Nov. 1977.
7. C. Shannon and W. Weaver, *The Theory of Communication*, Univ. of Illinois Press, Urbana, 1949.
8. D. A. Huffman, A method for the construction of minimum redundancy codes, *Proc. IEEE*, Sept. 1952.
9. F. deCoulon and M. Kunt, An alternative to run-length coding for black-and-white facsimilie, in *Proc. 1974 Int. Zurich Seminar on Digital Communications*.
10. T. S. Huang, Easily implementable suboptimum run-length codes, in *Proc. ICC, 1975, San Francisco*.

11. J. Capon, A probabilistic model for run-length coding of pictures, *IRE Trans. Inform. Theory*, Dec. 1959.
12. D. Preuss, Two-dimensional facsimile source coding based on a Markov model, *NTZ* **28**, H.10, 358–363.
13. T. S. Huang, Run-length coding and its extensions, in *Picture Bandwidth Compression* (T. S. Huang and O. J. Tretiak, Eds.), Gordon & Breach, New York, 1972.
14. R. N. Ascher and G. Nagy, A means for achieving a high degree of compaction on scan-digitized printed text, *IEEE Trans. Computers*, Nov. 1974.

Nonstationary Statistical Image Models (and Their Application to Image Data Compression)*

B. R. HUNT

Systems Engineering Department and Optical Sciences Center,
University of Arizona, Tucson, Arizona 85721

1. INTRODUCTION

It may not be immediately obvious, but statistical image models are frequently employed in some current procedures of digital image processing. The models are not usually made *explicit*, but are made *implicit* by the adoption of assumptions that incorporate certain model assumptions within them. To illustrate, consider the number of algorithms which employ the assumption that the image can be treated as a random process with wide-sense stationary properties. For example, image deblurring with a minimum-mean-square-error (MMSE) filter uses the wide-sense stationary assumption to derive the structure of the deblurring filter [1]. The parameters of a wide-sense stationary image model are readily specified, being the mean and autocorrelation function.

Some may object to the use of the term "model" to refer to the wide-sense stationary assumption and the associated mean and autocorrelation parameters. Nonetheless, it does satisfy most of the properties found in a textbook definition of the word "model," and we will consider the wide-sense stationary description of an image as a "model." Perhaps part of the objection to calling the wide-sense stationary description a "model" is that it is so extremely weak. This is a legitimate complaint and is at the root of our motivation for the work discussed herein. What is of interest is making this model less weak.

Little has been done in analyzing or processing images on the basis of assumptions other than a stationary process. What has been done has shown that favorable results come from abandoning the stationary process assumption. For example, Trussell developed an image deblurring algorithm that negated the assumed constancy of signal-to-noise ratio (SNR) throughout the image, and showed that superior deblurred images are produced [2]. Adaptive DPCM for image data compression allows the coefficients of the optimum predictor to vary within the image, and superior performance in data compression results [3]. In almost any situation, the virtue of abandoning the statistically stationary image

* This work was performed under the sponsorship of the U.S. Air Force Office of Scientific Research under Grant AFOSR-76-3024.

ISBN 0-12-597320-9

model can be understood if one recalls what an image that is truly stationary looks like. The most convenient example of a statistically stationary image is to tune a television set to channel where no station is broadcasting!

What is remarkable is that algorithms incorporating the stationary assumption produce usable results. Given that they work in the presence of clearly nonstationary data, we find the motivation for the ideas discussed in the remainder of this paper: to find transformations of a nonstationary image that will yield an image which satisfies the stationary model assumptions. We would then process the image by a stationary model algorithm and perform a transformation which is inverse to the original transformation to recreate the image.

2. CONVENTIONAL STATISTICAL IMAGE MODELS

The conventional statistical image model is simply stated. Let $f(x, y)$ be an image. Then the image is assumed to be described by the mean and autocorrelation statistics

$$\mu = E[f(x, y)], \qquad (1)$$
$$R(\xi, \eta) = E[f(x, y)f(x + \xi, y + \eta)],$$

where E denotes ensemble expectation. Occasionally the autocorrelation statistic will be replaced by the autocovariance

$$\Gamma(\xi, \eta) = E[f(x, y)f(x + \xi, y + \eta) - \mu^2] \qquad (2)$$

which is readily related to the autocorrelation.

Exactly how greatly an image can violate the conventional stationary assumptions is easily demonstrated. Given any image, a simple exercise is to calculate the mean of the image in blocks of $n \times n$ pixels. Ignoring the correlation between pixels, we would approximate the value of the mean in each block as a Gaussian random variable, whose variance is a function of the variance of the original pixels and the number of pixels, n^2. If the image were stationary, then paired comparison statistical tests of the means would show that no mean in any one block was statistically different from any of its neighbors (except for type I errors at the level of significance of the test, α). Performing this little experiment on any image invariably leads to failure. Almost all of the means in the blocks are statistically different from the others, even for very large values of the block size n. The reason why this occurs is obvious if one creates an image out of the means. Figure 1 is an image of size 128×128 (interpolated bilinearly to 512×512) and Fig. 2 is an image consisting of the means of Fig. 1 computed in 5×5 blocks (with the means of the 25×25 resulting data array bilinearly interpolated to 512×512). It is obvious that Fig. 2 is only a low-pass version of Fig. 1, with a recognizable relationship to Fig. 1. In local regions of the image the mean varies greatly, because the original image possessed great variations in the local value of optical intensity. Unfortunately this behavior is found in most images of any interest.

The local variation of the image mean is also seen in higher-order statistics of an image, e.g., the autocorrelation. Breaking the image up into blocks of size $n \times n$ and calculating the autocorrelation function within each block yields

FIG. 1. Original image.

autocorrelation functions which change within each block. The fact of locally changing autocorrelations is a prime reason for the development of block-adaptive methods of image data compression [3].

The failure of statistical stationary behavior for an image is also associated with another invalid assumption: the ergodic hypothesis. The derivation of the MMSE filter, for example, rests on the assumption of ensemble expectations to construct the correlation functions. However, an "ensemble of images," in the precise definition of the term, rarely exists. Common practice is to take the limited image sample (typically one image!) and carry out space-domain correlation function calculations. The correlation functions so calculated are then assumed

FIG. 2. Locally varying mean image.

to be valid in the MMSE filter. This is the essence of the ergodic hypothesis, interchanging space and ensemble averages. It is clearly invalidated by variability of the image statistics in space.

3. NONSTATIONARY STATISTICAL IMAGE MODELS

The development of nonstationary image models can be characterized through cases of increasing difficulty. The ranking we propose for the nonstationary models would be as follows:

Case 1: Nonstationary mean, stationary autocorrelation.
Case 2: Stationary mean, nonstationary autocorrelation.
Case 3: Both mean and autocorrelation nonstationary.

In symbols, the three cases would be characterized as

$$\text{Case 1:}\quad \text{Mean} = \mu_N(x, y),\quad \text{autocorrelation} = R_N(\xi, \eta); \tag{3}$$

$$\text{Case 2:}\quad \text{Mean} = \mu_N,\quad \text{autocorrelation} = R_N(x, y, \xi, \eta); \tag{4}$$

$$\text{Case 3:}\quad \text{Mean} = \mu_N(x, y),\quad \text{autocorrelation} = R_N(x, y, \xi, \eta); \tag{5}$$

where, as in Eqs. (1) and (2), (ξ, η) are lag variables in the correlation and (x, y) are coordinates in the image f. The symbolism in each case above includes the coordinates (x, y) to indicate that the specific statistic changes according to the image coordinates (x, y) *of the neighborhood* N in which the statistic is being calculated. Note the emphasis on the concept of calculating a statistic in a neighborhood. This is to emphasize that nonstationary statistics are defined in local neighborhoods of the points of an image. Since the breakdown of stationarity includes the loss of the ergodic assumption, then it is also necessary to specify the statistics in terms of spatial averages rather than ensemble averages. Thus, the neighborhood N defines the region over which spatial averaging takes place, and it is for this reason that we include the neighborhood N as part of the symbolism. Thus, in spatial averages we have

$$\mu_N(x, y) = \iint_N f(x, y)\,dx\,dy, \tag{6}$$

$$R_N(x, y, \xi, \eta) = \iint_N f(x, y) f(x + \xi, y + \eta)\,dx\,dy. \tag{7}$$

For cases of stationary behavior, of course

$$\mu_N = \mu_N(x, y),\qquad R_N(\xi, \eta) = R_N(x, y, \xi, \eta). \tag{8}$$

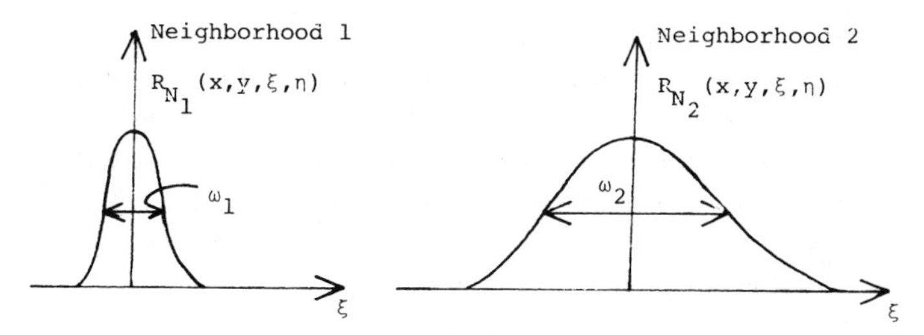

FIG. 3. Change in width of correlation function.

The mean being a single number, the breakdown of stationary behavior is readily described by tabulating the value of μ_N for each (x, y) coordinate pair of interest. However, since the autocorrelation is a function, the breakdown of stationarity is associated with the ways in which the correlation function itself may change for each (x, y) coordinate pair of interest. There are three principal *attributes* of the correlation function which can vary with location. The three attributes which we identify as capable of being space-variant are the following:

(1) *Energy*. The correlation function at $\xi = \eta = 0$ is (by definition) the mean-square energy of the image within the neighborhood N, i.e.,

$$R_N(x, y, 0, 0) = \sigma_N^2(x, y) + \mu_N^2(x, y), \tag{9}$$

where σ_N^2 is the variance of the neighborhood. The mean-square energy establishes the *vertical scale* of the correlation function.

(2) *Width*. The correlation function may change in width as a function of (x, y), which we illustrate in Fig. 3 with a one-dimensional plot along a single variable (ξ) for two neighborhoods.

(3) *Shape*. The correlation function may change in shape as a function of (x, y), which we illustrate in Fig. 4. Note that the widths at half-maximum are equal, $\omega_1 = \omega_2$, even though there are distinct differences in the shapes of the correlation functions.

Variation in image space of any of these three attributes would be sufficient to invalidate assumptions of stationary behavior. In the next section we consider methods of transforming an image so that the attributes become stationary.

We note that the choice of attributes for space variability is limited by our acceptance of stationary behavior as exemplified by "wide-sense" stationarity,

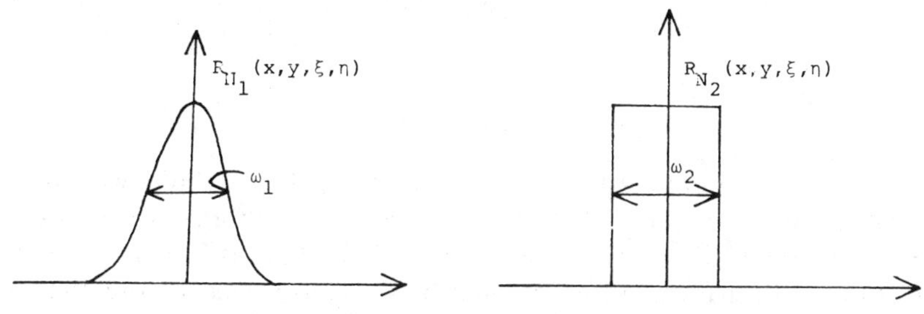

FIG. 4. Change in shape of correlation function.

i.e., the first two statistical moments of the random process. True stationarity involves all moments of a process, and would necessitate an infinite set of attributes changing in image space. Note that even with the three attributes chosen above it is possible to construct subattributes. For example, "shape" could be decomposed into subattributes such as monotonicity, convexity, compact versus extended shapes. We have not done so because of the prospects for normalizing shape, as we discuss in the next section.

4. TRANSFORMATION TO STATIONARY BEHAVIOR

For the obvious reasons of mathematical tractability, it is desirable to employ stationary image models. Thus, we are led to inquire about the prospects of transforming a nonstationary image model into a stationary one.

Suppose that Case 1 of the previous section prevails, i.e., stationary autocorrelation but nonstationary mean. The transformation

$$f_N(x, y) = f(x, y) - \mu_N(x, y) \tag{10}$$

will create an image which has a stationary (zero) mean *with respect to the neighborhood N*. Choice of a different neighborhood would require a different mean to be subtracted. A key question in the transformation of Eq. (10) is the relative invariance of the operation to the specific choice of N. Obviously, the invariance is dependent upon the size and shape of the neighborhood N and the specific image f as well; this is a topic which merits research.

Consider Case 2 discussed above, stationary mean and nonstationary autocorrelation. The breakdown in stationarity may be any one (or all) of the three phenomena discussed immediately above. A transformation to produce stationary behavior of the energy is straightforward:

$$f_N^{(1)}(x, y) = \frac{f(x, y)}{R_N(x, y, 0, 0)} \tag{11}$$

which yields an image normalized to unit energy with respect to the neighborhood N.

Treatment of the other two possibilities for variation in the autocorrelation is not so simple. The most difficult case is the third above, i.e., where the autocorrelation varies in shape. *In principle*, any autocorrelation shape can be synthesized, using the power spectrum relations:

$$\Phi_g(\omega_x, \omega_y) = |H(\omega_x, \omega_y)|^2 \Phi_f(\omega_x, \omega_y), \tag{12}$$

where g, h, f are related as

$$g(x, y) = h(x, y) ** f(x, y) \tag{13}$$

with h being the point-spread function of a filter. Φ_g, Φ_f are the power spectra of the associated processes, i.e., the Fourier transforms of the associated autocorrelation functions. The utilization of Eq. (12) is straightforward. If Φ_g represents the Fourier transform of the "standardized" autocorrelation function, which must be uniform throughout the image, then a digital filter which produces the standardized correlation can be constructed from the relation

$$|H(\omega_x, \omega_y)|^2 = \frac{\Phi_g(\omega_x, \omega_y)}{\Phi_f(\omega_x, \omega_y)}. \tag{14}$$

Equation (14) provides the *theoretical* grounds for "standardizing" the autocorrelation shapes.

Note in the previous paragraph we stated that "in principle" the above relations provide the "theoretical" grounds for transforming an image to a common autocorrelation throughout the frame. In practice we believe the prospects for using Eq. (14) are limited, for the following reasons. *First*, the quotient in (14) need not be well behaved; if Φ_g is more wide-band than Φ_f, then small denominator values can lead to gross amplification of any noise. A solution to this problem is to make the standard Φ_g more narrow-band than any spatial segments of the image. Since this implies low-pass filtering and loss of image detail, such a solution is considered to be unacceptable. *Second*, a transformation such as the above must be invertible (as we shall see below) and inversion of filters on images falls in the class of problems known as image restoration. This, in general, is a very difficult class of problems, and to willingly induce a transformation of this sort should be considered unwise [1].

Fortunately, there is reason to believe that the transformation to a standard autocorrelation/power-spectrum shape is unnecessary. Considerable studies of imagery for the purpose of adaptive data compression has shown that an image can be successfully treated as a single random process/autocorrelation model, with the parameters of the model varying in the image space. For example, adaptive DPCM is a successful data compression scheme because of the tendency of the image to fit a simple Markov process model (first- to third-order Markov), where the process parameters change within the image [3]. Therefore, we will assume that the necessity to normalize the autocorrelation function is not present. Instead, we will assume a simple first-order Markov model for the image process. The autocorrelation function of such a model is given as

$$R(\xi, \eta) = \sigma_f^2 \exp(-\rho(\xi^2 + \eta^2)^{\frac{1}{2}}) \tag{15}$$

which is a form chosen for rotational symmetry, i.e., no spatially preferential directions, and not spatially separable.

The assumed model for the image autocorrelation gives a space-variant form

$$R_N(x, y, \xi, \eta) = \sigma_{fN}^2(x, y) \exp(-\rho_N(x, y)(\xi^2 + \eta^2)^{\frac{1}{2}}), \tag{16}$$

in which the parameters varying with respect to the neighborhood N are the variance σ_{fN}^2 and the correlation parameter $\rho_N(x, y)$. Since we have discussed above a transformation to normalize energy (and hence variance), we can assume the simpler form

$$R_N(x, y, \xi, \eta) = \exp(-\rho_N(x, y)(\xi^2 + \eta^2)^{\frac{1}{2}}), \tag{17}$$

which leaves only the correlation parameter to vary with the neighborhood N and location (x, y) of N. Clearly, we are assuming a case in which the correlation

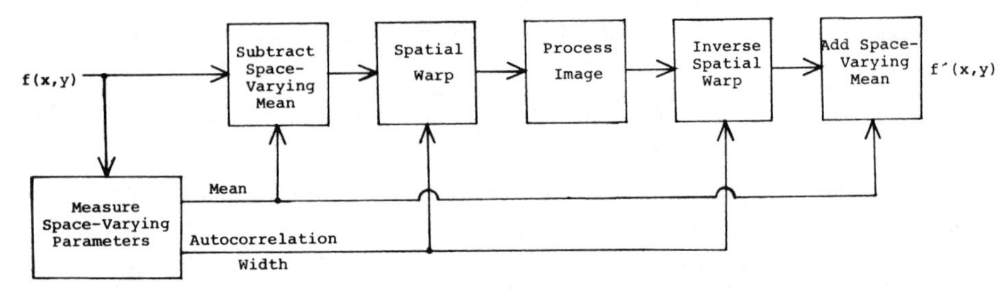

FIG. 5. A system for processing nonstationary images.

shape remains constant, with correlation width varying in space. (See the previous section.)

Under these assumptions, the problem of a transformation to render the image stationary becomes simple. One possibility is, of course, the filtering posed in Eq. (14). This we consider unacceptable, for the reason discussed above. Instead, we note the following. Since the correlation shape is assumed constant, then differences such as Fig. 3 can be equated through a scale factor. That is, the widths ω_1 and ω_2 in Fig. 3 are related as

$$\omega_1 = \alpha\omega_2. \tag{18}$$

A resampling (interpolation) of the image data to incorporate the scale factor α makes the correlations of equal width and produces stationary statistics.

We can now describe a transformation, given a first-order Markov model, to produce stationary image statistics. For neighborhoods of size N, calculate autocorrelation functions and estimate the parameters ρ_N of correlation for the neighborhoods. The changes in ρ_N are then used as control points in a least-squares fit for a spatial-warp polynomial [3]. The transformed image is obtained by interpolating the original image via the warp to produce an image with constant ρ parameter throughout.

5. A SYSTEM FOR PROCESSING NONSTATIONARY IMAGES

Figure 5 is a schematic diagram of a system that will carry out the processing of nonstationary images. The system incorporates the model of a nonstationary mean and nonstationary autocorrelation, with the nonstationary autocorrelation model being that of Eq. (17). The first stage of the model measures the spatial variation in model parameters. Using the spatial variation measurements, the successive stages correct for the nonstationary mean and autocorrelation. The nonstationary autocorrelation is corrected by a spatial warp. Following the correlations, the image is processed, through a given algorithm such as deblur or data compression, using stationary statistical assumptions. Then the transformations which created the stationary image are inverted, i.e., inverse spatial warp and reinsertion of the nonstationary mean.

6. EXAMPLES OF APPLICATIONS OF NONSTATIONARY MODELS

In the following we will present two different examples of the application of a nonstationary image model to particular image processing problems.

The first problem we consider is that of image restoration. Image restoration under optimum conditions usually incorporates assumptions that require stationary image statistics. For example, the most common image restoration method is the minimum-mean-square-error (MMSE) or Wiener filter. The filter is derived under an image formation model

$$g(x, y) = \int_{-\infty}^{\infty} \int_{-\infty}^{\infty} h(x - x_1, y - y_1) f(x_1, y_1) dx_1 dy_1 + n(x_1, y_1) \tag{19}$$

and is based upon the MMSE criterion

$$\underset{\{l(x,y)\}}{\text{Minimize}}\ E\{[f(x, y) - \hat{f}(x, y)]^2\}$$

where

$$\hat{f}(x, y) = \int_{-\infty}^{\infty} \int_{-\infty}^{\infty} l(x - x_1, y - y_1) g(x_1, y_1) dx_1 dy_1. \tag{20}$$

The solution of this problem yields the Fourier domain description of the MMSE filter as

$$L(\omega_x, \omega_y) = \frac{H^*(\omega_x, \omega_y)}{|H(\omega_x, \omega_y)|^2 + (\Phi_\eta(\omega_x, \omega_y)/\Phi_f(\omega_x, \omega_y))} \tag{21}$$

where Φ_η and Φ_f are the power spectra of noise and signal, respectively. The power spectra are defined, of course, only under spatially stationary statistical assumptions, and these are the assumptions required to solve the problem in the form of a convolution as in Eq. (20).

The solution of the problem in the form above requires only a simple model: the image formation model of Eq. (19) plus spatially stationary statistics. In particular, no assumption concerning the distribution of amplitude values of the image has been made. By making amplitude distribution assumptions we can significantly expand the sophistication of the image restoration model.

First, we represent the image formation model in terms of discrete operations. If we assume the image is sampled (at the appropriate Nyquist rate) then it is direct to show that the image formation model of Eq. (20) can be expressed as a vector-matrix operation

$$\mathbf{g} = [H]\mathbf{f} + \boldsymbol{\eta}, \tag{22}$$

where the matrix $[H]$ has special structure, e.g., block Toeplitz [1].

Equation (22) is a sampled form of a linear image formation model. It is possible to relax the assumption of linearity, in the case of image sensors which are nonlinear (e.g., photographic film). A model for image formation and recording by a nonlinear sensor thus has the model

$$\mathbf{g} = s\{[H]\mathbf{f}\} + \boldsymbol{\eta} \tag{23}$$

where $s\{\ \}$ is a point nonlinearity, such as the $D - \log E$ curve of a photographic film [1].

We have greatly complicated our image formation model. We now introduce an amplitude statistics model and allow for a major nonstationary model component. We will assume that image statistics of the vector **f** can be modeled by a multivariate Gaussian probability density function

$$p(\mathbf{f}) = K \exp(-(\mathbf{f} - \bar{\mathbf{f}})^T[R_f]^{-1}(\mathbf{f} - \bar{\mathbf{f}})), \tag{24}$$

where K is the standard normalizing constant, $[R_f]$ is the covariance matrix of the multivariate image process, and $\bar{\mathbf{f}}$ is the nonstationary mean of the image. The covariance matrix $[R_f]$ can be readily related to the correlation function of the process [1], under the assumption that **f** is stationary in correlation although not in the mean.

By adopting a solution criterion of finding the vector which maximizes the posterior probability density constructed from the prior density of Eq. (24), it is possible to construct an iterative solution to the restoration problem of Eq. (23). See [4] for details and examples of solutions. We note that the model of Eq. (23), incorporating nonlinear image formation/recording processes, cannot be directly solved in an optimum fashion (e.g., maximum a posteriori probability). It is the introduction of our Gaussian statistical model, with the nonstationary mean, which makes possible an optimum solution of the nonlinear problem.

It is relatively simple to construct the nonstationary mean $\bar{\mathbf{f}}$ in Eq. (24). Since the basic behavior of the nonstationary mean is to change with spatial location, then an estimate of the nonstationary mean can be constructed by forming a local average, e.g., convolve the sampled image **f** with the sampled point-spread function of a low-pass filter

$$\bar{\mathbf{f}} = [L]\mathbf{f}, \tag{25}$$

where $[L]$ is the matrix description for the convolution with the low-pass filter.

It is legitimate to inquire whether the simple process of Eq. (25) constructs a mean that has any relation to Gaussian assumptions. The following experiment is relevant. Construct an image which is the difference between the image and the nonstationary mean $\bar{\mathbf{f}}$ constructed as in Eq. (25), i.e.,

$$\mathbf{d} = \mathbf{f} - \bar{\mathbf{f}}. \tag{26}$$

Then tabulate the amplitude histogram of the difference image **d**. The amplitude histogram is surprisingly close to a Gaussian density function, as demonstrated in [9], for virtually any type of image.

Although the model of a Gaussian density for image amplitude statistics is convenient to simplify the mathematics and obtain a solution for Eq. (23), it is not a gross distortion for amplitude statistics of difference images constructed as in Eq. (26). Thus, the assumed Gaussian model, with nonstationary mean, is useful and reflects something actually observed in experiments with real data.

The above example reflects a simple image model for the problem of image restoration. We now consider another common problem in image processing: image data compression.

The importance of nonstationary image statistics has been long recognized in image data compression. The most powerful image data compression schemes are those known as adaptive, e.g., adaptive transform coding or adaptive DPCM coding. Adaptive compression schemes have the property of being able to vary

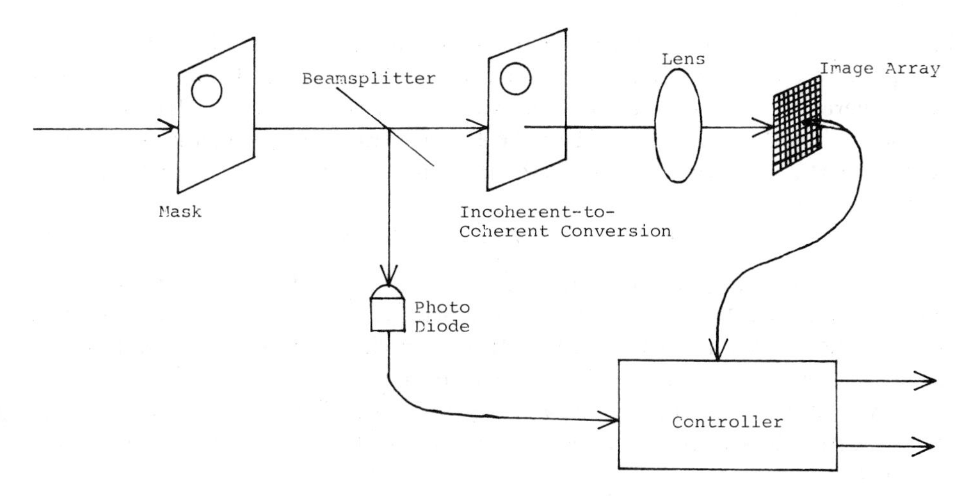

FIG. 6. Schematic for hybrid digital/optical system to measure space-varying parameters.

the important parameters of the compression algorithm as a function of localized behavior within the image. For example, an adaptive transform code will calculate the image transform in a given block, say 16 × 16 pixels, and then vary the assignment of code bits to the transform values as a function of the image structure within the block [3].

The general structure of the block diagram in Fig. 5 is suited for an image data compression system. The middle block in the diagram would be replaced by a transform or DPCM data compression system which had constant operating parameters, e.g., was spatially nonadaptive. Since the general structure of Fig. 5 is suited for an adaptive data compression system, we wish to examine a particular implementation. It is straightforward to envision hybrid digital/optical hardware that can implement the various stages in the block diagram at extremely high data rates. This is most advantageous since the spatial warp operation can be a very costly process when implemented by digital computation.

We consider first the box in Fig. 5 which derives all the critical information for the remainder of the system, i.e., the box with the function "measure space-varying parameters." Two parameters are the outputs of this box, the space-varying mean and the space-varying autocorrelation width. Both of these quantities must be measured in local image regions. Figure 6 shows the schematic outline of an optical system which can do this. The incoming image is intercepted by a mask. This is an electrooptic device capable of being spatially programmed for either full or zero transmission, for example, a Hughes liquid crystal or PLZT crystal. An aperture of full transmission is written onto the mask at the position in the image plane where it is desired to take a space-varying measurement; the rest of the mask is written at zero transmission. The beam-splitter diverts a portion of the masked image to a photodiode and the output of the photodiode is, by definition, the integrated or mean image intensity over the region selected by the mask.

To measure the other parameter, which is autocorrelation width, we use the Fourier transform properties of coherently illuminated lenses [6]. Since the input scene will almost always be observed in incoherent illumination, it is necessary to convert the masked portion of the input image from an incoherent to a coherent field. This can be done with a device such as the Hughes liquid crystal. A transform lens calculates the Fourier transform, which is sensed by a discrete array detector. The detector responds to intensity i.e., the detector output is

$$F_D(\omega_x, \omega_y) = F(\omega_x, \omega_y)F(\omega_x, \omega_y)^*, \tag{27}$$

where F is the complex field amplitude and F_D is the detected transform. The detector thus observes the Fourier transform of the autocorrelation function. Since we have assumed a first-order Markov model for the autocorrelation (see Eq. (17)), then the width of the Fourier spectrum observed on the detector is directly related to the parameter ρ_N in Eq. (17) [7]. Thus, the controller can use the measured data from the detector to calculate the space-varying autocorrelation width.

With the space-variant measurements completed the next step is to carry out the operations which adjust the space-variant image properties. First, we cannot directly subtract the space-variant mean, because an optical system such as described here could not deal with any negative light values which could arise. Instead we add to the image a quantity which is the difference between the space-variant mean and a bias sufficient to create everywhere positive light. Thus, the output of this adjustment would be

$$f_N(x, y) = f(x, y) - \mu_N(x, y) + b \tag{28}$$

which differs from the original in Eq. (10) only by the bias. Obviously, we would implement processing by adding the space-variant bias

$$f_N(x, y) = f(x, y) + b_N(x, y), \tag{29}$$

where

$$b_N(x, y) = b - \mu_N(x, y). \tag{30}$$

Finally, the spatial warp must be implemented. Warping is literally a "rubber-sheet" transformation. We conceive of the image as being on a sheet of rubber and find a mapping of coordinates to distort the sheet in some desired way. Let the coordinates of the original image be x, y and the coordinates of the warped image be x', y'. Then we assume a polynomial coordinate warp [13]

$$\begin{aligned} x' &= a_0 + a_1x + a_2y + a_3x^2 + a_4x^2 + a_5xy + \cdots, \\ y' &= b_0 + b_1x + b_2y + b_3x^2 + b_4y^2 + b_5xy + \cdots. \end{aligned} \tag{31}$$

The problem is to specify the polynomial order and the coefficients. Current practice has shown that a third-order polynomial is usually sufficient, and the coefficients can be calculated by least-squares techniques [3]. The calculation of least-squares coefficients requires control-point pairs, i.e., pairs in the original and warped images which are the same pixel. This is simple for our problem, since we only wish to change image scales to make the autocorrelation length constant in the warped image. Thus, we would make a two-dimensional map of the autocorrelation width, and from the map determine control-point pairs. For

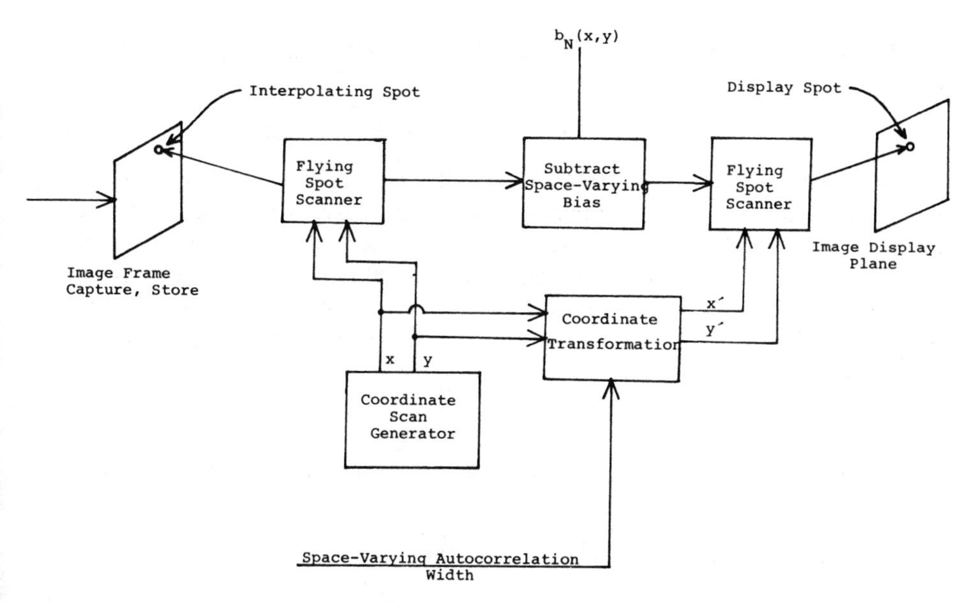

FIG. 7. System for electrooptic warping of imagery.

example, the variation in correlation length could be mapped to a constant value equal to the largest correlation length; such a choice would require interpolating extra pixels into image regions which possessed a smaller correlation length.

Digital computer implementation of the warp is a very costly process due to the interpolation calculations and the extensive I/O requirements resulting from the distortion of the image space. However, it can be implemented very easily by electrooptics. Interpolation is only a weighted sum of pixels in a region, and a flying-spot scan of an image plane can interpolate if the spot profile (apodization profile) is equal to the interpolator weights. Likewise, the mapping of coordinates in Eqs. (31) can be readily performed by either analog or digital hardware. As the scanning spot samples one image plane it extracts the interpolated pixel from coordinate x, y, the coordinates x, y are transformed into coordinates x', y' via Eqs. (31), and then x', y' are used to position a writing spot in a new plane.

Figure 7 shows a block-diagram schematic to carry out this process. The input scene is imaged on an electrooptic device which is capable of storing the image frame over the time required for processing. In the frame store a flying spot scans, under control of the coordinate generator, extracting from the spot read-out the interpolated image pixel values. The coordinate generator also passes values to a simple computer which generates new coordinate pairs in the warped plane, based on coefficients calculated from the autocorrelation width values. The new coordinate pairs drive a flying spot scan which writes the warped image in the output of an electrooptic image display. The value written to the output display is first modified by Eq. (28), to correct for the space-varying mean.

The image in the output display plane can now be the input to a bandwidth compression process, e.g., transform or DPCM, which is nonadaptive. In fact, the transform process can have fixed parameters and the warping system in Fig. 7

can be operated to adapt the image to the fixed parameters of the compression process.

7. CLOSING REMARKS

The introduction of nonstationary statistics for image models is a logical step, when we consider how many image processing analyses use stationary models in the face of certain nonstationarity in the image data. We have seen herein how such assumptions can be introduced into simple models and how applications of such models can be quickly derived. The exploitation of the models in special-purpose hardware is a prospect that could make such models of great utility for applications in real problems.

REFERENCES

1. H. C. Andrews and B. R. Hunt, *Digital Image Restoration*, Prentice–Hall, Englewood Cliffs, N.J., 1977.
2. H. J. Trussell and B. R. Hunt, Sectioned methods for image restoration, *IEEE Trans. Acoustics Speech Signal Processing* **ASSP-26**, 1978, 157–164.
3. W. K. Pratt, *Digital Image Processing*, Wiley, New York, 1978.
4. B. R. Hunt, Bayesian methods in nonlinear digital image restoration, *IEEE Trans. Computers* **C-26**, 1977, 219–229.
5. B. R. Hunt and T. M. Cannon, Nonstationary assumptions for Gaussian models of images, *IEEE Trans. Systems Man Cybernet.* **SMC-6**, 1976, 876–882.
6. J. W. Goodman, *Introduction to Fourier Optics*, McGraw–Hill, New York, 1968.
7. A. Papoulis, *The Fourier Integral and Its Applications*, McGraw–Hill, New York, 1964.

Markov Mesh Models

LAVEEN N. KANAL

Laboratory for Pattern Analysis, Department of Computer Science, University of Maryland, College Park, Maryland 20742

1. INTRODUCTION

Markov Mesh models presented in Abend, Harley, and Kanal [1] sought to incorporate spatial dependence in reducing the complexity of likelihood functions for image classification. This study of near neighbor spatial dependence anticipated the later interest in Markov Random Fields (MRF) in image modeling. Here, I summarize the Markov Mesh presentation as originally given in [1] and comment on some related references and developments of MRF models.

2. THE MARKOV MESH

Let $x_1, x_2, \ldots, x_n$ be a sequence of random variables assumed to have the first-order Markov chain property

$$p(x_k/x_1x_2\ldots x_{k-1}) = p(x_k/x_{k-1}) \tag{1}$$

for $k = 2, 3, \ldots, n$. Then

$$p(x_k/x_1x_2\ldots x_j) = p(x_k/x_j) \tag{2}$$

for all $j < k$. Also, for $k < n$,

$$p(x_k/x_1, x_2\ldots x_{k-1}, x_{k+1}\ldots x_n) = p(x_k/x_{k-1}, x_{k+1}) \tag{3}$$

so that any point is dependent on only its two nearest neighbors, one on either side. Similarly, for the rth order Markov chain,

$$p(x_k/x_1x_2\ldots x_{k-1}) = p(x_k/x_{k-r}\ldots x_{k-1}) \tag{4}$$

we have

$$p(x_k/x_1x_2\ldots x_{k-1}x_{k+1}\ldots x_n) = p(x_k/x_{k-r}\ldots x_{k-1}x_{k+1}\ldots x_{k+r}). \tag{5}$$

Extending the Markov chain methods to two dimensions gives the two-dimensional analog we called a Markov mesh.

The following definitions are used for the subsequent presentation.

(a) $X_{m,n}$ is an $m \times n$ array of discrete random variables, as shown in Fig. 1a.
(b) $x_{a,b}$ is the variable located in row a and column b of $X_{m,n}$.

ISBN 0-12-597320-9

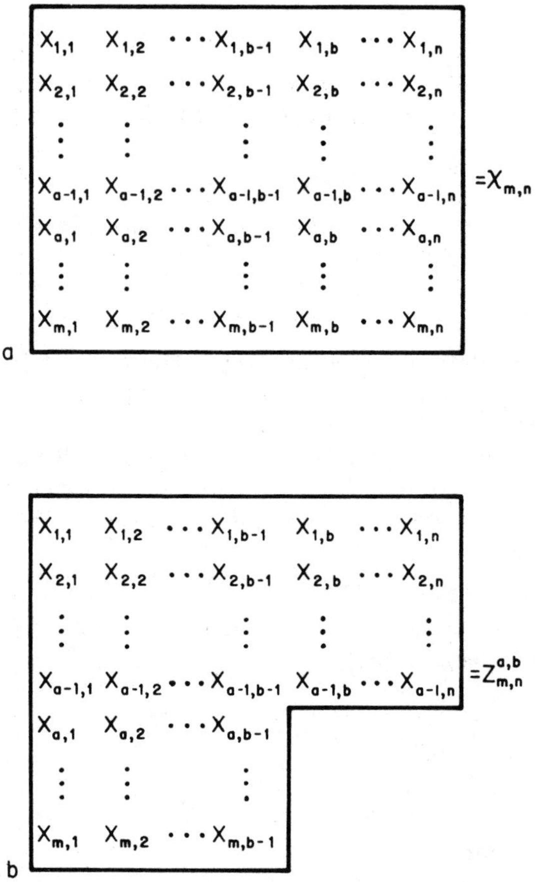

$$\begin{array}{cccccccc} x_{1,1} & x_{1,2} & \cdots & x_{1,b-1} & x_{1,b} & \cdots & x_{1,n} \\ x_{2,1} & x_{2,2} & \cdots & x_{2,b-1} & x_{2,b} & \cdots & x_{2,n} \\ \vdots & \vdots & & \vdots & \vdots & & \vdots \\ x_{a-1,1} & x_{a-1,2} & \cdots & x_{a-1,b-1} & x_{a-1,b} & \cdots & x_{a-1,n} \\ x_{a,1} & x_{a,2} & \cdots & x_{a,b-1} & x_{a,b} & \cdots & x_{a,n} \\ \vdots & \vdots & & \vdots & \vdots & & \vdots \\ x_{m,1} & x_{m,2} & \cdots & x_{m,b-1} & x_{m,b} & \cdots & x_{m,n} \end{array} = \chi_{m,n}$$

a

$$\begin{array}{cccccccc} x_{1,1} & x_{1,2} & \cdots & x_{1,b-1} & x_{1,b} & \cdots & x_{1,n} \\ x_{2,1} & x_{2,2} & \cdots & x_{2,b-1} & x_{2,b} & \cdots & x_{2,n} \\ \vdots & \vdots & & \vdots & \vdots & & \vdots \\ x_{a-1,1} & x_{a-1,2} & \cdots & x_{a-1,b-1} & x_{a-1,b} & \cdots & x_{a-1,n} \\ x_{a,1} & x_{a,2} & \cdots & x_{a,b-1} & & & \\ \vdots & \vdots & & \vdots & & & \\ x_{m,1} & x_{m,2} & \cdots & x_{m,b-1} & & & \end{array} = Z_{m,n}^{a,b}$$

b

FIG. 1. Definitions of $\chi_{m,n}$ and $Z_{m,n}{}^{a,b}$. (a) An $m \times n$ array of discrete random variables. (b) Sector to the left of or above $x_{a,b}$.

(c) $\chi_{c,d}$ is the rectangular array of all random variables $x_{i,j}$ with $i \leq c$ and $j \leq d$.

(d) $\chi_{c,d}{}^{a,b}$ is the rectangular array $\chi_{c,d}$ with the point $x_{a,b}$ deleted.

(e) $Z_{m,n}{}^{a,b}$ is the (nonrectangular) array of all random variables $x_{i,j}$ with $i < a$ or $j < b$; that is, all variables to the left of or above $x_{a,b}$. This is shown in Fig. 1b.

The assumption that the probability of $x_{a,b}$ conditional on $Z_{m,n}{}^{a,b}$ is equal to the probability of $x_{a,b}$ conditional on a limited set of near neighbors within the array $\chi_{a,b}{}^{a,b}$ defines a Markov mesh. For a third-order Markov mesh we have

$$p(x_{a,b}/Z_{m,n}{}^{a,b}) = p(x_{a,b}/x_{a-1,b}x_{a-1,b-1}x_{a,b-1}) \tag{6}$$

for all a,b with $1 \leq a \leq m$, $1 \leq b \leq n$. If we let $p(\chi_{a,b})$ denote the joint probability of all the variables in the array $\chi_{a,b}$, the third-order Markov mesh condition gives

$$p(\chi_{a,b}) = \prod_{i=1}^{a} \prod_{j=1}^{b} p(x_{i,j}/x_{i-1,j-1}x_{i-1,j}x_{i,j-1}), \tag{7}$$

where $x_{0,j}$ and $x_{i,0}$ are defined to be zero for all j and i. Corresponding to the result that in a first-order Markov chain

$$p(x_k/x_1 \ldots x_{k-1}x_{k+1} \ldots x_n) = p(x_k/x_{k-1}x_{k+1}), \tag{8}$$

the third-order Markov mesh assumption of Eq. (6) implies that

$$p(x_{a,b}/\chi_{m,n}{}^{a,b}) = p\left\{x_{a,b}\left|\begin{matrix} x_{a-1,b-1}x_{a-1,b}x_{a-1,b+1} \\ x_{a,b-1} \qquad x_{a,b+1} \\ x_{a+1,b-1}x_{a+1,b}x_{a+1,b+1}\end{matrix}\right.\right\}, \tag{9}$$

that is, the conditional probability of $x_{a,b}$ on the entire array $\chi_{m,n}$ less $x_{a,b}$, is the same as the probability of $x_{a,b}$ conditioned explicitly only on its eight nearest neighbors. The converse, in general, is not true; that is, Eq. (9) does not imply Eq. (6).

This third-order Markov mesh is a specific case. A general assumption of this type is

$$p(x_{a,b}/Z_{m,n}{}^{a,b}) = p(x_{a,b}/U_{a,b}), \tag{10}$$

where $U_{a,b}$ is a set of elements in the left upper sector, i.e., a subset of $\chi_{a,b}{}^{a,b}$. Using this assumption, the analysis presented in [1] can be generalized to show that

$$p(\chi_{m,n}) = \prod_{i=1}^{m}\prod_{j=1}^{n} p(x_{i,j}/U_{i,j}) \tag{11}$$

and

$$p(x_{a,b}/\chi_{m,n}{}^{a,b}) = p(x_{a,b}/Y_{a,b}), \tag{12}$$

where $Y_{a,b}$ is a minimum set of neighbors surrounding $x_{a,b}$, the configuration of $Y_{a,b}$ being determined by the set $U_{a,b}$. Figure 2 shows some of the *simpler* forms of $U_{a,b}$ and the resulting $Y_{a,b}$.

The results stated for the third-order Markov mesh are proved in the form of two theorems in our original paper [1].

3. SOME COMMENTS ON RELATIONS TO OTHER CONTEMPORARY AND LATER REFERENCES

In [1] noting that the form of $U_{a,b}$ for which the Markov mesh conditioning statement is

$$p(x_{a,b}/Z_{m,n}{}^{a,b}) = p(x_{a,b}/x_{a-1,b}x_{a,b-1}) \tag{13}$$

yields the expansion

$$p(\chi_{m,n}) = \prod_{i=1}^{m}\prod_{j=1}^{n} p(x_{i,j}/x_{i-1,j}x_{i,j-1}), \tag{14}$$

we commented that a search of the literature showed that Chow [2] had also considered neighbor dependence. Chow was interested in the case where a point $x_{i,j}$ is dependent on its four neighbors, $x_{i-1,j}x_{i,j-1}x_{i,j+1}x_{i+1,j}$ which he refers to as the north, west, east, and south neighbors, respectively. In Chow's Eq. (1), he

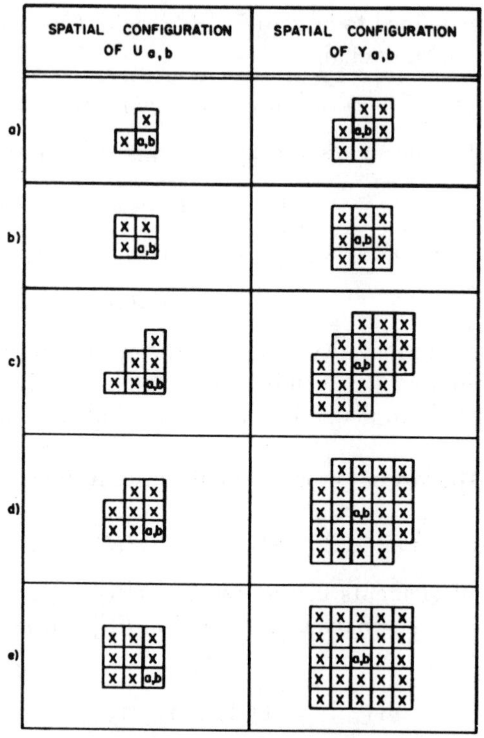

	SPATIAL CONFIGURATION OF $U_{a,b}$	SPATIAL CONFIGURATION OF $Y_{a,b}$
a)	X X a,b	X X X a,b X X X
b)	X X X a,b	X X X X a,b X X X X
c)	X X X X X a,b	X X X X X X X X X a,b X X X X X X X X X
d)	X X X X X X X a,b	X X X X X X X X X X X a,b X X X X X X X X X X X
e)	X X X X X X X X a,b	X X X X X X X X X X X X a,b X X X X X X X X X X X X

FIG. 2. Simple examples of $U_{a,b}$ and the resulting $Y_{a,b}$.

assumes exactly the same expansion as given in the last equation, as can readily be seen by substituting v for what we have called $\chi_{m,n}$ and $v_{i,j}$ for our $x_{i,j}$, etc. He then makes the statement "The general term in (1) includes only the north and west neighbors (above and to the left); the other two neighbors are not explicitly needed. The dependence propagates through the neighbors in this fashion." The development for the Markov mesh shows that, *using the expansion of Eq. (14)*, the smallest set of neighbors on which a variable is explicitly dependent is the set of six neighbors shown in Fig. 2a. By itself the expansion does not lead to the dependence of a point only on its four nearest neighbors.

In example 3 of Woods [6], the case of north–south–east–west dependence is derived by considering the Gaussian–Markov process generated by a difference equation driven by a zero mean Gaussian source. Woods concluded that "this formulation of discrete-space Markov field can be considered a generalization of that presented by Abend *et al.*, in that it allows north–south–east–west 'direct' dependence." The general Markov Mesh formulation we had in mind is that represented by Eq. (11). $Z_{m,n}{}^{a,b}$ was only one of many possible definitions of the "past" in the two-dimensional case and later on we did consider some alternate definitions. The material in [1] focused on the third-order Markov Mesh represented by Fig. 2b and Eq. (9) which seemed of greatest interest in our application. Even in this simple case as in later related investigations, we found the model more useful for the *generation* of images than in the estimation of image parameters for the classification of real images, for which other simpler procedures seemed to work equally well or better.

Following Woods [6] numerous papers were published on linear "spatial-temporal" and linear difference equation models for images, many appearing in the *IEEE Transactions on Information Theory, Communications, and Computers.*

Shortly after [1], through S. Sherman (who had been my teacher at the Moore School of Electrical Engineering, and had been working on Ising models, MRFs, etc.) I became aware of the literature on Ising models and related topics. The formulations presented in this literature up to that time gave me little hope that pursuing them would enable us to do any better on image modeling and classification than we did with our simple Markov mesh development.

The literature on MRFs has continued to grow with the recent past being well represented by the *Proceedings of the Conference on Spatial Patterns and Processes* (1978). Some of the work reported remains not far from our original Markov mesh development but other contributions describe potentially fruitful connections between Markov random fields and contingency tables and graphs (see Speed [5]). Current attempts to use MRFs as models of textured digital images (e.g., Hassner and Sklansky [3]) may have a better chance of producing useful results because of these connections. However, unlike the situation in molecular structure or crystal growth, the pixel representation of images may not be the best for modeling the dependencies occurring in real images.

ACKNOWLEDGMENT

The writing of this note was supported by NSF Grant Eng-7822159 from the Automation, Bioengineering and Sensing Program, Engineering Division, National Science Foundation.

REFERENCES

1. K. Abend, T. J. Harley, and L. N. Kanal, Classification of binary random patterns, *IEEE Trans. Inform. Theory* **IT-11**, 1965, 538–544.
2. C. K. Chow, A recognition method using neighbor dependence, *IRE Trans. Electronic Computers* **EC-11**, 1962, 683–690.
3. M. Hassner, and J. Sklansky, Markov random field models of digitized image texture, in *Proc. of the Fourth International Joint Conference on Pattern Recognition*, pp. 538–540, 1978.
4. S. Sherman, Markov random fields and Gibbs random fields, *Israel J. Math.* **13**, 1975, 92–102.
5. T. P. Speed, Relations between models for spatial data and contingency tables and Markov fields on graphs (*Proceedings of the Conference on Spatial Patterns and Processes*), *Suppl. Advan. Appl. Probability* **10**, 1978, 111–122.
6. J. W. Woods, Two-dimensional discrete Markovian fields, *IEEE Trans. Inform. Theory* **IT-18**, 1972, 232–240.

Univariate and Multivariate Random Field Models for Images *

R. L. KASHYAP

School of Electrical Engineering, Purdue University, West Lafayette, Indiana 47907

The paper introduces a special type of univariate random field model, the so-called periodic random field models, for describing *homogeneous* images (not necessarily isotropic). For modeling *nonhomogeneous* images, we introduce multivariate random field models. In the case of a univariate random field, *scalar autoregressive* representation on a lattice is used. In the case of a multivariate random field, *vector* or *multivariate* autoregressive representation on the lattice is discussed. For the univariate model, we discuss in detail the derivation of the expression for the probability density of the data, and the maximum likelihood estimation of the unknown parameters in the model. Using statistical decision theory we indicate the development of decision rules for the choice of the neighbors, the homogeneity of the data, etc. We discuss in some detail the information aspects of the data, such as compression of the data without effectively losing any information, and reconstruction of the data from the stored information. For the multivariate model we discuss only the problem of obtaining the probability density of the observations. In a subsequent paper, we will discuss further details of the multivariate model.

1. INTRODUCTION

We consider the modeling of two-dimensional images $\{y(i, j),\ 1 \leq i \leq N_1,\ 1 \leq j \leq N_2\}$ by means of a special class of discrete spatial time series of random fields.

We will consider univariate random fields with a *univariate* autoregressive representation on a lattice for characterizing a homogeneous two-dimensional image. Note that a univariate random field is quite different from a one-dimensional time series in image analysis [3–5]. We will represent nonhomogeneous images by a multivariate random field with a *multivariate* autoregressive representation on a lattice. The univariate random field given here is called "periodic" to distinguish it from other univariate random field representations used in image analysis and other spatial studies [1–6, 9]. The advantages of periodic random field models over the normal nonperiodic models are discussed in Section 2. Multivariate random fields have not been studied earlier.

The models considered here are very important from a theoretical point of view since they allow us to solve systematically many important problems such

* Partially supported by the National Science Foundation under Grant ENG-78-18271.

ISBN 0-12-597320-9

as testing whether an image is homogeneous or not, and the choice of the relevant set of neighbors. It is also important for many practical problems such as image compression and restoration, and classification.

Besides random field models, there are various other models for images, such as co-occurrence matrices [10]. One of the important advantages of the autoregressive random field models over the other models is that the former are regenerative, i.e., the random field model allows us to represent all the information in an $N \times N$ image by two sets of parameters, one set containing a small number of parameters having most of the information, and the second set containing N^2 parameters, the so-called residuals, having the remaining information. We can represent every residual using far fewer bits than the original image pixels without substantially sacrificing any accuracy. Using the residuals and the parameters of the first set, we can reconstruct a good approximation to the image. Using the parameters alone, we can generate the texture. This aspect is discussed in some detail in the paper.

The theory of univariate periodic random fields having an autoregressive representation is given in Section 2. The estimation of the parameters is given in Section 3. The decision rule for choosing the appropriate set of neighbors is given in Section 4. Tests for the homogeneity of an image are given in Section 5. The image compression and restoration aspects are discussed in Section 6. Multivariate random fields for modeling nonhomogeneous images are treated in Section 7.

2. PERIODIC UNIVARIATE RANDOM FIELD

In earlier studies [1, 2, 6, 9], an infinite two-dimensional image $\{y(s), s = (i, j), -\infty < i, j < \infty\}$ was represented by a (nonperiodic) random field obeying a univariate autoregressive process defined on a lattice. Here we will consider a *periodic* random field obeying a univariate autoregressive equation, the concept of "periodicity" being defined presently.

Consider the autoregressive equation in (2.1) for the random field

$\{y(s),\ s = (i, j),\ -\infty \leq i, j \leq \infty\}$:

$$y(s) + \sum_{k=1}^{m} \theta_k y(s + q_k) = \beta^{\frac{1}{2}} u(s), \qquad \forall s,\ q_k \in Q,\ \forall k = 1, 2, \ldots, m, \tag{2.1}$$

or equivalently

$$y(s) + \theta^T x(s) = \beta^{\frac{1}{2}} u(s), \tag{2.2}$$

where $Q = \{q_k = (q_{k1}, q_{k2}),\ k = 1, \ldots, m;\ q_k \neq (0, 0),\ q_{ki}$ are integers$\}$ is the effective neighborhood of influence for the pixel $s = (i, j)$,

$$\boldsymbol{\theta} = (\theta_1, \ldots, \theta_m)^T$$

is a vector of parameters, $\boldsymbol{\theta} \in R^m$,

$$x(s) = (y(s + q_1), \ldots, y(s + q_m))^T,$$

$\{u(s)\}$ is a Gaussian I.I.D. sequence with zero mean and unit variance. For

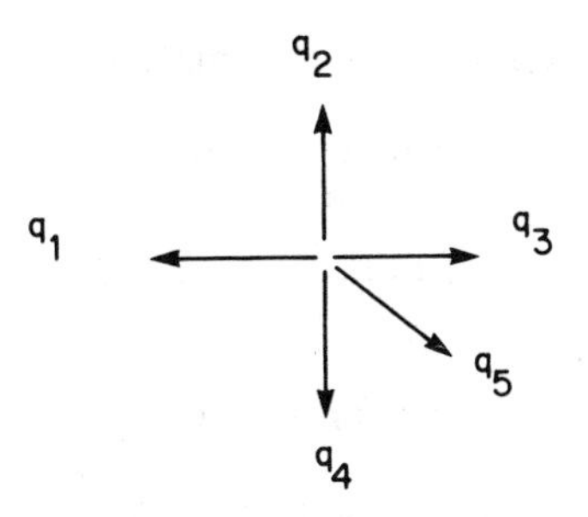

FIGURE 1

illustration, the set Q for the neighborhood in Fig. 1 is

$$Q = \{(-1, 0), (0, -1), (1, 0), (0, 1), (1, 1)\}, \tag{2.3}$$

The sequence $y(\cdot)$ in (2.1) is said to be *doubly periodic* if the infinite sequence $\{y(\cdot)\}$ is obtained by repeating the finite image $\{y(i, j), 1 \leq i \leq N_1, 1 \leq j \leq N_2\}$ an infinite number of times. One particular formula for repetition is

$$y(i + k_1N_1,\ j + k_2N_2) = y(i, j) \qquad \forall k_1, k_2 \text{ integers.} \tag{2.4}$$

(2.4) is not the only type of repetition; other formulas for repetition are given later.

Let $\Omega_0 = \{s = (i, j), 1 \leq i \leq N_1, 1 \leq j \leq N_2\}$. In view of (2.4), the finite image $\{y(s), s \in \Omega_0\}$ can be related to $\{u(s), s \in \Omega_0\}$ as in (2.5). Thus we do not need any other variables, the so-called initial conditions needed when representing *nonperiodic* autoregressive processes as discussed in all the current papers [1–6, 9].

$$\begin{bmatrix} A_1 & A_2 & \cdot\cdot & \cdot & A_{N_1} \\ A_{N_1} & A_1 & A_2 & \cdot & A_{N_1-1} \\ \cdot & \cdot & \cdot & & \cdot \\ \cdot & \cdot & \cdot & & \cdot \\ A_2 & & & & A_1 \end{bmatrix} \begin{bmatrix} \mathbf{y}_1 \\ \mathbf{y}_2 \\ \cdot \\ \cdot \\ \mathbf{y}_{N_1} \end{bmatrix} = \beta^{\frac{1}{2}} \begin{bmatrix} \mathbf{u}_1 \\ \cdot \\ \cdot \\ \cdot \\ \mathbf{u}_{N_1} \end{bmatrix}, \tag{2.5}$$

where

$$\mathbf{y}_i = (y(i, 1), \ldots, y(i, N_2))^T,$$
$$\mathbf{u}_i = (u(i, 1), \ldots, u(i, N_2))^T,$$

$A_1, A_2, \ldots$ are all $(N_2 \times N_2)$ circulant matrices [11] depending on the parameter $\boldsymbol{\theta}$ and on Q. When the set Q is as in (2.3), the matrices $A_1, A_2, \ldots$ have the form

$$\begin{aligned} A_i &= 0 \qquad \forall i \neq 1, 2, N_1, \\ A_1 &= \text{Circulant } (1, \theta_3, 0, \ldots, 0, \theta_1), \\ A_2 &= \text{Circulant } (\theta_4, \theta_5, 0, \ldots, 0), \\ A_{N_1} &= \text{Circulant } (\theta_2, 0, \ldots, 0), \end{aligned} \tag{2.6}$$

where Circulant $(\alpha_1, \ldots, \alpha_n)$ is the $n \times n$ matrix

$$\begin{bmatrix} \alpha_1, \alpha_2, \ldots, \alpha_n \\ \alpha_n, \alpha_1, \ldots, \alpha_{n-1} \\ \vdots \\ \alpha_2, \alpha_3, \ldots, \alpha_n, \alpha_1 \end{bmatrix}$$

We can rewrite (2.5) as

$$B(\boldsymbol{\theta})y = \beta^{\frac{1}{2}}\mathbf{u}(s), \tag{2.7}$$

where

$$\mathbf{y}^T = (\mathbf{y}_1{}^T, \ldots, \mathbf{y}_{N_1}^T),$$
$$\mathbf{u}^T = (\mathbf{u}_1{}^T, \ldots, \mathbf{u}_{N_1}^T),$$

$B(\theta)$ = Block Circulant $(A_1, A_2, \ldots, A_{N_1})$ is an $N_1N_2 \times N_1N_2$ matrix. The expression for the determinant of $B(\boldsymbol{\theta})$ is given in Theorem 1; the proof is in Appendix 1.

THEOREM 1.

$$\begin{aligned} \textit{Det. } B(\theta) &= \prod_{i=0}^{N_1-1} \prod_{j=0}^{N_2-1} A(z_1{}^i, z_2{}^j, \boldsymbol{\theta}) \\ &= \prod_{i=0}^{N_1-1} \prod_{j=0}^{N_2-1} \|A(z_1{}^i, z_2{}^j, \boldsymbol{\theta})\|, \end{aligned}$$

where $\|\cdot\|$ *indicates the modulus of the complex quantity*

$$A(z_1, z_2, \boldsymbol{\theta}) \triangleq 1 + \sum_{k=1}^{m} \theta_k z_2^{q_{k1}} z_1^{q_{k2}}, \qquad q_k = (q_{k1}, q_{k2}) \in Q,$$
$$z_i = exp[\sqrt{-1}\, 2\pi/N_i], \quad i = 1, 2.$$

Since the sequence $u(\cdot)$ is Gaussian and independent, (2.7) yields the following expression for the probability density of $\mathbf{y}$:

$$\begin{aligned} p(\mathbf{y} \mid \boldsymbol{\theta}, \beta) &= [\text{Det } B(\theta)] \prod_{s \in \Omega_0} (1/2\pi\beta)^{\frac{1}{2}} \exp[-(1/2\beta)(y(s) + \boldsymbol{\theta}^T\mathbf{x}(s))^2] \\ &= \prod_{s=(i,j)\in\Omega_0} (S(z_1{}^i, z_2{}^j, \boldsymbol{\theta}))^{-\frac{1}{2}} (1/2\pi\beta)^{\frac{1}{2}} \exp[-1/2\beta) \\ &\qquad \times [(y(s) + \boldsymbol{\theta}^T x(s))^2], \end{aligned} \tag{2.8}$$

where

$$S(z_1, z_2, \boldsymbol{\theta}) = \frac{1}{\|A(z_1, z_2, \boldsymbol{\theta})\|^2}. \tag{2.9}$$

Comment 1. βS is the spectral density of $y(\cdot)$ defined as the DFT of the bivariate correlation function.

Comment 2. We mentioned earlier that the formula (2.4) for defining the periodic repetition is not the only one. Another possible and useful formula is

$$\begin{aligned} y_{i,N_2+j} &= y_{i+1,j}, & i &= 1, \ldots, N_1 - 1, \forall j, \\ y_{N_1,N_2+j} &= y_{1,j} & &\forall j, \\ y_{i+N_1,j} &= y_{i,j} & &\forall i, j. \end{aligned} \tag{2.10}$$

In this case, the vectors $\mathbf{y}$, $\mathbf{u}$ defined in (2.7) are related by

$$\mathbf{B}_1(\theta)\mathbf{y} = \beta^{\frac{1}{2}}\mathbf{u}, \tag{2.11}$$

where $\mathbf{B}_1(\theta)$ is a matrix depending on θ and Q. When Q is as in (2.3), $B_1(\theta)$ has the form

$$B_1 = \text{Circulant } (1, \theta_3, 0, 0, \ldots, 0, \underset{\substack{\downarrow \\ N_2+1}}{\theta_4}, \theta_5, 0, \ldots, 0, \underset{\substack{\downarrow \\ (N_1N_2-N_2+1)}}{\theta_2}, 0, \ldots, 0, \underset{\substack{\downarrow \\ N_1N_2}}{\theta_1}),$$

and $\quad N_1N_2 \times N_1N_2$ matrix.

The determinant B_1 needed for the probability distribution can be obtained from the theory of circulants [11]:

$$\text{Determinant } B_1(\theta) = \prod_{k=0}^{N_1N_2-1} (1 + \theta_3\alpha_k + \theta_4\alpha_k^{N_2} + \theta_5\alpha_k^{(N_2+1)} + \theta_2\alpha_k^{(N_1N_2-N_2)} + \theta_1\alpha_k^{(N_1N_2-1)}), \qquad \text{where} \qquad \alpha_k = \exp[\sqrt{-1}\, 2\pi k/N_1N_2].$$

The form in (2.11) is useful for image compression and regeneration studies. This aspect will be discussed in a later paper.

Comment 3. Note that $y(\cdot)$ is not *in general* an *ordinary* 2-dimensional Markov field with memory Q, i.e.,

$$p(y(s_1) \mid \text{all } y(s),\, s \neq s_1) \neq p(y(s_1) \mid y(s_1 + q),\, q \in Q).$$

We can show that in general

$$p(y(s_1) \mid \text{all } y(s),\, s \neq s_1) = p(y(s_1) \mid y(s_1 + q),\, q \in Q_1)$$

where Q_1 is a set having Q as its subset, usually $Q \subset Q_1$. For instance, if

$$\begin{aligned} Q &= \{(0, 1), (1, 0), (0, -1), (-1, 0)\}, \\ Q_1 &= \{(i, j) : |i|;\, |j| \leq 2,\, (i, j) \neq (0, 0)\}. \end{aligned}$$

Random field models (2.1) in which (2.12) is indeed satisfied are called one-sided models since the corresponding field $\{y(s)\}$ is equivalent to a one-dimensional time series

$$p(y(s_1) \mid \text{all } y(s),\, s \neq s_1) = p(y(s_1) \mid y(s_1 + q),\, q \in Q). \tag{2.12}$$

There is extensive discussion of such one-sided models in the literature [3–5].

Comment 4. The papers [1, 2, 6, 9] consider the random field in (2.1) without using the periodicity condition in (2.4). The Jacobian of the transformation from $\mathbf{u}$ to $\mathbf{y}$ given in [1] is correct only for infinite images and is only approximate for finite images. Consequently the likelihood expression in [1] is only asymptotically correct. The likelihood expression given in [2] is also only asymptotically true and it is a much coarser approximation than that in [1]. Note that the likelihood expression given here is exact for the given $(N_1 \times N_2)$ image, regardless of N_1 and N_2.

Comment 5. In defining $A(z_1, z_2, \theta)$ in Theorem 1 the vectors q_k and θ_k should be the same as in the defining equation (2.1). The exponents are q_{k1} for z_2 and q_{k2} for z_1 because we are dealing with an $N_1 \times N_2$ image. If we were dealing with an $N_2 \times N_1$ image, $A(z_1, z_2, \theta)$ would read as follows:

$$A(z_1, z_2, \boldsymbol{\theta}) = 1 + \sum_{k=1}^{m} \theta_k z_1^{q_{k1}} z_2^{q_{k2}}, \qquad q_k = (q_{k1}, q_{k2}) \in Q.$$

3. PARAMETER ESTIMATION

The unknown parameters $(\boldsymbol{\theta}, \beta)$ in the model (2.1) can be estimated using the finite image data $\{y(s),\ s \in \Omega_0\}$ using a standard procedure such as maximum likelihood. In this procedure, the estimates θ^* and β^* of θ and β in (2.1) are obtained by maximizing the probability density of the observation set with respect to θ and β. $\boldsymbol{\theta}^*$ and β^* have the following expressions. Let

$$\Omega_0 = \{s = (i, j) < 1 \leq i \leq N_1,\ 1 \leq j \leq N_2\}.$$

Then

$$\ln p(y(s),\ s \in \Omega_0 \,|\, \boldsymbol{\theta}, \beta) = -\left(\frac{N_1 N_2}{2}\right) \ln(2\pi\beta) - \sum_{s \in \Omega_0} (y(s) + \theta^T x(s))^2 / 2\beta$$
$$+ (1/2) \sum_{s=(i,j) \in \Omega_0} \ln \| A(z_1^{\,i}, z_2^{\,j}, \boldsymbol{\theta}) \|^2.$$

Maximizing $\ln p$ with respect to $\boldsymbol{\theta}$ and β yields the following estimates:

$$\theta^* = \text{Argument} \left[\underset{\theta \in R^m}{\text{Minimum}}\ L(\theta) \right],$$
$$\beta^* = (1/N_1 N_2) \sum_{s \in \Omega_0} (y(s) + (\theta^*)^T x(s))^2,$$

where

$$L(\theta) = N_1 N_2 \ln \left[\sum_{s \in \Omega_0} (y(s) + \theta^T x(s))^2 \right] - \sum_{s \in \Omega_0} \ln \| A[z_1^{\,i}, z_2^{\,j}, \boldsymbol{\theta} \|^2.$$

The first term in $L(\theta)$ has a minimum at

$$\bar{\boldsymbol{\theta}} = \left[\sum_{s \in \Omega_0} \mathbf{x}(s) \mathbf{x}^T(s) \right]^{-1} \sum_{s \in \Omega_0} \mathbf{x}(s) y(s).$$

One can minimize $L(\boldsymbol{\theta})$ with respect to $\boldsymbol{\theta}$ using a standard gradient or Newton–Raphson procedure with $\boldsymbol{\theta}$ as the starting point. The available numerical results [1] indicate that θ^* is often close to $\bar{\theta}$.

The estimates $\boldsymbol{\theta}^*$ and β^* are strongly consistent and the mean square value of the difference between the estimate and the true value is of the order $O(1/N_1 N_2)$.

4. CHOICE OF APPROPRIATE NEIGHBORS

An important problem in all image models is the choice of the appropriate set of neighbors for each pixel. Usually this problem is tackled by ad hoc methods or simulation. However, in our case, the choice of the neighbors can be posed as a standard statistical decision problem and a decision rule for the choice can be derived by standard decision-theoretic techniques.

Suppose we had three sets Q_1, Q_2, Q_3 corresponding to the following three choices for the neighborhood:

$$\begin{aligned} Q_1 &= \{(1, 0), (0, 1), (-1, 0), (0, -1)\} \\ &= \{\text{The set of east, south, west, north neighbors}\}, \\ Q_2 &= \{(0, -1), (-1, 0)\} \\ &= \{\text{north and west neighbors}\}, \\ Q_3 &= \{(0, 1), (0, -1), (-1, 0), (1, 0), (1, 1)\} \\ &= \{\text{south, north, west, east, southwest neighbors}\}. \end{aligned}$$

Corresponding to each set Q_i, we will write the corresponding autoregressive equation E_i, $i = 1, 2, 3$:

$$\begin{aligned} &E_i\colon y(s) + \boldsymbol{\theta}_i^T \mathbf{x}_i(s) = \beta_i^{\frac{1}{2}} u(s), && i = 1, 2, 3, \\ &\boldsymbol{\theta}_i = (\theta_{i1}, \theta_{i2}, \ldots, \theta_{im_i})^T, && \theta_{ij} \neq 0 \;\forall j = 1, \ldots, m_i, \\ &m_1 = 4, \; m_2 = 3, \; m_3 = 5, \\ &\mathbf{x}_i(s) = \text{vector of neighbors of } y(s) \text{ given by } Q_i. \end{aligned}$$

Note that the $\boldsymbol{\theta}_i$ are unknown. We only know that $y(\cdot)$ obeys one of the given equations E_i. Note that the equations E_i are mutually exclusive.

We can use standard Bayesian methods to obtain a decision rule [8, 12]. The simplified decision rule is given below. Compute the statistics C_k, $k = 1, 2, 3$,

$$C_k = N_1 N_2 \ln \beta^*_k - \sum_{s=(i,j)\in\Omega_0} \ln\|A_k(z_1{}^i, z_2{}^j, \boldsymbol{\theta}^*_k)\|^2 + m_k \ln (N_1 N_2),$$

where β^*_k, $\boldsymbol{\theta}^*_k$ are the M.L. estimates of the β_k, $\boldsymbol{\theta}_k$ in the model E_k based on the given data, computed as in the previous section.

Decision rule [8]. Choose the set Q_k such that C_{k*} has the smallest value among C_1, C_2, C_3.

It is possible to show that the decision rule is strongly consistent, i.e., the probability of error in the decision tends to zero as N tends to infinity. Further, the decision rule approximately minimizes the average probability of error. The details can be found in [8].

Whittle [1] and Larimore [2] have also considered this problem. However, in [1], only two models can be compared at one time using significance testing methods. This can lead to an intransitive decision [12]. Akaike's AIC rule which is quite arbitrary and has no consistency properties is used in [2].

5. SEGMENTATION OF AN IMAGE

Often an entire image such as a photograph of a human face or an airport is not homogeneous and, as such, a single model like (2.1) cannot be fitted to the entire image, whatever may be the set Q. Hence, we have to segment the image into blocks and fit a separate model to each block. Clearly we want each block to be as large as possible. Hence we need a test to determine whether two given blocks of image data together are homogeneous. If they are homogeneous, then the two blocks can be merged into a single block. Otherwise, we have to treat them separately. The problem of testing whether the two blocks together are homogeneous or not can be posed as a standard problem in statistical decision theory and the decision rule can be obtained in a routine way. Using decision theoretic methods we will give the decision rule below.

Let $\Omega_{12} = \Omega_1 \cap \Omega_2$, size of Ω_1 = size of $\Omega_2 = N \times N$. The two mutually exclusive hypotheses are H_1 and H_2. H_1: The image $y(s)$, $s \in \Omega_{12}$, is homogeneous and is described by the difference equation

$$y(s) + \boldsymbol{\theta}_{12}^T x(s) = \beta_{12}^{\frac{1}{2}} u(s), \qquad s \in \Omega_{12}, \tag{5.1}$$

where $\boldsymbol{\theta}_{12}$ is a m_{12} vector of unknown parameters, none of which is zero. H_2: The images $\{y(s), s \in \Omega_1\}$ and $\{y(s), s \in \Omega_2\}$ are individually homogeneous obeying (5.2) and (5.3), respectively:

$$y(s) + \boldsymbol{\theta}_1{}^T \mathbf{x}_1(s) = \beta_1^{\frac{1}{2}} u(s), \qquad s \in \Omega_1, \tag{5.2}$$

$$y(s) + \boldsymbol{\theta}_2{}^T \mathbf{x}_2(s) = \beta_2^{\frac{1}{2}} u(s), \qquad s \in \Omega_2, \tag{5.3}$$

where the dimensions of $\boldsymbol{\theta}_i = m_i$, $\boldsymbol{\theta}_1$, $\boldsymbol{\theta}_2$ are unknown but none of their components are zero. Note that (5.1), (5.2), (5.3) are mutually exclusive since

$$x_{12}(s) \neq \mathbf{x}_1(s) \neq x_2(s).$$

Using Bayesian considerations, we can derive the following decision rule. Let $(\theta^*_{12}, \beta^*_{12})$, (θ^*_1, β^*_1), and (θ^*_2, β_2) be the M.L. estimates of the corresponding parameters obtained from the data $\{y(s), s \in \Omega_{12}\}$, $\{y(s), s \in \Omega_1\}$, and $\{y(s), s \in \Omega_2\}$, respectively.

Compute the following statistics:

$$C_{12} = 2N^2 \ln \beta^*_{12} - \ln \sum_{(i,j) \in \Omega_{12}} \|A_{12}(z_1{}^i, z_2{}^j, \theta^*_{12})\|^2 + m_{12} \ln 2N^2,$$

$$C_k = N^2 \ln \beta^*_k - \ln \sum_{(i,j) \in \Omega_k} \|A_k(z_1{}^i, z_2{}^j, \boldsymbol{\theta}^*_k)\| + m_k \ln N^2, \qquad k = 1, 2$$

where

$$A_{12}(z_1{}^i, z_2{}^j, \boldsymbol{\theta}_{12}) = 1 + \sum_{k=1}^{m_{12}} \theta_{12k} z_1^{q_{k1}} z_2^{q_{k2}}, \qquad \text{where } q_k = (q_{k1}, q_{k2}) \in Q_{12},$$

$$A_l(z_1{}^i, z_2{}^j, \boldsymbol{\theta}_l) = 1 + \sum_{k=1}^{m_i} \theta_{lk} z_1^{q_{k1}} z_2^{q_{k2}}, \qquad \text{where } q_k = (q_{k1}, q_{k2}) \in Q_i,\ l = 1, 2,$$

Q_{12}, Q_1, and Q_2 are the neighborhood sets associated with $x_{12}(s)$ and $x_1(s)$ and $x_2(s)$, respectively.

Decision rule. Choose

$$H_1 \quad \text{if} \quad C_{12} < (C_1 + C_2),$$
$$H_2 \quad \text{if} \quad C_{12} > (C_1 + C_2).$$

One can show that the decision rule is consistent. We can use a similar rule to compare whether a single model fits r blocks or r separate models are needed for the r blocks.

6. IMAGE COMPRESSION AND RESTORATION

One of the important advantages of the random field model is that it is a generative model, i.e., suppose we assume a model for the image as in (2.1), obtain estimates $\boldsymbol{\theta}^*$ and β^* of the parameters θ and β and construct the residuals $\bar{u}(*)$:

$$\bar{u}(s) = (y(s) + (\boldsymbol{\theta}^*)^T x(s))/(\beta^*)^{\frac{1}{2}}, \qquad s \in \Omega.$$

Then all the information contained in $\{y(s),\ s \in \Omega\}$ is also contained in the set $\{\bar{u}(s),\ s \in \Omega\}$ and $\boldsymbol{\theta}^*$ and β^*, since it is possible to recover $\{y(s),\ s \in \Omega\}$ from $\{\bar{u}(s),\ s \in \Omega\}$ and θ^* by Eq. (6.1) which is derived from (2.7), $B(\theta)$ being defined in (2.7)

$$\mathbf{y} = [B(\theta^*)]^{-1}\beta^*\bar{\mathbf{u}},$$
$$\bar{\mathbf{u}}^T = [\bar{\mathbf{u}}_1{}^T, \ldots, \bar{\mathbf{u}}_{N_1}{}^T], \qquad \bar{u}_i{}^T = (\bar{u}(i, 1), \ldots, \bar{u}(i, N_2)). \tag{6.1}$$

If the given image is homogeneous and really obeys the model in (2.1) as indicated by the tests in Section 5, then $\{\bar{u}(\cdot)\}$ is an I.I.D. sequence. Thus the information contained in the original image $\{y(s),\ s \in \Omega\}$ is divided into two parts, namely the low-dimensional vector (θ^*, β^*) which contains most of the information, and the residual sequence $\{\bar{u}(\cdot)\}$ which has much less information in it.

The above statement forms the basis for compressing the images for storage without effectively losing any information. Suppose the image is $N \times N$ and each pixel needs 8 bits. If we are interested in reconstructing only the texture of the image, then we need store only the coefficient vector $\boldsymbol{\theta}^*$ and β^*. Using these and a vector $\bar{u}$ obtained from a standard Gaussian number table, we can compute $\bar{y}$ from

$$\bar{y} = [B(\theta^*)]^{-1}\beta^*\bar{u},$$
$$\text{Dimension of } \bar{u} = N_1N_2. \tag{6.2}$$

However, if we are interested in actually storing the image in some sort of archival storage, then we want to compress the data into fewer bits than the original set of $8N_1N_2$ (assuming 8 bits per pixel), say $N_0 \leq 8N_1N_2$, such that the reconstructed image using these N_0 bits will be practically indistinguishable from the original picture. We know that the residual vector $\bar{u}^*$ and (θ^*, β^*) have *all* the information in $\{y(s),\ s \in \Omega\}$. Since $\bar{u}^*$ is made of mutually independent components, every component in it requires fewer than 8 bits for storage without sacrificing any accuracy at all.

In [3], many examples of compression and reconstruction are given using one-dimensional random field models. In such a case, an image which needs 8 bits per pixel can be compressed into about 1.7 bits per pixel without substantially losing any accuracy. All the examples in [3] are 256×256 images divided into 16×16 blocks, each block being assumed to be homogeneous. Additional compression seems to be possible by using the sophisticated models mentioned in this paper.

7. MULTIVARIATE RANDOM FIELD

We have already mentioned that an image representing an entire human face or an outdoor scene is unlikely to be homogeneous and hence cannot be adequately represented by an autoregressive equation of type (2.1), whatever may be the finite set Q. In this section, we show that an entire *nonhomogeneous image* can be adequately represented by a two-stage model involving both univariate and multivariate random fields, described below. Such a two-stage model is useful in many classification and texture generation problems.

The $N_{01} \times N_{02}$ original image is divided into $r_1 r_2$ blocks labeled (i, j), $i = 1, \ldots, r_1$; $j = 1, \ldots, r_2$, such that the (i, j)th block is

$$(i, j) \text{ block} = \{y(s),\ s \in \Omega_{ij}\},$$

$$\Omega_{i_1 j_1} \cap \Omega_{i_2 j_2} = 0 \qquad \text{if } i_1 \neq i_2 \text{ or } j_1 \neq j_2,$$

$$\Omega_0 = \bigcup_{i=1}^{r_1} \bigcup_{j=1}^{r_2} \Omega_{i,j}. \tag{7.1}$$

Each block is assumed to be sufficiently small so that it is homogeneous. We will fit a separate univariate model for each block, the model for the (i, j) block being

$$y(s) + \boldsymbol{\theta}_{i,j}^T \mathbf{x}_{i,j}(s) = \beta_{ij}^{\frac{1}{2}} u(s), \qquad s \in \Omega_{ij}, \tag{7.2}$$

where $\boldsymbol{\theta}_{i,j} = (\theta_{i,j,1}, \ldots, \theta_{i,j,m})^T$. Of course $\boldsymbol{\theta}_{i,j} = 1 = 1, \ldots, r_1$; $j = 1, \ldots, r_2$ are unknown and have to be estimated from the corresponding block data.

It is clear that the vectors $\theta_{i,j}$, $i = 1, 2, \ldots$ and $j = 1, 2, \ldots$ are not completely independent of one another. Hence we can regard the vector sequence $\{\boldsymbol{\theta}_{i,j}\}$ as a *vector* or multivariate random field just as the scalar sequence $\{y(\cdot)\}$ constitutes a univariate random field. As before, we can make the field $\{\theta_{i,j}\}$ periodic by introducing the following conditions:

$$\boldsymbol{\theta}_{i,r_2+k} = \boldsymbol{\theta}_{i,k}, \qquad \boldsymbol{\theta}_{i-k,j} = \boldsymbol{\theta}_{i-k+r_1,j}, \qquad k = 0, \pm 1, \pm 2, \text{ etc.} \tag{7.3}$$

Just as $y(\cdot)$ can be described by a univariate autoregressive representation on a lattice, the vector field $\{\theta_{i,j}\}$ can be represented by a *vector* autoregressive representation defined on a lattice. If we want $\boldsymbol{\theta}_{i,j}$ to depend only on the four neighbors (east, west, north, south), then the corresponding autoregressive representation is

$$\boldsymbol{\theta}_{i,j} + \boldsymbol{\delta} + \boldsymbol{\alpha}_1 \boldsymbol{\theta}_{i-1,j} + \boldsymbol{\alpha}_2 \boldsymbol{\theta}_{i,j-1} + \boldsymbol{\alpha}_4 \theta_{i,j+1} + \alpha_3 \theta_{i+1,j} = V_{i,j}, \tag{7.4}$$

where $\boldsymbol{\delta} = m$-vector, $\boldsymbol{\alpha}_1, \boldsymbol{\alpha}_2, \boldsymbol{\alpha}_3, \boldsymbol{\alpha}_4 = m \times m$ matrices and $\{V_{i,j}, i, j = 1, 2, 3, \ldots\}$ is an independent sequence with Gaussian distribution, zero mean, and covariance matrix S. Of course the vectors and matrices $\boldsymbol{\delta}$, $\boldsymbol{\alpha}_1$, $\boldsymbol{\alpha}_2$, etc., are unknown and have to be estimated from the given numerical values of the vectors $\{\boldsymbol{\theta}_{i,j}, i, j = 1, 2, 3, \ldots\}$.

We will obtain an expression for the probability density of $\{y(s), s \in \Omega_0\}$. This can be done in two stages. First we obtain the probability density of

$$p(y(s) \mid s \in \Omega_0 \mid \theta_{i,j}, i, j = 1, 2, \ldots).$$

Then we obtain the probability density of $\{\theta_{i,j}\}$ given δ, α_1, α_2, α_3, α_4. Then we can find the density $p(y(s) \mid s \in \Omega_0; \delta, \alpha_1, \ldots, \alpha_4, S)$ by integration. This density is used for estimation of the unknown parameters δ, α_1, $\ldots$, α_4, S. We have to stress here that $\theta_{i,j}$, i, $j = 1, 2, 3, \cdots$ are unknown. We can only obtain their estimates for any finite image and not recover them exactly.

The various blocks of images $\{y(s), s \in \Omega_{ij}\}$ are conditionally independent of one another given the corresponding parameter set $\{\theta_{i,j}\}$. Hence

$$p[y(s), s \in \Omega_0 = \bigcup_{i,j}\bigcup \Omega_{i,j} \mid \theta_{i,j}, \beta_{ij}, i = 1, \ldots, r_1; j = 1, \ldots, r_2]$$

$$= \prod_{i=1}^{r_1} \prod_{j=1}^{r_2} p(y(s), s \in \Omega_{ij} \mid \theta_{i,j}, \beta_{i,j}). \tag{7.5}$$

To write the probability density of $\{\theta_{ij,}\}$ we will rewrite (7.4) in the form of a partitioned matrix just like (2.5):

$$\begin{bmatrix} A_1 & A_2 \cdots A_{r_1} \\ A_{r_1} & A_1 \cdots A_{r_1-1} \\ \vdots & \\ A_2 & A_1 \end{bmatrix} \begin{bmatrix} \boldsymbol{\theta}^{(1)} \\ \vdots \\ \boldsymbol{\theta}^{(r_1)} \end{bmatrix} = \begin{bmatrix} V_{1,1} \\ \vdots \\ V_{r_1,r_2} \end{bmatrix} - \begin{bmatrix} \boldsymbol{\delta} \\ \boldsymbol{\delta} \\ \vdots \\ \boldsymbol{\delta} \end{bmatrix}, \tag{7.6}$$

where $\boldsymbol{\theta}^{(k)} = (\boldsymbol{\theta}_{k,1}^T, \ldots, \theta_{k,r_2}^T)^T$,

$$\begin{aligned} A_i \quad & \text{are all } mr_2 \times mr_2 \text{ matrices}, \\ A_1 &= \text{Block Circulant } (I, \boldsymbol{\alpha}_3, 0, \ldots, 0, \boldsymbol{\alpha}_1), \\ A_2 &= \text{Block Circulant } (\boldsymbol{\alpha}_4, 0, \ldots, 0), \\ A_{r_1} &= \text{Block Circulant } (\boldsymbol{\alpha}_2, 0, \ldots, 0), \\ A_i &= 0 \quad \forall \quad i \neq 1, 2, r_1. \end{aligned}$$

We can rewrite (7.6) as

$$\begin{aligned} B_2\boldsymbol{\theta} &= \mathbf{V} - \gamma, \\ \boldsymbol{\theta} &= ((\boldsymbol{\theta}^{(1)})^T, \ldots, (\theta^{(r_1)})^T)^T \\ &= (\boldsymbol{\theta}_{11}^T, \boldsymbol{\theta}_{12}^T, \ldots, \boldsymbol{\theta}_{r_1,r_2}^T)^T, \\ V &= (\mathbf{V}_{11}^T, \mathbf{V}_{12}^T, \ldots, V_{r_1,r_2}^T)^T, \\ V^T &= (\delta^T, \delta^T, \ldots, \delta^T). \end{aligned} \tag{7.7}$$

Hence,

$$p(\theta_{i,j}, 1 \le i \le r_1, 1 \le j \le r_2 | \boldsymbol{\delta}, \boldsymbol{\alpha}_1, \boldsymbol{\alpha}_2, \boldsymbol{\alpha}_3, \boldsymbol{\alpha}_4, \mathbf{S})$$

$$= [\text{Det. } B_2(\boldsymbol{\theta})] \prod_{i=1}^{r_1} \prod_{j=1}^{r_2} \frac{1}{(2\pi)^{m/2}} (\text{Det. } S)^{-\frac{1}{2}} \exp[-(1/2)\|\theta_{i,j} + \delta + \alpha_1\theta_{i-1,j} + \alpha_2\theta_{i,j-1} + \alpha_3\theta_{i+1,j} + \alpha_4\theta_{i,j+1}\|_{S^{-1}}^2]. \quad (7.8)$$

Hence,

$$p(y(s), s \in \Omega_0 | \boldsymbol{\delta}, \boldsymbol{\alpha}_1, \boldsymbol{\alpha}_2, \boldsymbol{\alpha}_3, \boldsymbol{\alpha}_4, \mathbf{S}) = \int d[\prod_i \prod_j \theta_{i,j}, \beta_{ij}] p(y(s), s \in \Omega_{ij} | \theta_{i,j}, \beta_{i,j})) p(\theta_{i,j}, i = 1, \ldots, r_1, j = 1, \ldots, r_2 | \boldsymbol{\delta}, \boldsymbol{\alpha}_1, \boldsymbol{\alpha}_2, \boldsymbol{\alpha}_3, \boldsymbol{\alpha}_4, \mathbf{S}), p(\beta_{i,j}, i = 1, \ldots, r_1, j = 1, \ldots, r_2). \quad (7.9)$$

Let $\theta_{i,j}^*$, β_{ij}^* be the estimates of $\theta_{i,j}$ and $\beta_{i,j}$ based on $\{y(s), s \in \Omega_{ij}\}$. Then using asymptotic integration, we can simplify (7.9) into

$$p(y(s), s \in \Omega_0 | \boldsymbol{\delta}, \boldsymbol{\alpha}_1, \ldots, \boldsymbol{\alpha}_4, S) = g(y(s), s \in \Omega_0) \times p(\theta_{i,j}^*, i = 1, \ldots, r_1; j = 1, \ldots, r_2 | \boldsymbol{\delta}, \ldots, \boldsymbol{\alpha}_1, \boldsymbol{\alpha}_4, \mathbf{S}) \times p(\beta_{i,j}^*, i = 1, \ldots, r_1; j = 1, \ldots, r_2), \quad (7.10)$$

where $g(y(s), s \in \Omega_0)$ is a functi.n independent of δ, α_1, α_2, α_3, α_4, S. The highlight of (7.10) is that it involves only $\theta_{i,j}^*$ and $\beta_{i,j}^*$. Note that $\theta_{i,j}^*$, $\beta_{i,j}^*$, $i = 1, \ldots, r_1$, $j = 1, \ldots, r_2$ serve as the sufficient statistics of the entire data $\{y(s), s \in \Omega_0\}$ and contain all the relevant information about the image for various purposes such as the estimation of the parameters $\boldsymbol{\delta}$, $\boldsymbol{\alpha}_i$ or classification of the image.

In a subsequent paper, we will consider the applications of the multivariate random field model for image classification, compression, regeneration, etc. All these applications can be carried out in a manner similar to the corresponding operations on the univariate random field done in Sections 2–6.

8. CONCLUSIONS

We have introduced a new type of univariate random field model, the so-called periodic random field model with scalar autoregressive structure for describing a two-dimensional homogeneous image. We discussed the various applications of the model such as image compression, decision rules for the choice of the appropriate neighborhood, decision rules for testing the homogeneity of two blocks of data. We also discussed the advantages of the periodic random field model over the nonperiodic random field models mentioned in the literature earlier.

We also are considering the modeling of two-dimensional *nonhomogeneous* image. We introduced a two-stage model. In the first stage, the image is divided into a large number of segments. In the second stage, the relationship between the various segments is described by a multivariate random field model.

APPENDIX 1

Proof of Theorem 1

We will prove Theorem 1 for a general $N_1 \times N_2$ image for the particular case of Q being as in Fig. 1. We can easily generalize the theorem to other neighborhood sets.

LEMMA 1 [11]. *Consider an $n \times n$ circulant matrix*

$$A = \text{Circulant } (a_1, \ldots, a_n).$$

The eigenvalues of A are

$$\left(\sum_{i=1}^{n} a_i \alpha^{k(i-1)}\right),$$

$k = 1, \ldots, n$, where $\alpha = \exp[\sqrt{-1}\, 2\pi/n]$. *The eigenvector corresponding to the eigenvalue $\sum a_i \alpha^{k(i-1)}$ is $(1, \alpha^k, \alpha^{2k}, \ldots, \alpha^{(n-1)k})$.*

The Lemma can be proved by inspection.

LEMMA 2. *Consider the $N_2 \times N_2$ circulant A_i, $i = 1, \ldots, N_1$:*

$$A_i = \text{circulant } (a_{i1}, a_{i2}, \ldots, a_{iN_2}).$$

Let $\lambda_{i1}, \lambda_{i2}, \ldots, \lambda_{iN_2}$ be the eigenvalues of A_i where

$$\lambda_{ik} = a_{i1} + a_{i2} z_2{}^k + a_{i3} z_2^{2k} + \ldots + a_{iN_2} z_2^{(N_2-1)k}, \quad z_2 = \exp[\sqrt{-1}\, 2\pi/N_2].$$

Let B = Block circulant matrix $(A_1, A_2, \ldots, A_{N_1})$, (an $N_1N_2 \times N_1N_2$ matrix). Then μ_{ij}, $i = 1, \ldots, N_2$, $j = 1, \ldots, N_1$ are the eigenvalues of B:

$$\mu_{ij} = \lambda_{1i} + \lambda_{2i} z_1{}^j + \lambda_{3i} z_1^{2j} + \ldots + \lambda_{N_1 i} z_1^{(N_1-1)j}, \quad z_1 = \exp[\sqrt{-1}\, 2\pi/N_1].$$

The Lemma can be proved by inspection.

In the present case, with Q as in Fig. 1,

$$A_i = 0, \qquad \forall\, i \neq 1, 2, N_1.$$

Hence

$$\begin{aligned} \mu_{ij} &= \lambda_{1i} + \lambda_{2i} z_1{}^j + \lambda_{N_1 i} z_1^{(N_1-1)j} \\ &= \lambda_{1i} + \lambda_{2i} z_1{}^j + \lambda_{N_1 i} z_1^{-j}. \end{aligned}$$

Using the definition of A_i in (2.6), and the formula for λ_{ij} from Lemma 2

$$\begin{aligned} \lambda_{1k} &= 1 + \theta_3 z_2{}^k + \theta_1 z_2^{k(N_2-1)} \\ &= 1 + \theta_3 z_2{}^k + \theta_1 z_2^{-k}, \\ \lambda_{2k} &= \theta_4 + \theta_5 z_2{}^k, \\ \lambda_{N_1 k} &= \theta_2. \end{aligned}$$

Hence B has the eigenvalues

$$\mu_{ij} = (1 + \theta_3 z_2{}^i + \theta_1 z_2^{-i}) + (\theta_4 + \theta_5 z_2{}^i) z_1^{+j} + \theta_2 z_1^{-j}.$$

By definition

$$A(z_1, z_2, \boldsymbol{\theta}) = 1 + \theta_1 z_2^{-1} + \theta_3 z_2 + \theta_4 z_1 + \theta_2 z_1^{-1} + \theta_5 z_1 z_2.$$

Hence

$$\text{Det. } B(\theta) = \prod_{i=1}^{N_2} \prod_{j=1}^{N_1} \mu_{ij}$$

$$= \prod_{i=1}^{N_2} \prod_{j=1}^{N_1} (A(z_1{}^j, z_2{}^i, \theta)).$$

REFERENCES

1. P. Whittle, On stationary processes in the plane, *Biometrika* **41**, 1954, 434–449.
2. W. E. Larimore, Statistical inference on stationary random fields, *Proc. IEEE* **65**, 1977, 961–970.
3. E. J. Delp, R. L. Kashyap, and O. R. Mitchell, Image data compression using autoregressive time series models, *Pattern Recognition* **11**, 1979, 313–323.
4. J. T. Tou *et al.*, Pictorial texture analysis and synthesis, *Proc. 3rd Int. Conference on Pattern Recognition, Nov. 1976.*
5. B. H. McCormick and S. N. Jayaramamurthy, Time series models for texture synthesis, *Internat. J. Comput. Inform. Sci.* **3**, 1974, 329–343.
6. M. S. Bartlett, *The Statistical Analysis of Spatial Patterns*, Chapman and Hall, London, 1975.
7. E. J. Delp and O. R. Mitchell, Some aspects of moment preserving quantizers, in *IEEE Communication Society Int. Conf., 1979.*
8. R. Chellappa, R. L. Kashyap, and N. Ahuja, Decision rules for choice of appropriate neighbors, Tech. Rept. 802, Computer Science, Univ. of Maryland, August 1979.
9. J. E. Besag, On the correlation structure of some 2-dimensional stationary processes, *Biometrika* **59**, 1972, 43–48.
10. R. M. Haralick, K. Shanmugam, and I. Dinstein, Textural features for image classification, *IEEE Trans. Systems Man Cybernet.* **SMC-3,** Nov. 1973, 610–621.
11. R. Bellman, *Theory of Matrices*, 1960.
12. R. L. Kashyap, A Bayesian comparison of different classes of models using empirical data, *IEEE Trans. Automatic Control* **AC-22,** 1977, 715–727.
13. J. S. Weszka, C. R. Dyer, and A. Rosenfeld, A comparative study of texture measures for terrain classification, *IEEE Trans. Systems Man Cybernet.* **SMC-6,** April 1976, 269–285.
14. K. Deguchi and I. Morishita, Texture characterization and texture based image partitioning using two-dimensional linear estimation techniques, *IEEE Trans. Computers* **C-27,** August 1978, 739–765.
15. R. L. Kashyap, Two-dimensional autoregressive models for images: Parameter estimation and choice of neighbors, in *IEEE Computer Society Workshop on Pattern Recognition and Artificial Intelligence, April 1978.*

Image Models in Pattern Theory *

DONALD E. MCCLURE

Division of Applied Mathematics, Brown University, Providence, Rhode Island 02912

The pattern theory developed by U. Grenander provides a unified framework for the description of observed images in terms of algebraic and probabilistic processes that generate the images. The main elements of the general theory are outlined and specific image models are related to the general theory.

1. INTRODUCTION

The theory of regular structures—pattern theory—developed by Ulf Grenander is a unified framework for the description of observed images in terms of algebraic and probabilistic processes that generate the images. The purpose of this paper is to give an overview of certain aspects of pattern theory that are useful in the description and analysis of pictorial data. The first part outlines the basic elements of the theory; the second describes how familiar image models, for example, continuous and discrete random fields, fit within the framework of the theory. Then, results for three specific examples of problems in image analysis are reviewed. The three examples are concerned with (i) optimal spatial quantization for transforming continuous pictures into discrete ones, (ii) a characterization of certain discrete random fields on a square lattice, and (iii) restoration of the image of a planar convex set which is incompletely observed.

2. ELEMENTS OF PATTERN THEORY

The fundamentals of pattern theory are developed in depth by Grenander [3]. The viewpoint adopted in [3] and the companion volume [5] is that the analysis of structure or regularity in observed patterns should be guided by a unified theory of pattern formation. The theory should describe how patterns are synthesized from atoms, indecomposable building blocks, that are combined according to prescribed rules for regularity. The deformations that link the mathematical model of idealized images to the model for observable images should be included in the theory so that the complete history of formation of an observable image can be modeled within a single theoretical framework.

Four main elements of the theory are used to describe the synthesis of observed

* Research supported in part by the National Science Foundation through Grant MCS 76-07203 and by the Air Force Office of Scientific Research through Grant 78-3514.

ISBN 0-12-597320-9

images. These are (1) *generators*, the basic building blocks; (2) *combinatory rules*, the structural framework for construction of images; (3) *identification rules*, an equivalence relation that distinguishes observable from unobservable features of images; and (4) *deformations*, mappings that associate the idealized images of the formal model with the actual images that are accessible to the observer.

The analysis of an observed pattern involves relating the observable image to its history of formation; for example, inferring (i) the generators of which it is composed, (ii) the particular combinatory rules used to connect those generators, and (iii) a possible deformation to which the model image was subjected in order to produce the observed image. Analysis is thus the inverse problem of pattern synthesis, which starts with the four basic elements of the theory and deduces properties of the resulting images.

An overview of the four main elements of pattern theory will allow us to relate specific cases to the general theory.

Generators

The generators are the elements of a set G. For modeling pictorial data, one frequently used set of generators is a family of functions g mapping a background space, e.g., the unit square Q or a subset of Q, into a contrast space such as the set of reals $\mathbf{R}$ or an interval subset of $\mathbf{R}$; the value $g(x, y)$ is identified with the gray level at the point (x, y) in the background. Another choice we use for one of the examples described below is to take G to be a set of closed half-planes,

$$G = \{g_{(\theta,\delta)} : 0 \leq \theta < 2\pi, \ -\infty < \delta < \infty\}, \tag{1a}$$

where

$$g_{(\theta,\delta)} = \{(x, y) \in \mathbf{R}^2 : x \cos\theta + y \sin\theta \leq \delta\}. \tag{1b}$$

This set of generators is useful for building binary images, black–white pictures, of highly structured regions in the plane.

The set G may be composed of pairwise disjoint *generator classes* G^α, where α takes values in an index set A. Generators from the same class are regarded as qualitatively alike and generators from distinct classes are considered unlike. For example, the set G in Eq. (1a) admits the decomposition $G = \bigcup_\theta G^\theta$, where the index set is $[0, 2\pi)$ and $G^\theta = \{g_{(\theta,\delta)} : -\infty < \delta < \infty\}$; the boundaries of the half-planes in a class G^θ have a common orientation.

The schema in Fig. 1 is used to depict a generator, illustrating its internal structure and the ways it can be combined with other generators.

Generators have *attributes* $a(g)$ whose values may be integers, real values, finite-dimensional vectors, or elements of more general sets. For example, among the attributes of the half-plane generators of Eq. (1b) are orientation θ and the scalar δ associated with distance from the origin.

Besides its attributes, a generator may have an *identifier*, simply a name, to distinguish it from other generators used to construct an image. The use of identifiers permits the use of identical copies of the same element of G in the formation of an image; the unique identifiers distinguish the separate copies.

Finally, a generator has *bonds* that determine how sets of generators may be interconnected. The bonds may be oriented, as in-bonds or as out-bonds, and the orientation will in turn induce a direction on connections between the bonds of two generators. The semicircles in the diagram of Fig. 1 depict bonds. The total number of bonds of a generator g is its *arity*, denoted $\omega(g)$, which may be any cardinal number. The arity is the sum of the in-arity $\omega_{\text{in}}(g)$ and the out-arity $\omega_{\text{out}}(g)$, which count the numbers of in-bonds and out-bonds of g, respectively. The generator in Fig. 1 has arity 4, with $\omega_{\text{in}}(g) = 3$ and $\omega_{\text{out}}(g) = 1$.

Attached to each bond of g is a *bond value*, commonly denoted by β, which is used as described below to determine whether a bond of one generator is compatible with another bond of a second generator.

At this level in the mathematical formalism, we may wish to introduce a set S of *similarity transformations* acting on G. Similarity transformations defined initially on G will be extended through the other three levels of the theoretical hierarchy to define similarity transformations on ideal and observable images. For example, if the images in a particular model are to be planar subsets built from the generators of Eq. (1) and if two images are regarded as equivalent if one is a translate of the other, then it is convenient to take S to be the group of translations acting on the elements of G. In general, the only restrictions imposed on a set S of similarity transformations are that (i) S is a semigroup with identity, (ii) members of S map every generator class G^{α} into itself, and (iii) the mappings in S do not alter the bonds of any generator; that is, out-bonds remain as out-bonds and in-bonds remain as in-bonds (though bond values may be changed).

Combinatory Rules and Configurations

Given a set $\{g_{\nu}\}$ of generators, connections can be established between out-bonds of the generators and in-bonds of other generators in the set. The interconnected set of generators is a *configuration* c. The set of all configurations is denoted $\mathfrak{C}$. Examples of schemata for configurations are shown in Fig. 2.

A configuration has both *content* and *structure*. The content of c is simply the subset of G out of which it is formed. The structure of c, commonly denoted $\sigma =$ structure (c), is determined by the connections set up between the bonds within c. We can represent σ by a directed graph whose nodes are the set of bonds of all the generators in c and whose edges lead from an out-bond to the in-bond, if any, to which the out-bond is linked. It is convenient to use the incidence matrix of the directed graph as a canonical representation of σ.

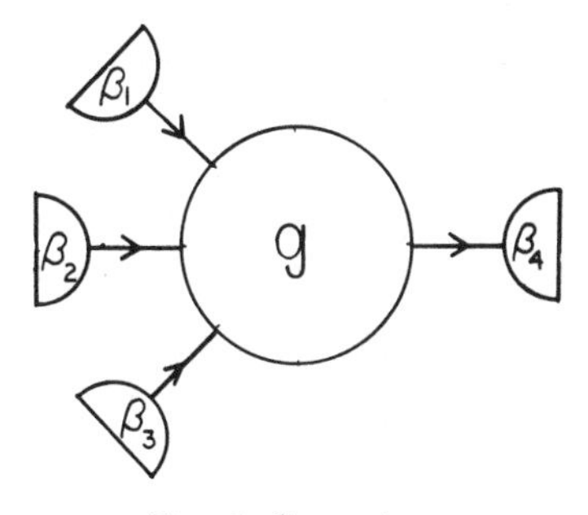

FIG. 1. Generator *g*.

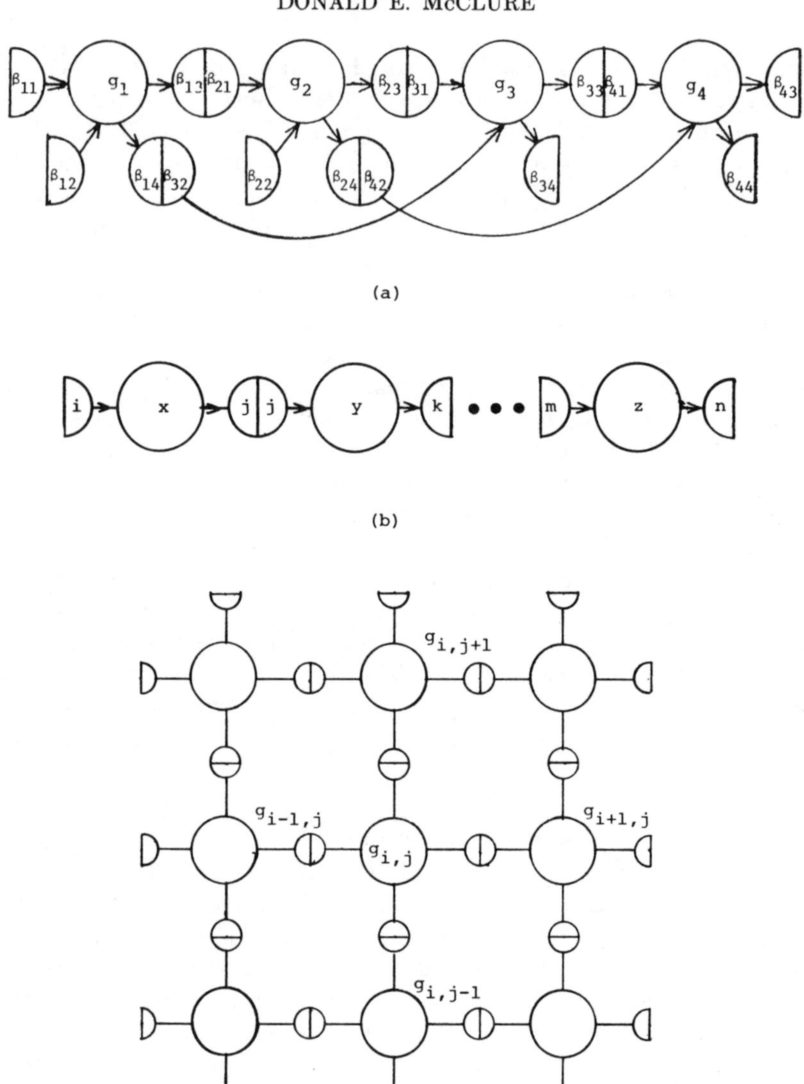

FIG. 2. Regular configurations. (a) General structure Σ. (b) Σ = LINEAR. (c) Σ = SQUARE LATTICE.

A system of *combinatory rules* $\mathcal{R}$ singles out the *regular configurations* from the set $\mathcal{C}$. $\mathcal{R}$ is defined by (i) a set Σ of admissible structures and (ii) a relation ρ between pairs of bond values. A configuration c is regular according to the system of rules $\mathcal{R} = (\Sigma, \rho)$ if (i) $\sigma \in \Sigma$, and (ii) the relation $\beta\rho\hat{\beta}$ is TRUE between the bond value β of the out-bond and the bond value $\hat{\beta}$ of the in-bond, for every connected pair of out-bonds and in-bonds. Σ decides where it is permissible to establish connections between bonds and, independently, ρ decides whether or not a pair of linked bonds is compatible. The set of regular configurations is denoted $\mathcal{C}(\mathcal{R})$.

For illustration, consider the following examples.

(I) Σ = MONATOMIC. This is the simplest possible structure. Each regular configuration consists of a single isolated generator. Thus, $\mathcal{C}(\mathcal{R}) = G$. A bond relation ρ plays no role.

(II) Σ = LINEAR, ρ = EQUAL. Each generator has arity 2, with one in-bond and one out-bond. Suppose the content of c is finite. The content can be linearly ordered and the out-bond of the ith generator g_i is linked to the in-bond of g_{i+1}. If the in- and out-bond values of g_i are denoted $\beta_{i,1}$ and $\beta_{i,2}$, then ρ requires that $\beta_{i,2} = \beta_{i+1,1}$ for each i.

Figure 2b depicts a regular configuration of this type modeling a substring of a grammatical sentence from a finite-state language. The generator with bond values i and j and with attribute x corresponds to a transition in a finite-state machine from state i to state j with an output of word x. A complete sentence is identified with a regular configuration whose initial in-bond has the value of the initial state and whose terminal out-bond has the value of a final state.

(III) Σ = SQUARE LATTICE, ρ = EQUAL. A finite regular configuration of this type is depicted in Fig. 2c. This structure is the basis of models for discrete pictures introduced later.

Identification Rules and Images

Two different configurations, c and $\hat{c}$, may be indistinguishable to an observer if he can observe only certain features of c and cannot obtain complete knowledge of its structure. For instance, in the example of a finite-state language one might see only the strings of terminal symbols in a grammatical sentence and not see the actual productions or sequence of state transitions within a finite-state machine that generated the sentence. If the finite-state machine is nondeterministic then it may be possible for distinct configurations c and $\hat{c}$ both to correspond to the same string of terminal symbols. Then c and $\hat{c}$ are equivalent in the eyes of the observer.

In the general theory we assume that there is an equivalence relation R defined between configurations in $\mathcal{C}(\mathcal{R})$. There are mild restrictions imposed on R, details of which need not concern us here; see [3]. Briefly, the restrictions require that (i) if two configurations c and $\hat{c}$ are equivalent modulo R, denoted $cR\hat{c}$, then the external (i.e., unconnected) bonds of c and $\hat{c}$ are the same, and (ii) if s is any similarity transformation defined on $\mathcal{C}(\mathcal{R})$ and if $cR\hat{c}$, then $s(c)Rs(\hat{c})$.

The equivalence relation R induces a partition of $\mathcal{C}(\mathcal{R})$ into equivalence classes. The set of equivalence classes $\mathcal{I} = \mathcal{C}(\mathcal{R})/R$ is called the algebra of pure or ideal images. The set $\mathcal{I}$ inherits its algebraic structure from the combinatory rules that dictate how generators and configurations can be combined. It is possible to extend similarity transformations s defined on $\mathcal{C}(\mathcal{R})$ to well-defined similarity transformations on $\mathcal{I}$.

In pattern theory, the term *image* refers to an arbitrary element I of the image algebra $\mathcal{I}$. A broader meaning is given to the term than is usual in the area of "image" processing. In the examples that follow, I shall focus on image algebras for the modeling of pictorial data and the subtleties of the terminology should cause no problems.

Deformations

The last main ingredient in the models of pattern theory is the deformation mechanism that relates observable to ideal images. A *deformation* is a mapping d from $\mathcal{I}$ to a set $\mathcal{I}^{\mathfrak{D}}$. The elements $I^{\mathfrak{D}}$ of $\mathcal{I}^{\mathfrak{D}}$ are called deformed images. They model the images that can actually be observed.

There may be many possible deformations d. The full set of them is denoted by $\mathfrak{D}$. Often the particular deformation d applied to an ideal image is not known in advance and it is chosen at random, according to a probability measure on the set $\mathfrak{D}$.

The set $\mathcal{I}^{\mathfrak{D}}$ can be, but generally is not, the same as $\mathcal{I}$. Whether or not the algebraic structure of $\mathcal{I}$ and similarity transformations defined on $\mathcal{I}$ can be extended to $\mathcal{I}^{\mathfrak{D}}$ relies intimately on the specific deformations. Some general results for these questions are developed by Grenander [3].

Commonly, deformations destroy information and corrupt the structure of the ideal images. The problem of *image restoration* is concerned with recovering an ideal image I as completely as possible, given the observation of the deformed version dI. If the deformation d is random and if something is known about its probability distribution, then basic statistical methods, such as methods of maximum likelihood or moments, can guide the development of image restoration algorithms. A concrete example is described in the last section.

Deformations that are encountered in picture processing arise from discretization of a continuous gray scale into a finite set of allowable gray levels, spatial quantization of continuous backgrounds into a finite number of discrete pixels, measurement error—for example, additive noise—in the observation of signals, incomplete observability of complex images, nonlinear distortions of backgrounds and gray scales, and so on. I shall discuss two of these deformation mechanisms in greater depth in the examples that follow.

There are several ways that probability measures can be defined or induced on an image algebra. This is an important aspect of the general models, in addition to the four basic elements outlined above, when we wish to quantify the likelihood or frequency of occurrence of either pure or deformed images. A probability measure may be defined directly on a σ-algebra of subsets of $\mathcal{I}$, neglecting for now the mathematical technicalities involved, which depend on properties of the set $\mathcal{I}$. In some cases it is natural to start with a probability measure on a σ-algebra of subsets of G, then assume that generators are sampled independently from G according to this measure and that their probability distributions are conditioned by the constraints of Σ and the bond relation ρ when regular configurations are formed. This will in turn induce probability measures on $\mathcal{C}(\mathcal{R})$ and on $\mathcal{I}$. A specific example of this method of constructing a probability measure on an image algebra is discussed in the last section.

3. GENERAL IMAGE MODELS

Familiar and widely used image models such as continuous random fields and digital rectangular arrays can be cast in the framework of pattern theory. I shall describe several cases and attempt to reinforce with familiar models what some of the possible interpretations for the basic elements of pattern theory are—generators, bonds, combinatory rules, configurations, and so on. The first two cases are models for continuous pictures on a continuous background, the next two cases model subregions of a continuous background, and the last two cases model digital pictures.

(a) Continuous gray-level pictures on a continuous background can be modeled by schemes of varying complexity. The simplest model has an almost trivial set Σ of admissible structures for regular configurations. Still, the model is adequate for many kinds of analysis, as the first example in the last section shows.

We identify the picture background with the unit square Q in $\mathbf{R}^2$. The brightness or gray level of a picture at a point (x, y) in Q can be described by a real value $g(x, y)$, usually assumed nonnegative. Thus a single picture is associated with a real-valued function $g: Q \to \mathbf{R}$. A family of pictures is determined by a set $G \subseteq \mathbf{R}^Q$ of functions on Q. We take the generators to be such a set. Each generator will have arity zero—no bonds—and Σ = MONATOMIC is the set of admissible structures. Then every regular configuration consists of a single generator (function). If the identification rule R for equivalent configurations is "equality of functions," then $\mathcal{I} = \mathcal{C}(\mathcal{R}) = G$.

A random field model can be constructed around this image algebra by first specifying (by standard techniques) a σ-algebra $\mathcal{G}$ of subsets of G. Then we can either specify a probability measure P_G on $\mathcal{G}$ or, equivalently, introduce an arbitrary probability space $(\Omega, \mathcal{F}, P)$ and a measurable function Γ from Ω to G. For each $\omega \in \Omega$, the sample function $g = \Gamma(\omega)$ is a realization of a random image.

(b) If the image algebra $\mathcal{I}$ is a vector space of real-valued functions on Q, for example, $L_2(Q)$, then it is natural to take as generators a set G of functions that span $\mathcal{I}$. We set

$$G = \bigcup_{\alpha \in A} G^\alpha,$$

where the generator class G^α consists of all scalar multiples of a fixed function g^α:

$$G^\alpha = \{cg^\alpha : c \in \mathbf{R}\}. \tag{2}$$

Regular configurations will be formed by selecting a subset of G, choosing at most one generator from each G^α, and summing the selected generators. The restrictions on the selection of generators can be enforced by taking the arity of each generator to be the same as the cardinality of the index set A, $\omega(g) = \text{card } A$, so that each generator in a configuration c has a bond to link with every other generator in c. This is the structure identified as Σ = FULL. The orientation of the bonds does not matter in this model; we can regard each bond as being bidirectional, both an in-bond and an out-bond. If $g \in G^\alpha$, then the bond value of every bond of g is α, its class index. The bond relation ρ = UNEQUAL then assures that all the generators in a regular configuration come from distinct classes.

As in example (a) above, we take the identification rule R to be "equality of functions" if two configurations are deemed different when the pictures with which they are associated are different. If we require the set of generator classes

$\{G^\alpha: \alpha \in A\}$ to be a linearly independent family of subsets of $\mathcal{C}(\mathcal{R})$, then distinct regular configurations will automatically correspond to distinct functions on Q and we will have $\mathcal{I} = \mathcal{C}(\mathcal{R})$.

Two special instances of this model deserve special mention. If one is doing Fourier analysis of images on the background Q, then it is natural to imbed $\mathcal{I}$ in the set of square-integrable complex-valued functions on Q, take $A = Z^2$, the infinite lattice of integer pairs, and define the generator classes by Eq. (2) and

$$g^{(m,n)}(x, y) = \exp\{2\pi i(mx + ny)\}.$$

If one is computing orthogonal expansions of a random field with expected value zero and known continuous covariance function $r((x, y), (\hat{x}, \hat{y}))$, then it is natural to set $A = Z^+$, the positive integers, and take g^α to be an eigenfunction of the kernel r. Specifically, g^α is chosen as a nontrivial solution of

$$\lambda_\alpha g^\alpha(x, y) = \int_Q\int r((x, y), (\hat{x}, \hat{y}))g^\alpha(\hat{x}, \hat{y})d\hat{x}d\hat{y},$$

where $\{\lambda_\alpha\}_{\alpha=1}^\infty$ is the full set of eigenvalues of r. Regular configurations then correspond directly to the Karhunen–Loève expansion of the random field. See Rosenfeld and Kak [9] for a discussion of these image representations.

(c) To model images of planar regions, black-and-white images that divide a background into complementary parts, we can use models with simple structure and large sets of generators or models with richer structure and relatively small sets of generators. The alternatives are analogous to the two types of models for gray-level pictures in (a) and (b).

In particular, suppose the regions to be modeled are compact convex subsets of the plane. The model with simplest structure sets $G = \mathcal{K}$, the family of compact convex planar sets, Σ = MONATOMIC and R as "equality of sets." Then, as in (a) above, $\mathcal{I} = \mathcal{C}(\mathcal{R}) = G$. Artstein and Vitale [1] used a random set model constructed around this image algebra to analyze limiting shapes for sequences of random sets.

(d) Alternatively, we can base a model for $\mathcal{K}$ on the generators defined by Eqs. (1). The generators G are partitioned into generator classes G^θ, $0 \leq \theta < 2\pi$, where all half-spaces in G^θ have a common orientation with their outward normal vectors in direction θ.

Each generator has arity $\omega(g) = 2^{\aleph_0}$, with one bidirectional bond for every generator class. The bond value of each bond of g is θ, its class index. Regular configurations are formed by taking generators from some or all of the G^θ, combining them with structure Σ = FULL and testing the compatibility of their bonds with the relation ρ = UNEQUAL.

Each configuration is viewed as the intersection of the generators in its content. Thus the regular configurations determine closed convex sets in $\mathbf{R}^2$. As in (c), we use the equivalence relation "equality of sets" for the identification rule R. But now there is a many-to-one correspondence between $\mathcal{C}(\mathcal{R})$ and $\mathcal{I}$ because distinct sets of generators can have the same intersection.

This model is used in the image reconstruction example below. It was used previously by McClure and Vitale in [6], where we studied how closely an arbitrary image can be approximated by regular configurations with finite content, that is, polygons.

(e) In models for digital pictures, a structure Σ that embodies the spatial arrangement of pixels is appropriate. The finite square lattice structure is used to model images on an $N \times N$ array of elementary subregions of the background Q.

A single generator will describe the brightness or gray-level of the image on a single pixel. If the global background is the unit square Q, then an individual pixel is identified with a subsquare with sides of length $h = 1/N$. The generators are defined by

$$g_{i,j}(x, y) = a_0 I_{i,j}(x, y), \tag{3}$$

where $I_{i,j}$ is the indicator function of the (i, j)th pixel

$$\begin{aligned} I_{i,j}(x, y) &= 1 \qquad \text{for} \quad (i-1)h \le x \le ih, \quad (j-1)h \le y \le jh \\ &= 0 \qquad \text{otherwise} \end{aligned}$$

and a_0 is any value in the set $Y \subseteq \mathbf{R}$ of allowable gray levels. Then

$$G = \{g_{i,j}: 1 \le i \le N, 1 \le j \le N, a_0 \in Y\}.$$

Each generator has four bidirectional bonds corresponding to the edges of its pixel. Specifically, the bond values of $g_{i,j}$ are

$$\begin{aligned} \beta_{(i+),j} &= \{(x, y): x = ih, (j-1)h \le y < jh\}, \\ \beta_{(i-),j} &= \{(x, y): x = (i-1)h, (j-1)h \le y < jh\} \end{aligned} \tag{4}$$

with analogous definitions for the top and bottom edge sets $\beta_{i,(j+)}$ and $\beta_{i,(j-)}$ of the pixel.

Regular configurations are formed by combining N^2 generators by the structure Σ = SQUARE LATTICE. The right-edge bond of $g_{i,j}$ is linked to the left-edge bond of $g_{(i+1),j}$, and so on. The bond relation ρ = EQUAL assures that the pixels are contiguous. The regular configurations can be regarded as $N \times N$ arrays whose entries assume values in Y, or, using the correspondence defined by Eq. (3) between an array entry a_0 and a function on Q, we can regard a regular configuration as a piecewise constant function on the background Q. Normally the identification rule R is "equality of functions" and consequently $\mathcal{I} = \mathcal{C}(\mathcal{R})$.

(f) Small variations on model (e) describe smoother digital pictures than those that assume different constant values on separate pixels. For instance, enlarge the set of generators to contain the piecewise linear functions

$$g_{1,1}(x, y) = (a_0 + a_1 x + a_2 y + a_3 xy) I_{1,1}(x, y)$$

and

$$g_{i,j}(x, y) = g_{1,1}(x - (i-1)h, y - (j-1)h),$$

where the coefficients a_0, a_1, a_2, a_3 are arbitrary values in a set $Y \subseteq \mathbf{R}$. Change the bond values of Eq. (4) to

$$\beta_{(i+),j} = \{(x, y, g_{i,j}(x, y)): x = ih, (j-1)h \le y < jh\}$$

and similarly for $\beta_{(i-),j}$, $\beta_{i,(j+)}$, and $\beta_{i,(j-)}$. Then the same structure Σ, bond relation ρ, and identification rule R as those in (e) will lead to the image algebra of continuous piecewise linear functions on the partitioned square Q.

4. EXAMPLES

Three examples are described in order to amplify the discussion of some of the models introduced above and to state substantive results which have been obtained in three problem areas. The first example focuses on the analysis of image deformations due to quantization, the second is concerned with the characterization of certain probability measures on the types of regular configurations used to model digital images, and the last example shows how a complete model for image formation along with classical statistical principles can yield algorithms for image restoration. The choice of examples is necessarily limited and many others can be found in Grenander [3, 5].

Optimal Spatial Quantization

Invariably, when pictures on a continuous background with a continuous gray scale are digitized, they are deformed by the quantization of the background and by the discretization of the gray scale. Deformations due to discretization of the gray scale and methods of minimizing these deformations are fairly well understood; see, for example, Rosenfeld and Kak [9]. More recently, we have developed some results that characterize optimal methods of quantizing the background. Analysis of optimal background quantization is more involved than the analysis of optimal gray-scale quantization; quantizing the two-dimensional background entails an interesting geometric aspect which quantizing a one-dimensional gray scale does not. Proofs of some of the results cited here are given in McClure [7]. The results will be published in full generality elsewhere [8].

Commonly a picture background is quantized into rectangular pixels of identical shapes and sizes arranged in a rectangular lattice. Occasionally, pictures are sampled at points of an equilateral triangular lattice, corresponding to regular hexagonal pixels. We are led to ask whether one of these schemes is preferable to the other in terms of preserving greater fidelity between an original continuous image and its deformed, digitized version. We can obtain answers to a more general question: Among all possible ways of partitioning the background Q into L polygonal pixels of arbitrary shape and size, can we characterize the method, if any, which minimizes the average distance between a family of continuous images and their digitized versions?

To make the formulation of the question precise, we shall assume models of the type introduced in the previous section for the family of continuous images and for the discrete images. The random field model of case (a) is used for the continuous images. We shall impose some mild restrictions on the set G of generators and on the structure of the random field below. $\mathcal{I}$ will denote this image algebra of continuous pictures.

For the discrete images we adopt a model which is patterned after, but is considerably more general than, case (e) in the previous section. The generators are identified with functions that assume a constant value on an arbitrary convex polygonal subset of Q. The structures Σ of regular configurations are associated, as in case (e), with the contiguity relationships among the cells of a finite partition (quantization) of Q.

The generators are defined by

$$g_S(x, y) = a_0 I_S(x, y),$$

where S is an arbitrary convex polygonal subset of Q, I_S is the indicator function of S, and a_0 is any real value. The generator g_S has a bond for each edge of S and the associated bond value is the set of points on that edge, in analogy with Eqs. (4).

Regular configurations are formed from sets of L generators $\{g_{S_k}\}$ associated with the subsets S_k in a partition $\mathcal{S} = \{S_k\}_{k=1}^{L}$ of Q. Two bonds between g_{S_k} and g_{S_j} are linked if they correspond to a common edge of S_k and S_j. The bond relation is ρ = EQUAL. Figure 3 illustrates the correspondence between a partition $\mathcal{S}$ and the associated structure of a regular configuration. Since structures of regular configurations correspond to planar graphs, we refer to the set of admissible structures as Σ = PLANAR. The image algebra associated with a fixed partition $\mathcal{S}$ of Q is denoted $\mathcal{g}^{\mathcal{S}}$ and the image algebra associated with *all* such partitions of Q into L subsets is denoted by $\mathcal{g}^{L}$.

Now let $\mathcal{S}$ be an arbitrary partition of Q into L subsets. A mapping (deformation mechanism) D from $\mathcal{g}$ into $\mathcal{g}^{\mathcal{S}}$ is defined by

$$Dg(x, y) = \sum_{k=1}^{L} g(x_k, y_k) I_{S_k}(x, y),$$

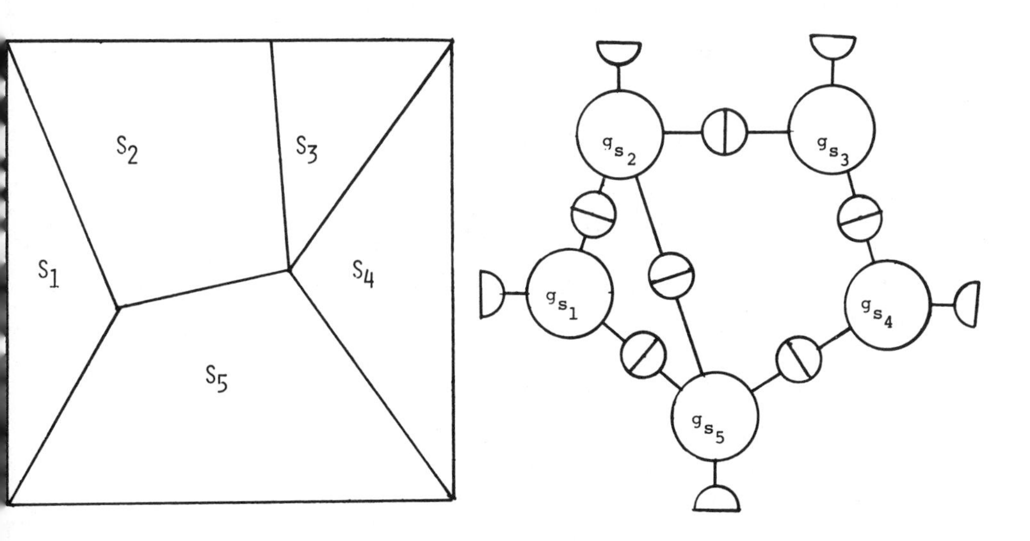

FIG. 3. (a) Partition $\mathcal{S}$ of Q. (b) Corresponding regular configuration with PLANAR structure.

where (x_k, y_k) is the centroid of S_k. Discrete images are formed by sampling continuous images at the centroids of the cells S_k. An average squared distance between $\mathcal{G}$ and $\mathcal{G}^{\mathcal{S}}$ is

$$\delta(\mathcal{G}, \mathcal{G}^{\mathcal{S}}) = \int_{\Omega} \left\{ \int_Q \int [g_\omega(x, y) - Dg_\omega(x, y)]^2 dxdy \right\} P(d\omega),$$

where Ω is the sample space and P is the probability measure in the definition of the random field. The average squared distance between $\mathcal{G}$ and $\mathcal{G}^L$ is defined by

$$\delta(\mathcal{G}, \mathcal{G}^L) = \inf_{\mathcal{S}} \delta(\mathcal{G}, \mathcal{G}^{\mathcal{S}}) \tag{5}$$

with the infimum taken over all partitions $\mathcal{S}$ of Q into L polygonal subsets.

Results obtained in [7, 8] describe the asymptotics of (5) for large L and characterize partitions $\mathcal{S}^*$ which are optimal in the sense that

$$\delta(\mathcal{G}, \mathcal{G}^{\mathcal{S}^*}) = \delta(\mathcal{G}, \mathcal{G}^L). \tag{6}$$

The results are expressed in terms of the 2×2 matrix-valued function

$$H(x, y) = \int_{\Omega} [\nabla g_\omega(x, y)][\nabla g_\omega(x, y)]^t P(d\omega). \tag{7}$$

Here ∇g_ω is the column vector with components $\partial g_\omega / \partial x$ and $\partial g_\omega / \partial y$. We must assume that, with probability 1, the sample functions g_ω of the random field have continuous partial derivatives and the integral in (7) is finite. We also assume for convenience that H is positive definite in Q; it is necessarily nonnegative definite.

These assumptions imply that for large L,

$$\delta(\mathcal{G}, \mathcal{G}^L) = \frac{1}{L} \frac{5(3)^{\frac{1}{2}}}{54} \left[\int_Q \int [\det H(x, y)]^{\frac{1}{4}} dxdy \right]^2 + o\left(\frac{1}{L}\right).$$

This result then leads to characterizations of the structure of optimal partitions $\mathcal{S}^*$ satisfying Eq. (6). The results say that optimal partitions necessarily have structures such as those depicted in Fig. 4. Figure 4a depicts an optimal partition when the matrix H is a constant multiple of the identity matrix. Figure 4b illustrates an optimal partition when H is given by

$$H(x, y) = f(x, y) \begin{bmatrix} 1 & 0 \\ 0 & 1 \end{bmatrix}, \tag{8}$$

where f is positive on Q. The particular function f for which Fig. 4b depicts an approximately optimal partition is piecewise constant on Q, assuming a value four times as large in the central square as in the surrounding ring.

Equation (8) expresses a weak condition of isotropy on the random field. If (8) holds, then the matrix H is invariant under rotations of the coordinate system. Since H is defined by the gradient of the random field, we use the term *differentially isotropic* to describe a random field satisfying (8). An isotropic random field with differentiable sample functions will be differentially isotropic; hence (8) is weaker than the normal condition for isotropy.

In brief, the necessary conditions for optimality of $\mathcal{S}^*$, assuming L is large and the random field satisfies (8), are: (i) the proportion of hexagons in $\mathcal{S}^*$ approaches 1 as $L \to \infty$, (ii) the shapes of almost all hexagons in $\mathcal{S}^*$ converge to the shape of a regular hexagon as $L \to \infty$, and (iii) the area of an arbitrary cell S_k in $\mathcal{S}^*$ is inversely proportional to $[f(x_k, y_k)]^{\frac{1}{2}}$, where (x_k, y_k) is the centroid of S_k. Thus, the points where the continuous images are sampled should be arranged locally in an equilateral triangular lattice and the spacing between lattice points will vary as $f^{-\frac{1}{4}}$. When the isotropy assumption is dropped, then the condition that the cells S_k be approximately regular in shape is replaced by the condition that the linearly transformed cells $P(x_k, y_k)S_k$ be approximately regular, where $P(x, y)$ is any square root of the matrix $H(x, y)$; $P^tP = H$.

These characterizations of optimal partitions translate directly into methods of constructing approximately optimal spatial quantizations. Details are given in [7].

Probabilities on Regular Configurations

Recently Thrift [10] proved some general, strong results that characterize probability measures induced on spaces $\mathcal{C}(\mathcal{R})$ of regular configurations. More specific results of the same type were reported by Grenander [3]. The measure is induced on $\mathcal{C}(\mathcal{R})$ in the following steps: (i) a probability measure P_G is prescribed for the set G of generators; (ii) subsets of generators $\{g_\nu\}$ are sampled independently from G, distributed according to P_G; (iii) then the joint distributon of $\{g_\nu\}$ is conditioned by the structural constraints Σ and the bond relation ρ that define the set $\mathcal{C}(\mathcal{R})$. The basic problem is to understand how the bond structure (Σ, ρ) affects the stochastic dependence among the generators in a regular configuration.

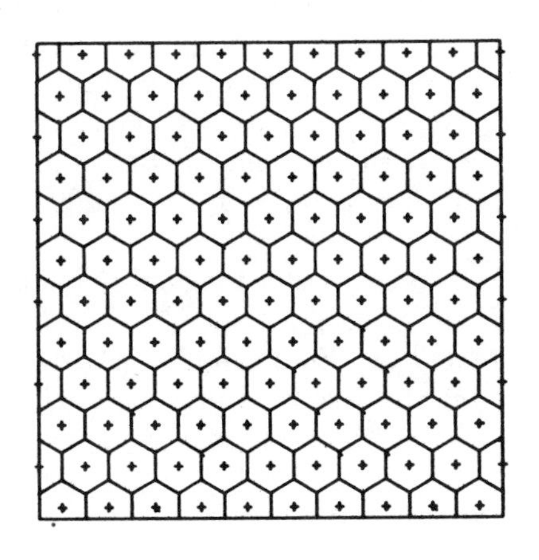

(a)

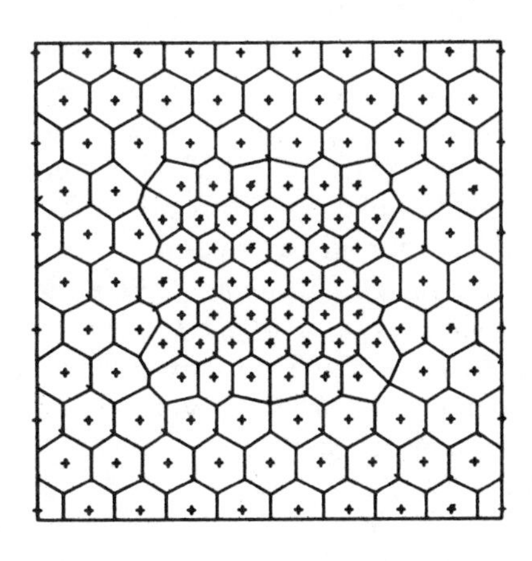

(b)

FIG. 4. Approximately optimal partitions $\mathcal{S}^*$. (a) Constant f. (b) Nonuniform f.

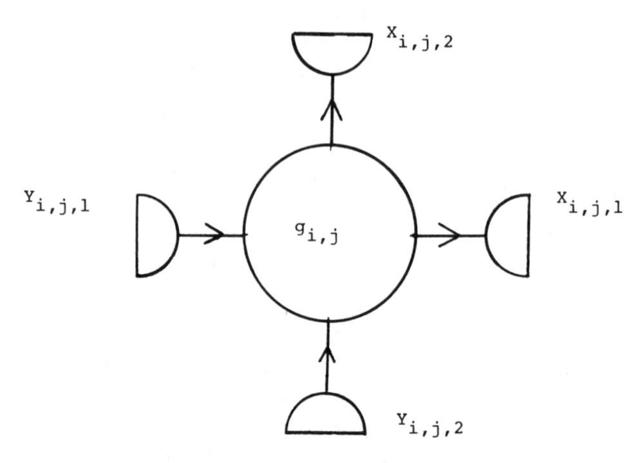

FIG. 5. Generator at lattice point (i, j).

Thrift obtained solutions for this problem for a variety of situations, including all of the structures depicted in Fig. 2 and others. In all cases the generators are distributed according to a finite-dimensional Gaussian distribution P_G and the bond relation is ρ = EQUAL. I shall describe Thrift's results for the special case related to cases (e) and (f) of the previous section, the models for digital pictures.

Each generator, as shown in Fig. 5, has arity 4, with two in-bonds and two out-bonds. The orientations of the bonds do not play an essential role in the interpretation of the model or in the results proved, but the initial assignment of orientations facilitates the analysis. The bond values of the out-bonds of a generator $g_{i,j}$ are denoted $X_{i,j,k}$ for $k = 1$ or 2, and these are associated with the East and North bonds as shown in the figure. Similarly, bond values $Y_{i,j,k}$ for $k = 1$ and 2 are associated with the West and South in-bonds.

For regular configurations, the generators $\{g_{i,j}\}_{i,j=-\infty}^{\infty}$ are combined in an *infinite* SQUARE LATTICE structure Σ. Actually in Thrift's analysis the infinite structure is approximated by a sequence of finite structures, so there is a connection to the finite lattice models introduced before. Initially, the vectors $(X_{i,j,1}, X_{i,j,2}, Y_{i,j,1}, Y_{i,j,2})$ for $-\infty < i, j < \infty$ are independent and identically distributed with a four-dimensional Gaussian distribution having mean 0 and covariance matrix H^{-1}. Then the generators are combined and the distributions are conditioned by the bond relation; in particular, $X_{i,j,1} = Y_{i+1,j,1}$ and $X_{i,j,2} = Y_{i,j+1,2}$ for all pairs i and j.

After the conditioning, there are essentially two degrees of freedom remaining for each generator $g_{i,j}$. Knowing the pair of values $X_{i,j,1}$, $X_{i,j,2}$ for all i and j tells us all of the bond values for all of the generators. So a complete characterization of the probability measure induced on the regular configurations will follow from a characterization of the vector-valued discrete-index random field $\mathfrak{X}_{i,j} = (X_{i,j,1}, X_{i,j,2})$.

Thrift proves that $\mathfrak{X}_{i,j}$ can be represented by an autoregression. The proof involves a very intricate analysis of the factorization of polynomials in two complex variables whose coefficients are 2×2 matrices determined by the covariance matrix H^{-1}. The polynomials enter into expressions for the spectral density matrix for the process $\mathfrak{X}_{i,j}$. Out of the factorization results comes a clear understanding of what the *past* of each $\mathfrak{X}_{i,j}$ is, that is, of which other values $\mathfrak{X}_{k,l}$ of the process enter into the autoregression for $\mathfrak{X}_{i,j}$. In this square lattice model the past of each $\mathfrak{X}_{i,j}$ is infinite, even though the original conditioning depends only on nearest neighbors in the lattice. In particular, the past of $\mathfrak{X}_{0,0}$ is $\{\mathfrak{X}_{i,j}: i < 0 \text{ or } [i = 0 \text{ and } j < 0]\}$, a "nonsymmetric half-plane" like the one found in autoregressive representations derived by Ekstrom and Woods [2]. Only terms identified with the *boundary* of this past actually enter the autoregression in the Gaussian model. The boundary of the past of $\mathfrak{X}_{0,0}$ is contained in $\{\mathfrak{X}_{i,j}: [i = -1 \text{ and } j \geq 0] \text{ or } [i = 0 \text{ and } j < 0]\}$.

The viewpoint and the theoretical framework of pattern theory made a substantial contribution to the analysis carried out by Thrift by providing the setting in which the very general characterization problems could be clearly formulated and studied.

Image Restoration

The last example develops an algorithm for the restoration of images of planar convex regions which are deformed by a mechanism of incomplete observation. The ideal images are modeled by the image algebra $\mathcal{I}$ of case (d) in the previous section, where the generators are half-planes.

Let K denote a pure image, that is, a convex subset of $\mathbf{R}^2$. The observer cannot see K directly, but instead can see a realization $K^{\mathfrak{D}}$ of an inhomogeneous Poisson process with constant intensity $\lambda(x, y) = \lambda_1$ within K and constant intensity $\lambda(x, y) = \lambda_2 < \lambda_1$ in K^c. For the moment we assume that λ_1 and λ_2 are known. Such a deformation mechanism might model the detection of radioactive disintegrations when the decaying isotopes have a higher concentration inside K than outside K. The deformation mechanism is studied by Grenander in [4, 5]. An example of pure and deformed images is shown in Fig. 6.

A nonparametric statistical method of estimating the ideal image K from its deformed counterpart $K^{\mathfrak{D}}$ is proposed by Grenander [4]. We describe a modification of that method here. The modification assumes a bit less a priori knowledge about the ideal image than the algorithm in [4].

The goal of the reconstruction is to find an element of the image algebra $\mathcal{I}$ with which the random point set $K^{\mathfrak{D}}$ is most compatible. "Compatibility" is quantified by likelihood and the principle of maximum likelihood guides the reconstruction. With the given structure Σ and bond relation ρ for $\mathcal{I}$, it suffices to infer from $K^{\mathfrak{D}}$ a set of generators $\{g^\theta\}$ for which the associated regular configuration is a good estimate of K. Searching for the right generators g^θ is the same as searching for the support planes (lines) of the convex body K.

Consider a fixed angle θ (Fig. 6) and let us focus on the estimation of the support planes g^θ and $g^{\theta+\pi}$ having outward normals at angles θ and $\theta + \pi$ relative to a fixed direction, the horizontal axis. We shall sweep a line segment of length L, with normal at angle θ, across K. We must know the approximate location of K to do this. Let u denote a coordinate in the direction θ (Fig. 6); the line segment will move across an interval from $u = a$ to $u = b$ which contains K. The segment has a nonempty intersection with K when u is in $[\xi_0, \xi_1]$.

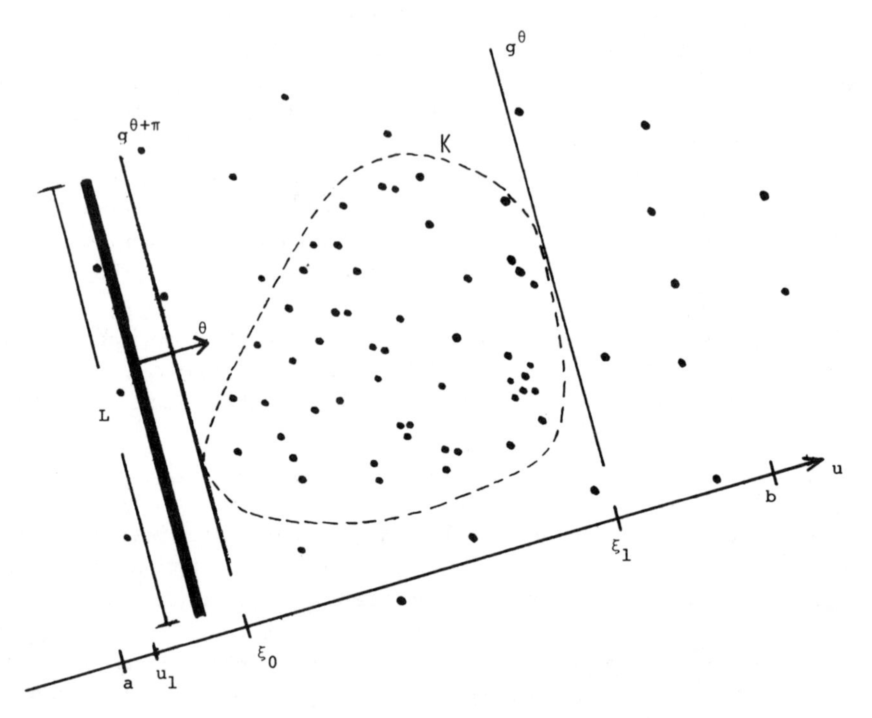

FIG. 6. Ideal image K and deformed image $K^{\mathfrak{D}}$.

The sweeping segment is used to extract a relevant one-dimensional sample from the two-dimensional point set $K^{\mathfrak{D}}$. The Poisson points that the segment intersects are projected onto the u-axis. Let $u_1, u_2, \ldots, u_N$ denote the projected points in $[a, b]$. The set $\{u_k\}_{k=1}^N$ is a realization of a one-dimensional inhomogeneous Poisson process with intensity function

$$\lambda(u) = \lambda_1 l(u) + \lambda_2(L - l(u)),$$

where $l(u)$ is the length of the intersection of K and the segment positioned at u. Since K is convex, l is unimodal on $[a, b]$. The function $\lambda(u)$ has the same qualitative behavior. It equals the constant λ_2 on the intervals $[a, \xi_0)$ and $(\xi_1, b]$ where the segment is disjoint from K, and within $[\xi_0, \xi_1]$ it is first nondecreasing up to a point M and then nonincreasing from M to ξ_1.

The estimation of $\lambda(u)$ from the sample $\{u_k\}_{k=1}^N$ begs for the application of the maximum likelihood methods for estimating monotone and unimodal density functions. The present problem is different in some ways from the density estimation problems since the sample is Poisson and not of fixed size. Nevertheless, Grenander shows that the maximum likelihood methods for estimating monotone and unimodal functions carry over to this setting.

The likelihood function is

$$\mathfrak{L}(\lambda) = \prod_{k=1}^N \lambda(u_k) \exp\left\{-\int_a^b \lambda(u)du\right\} \tag{9}$$

and it is to be maximized with respect to its dependence on λ where (i) λ is unimodal on $[a, b]$ and (ii) the mode M of λ is unknown. The solution can be found by the following method.

Step 1. For each m between 1 and N compute the function λ_m which is nondecreasing on $[a, u_m]$, is nonincreasing on $[u_m, b]$, and maximizes (9). An effective algorithm for computing λ_m is given in [4].

Step 2. Compute $\mathfrak{L}(\lambda_m)$ for each m and select $\hat{m}$ for which

$$\mathfrak{L}(\lambda_{\hat{m}}) = \max_{1 \leq m \leq N} \mathfrak{L}(\lambda_m).$$

The maximum likelihood estimate of λ is $\lambda_{\hat{m}}$.

Now we can use $\lambda_{\hat{m}}$ to estimate the points ξ_0 and ξ_1 that determine g^θ and $g^{\theta+\pi}$. Choose a positive constant ϵ whose size will depend on λ_2 and λ_1. Then solve for ξ_0^* and ξ_1^* as the smallest and largest values of u that satisfy

$$\lambda_{\hat{m}}(u) \geq \lambda_2 + \epsilon.$$

The support planes g^θ and $g^{\theta+\pi}$ are then estimated from ξ_1^* and ξ_0^*.

Finally, the set K is estimated by finding the estimates of the support planes for a set of directions $\{\theta_\nu\}$ and forming the intersection K^* of the associated half-planes.

The known consistency results and distribution theory for maximum likelihood estimation of a unimodal density function with unknown mode [11] can probably be carried over to this setting to prove a consistency result for K^*, assuming $\lambda_1 \to \infty$ and λ_2/λ_1 is bounded away from one. Also, since the estimator $\lambda_{\hat{m}}$ does not use a priori knowledge of λ_1 and λ_2, it may be possible to adapt the method of restoring K to the case when λ_1 and λ_2 are unknown.

REFERENCES

1. Z. Artstein and R. A. Vitale, A strong law of large numbers for random compact sets, *Ann. of Probability* **3**, 1975, 879–882.
2. M. P. Ekstrom and J. W. Woods, Two-dimensional spectral factorization with applications in recursive digital filtering, *IEEE Trans. Acoust. Speech and Signal Processing* **ASSP-24**, 1976, 115–128.
3. U. Grenander, *Lectures in Pattern Theory I: Pattern Synthesis*, Applied Mathematical Sciences Series, Vol. 18, Springer-Verlag, New York, 1976.
4. U. Grenander, *Restoration of Convex Sets with $\mathfrak{D}_3$-Deformations*, Reports on Pattern Analysis No. 62, Division of Applied Mathematics, Brown University, 1977.
5. U. Grenander, *Lectures in Pattern Theory II: Pattern Analysis*, Applied Mathematical Sciences Series, Vol. 24, Springer-Verlag, New York, 1978.
6. D. E. McClure and R. A. Vitale, Polygonal approximation of plane convex bodies, *J. Math. Anal. Appl.* **51**, 1975, 326–358.
7. D. E. McClure, *Characterization and Approximation of Optimal Plane Partitions*, Reports on Pattern Analysis No. 36, Division of Applied Mathematics, Brown University, 1976.
8. D. E. McClure, Quantization of random fields, to appear.
9. A. Rosenfeld and A. C. Kak, *Digital Picture Processing*, Academic Press, New York, 1976.
10. P. Thrift, *Autoregression in Homogeneous Gaussian Configurations*, Ph.D. dissertation, Division of Applied Mathematics, Brown University, 1979.
11. E. J. Wegman, Maximum likelihood estimation of a unimodal density, II, *Ann. Math. Statist.* **41**, 1970, 2169–2174.

A Survey of Geometrical Probability in the Plane, with Emphasis on Stochastic Image Modeling

R. E. MILES

Department of Statistics, Institute of Advanced Studies, Australian National University, P.O. Box 4, Canberra, A.C.T. 2600, Australia

An attempt is made to isolate key elements and themes of planar geometrical probability, both within a bounded region and over the whole plane, at a suitable level for those conversant with elementary continuous probability theory. Emphasis is placed upon possible stochastic models for planar image phenomena of a random character.

1. INTRODUCTION

Doubtless some of you are aware of the birth of Geometrical Probability in 1777, when the famed 18th century French naturalist Buffon published his work [10] on the so-called "needle problem," concerned with the probability distribution of the number of needle–line intersections when a needle is thrown "at random" onto a parallel grid of equispaced lines. It seems significant that, although Buffon had earlier considered similar fairground games relating to the intersections of coins thrown at random onto square, triangular, and hexagonal tessellations [38, pp. 346–347], this was the first consideration of *random orientation*. The bicentenary of Buffon's publication was recently celebrated by symposia held in Armenia and Paris [28].

About a century was to elapse before the next major development, due to the Englishman Crofton, who obtained the first of the elegant results for which the discipline is noted. The next really substantial development took place in Hamburg in the years preceding the Second World War, when Blaschke and his school established the intimately related Integral Geometry on a thorough and systematic basis. Understandably, that vigorous development was halted, but the story since then is one of ever increasing activity.

Academically, Geometrical Probability has been a "Cinderella" field, overshadowed and separate from the mainstream of Probability and Statistics, but with definite links via Integral Geometry with Differential Geometry and Algebraic and Differential Topology. Throughout its history, it seems to have attracted isolated ventures by distinguished mathematicians or scientists from other areas, early names besides Buffon being Cauchy, Euler, Laplace, Lebesgue, and Minkowski, possibly attracted by its simply stated yet difficult problems [28]. Nowadays there are signs of an ever-increasing interest by probabilists and

ISBN 0-12-597320-9

statisticians (especially in England), emerging from the ordered "line and time" into higher dimensions in a nontrivial geometrical manner, and making heavy use of the computer in assaults on hitherto impregnable problems. Without a doubt, no more is Geometrical Probability "little more than a plaything" (J. L. Coolidge, 1928)—see Blaschke [9, Foreword to Second Part]!

For anyone familiar with elementary continuous probability theory (to the level of [31], say) and with a spare week available, I cannot recommend any better introduction to Geometrical Probability than Kendall and Moran's 1963 monograph [1], together with the four subsequent reviews of recent work [2–5]. These works illustrate especially well the manifold practical applications. As a perfect supplement to these, for browsing and reference rather than reading, I would recommend the recent encyclopedic treatise by Santaló on Integral Geometry [6].

This paper is an attempt to present in brief compass what I personally regard to be the key elements and themes of classical planar Geometrical Probability, with emphasis on stochastic planar image modeling. While it presupposes a good working knowledge of continuous probability theory, including Poisson processes in time or on a line, the geometry itself involved is quite elementary. Naturally in a paper of this length much of importance is omitted. Even with a bias toward personal interests, I feel the picture is a fairly unified one that might serve as an appetizer before references [1–5]. I also hope that it may help to strengthen the currently somewhat tenuous links of Geometric Probability with Pattern Recognition, Image Modeling and Analysis, Picture Processing, and so forth.

2. MOBILE GEOMETRICAL OBJECTS

We shall be largely concerned with mobile geometrical objects in the Euclidean plane $\mathcal{R}^2$. Thus, if $\bar{X}$ is an arbitrary subset of $\mathcal{R}^2$ we consider $\mathcal{E}(\bar{X})$, the set of all congruent subsets $X = e\bar{X}$ of $\mathcal{R}^2$ obtained from $\bar{X}$ by Euclidean (or rigid body) motion e, i.e., a rotation followed by a translation. $\bar{X}$ will often be a simple subset like a point, segment, ray, (infinite) line, disk, ellipse, square, or rectangle.

The most general $\bar{X}$ we shall consider is a *domain*, a bounded region of $\mathcal{R}^2$ limited (internally and externally) by a finite set of closed curves without double points. Each such curve may be (topologically) oriented by the property that, as it is traversed in a positive direction, the interior of the domain lies on the left. We suppose each bounding curve is smooth in the sense that it is specified by the tangent orientation $\phi(s)$ at distance s along the curve, with ϕ differentiable except at a finite number of points of discontinuity, corresponding to cusps, each of any angle in $(-\pi, 0) \cup (0, \pi]$. At points at which it exists, $\kappa = \kappa(s) = d\phi(s)/ds$ is the (local) *curvature*, the rate of turning of the tangent vector. The total curvature of a closed boundary curve is the net total variation of ϕ, i.e., $\pm 2\pi$, and the *total curvature* C of a domain is the sum of the total curvatures of all of its bounding curves. Thus a domain comprises an at most finite union of disconnected components. Especially important and useful are convex such domains, on which $\phi(s)$ is monotone, with $C = +2\pi$, and for which we write K rather than X. The term "object" is used for any set of $\mathcal{R}^2$, be it domain, point, or line.

Parameterization. A handy way of thinking of mobile objects is to imagine a transparent rigid mobile plane $\bar{\mathcal{R}}^2$ on which $\bar{X}$ has been inscribed. The position of $X = e\bar{X}$ is determined by the position in $\mathcal{R}^2$ of an arbitrary ray fixed in $\bar{\mathcal{R}}^2$, e.g., the nonnegative x-axis $\bar{0}\bar{x}$. Suppose that, relative to $\mathcal{R}^2$, $\bar{0}$ has coordinates (x, y) and $\bar{0}\bar{x}$ makes an angle θ (increasing anticlockwise, say) with $0x$. Then the position of $\bar{\mathcal{R}}^2$, and hence that of X, is determined by the triple $v = (x, y; \theta)$. Structurally, the position v decomposes into the *location* $\xi = (x, y)$ and the *orientation* θ. Of the elementary sets mentioned above, this full set of three parameters is required to determine the position of segments, rays, ellipses, squares, and rectangles. However, due to invariance under rotations through $\pi/2$ or π about their centroids, the ranges of θ are reduced from the usual 2π down to π, 2π (no reduction), π, $\pi/2$, and π, respectively. When there is rotational symmetry (i.e., full rotational invariance), as for points and disks, only the two locational parameters x, y are required—rotationally symmetric objects have no orientation. The other type of symmetry we need consider is translational symmetry, as for (infinite) lines, strips, and half-planes. Here the best parameterization is $\lambda = (\theta, p)$, the polar coordinates of the foot of the perpendicular from 0 to a line parallel to, and fixed relative to, the object. θ, p represent orientation and location, respectively. In addition to these symmetries, there are also periodic infinitely extensive objects like point lattices and periodic polygonal tessellations. They are parameterized by v, with the range of ξ typically reduced from an unbounded to a bounded one.

One may think of mobile geometric objects as "suspended" on one of the three following basic types of mobile "frameworks" (as various types of coat may be hung on a coathanger):

(a) arrow $v = (x, y; \theta)$ (i.e., ray, or vector, or whatever).
(b) particle $\xi = (x, y)$ (we consistently use this term, rather than the overworked "point").
(c) line $\lambda = (\theta, p)$ (undirected, corresponding to range $[0, \pi) \otimes \mathcal{R}^1$).

Actually, an arrow can alternatively be naturally expressed as $(\theta, p; \alpha)$ where (θ, p) is as in (c), and α represents location along the line. There may be some didactic advantage in distinguishing between "primary theory" relating to the three basic frameworks (a)–(c), and "secondary theory" relating to objects suspended on such frameworks. Secondary objects are parameterized (or, more accurately, specified) by

(a) v together with some specification of $\bar{X}$ relative to $\bar{0}\bar{x}$.
(b) ξ together with the indicator function $I(r)$ on $[0, \infty)$ indicating whether the circle of center ξ and radius r, belongs to the set. (That is, $I(r) = 1$ if so, and 0 otherwise.)
(c) λ together with the indicator function $J(q)$ on $\mathcal{R}^1$ indicating whether the parallel line at distance q from λ belongs to the set.

Of course, $\bar{X}$, I, and J may simplify drastically and reduce to a finite number of parameters, as for the simple objects mentioned above. In the ensuing theory, v, ξ, and λ are always random, whereas $\bar{X}$, I, and J may be "just about anything": deterministic (i.e., nonrandom) or random. This is because, to get a random model, we only require randomness at either the primary or the secondary stage; and because randomness at the primary stage yields more interesting models.

3. MEASURES AND INVARIANT MEASURES

Let us write generally h for any of the two- or three-tuples υ, ξ, and λ, dh for the corresponding differential element (e.g., $d\lambda = d\theta dp$), and $\mathcal{H}$ for the corresponding range space. To proceed further, we shall need to define measures on the usual σ-algebras of subsets H of $\mathcal{H}$. For the (absolutely continuous) measure

$$\mathfrak{M}(H) = \int_H g(h)dh$$

corresponding to the continuous nonnegative density g, we know that relative to another Cartesian coordinate frame in $\mathcal{R}^2$ (corresponding quantities relative to this latter frame bearing a prime)

$$\mathfrak{M}(H) = \int_{H'} g(h(h'))\,|\partial h(h')/\partial h'|\,dh', \tag{1}$$

by the usual "Jacobian" transformation rule. Imposing the natural and fundamental condition that $\mathfrak{M}$ be invariant under all such Euclidean motions of the coordinate frame in $\mathcal{R}^2$,

$$\mathfrak{M}(H) = \mathfrak{M}(H') = \int_{H'} g(h')dh' \qquad \forall\, e \in \mathcal{E}:$$

Hence $\mathfrak{M}$ is invariant iff

$$g(h') = g(h(h'))\,|\partial h(h')/\partial h'| \qquad \forall\, e \in \mathcal{E}.$$

It turns out that this relation is satisfied by $g(h) =$ constant in each of the cases $h = \upsilon, \xi, \lambda$, and that these densities and the resulting measures are the unique such *invariant densities* and *measures*. For the mathematical background to the theory of such "Haar" measures, see Nachbin [29]. They are, with the arbitrary constant factor chosen in the natural way as unity:

$$\text{arrows:} \quad \int_H d\upsilon \equiv \iiint_H dx\,dy\,d\theta;$$

$$\text{particles:} \quad \int_H d\xi \equiv \iint_H dx\,dy;$$

$$\text{lines:} \quad \int_H d\lambda \equiv \iint_H d\theta\,dp.$$

Each is simple Lebesgue measure in its respective space, this space being $\mathcal{R}^2$ itself for particles. They measure in a very natural sense the "quantity" of objects h within the object-set H. Alternatively, writing $g(\ )$ for the invariant density of its argument

$$g(v) = g(\xi) = g(\lambda) = 1$$

(f is also used in this concise "nonfunctional" manner below). Relative to other parameterizations h^*, we have

$$\mathfrak{M}(H) = \int_{H^*} g(h(h^*)) \,|\, \partial h(h^*)/\partial h^* \,|\, dh^*$$

(cf. (1))—in other words

$$g(h^*) = g(h) \,|\, \partial h(h^*)/\partial h^* \,|.$$

As examples, the invariant density of particles relative to polar coordinates (r, θ) is $g(r, \theta) = r$, and that of lines relative to their intercepts u, v with the x- and y-axes is $g(u, v) = uv/(u^2 + v^2)^{\frac{3}{2}}$. The constancy of $g(v)$, $g(\xi)$, and $g(\lambda)$ makes v, ξ, and λ natural parameterizations. In practice the domain H of integration almost always corresponds to a *hitting set*

$$[h \uparrow X] \equiv \{h : h \uparrow X \subset \mathbb{R}^2\},$$

where we write $\uparrow$ for "hits": $h \cap X \neq \phi$. Examples are (with $\mathfrak{M}$ now denoting invariant measure):

(a) If $D(a)$, $D(b)$ $(0 < a < b)$ are concentric disks in $\mathbb{R}^2$, then $\mathfrak{M}\{$arrows issuing from $D(b) - D(a)$ and, when extended, hitting $D(a)\}$

$$= 2\pi b^2 \sin^{-1}(a/b) + 2\pi a(b^2 - a^2)^{\frac{1}{2}} - \pi^2 a^2,$$

which $\to 0$ as $a \to 0$, and $\to \infty$ as $b \to \infty$.

(b, c) With A, B denoting area and perimeter, respectively,

$$\mathfrak{M}(\xi \in \text{domain } X) = A(X), \tag{2}$$

and

$$\mathfrak{M}(\lambda \uparrow \text{convex domain } K) = B(K). \tag{3}$$

(2) is obvious, unlike (3) which asserts that the "quantity" of lines of $\mathbb{R}^2$ intercepting a convex domain equals its perimeter. What we have so far sketched is classical IG (*integral geometry*), which is characteristically concerned with such invariant densities and measures, especially interrelations between them.

Invariance under translations only. If we only impose the condition that the measures be invariant with respect to all translations of coordinate frames in $\mathbb{R}^2$, $g(\xi)$ is unaltered, unlike $g(v)$ and $g(\lambda)$ which are no longer unique. In fact, if Θ denotes an arbitrary measure on the appropriate range of θ, then invariant densities are of the forms

$$\text{(a) } dxdy\, \Theta(d\theta)$$

and

$$\text{(c) } \Theta(d\theta)dp,$$

respectively. Broadly speaking, IG requires at least translational invariance;

with rotational invariance too—the isotropic case—there is the bonus of a battery of elegant formulas. For reasons of space we shall not be much concerned with the anisotropic case, although much of our theory has a useful anisotropic version [27].

4. A SINGLE RANDOM MOBILE OBJECT

We have established in Sections 2 and 3 a natural framework with respect to which random primary, and hence secondary, objects may be defined. With Section 2 as basis, to *any* probability distribution with range $\mathcal{H}$, there corresponds a *random object*, which in principle may be generated repeatedly (and independently), just as a coin may be tossed repeatedly to produce a "Bernoulli sequence of trials." However, this yields an embarrassingly large collection of possible random objects!

Consider the subclass of these distributions with density proportional to the invariant density $g(h)$ over $[h \uparrow X]$ for a domain X. Here h is either ξ or λ, not υ, because "$\upsilon \uparrow H$" is as yet undefined (this is remedied below by means of a secondary hitting condition). By (2), (3) the corresponding probability densities are

$$f(\xi) \doteq 1/A(X) \qquad [\xi \in X] \tag{4}$$

and

$$f(\lambda) \doteq 1/B(K) \qquad [\lambda \uparrow K] \tag{5}$$

($\doteq$ means that the density is zero on the residue of the relevant parameter space $\mathcal{H}$). By the properties of invariance discussed in Section 3, both of these distributions are *uniform* in the sense that each hitting ξ or λ is "equally likely to be chosen." Consequently, for a random particle ξ and a random line λ with densities $p(\xi)$ on $[\xi \in X]$ and $p(\lambda)$ on $[\lambda \uparrow K]$, respectively, these densities also measure in a natural (Radon–Nikodym) sense the *relative frequencies* of ξ and λ. This property persists, when we drop the hitting condition, for densities $p(h)$ ($h = \xi, \lambda$, and now υ!) on $\mathcal{H}$. Note that none of the invariant densities can be normalized into probability densities over their entire ranges $\mathcal{H}$.

Classical GP (*geometrical probability*) restricts itself largely to consideration of uniformly distributed objects, such as those specified by (4), (5). This is because uniformity is a very natural, unique, and simply described type of randomness, that occurs often in practice, and which gives rise to a rather elegant theory. Formally, GP bears the same relation to IG that probability theory does to measure theory: the former results from the latter on normalizing. Beyond this, both GP and probability theory are largely inspired by real-life phenomena, whereas the other two disciplines are of a more abstract mathematical nature. Their close interrelations yet different emphases have been of considerable mutual benefit.

Random particle in domain. We term a random particle $\xi = (x, y)$ with density (4) a *UR* (uniform random) *particle in* X. Its distribution is completely determined by knowledge of the domain X. The marginal density of x is in general nonuniform:

$$f(x) = L(X \cap \lambda_x)/A(X),$$

where λ_x is the line parallel to the y-axis with abscissa x and L denotes length (similarly for $f(y)$); whereas the conditional density of y given x, which exists only for x with $L(X \cap \lambda_x) > 0$, is uniform:

$$f(y|x) \doteq 1/L(X \cap \lambda_x), \qquad X \cap \lambda_x$$

(similarly for $f(x|y)$). Note the validity of the standard probability relation $f(x, y) = f(x)f(y|x) = f(y)f(x|y)$. UR ξ in X generates further random variables—for example, d, the shortest distance between ξ and a point of the bounding curves of X. If the range of d is $(0, d_0)$ then the probability density of d at $0+$ is $B(X)/A(X)$ and, for sufficiently general X, at $d_0 - \epsilon$ is $\sim c\epsilon$ as $\epsilon \downarrow 0$, where c depends on the geometry of X.

Stemming from invariance is the fundamental characterizing

Hitting property for UR particles. For a UR particle ξ in X, the probability $P(\xi \uparrow X' \subset X) = A(X')/A(X)$. Moreover, given $\xi \uparrow X'$, ξ is (conditionally) UR in X'.

Random line through convex domain. We term a random line $\lambda = (\theta, p)$ with density (5) an *IUR* (*isotropic uniform random*) *line through* K. Integration of (5) shows that the marginal density

$$f(\theta) = W_K(\theta)/B(K) \qquad [0, \pi),$$

where $W_K(\theta)$ is the relevant width or "caliper diameter" of K; it follows that

$$f(p|\theta) = f(\theta, p)/f(\theta) \doteq 1/W_K(\theta) \qquad [p : \lambda \uparrow K] \tag{6}$$

for all θ (hence the description "uniform"). Unlike the symmetric case of (x, y), there is a natural order here, viz., *first* θ, *then* p, corresponding to the basic conditional relation (6). The reverse quantities $f(p)$ and $f(\theta|p)$ are complicated, depending much upon the arbitrary p coordinates adopted for each θ, and are of little use or interest.

Perhaps the most obvious associated random variable is the length L of the random *secant* (or chord) $\lambda \cap K$. We shall now determine the mean value or expectation of L. Following the natural order, the conditional uniformity of p given θ yields the conditional expectation

$$E(L|\theta) = A(K)/W_K(\theta),$$

so that, using a standard probability relation, the expectation

$$E(L) = \int E(L|\theta)f(\theta)d\theta = \pi A(K)/B(K).$$

Stemming from invariance is the fundamental characterizing

Hitting property for IUR lines. For λ IUR $\uparrow K$, $P(\lambda \uparrow K' \subset K) = B(K')/B(K)$. Moreover, given $\lambda \uparrow K'$, it is (conditionally) IUR $\uparrow K'$.

More generally we may define IUR lines through nonconvex domains X, but the density (5) becomes more complicated. In fact, in several places below we have the option of using domains X or convexes K; for simplicity we shall usually use convexes, even though a corresponding theory for domains exists and is not indicated in the text.

Anisotropic ΘUR *lines.* The random line just discussed is termed "isotropic" because it derives from the isotropic invariant density $d\theta dp$, but the marginal density $f(\theta)$ is isotropic only if K is of constant width. More generally, similarly normalizing the translationally invariant density element $\Theta(d\theta)dp$ over $[\lambda \uparrow X]$, we get an anisotropic ΘUR *line through* X, whose marginal θ distribution can, perversely, be isotropic (when $\Theta(d\theta) = d\theta/W_X(\theta)$)! Such lines can be useful to model anisotropic structures. The usual hitting property holds, but $P(\lambda \uparrow X' \subset X)$ now depends on the orientation of X' within X.

Random domain hitting fixed domain. To give meaning to "$\upsilon \uparrow X$", we use hitting conditions for domains suspended on the arrow υ. Thus, for two domains X_0, $\bar{X}_1$ we define IUR X_1 *hitting* X_0, by

$$f(\upsilon_1) \doteq 1 \bigg/ \int_{01} d\upsilon_1,$$

where X_0 is fixed, X_1 is mobile suspended on υ_1, and $01 \equiv \{\upsilon_1 \colon X_1 \uparrow X_0\}$. Geometrically, given $X_1 \uparrow X_0$, $X_0 \cap X_1$ is almost surely (i.e., with probability 1) also a domain. Writing $*_i \equiv *(X_i)$, $*_{01} \equiv *(X_0 \cap X_1)$ $(* = A, B, C)$, we have the fundamental set of IG relations

$$\int_{01} A_{01} d\upsilon_1 = 2\pi A_0 A_1, \tag{7}$$

$$\int_{01} B_{01} d\upsilon_1 = 2\pi(A_0 B_1 + B_0 A_1), \tag{8}$$

$$\int_{01} C_{01} d\upsilon_1 = 2\pi(A_0 C_1 + B_0 B_1 + C_0 A_1), \tag{9}$$

the last of which, the most fundamental, is Blaschke's formula [6, Sect. I.7.4]. These are extraordinarily general results, due to the generality of X_0, X_1; for example, (2) is a limiting case of (9) in which X_1 is a disk whose radius decreases to zero. A, B, C have length dimensions 2, 1, 0, respectively.

For a useful GP, we need a general formula for the value of $\int_{01} d\upsilon_1$, with which to normalize. This is only possible when C_{01} is constant, corresponding essentially (but not necessarily!) to the *convex* case $X_0 = K_0$, $X_1 = K_1$. Then, since convexity is preserved on intersection, (9) becomes

$$\int_{01} d\upsilon_1 = 2\pi\{A_0 + (2\pi)^{-1} B_0 B_1 + A_1\} \equiv 2\pi\omega(K_0, K_1) \tag{10}$$

in our later notation. Essentially, (3) is the limiting case of (10) in which K_1 is a segment whose length increases to ∞. Two immediate results in the convex GP case are

$$E(A_{01}) = \int_{01} A_{01} d\upsilon_1 \bigg/ \int_{01} d\upsilon_1$$

and

$$E(B_{01}) = \int_{01} B_{01} dv_1 \Big/ \int_{01} dv_1,$$

the values of which follow from (7)–(9).

Again, we have the fundamental characterizing

Hitting property for IUR convex domains. For K_1 IUR $\uparrow K_0$,

$$P(K_1 \uparrow K_0' \subset K_0) = \omega(K_0', K_1)/\omega(K_0, K_1).$$

Moreover, given $K_1 \uparrow K_0'$, it is (conditionally) IUR $\uparrow K_0'$. (Similarly where X replaces K, but the value of the probability is not then known.)

Stochastic construction of random mobile objects. To simulate images comprising random mobile objects, an efficient method of generation is desirable. This is achieved by utilizing the relative simplicity of generating UR particles in *rectangles* aligned with the axes in any $\mathcal{R}^i$ by "random numbers," in conjunction with the fundamental hitting properties. Thus to generate an IUR $X \uparrow X_0$ we proceed as follows:

(i) With T_i representing an aligned rectangle in $\mathcal{R}^i$, select $T_3 = T_2 \otimes [0, 2\pi)$ such that it contains $[v: X \uparrow X_0]$.

(ii) Construct a UR particle in T_3, and accept the resulting random X if it hits X_0; if not, repeat until an X does hit X_0—this is then IUR $\uparrow X_0$.

For efficiency, (a) ξ should be "central" in X, e.g., its centroid; and (b) given (a), T_2 should be chosen as small as possible "around X_0," to reduce the average number of repetitions required. Simpler stochastic constructions of a UR particle in X or an IUR line through X utilize the minimal rectangle and disk, respectively, $\supset X$.

5. TWO OR MORE RANDOM MOBILE OBJECTS

Next consider two such random objects, which for simplicity are assumed *independent.* (Our philosophy is to proceed to more complex models in the least painful way.) IG-wise, this means looking at the corresponding *product*-invariant density, on the corresponding *product* range space.

We shall now examine quite closely a most instructive special case, since it very effectively demonstrates features which occur repeatedly in this type of model building. Consider two independent UR particles ξ, η in a convex domain K (with $A(K) > 0$). Their joint probability density is

$$f(\xi, \eta) \doteq A(K)^{-2}, \qquad K \otimes K.$$

Almost surely ξ, η do not coincide, so that almost surely a unique line λ through ξ and η exists. A natural question is how the distribution of the random line λ relates to that of a basic IUR line hitting K, especially since both have the same range space $[\lambda \uparrow K]$. Standard IG Jacobian methods [6, Sect. I.4.1] give the invariant density relation

$$d\xi d\eta = d\lambda \cdot |\alpha - \beta| \, d\alpha d\beta, \tag{11}$$

where α, β are the coordinates of ξ, η, respectively, relative to a linear Cartesian coordinate frame on λ, so that $|\alpha - \beta| = R$, the distance between ξ and η. We use a dot to emphasize the essential order: first λ, then α, β. Multiplying both sides of (11) by R^i and integrating totally

$$\int_K \int_K R(\xi, \eta)^i d\xi d\eta = \int_{[\lambda\uparrow K]} \int_a^{a+L(\lambda\cap K)} \int_a^{a+L(\lambda\cap K)} |\alpha - \beta|^{i+1} d\alpha d\beta d\lambda \quad (12)$$

for some a,

$$= \{2/(i+2)(i+3)\} \int_{[\lambda\uparrow K]} L(\lambda \cap K)^{i+3} d\lambda,$$

valid for all real $i > -2$. Dividing by $A(K)^2$, we obtain the corresponding GP result

$$E_{\text{two independent UR particles in } K}(R^i)$$
$$= \{2B(K)/(i+2)(i+3)A(K)^2\} E_{\text{IUR line through } K}(L^{i+3}) \qquad (i > -2).$$

This is a typical IG/GP moment relationship, in which an explicit value is only known for the special case $i = 0$, viz.,

$$E(L^3) = 3A(K)^2/B(K) \qquad \forall \text{ convex } K$$

—"Crofton's second theorem" [12]. By specializing K, the remaining explicit values and even the probability density itself may sometimes be determined, for example for a disk [1, pp. 41–42]. In fact, the relation between the moment generating functions of R and L may be similarly obtained, by considering instead $\int\int e^{-t_1 R} d\xi d\eta$; most generally, we could consider $\int\int \chi(\xi, \eta) d\xi d\eta$ for arbitrary integrable χ.

Let us turn from moments to distributions. Dividing (11) by $A(K)^2$ yields the joint probability density

$$f(\lambda; \alpha, \beta) = f(\lambda) \cdot f(\alpha, \beta | \lambda)$$
$$\doteq A(K)^{-2} \int_0^L \int_0^L |\alpha - \beta| d\alpha d\beta \cdot |\alpha - \beta| \Big/ \int_0^L \int_0^L |\alpha - \beta| d\alpha d\beta$$

which, by the second equality of (12) with $i = 0$,

$$= L(\lambda \cap K)^3 / 3A(K)^2 \cdot 3|\alpha - \beta| / L(\lambda \cap K)^3.$$

Thus the distribution of λ is that of an IUR line through K *weighted* by the cube of the secant length; while the conditional joint distribution of α, β given λ is symmetric about the secant center. Whereas ξ, η are independent, α, β given λ are dependent, with a uniform joint distribution weighted by $|\alpha - \beta|$. Taking a in (12) to be zero, the marginal density

$$f(\alpha | \lambda) = \int_0^L f(\alpha, \beta | \lambda) d\beta \doteq 3[\{\alpha - (L/2)\}^2 + (L/2)^2]/L^3 \qquad (0 \le \alpha \le L),$$

symmetrical about $L/2$.

An equally instructive case that the reader may care to pursue is that of two independent IUR lines through K. Here one uses the basic IG density relation

$$d\lambda_1 d\lambda_2 = d\zeta \,|\sin \overline{\theta_1 - \theta_2}|\, d\theta_1 d\theta_2,$$

where $\zeta = \lambda_1 \cap \lambda_2$ and θ_1, θ_2 are the orientations of λ_1, λ_2 [6, Sect. I.4.3]. With good fortune, "Crofton's first theorem" [1, p. 63] is recovered.

There is no better introduction to the elements of continuous probability theory than these two examples, yet few teachers of probability theory are probably aware of them.

Useful extensions reexpress the product-invariant density of sets of three particles and of three lines. First we need to define

$$\begin{aligned}\sigma(\theta_1, \theta_2, \theta_3) &\equiv \sin(\theta_2 - \theta_1) + \sin(\theta_3 - \theta_2) + \sin(\theta_1 - \theta_3) \\ &= 4 \sin \tfrac{1}{2}(\theta_1 - \theta_2) \sin \tfrac{1}{2}(\theta_2 - \theta_3) \sin \tfrac{1}{2}(\theta_3 - \theta_1) \\ &\qquad\qquad (0 \le \theta_1, \theta_2, \theta_3 < 2\pi),\end{aligned}$$

which is $\pm$ twice the area of the triangle whose vertices are the points on the unit circle with angular coordinates θ_1, θ_2, θ_3. Then for three particles (with for brevity $\phi_* \equiv (\phi_1, \phi_2, \phi_3)$)

$$d\xi_1 d\xi_2 d\xi_3 = R^3 |\sigma(\phi_*)|\, d\zeta dR d\phi_1 d\phi_2 d\phi_3, \tag{13}$$

where ζ, R are the center and radius of the circumcircle through ξ_*, and ϕ_* are the angular coordinates of ξ_* with respect to ζ [22, Relation (4.11)]. This new parameterization is "structural," in the sense that $\zeta \sim$ location, $R \sim$ size, and (say) $\phi_1 \sim$ orientation, $\phi_2 - \phi_1$, $\phi_3 - \phi_1 \sim$ shape. For three lines

$$d\lambda_1 d\lambda_2 d\lambda_3 = |\sigma(\psi_*)|\, d\eta dI d\psi_1 d\psi_2 d\psi_3,$$

where η, I are the center and radius of the incircle of the triangle with sides λ_*, and ψ_* are the angular coordinates of the *directed* perpendiculars from η to the lines [25, Sect. 12]. The parameterization $(\eta, I; \psi_1, \psi_2, \psi_3)$ is clearly also structural.

Two or more random domains. Suppose X_0 is a fixed, and X_1, X_2 are mobile, domains. Then, writing $012 \equiv [X_0 \cap X_1 \cap X_2 \ne \phi]$, $*_{012} \equiv *(X_0 \cap X_1 \cap X_2)$, etc., we have

$$\begin{aligned}\iint_{012} C_{012} dv_2 dv_1 &= \int_{01} \int_{X_2 \uparrow X_0 \cap X_1} C_{012} dv_2 dv_1 \\ &= 2\pi \int_{01} (A_{01} C_2 + B_{01} B_2 + C_{01} A_2) dv_1 \qquad \text{by (9)} \\ &= (2\pi)^2 \Big[\Big(\prod_0^2 A_i\Big)\Big\{\sum_0^2 (C_i/A_i)\Big\} + \Big(\prod_0^2 B_i\Big)\Big\{\sum_0^2 (A_i/B_i)\Big\}\Big] \\ &\qquad\qquad \text{by (7)–(9).}\end{aligned} \tag{14}$$

Similarly,

$$\iint_{012} A_{012} dv_2 dv_1 = (2\pi)^2 \prod_0^2 A_i \tag{15}$$

and

$$\iint_{012} B_{012} dv_2 dv_1 = (2\pi)^2 \left(\prod_0^2 A_i\right) \left\{\sum_0^2 (B_i/A_i)\right\}. \tag{16}$$

Again, lacking a formula for $\iint_{012} dv_2 dv_1$ for general domains, to switch from IG to GP we must take all domains convex. Then

$$P(012) = \iint_{012} dv_2 dv_1 \Big/ \int_{01} dv_1 \int_{02} dv_2,$$

$$E(A_{012}|012) = E(A_{012})/P(012)$$

$$= \iint_{012} A_{012} dv_2 dv_1 \Big/ \iint_{012} dv_2 dv_1,$$

and

$$E(B_{012}|012) = E(B_{012})/P(012)$$

$$= \iint_{012} B_{012} dv_2 dv_1 \Big/ \iint_{012} dv_2 dv_1,$$

with the explicit values of all the integrals being given in (9) and (14)–(16).

In fact, such iteration may be continued (inductively). The full GP form of the results, relating to independent IUR mobile convex domains $K_1, \ldots, K_n$ each hitting the fixed convex domain K_0, is

$$P(0\cdots n) = \omega_{0\cdots n} \Big/ \prod_1^n \omega_{0i}, \tag{17}$$

$$E(A_{0\cdots n}|0\cdots n) = E(A_{0\cdots n})/P(0\cdots n) = \left(\prod_0^n A_i\right)/\omega_{0\cdots n}, \tag{18}$$

$$E(B_{0\cdots n}|0\cdots n) = E(B_{0\cdots n})/P(0\cdots n) = \left(\prod_0^n A_i\right)\left\{\sum_0^n (B_i/A_i)\right\}/\omega_{0\cdots n}, \tag{19}$$

where

$$0\cdots n \equiv \left[\bigcap_0^n K_i \neq \emptyset\right]$$

and

$$\omega_{0\cdots n} \equiv \left(\prod_0^n A_i\right)\left\{\sum_0^n (1/A_i) + (2\pi)^{-1} \sum_{0\le i<j\le n}\sum (B_i B_j/A_i A_j)\right\} \qquad (n \geq 1).$$

Taking note of the manifold possible choices of n and each K_i, this constitutes a quite extraordinarily wide collection of formulas, which includes many of the standard formulas of GP as special or limiting cases. The limit $n \to \infty$ is also of interest [21]. Equation (17) has strong claim to be the single most important formula in planar GP, yet is little known. The multidimensional form of these formulas is given by Streit [37].

We write $\{K_1, \ldots, K_n \uparrow K_0\}$ for the portion of the model just discussed which lies within K_0, and describe it as the "hit" model, reflecting its definition. It, and the more general hit model $\{X_1, \ldots, X_n \uparrow X_0\}$, constitute basic stochastic models for images, which may well be taken as null hypotheses in any statistical study. It is homogeneous and isotropic within K_0 in the sense that the realization within any $K_0' \subset K_0$ does not depend upon the position of K_0' within K_0 ($\forall K_0' \subset K_0$). Usually, in model building, each of $K_1, \ldots, K_n$ is supposed very much smaller than K_0, with n reasonably large, even though there is no limit on the relative sizes of $K_0, \ldots, K_n$. The great variation in choice of the K_i allows the modeling of a wide variety of textures. Special or limiting cases are n independent UR particles in K_0, and n independent IUR secants of K_0. In turn, $\{K_1, \ldots, K_n \uparrow K_0\}$ is a basic framework on which we may suspend all manner of "tertiary" detail. For example, we may define the planar stochastic process

$$u(\xi) = \sum_{i:\, \xi \in K_i} v_i \qquad [\xi \in K_0],$$

where v_i are arbitrary scalars or random variables. If each v_i is unity, the set $\{\xi\colon u(\xi) = j\}$ is the j-covered subset of K_0. Further, we may choose the K_i $(i = 1, \ldots, n)$ themselves to be random from some convex set distribution. To drop isotropy while retaining homogeneity we simply locate the mobile sets by $dxdy\, \Theta(d\theta)$ rather than $dv = dxdyd\theta$; iteration is still possible, although the resulting formulas are more complicated [27, Relation (15)].

6. TWO STANDARD TECHNIQUES

(i) *Discrete indicators.* Let M be the total number of domain–domain contacts of $\{K_1, \ldots, K_n \uparrow K_0\}$ occurring *within* K_0, i.e., the number of the triples $0, j, k$ for which the event $0jk$ occurs. Thus $0 \leq M \leq \binom{n}{2}$. For simplicity only, suppose $K_1 = \ldots = K_n = K$. M is a random variable, and we now determine its first two moments. We have the indicator representation

$$M = \sum_{0 \leq j < k \leq n} \sum I_{jk},$$

where the random variable $I_{jk} = 1$ if $0jk$ occurs and 0 otherwise. The $\binom{n}{2}$ I_{jk} are by no means independent, and we rely heavily upon the basic expectation relation $E(\sum u_i) = \sum E(u_i)$, which holds whether or not the random variables u_i are independent. Thus the mean value

$$E(M) = \sum_{0 \leq j < k \leq n} \sum E(I_{jk})$$

$$= \binom{n}{2} P(012) \equiv \binom{n}{2} p,$$

say, which is known from (17). The second-order moment

$$E(M^2) = E\Big(\sum_{0 \leq j < k \leq n} \sum I_{jk} + \sum_{j=1}^{n} \sum_{k=1}^{n} \sum_{l=1}^{n}{}' I_{jk} I_{jl} + \sum_{j=1}^{n} \sum_{k=1}^{n} \sum_{l=1}^{n} \sum_{m=1}^{n}{}' I_{jk} I_{lm}\Big),$$

where the primes mean that in the summations no pair of j, k, l, m is allowed to be equal,

$$= \binom{n}{2} p + 2n \binom{n-1}{2} pq + \binom{n}{2}\binom{n-2}{2} p^2$$

with $q \equiv P(0jl \mid 0jk)$. Thus the variance

$$\begin{aligned} \text{Var } M &= \tfrac{1}{2}n(n-1)p\{1 + 2(n-2)q - (2n-3)p\} \\ &\sim p(q-p)n^3 \end{aligned}$$

as $n \to \infty$, if $p \neq q$ (hence $q \geq p$!). On the other hand, if the size of K decreases to zero, then the chance of K_j being "well interior" to K_0, given $0jk$, tends to 1, so that $q \downarrow p$.

In fact, the nonequality of p, q may be regarded as an "edge effect," which may be eliminated as follows, at the cost of drastically restricting the possible shapes of K_0; imposing a size limitation on $K_1, \ldots, K_n$ in relation to K_0; and introducing a high dependence between distant border regions. Take (a) K_0 to be "space-filling," e.g., as the rectangle $\{(x, y): 0 \leq x < a, 0 \leq y < b\}$, and (b) $\bar{K}_1, \ldots, \bar{K}_n$ such that every sum of diameters for pairs of different $\bar{K}_i$ is exceeded by min (a, b). Define K_i' corresponding to $\bar{K}_i$ to be the random domain congruent to $\bar{K}_i$ obtained by taking v uniform in $K_0 \otimes [0, 2\pi)$. Next let K_i'' be the doubly infinite lattice of convexes corresponding to K_i', i.e., $\bigcup t_{jk}K_i'$ where t_{jk} denotes a translation by (ja, kb). Choose $K_1'', \ldots, K_n''$ independently in this way, and finally define the "periodic" model $\{K_1, \ldots, K_n \downarrow K_0\}$ to be the intersection of K_0 with the union of these n random lattices. Note that K_0 may intersect each K_i'' in one, two, or even four disjoint convex sets, and that the model is homogeneous, since the intersection of tK_0 with $K_1'', \ldots, K_n''$ is stochastically equivalent to $\{K_1, \ldots, K_n \downarrow K_0\}$ for all translations t. Topologically the rectangle K_0 has become a torus.

After the effort, the reward: taking note of condition (b), (17)–(19) extend, with $0 \cdots n$ replaced by $1 \cdots n$, i.e.,

$$\begin{aligned} P(1 \cdots n) &= (ab)^{1-n}\omega_{1\cdots n}, \\ E(A_{1\cdots n} \mid 1 \cdots n) &= E(A_{1\cdots n})/P(1 \cdots n) = \Big(\prod_1^n A_i\Big)/\omega_{1\cdots n}, \\ E(B_{1\cdots n} \mid 1 \cdots n) &= E(B_{1\cdots n})/P(1 \cdots n) = \Big(\prod_1^n A_i\Big)\Big\{\sum_1^n (B_i/A_i)\Big\}/\omega_{1\cdots n}. \end{aligned} \tag{20}$$

Moreover, with $K_1 = \ldots = K_n = K$,

$$\begin{aligned} q &= P(jl \mid jk) = P(jl) = p \\ &= (4\pi A + B^2)/2\pi ab, \end{aligned}$$

so that

$$\text{Var } M = \tfrac{1}{2}n(n-1)p(1-p).$$

Hence, in the periodic model, the mean and the variance of M coincide with those for a Bernoulli sequence of $\binom{n}{2}$ independent trials of probability p (dependence begins to rear its head with the third order moments). In summary, the variance of M is of $O(n^3)$ and $O(n^2)$ in the hit and periodic models, respectively. In fact, asymptotic normality of M as $n \to \infty$ may be proved in both cases by the method of moments, examining the dominant terms in expressions derived as above [17].

A similar theory may be developed to explore the moments of M_i ($i \geq 3$)—the number of i-subsets of $K_1, \ldots, K_n$ having a nonvoid intersection with K_0 (this last condition is unnecessary in the periodic model). The first-order moments drop out of (17) and (20), but higher-order moments are affected by dependence problems.

(ii) *Continuous indicators.* Let U be the subarea of K_0 not covered by any of $K_1, \ldots, K_n$. It has the indicator representation

$$U = \int_{K_0} I(\xi) d\xi,$$

where $I(\xi)$ indicates whether ξ is uncovered. Since an expectation is essentially an integral, Fubini's theorem applies [34], to yield

$$E(U) = \int_{K_0} E\{I(\xi)\} d\xi$$

$$= \int_{K_0} P(\xi \text{ is uncovered}) d\xi$$

which, by the homogeneity in both hit and periodic models,

$$= A_0 P(\text{an arbitrary point of } K_0 \text{ is uncovered})$$

$$= A_0 \prod_1^n \{1 - (A_i/\omega_{0i})\} \qquad \text{(hit model)}$$

$$= A_0 \prod_1^n \{1 - (A_i/A_0)\} \qquad \text{(periodic model).}$$

The second-order moment

$$E(U^2) = E \int_{K_0} \int_{K_0} I(\xi) I(\eta) d\xi d\eta$$

$$= \int_{K_0} \int_{K_0} P(\xi, \eta \text{ both uncovered}) \, d\xi d\eta$$

$$= \int_{K_0} \int_{K_0} P(\xi \text{ uncovered}) \, P(\eta \text{ uncovered} \mid \xi \text{ uncovered}) \, d\xi d\eta.$$

The main problem here is the evaluation of the conditional probability. Edge effects make this difficult in the hit model, so we turn to the periodic model in the simplest case, with each K_i a disk of radius $r < \min(a/4, b/4)$. Then

$$P(\xi \text{ uncovered}) = \{1 - (\pi r^2/ab)\}^n$$

and

$$P(\eta \text{ uncovered} \mid \xi \text{ uncovered}) = [1 - \{l(s)/(ab - \pi r^2)\}]^n,$$

where $s = |\xi - \eta|$ and $l(s)$ is the subarea of the disk with center η and radius r *not* contained in the disk with center ξ and radius r (thus $l(s) = \pi r^2$ for $s \geq 2r$). It follows that

$$E(U^2) = (ab)^{1-n}\left[2\pi \int_0^{2r} \{ab - \pi r^2 - l(s)\}^n s ds + (ab - \pi r^2)(ab - 2\pi r^2)^n\right].$$

Just as for the discrete indicators, this method may be extended to write down integral expressions for higher-order moments of U, but the integrations involved are formidable. The moments of the covered area C follow at once from the relation $C = A_0 - U$, and the method extends to finding the moments of the i-covered area C_i $(i \geq 2)$.

7. POISSON MODELS

Both the hit and periodic models postulate a fixed number n of random mobile objects in (or straddling) the fixed domain of interest K_0. Both possess homogeneity properties within K_0, but outside the hit model is inhomogeneous and the periodic model although homogeneous is of little interest, being simply a periodic replica of the realization within K_0. Both models exhibit dependence of all separated regions, a dependence that, at least for the hit model, decreases as the separation increases. We now show how to construct a related model which is continued homogeneously and nontrivially outside K_0 and which possesses such "complete independence" properties, including the independence of all sufficiently separated regions, that it has great claim to be a "yardstick" model.

Suppose m is an integer >1, write mK_0 for the dilated domain $\{m\xi \colon \xi \in K_0\}$, and suppose $K_0 \subset$ interior of mK_0 (arranged, for example, by taking the origin at the centroid of K_0). Consider the hit model $\{K_1, \ldots, K_{m^2n} \uparrow mK_0\}$, with $\bar{K}_1 = \ldots = \bar{K}_{m^2n} = \bar{K}$. We investigate the limiting behavior of the intersections of $K_1, \ldots, K_{m^2n}$ with K_0 as $m \to \infty$. Equivalently in υ-space, we have m^2n independent UR particles in $[\upsilon \colon K \uparrow mK_0]$. By (10), the number of these lying in $[\upsilon \colon K \uparrow K_0]$ has a binomial $\{m^2n, \omega(K_0, K)/\omega(mK_0, K)\}$ distribution which tends, as $m \to \infty$, to a Poisson $\{n\omega(K_0, K)/A_0\}$ distribution. In fact, in any bounded region of υ-space we get in the limit a three-dimensional Poisson particle process of intensity (i.e., average number per unit volume) $n/2\pi A_0$. The same limit results from corresponding limits of the periodic model. Both are essentially examples of the classical Poisson limit of a binomial distribution [14, Sect. VI.5].

I trust the reader is familiar with the standard properties of Poisson (or "completely random") particle processes on a line or in higher dimensions. Here, for example, the chance of a particle occurring in the volume element $d\upsilon$ is $(n/2\pi A_0)d\upsilon$; the number of particles occurring in a domain of υ-volume V has a Poisson $(nV/2\pi A_0)$ distribution; given the number in V is m, then these m are (conditionally) UR particles in V; and the numbers of particles in disjoint υ-domains are mutually independent. We write below **P** for a homogeneous Poisson particle process (of intensity ρ if not otherwise stated), and $\mathbf{P}\{\uparrow K_0\}$ for its restriction to K_0.

Following the line of least resistance, we consider directly the *actual limit*, viz., *Poisson convex domain model*. For a convex domain $K \subset \mathbb{R}^2$, $\mathbf{P}\{K\}$ is the convex domain process in $\mathbb{R}^2$ corresponding to a **P** of intensity $\rho/2\pi$ in υ-space;

$$\mathbf{P}\{K \uparrow K_0\} \equiv K_0 \cap \mathbf{P}\{K\}$$

is the bounded model comparable to the hit and periodic models already discussed.

$\mathbf{P}\{K\}$ is homogeneous and isotropic in $\mathbb{R}^2$, with intensity ρ domains per unit area. The realizations of $\mathbf{P}\{K\}$, and hence of $\mathbf{P}\{K \uparrow K_0\}$, within disjoint domains, each pair of domains being separated by at least the diameter of K, are mutually independent.

The Poisson structure yields the basic

Stochastic construction of $\mathbf{P}\{K \uparrow K_0\}$.

(i) Generate a Poisson random variable m, with mean value

$$\mu = \rho\omega(K_0, K).$$

(ii) Generate

$$\{\underbrace{K, \ldots, K}_{m}\uparrow K_0\}.$$

Since $\mathbf{P}\{K \uparrow K_0\}$ is a Poisson mixture of the hit model, moments for the hit model extend by the standard probability formula

$$Eu = EE(u \mid m) = \sum_{m=0}^{\infty} E(u \mid m)(e^{-\mu}\mu^m/m!).$$

One conceivable drawback of $\mathbf{P}\{K \uparrow K_0\}$ is that there is positive probability $e^{-\mu}$ that it is vacuous; however, this becomes academic when μ is chosen large enough.

The intersection of an arbitrary infinite straight line λ with $\mathbf{P}\{K\}$—a Poisson random segment process on the line—is of some interest. Since the probability that an arbitrary segment of length L in $\mathbb{R}^2$ hits no domain of $\mathbf{P}\{K\}$ is $\exp\{-\rho(A + \pi^{-1}BL)\}$, the conditional probability that segment extensions L_1 and L_2 at its two ends are also uncovered, given the segment is uncovered, equals $\exp\{-\pi^{-1}\rho B(L_1 + L_2)\}$. It follows that the uncovered intervals of λ are mutually independent exponential $(\pi^{-1}\rho B)$ random variables. The covered intervals too are mutually independent, and conform to a distribution which may more properly be regarded as the concern of queuing theory! In fact, the alternating covered and uncovered intervals of λ form an "alternating renewal process" [11].

Special and limiting cases of $\mathbf{P}\{K \uparrow K_0\}$ occur when K is a particle and when K is a segment of increasing length (balanced by decreasing ρ)—for properties of the resulting Poisson point and line processes, see [20, 22, 25].

More general Poisson models. In fact, to model any image data, one may "suspend" onto each Poisson arrow just about any deterministic (as the K's above) or randomly sampled set. A rather general suspended set would be one, X say, sampled from some random set distribution with centroid at the origin of $\mathbb{R}^2$; after sampling, X is suspended on the arrow $\upsilon = (\xi, \theta)$ by rotation through θ followed by translation through ξ. Examples are fixed or random domains and fixed or random particle clusters. Such suspensions are only limited by the imagination of the modeler, or by suspensions that can be satisfactorily simulated, e.g., by a computer. Unfortunately, theoretical results are rather limited: apart from the convex theory sketched here, the only other theory relates to point cluster processes [30]. The homogeneity and isotropy in the above models may be relaxed, by taking the underlying Poisson particle process inhomogeneous, i.e., with nonconstant intensity $\rho(\xi)$, and selecting anisotropic suspended sets.

8. EXPECTED NUMBERS AND ERGODIC DISTRIBUTIONS OF n-FIGURES

Define an *n-figure* to be an ordered set of n mobile objects (examples have of course already been considered). The choice in our models of primary frames with independent uniform distributions enables the IG of n-figures to be applied with advantage.

Consider the Poisson particle process **P**. By its mutual independence within disjoint domains, we have that the probability of getting a particle in each of the area elements $d\xi_1$, $d\xi_2$, $d\xi_3$ is

$$P/E(d\xi_*) = \prod (\rho d\xi_i);$$

here we have written P/E to indicate that this is also the *expected number* of such 3-figures. By (13) this may be reexpressed as

$$P/E(d\zeta, dR, d\phi_*) = \rho^3 R^3 |\sigma(\phi_*)| d\zeta dR \prod d\phi_i.$$

Integrating totally with respect to $\zeta \in K_0$, we get

$$P/E(K_0, dR, d\phi_*) = \rho^3 R^3 |\sigma(\phi_*)| A_0 dR \prod d\phi_i. \tag{21}$$

Integration may be continued, leading to "P/E" relations, provided at least one differential is retained; without differentials, we have simply the expected number of 3-figures with circumcenter $\in K_0$ satisfying whatever condition is imposed. For example, we may integrate (21) to obtain $P/E(K_0, R \leq R_0, d\phi_*)$ and then again to obtain $E(K_0, R \leq R_0)$. Such formulas can be quite revealing for structural parameterizations like this. The useful examples generally relate to n-figures of more basic objects, like particles and lines, rather than arrows and convex domains.

Hitting conditions. The independence properties of Poisson processes may be exploited even further, by restricting attention to n-figures of particular types determined by a hitting condition. A parallel theory to that holding without the hitting condition may be developed. For example, restricting attention to 3-figures whose circumdisks each contain precisely k other Poisson particles (besides the three peripheral particles), each of the above relations which contain a dR must be multiplied by the corresponding Poisson probability $\exp(-\pi\rho R^2)(\pi\rho R^2)^k/k!$.

Large domains. In fact, stochastic "ergodic" convergence, to the corresponding mean values, may be generally proved for the numbers of 3-figures with $\zeta \in$ the disk with center 0 and radius r, and with R and ϕ_* satisfying arbitrary conditions, as $r \to \infty$. Thus, for large K_0, the number of 3-figures within size limits $(dR, d\phi_*)$ and with empty circumdisk is asymptotically

$$\rho^3 R^3 e^{-\pi\rho R^2} |\sigma(\phi_*)| A_0 dR \prod d\phi_i. \tag{22}$$

Total integration yields the total number of such empty disks as asymptotically

$$12\rho A_0. \tag{23}$$

Due to the ordering, each circumdisk is counted 3! times, so that the intensity of such empty circumdisks in $\mathcal{R}^2$ is 2ρ, twice that of the base particles. Equation (22) may be normalized by (23), yielding the "ergodic" probability density of the empty circumdisks generated by **P** as

$$f(R, \phi_*) = 2(\pi\rho)^2 R^3 e^{-\pi\rho R^2} \cdot (24\pi^2)^{-1} |\sigma(\phi_*)|. \tag{24}$$

In this distribution R and ϕ_* are independent, while the ϕ_* are dependent, with marginals given in [22, Relations (4.18)–(4.20)]. In a natural sense, this is the distribution of a "uniform random" such circumdisk from **P**.

Often as here, a hitting condition is required to enable normalization to a proper ergodic probability distribution. The ergodic distributions of n-figures for underlying Poisson flat processes in $\mathcal{R}^d$ are systematically developed in [23, 26]. For *any* stochastic model comprising an aggregate of objects, the generated aggregates of n-figures ($n = 2, 3, \ldots$) may well be useful for characterizing and analyzing the resulting texture.

9. ASSORTED TOPICS

9.1. *Chains and percolation.* We have just seen how, for Poisson particles, given n-figures recur throughout $\mathcal{R}^2$ with a given intensity, at the same time conforming to certain ergodic distributions, normalizable or not. One particular consequence of suspending domains K onto the particles of **P** to form $\mathbf{P}\{K\}$ is that for each particle n-figure the corresponding domains may or may not form an *n-chain*, i.e., a connected set in which every point pair may be joined by a continuous path lying entirely within the set. It is natural to restrict attention to *isolated* n-chains—those not hit by other domains. This implies the existence of a "taboo" region around the corresponding n particles in υ-space, which must contain no particles. Unfortunately the probabilities for chains, even for the simplest case of equal disks, rapidly become intractable as n increases. Writing p_n ($n = 1, 2, \ldots$) for the probability that an arbitrary domain of $\mathbf{P}\{K\}$ belongs to an isolated chain of length n (the homogeneity in $\mathcal{R}^2$ makes this definition possible), we have that

$$\zeta \equiv 1 - \sum_{n=1}^{\infty} p_n$$

is the probability that an arbitrary domain belongs to a chain of infinite length. It is a remarkable fact that there exists a critical value $\rho_c(K)$ of ρ such that

$$q \begin{Bmatrix} = \\ > \end{Bmatrix} 0 \qquad \text{according as} \qquad \rho \begin{Bmatrix} < \\ > \end{Bmatrix} \rho_c(K).$$

This phenomenon is described as *percolation*, since a percolating flow from domain to domain *throughout* $\mathbb{R}^2$ is or is not possible, depending almost surely upon the value of ρ in relation to that of $\rho_c(K)$. Simple dimensional argument tells us that $\rho_c(lK) = l^{-2}\rho_c(K)$ so that essentially ρ_c is defined for the class of domain shapes. In the case of disks of unit radius simulation studies suggest that $\rho_c \sim 0.30$ [35]. Although strictly speaking the percolation phenomenon depends for its existence upon a model extending throughout $\mathbb{R}^2$, nevertheless its effects should manifest themselves within sufficiently large domains K_0.

9.2. *Random polygonal tessellations.* A *tessellation* of X is a division of X into nonoverlapping cells which cover X. There are several natural interesting homogeneous random polygonal tessellations of $\mathbb{R}^2$ with Poisson bases, in which each cell is a bounded convex polygon. All distributions and moments mentioned here are "ergodic" in the abovementioned limit sense.

(i) The lines of a homogeneous Poisson line process with orientation distribution Θ on $[0, \pi)$, corresponding to a **P** of intensity $\tau\Theta$ in the strip $[0, \pi) \otimes \mathbb{R}^1$, tessellate $\mathbb{R}^2$. Two general results are that the mean number of vertices (or sides) per polygon is four, and the distribution of polygon in-radii is exponential with mean τ [20, 25].

(ii) (a) Almost surely each point of $\mathbb{R}^2$ has a nearest particle of **P**. For a given particle, the set of points of $\mathbb{R}^2$ having this particle as nearest particle almost surely constitutes a convex polygon, and the aggregate of such polygons tessellate $\mathbb{R}^2$. This is the Voronoi (sometimes Dirichlet) tessellation corresponding to the underlying particle aggregate. Since almost surely three sides meet at each vertex, the mean number of vertices (or sides) per polygon is six.

(ii) (b) Joining all particle pairs of **P** whose associated polygonal cells have a common side yields the random triangular Delaunay tessellation of $\mathbb{R}^2$, which is in a sense dual to the Voronoi tessellation. Since geometrical identity ensures that each Voronoi vertex is a Delaunay circumcenter, the complete ergodic distribution for Delaunay triangles is given by (24) above. That distribution also governs the joint orientation of the three sides meeting in a random Voronoi vertex.

Other topologically similar random tessellations with three sides meeting at each vertex arise from

(a) considering points of $\mathbb{R}^2$ with the same nearest n particles of **P** ($n = 2, 3, \ldots$) [22]; and

(b) plane sections of higher order Voronoi tessellations [24].

A stochastic process arises from each random tessellation by associating a random variable with each cell. This includes random colorings as a special case. The line-generated tessellations (i) may be given a black–white "checkerboard" pattern by randomly coloring an arbitrary cell black or white and requiring that at every vertex the colors of the associated four cells alternate as the vertex is encircled.

We have used Poisson bases for these superstructures, but clearly any line-aggregate or point-aggregate base would suffice.

9.3. *Clustering.* The phenomenon of "clustering" of the particles of stochastic point processes is a notion that I make no attempt to *define* precisely—hence the use of "quotes" below. The basic Poisson particle process **P** possesses a certain degree of inherent "clustering," as is apparent to anyone who has ever generated independent UR points in a rectangle. By suspending onto each of its particles an aggregate of subparticles, of linear extent sufficiently small relative to $\rho^{-\frac{1}{2}}$, the length scale of **P**, one may generate as highly "clustered" a subparticle process as desired. This is "superclustering" relative to **P**. Here, thanks to the constructive procedure, "cluster" is well defined. Such models have been used for the "clustering" of galaxies on the celestial sphere [30].

To generate "subclustered" particle processes relative to **P** is rather more difficult. One simple model which achieves this, due to Matérn [19, Sect. 3.6], runs as follows. In the lamina $0 \leq z \leq 1$ of $\mathcal{R}^3$ generate a three-dimensional **P** and center at each of its particles opaque disks of radius $r/2$ parallel to the xy-plane. Looking down on the system along lines parallel to the z axis, from the z positive end, and taking account of the obscuring effect of the disks, one sees an aggregate of complete and partial disks. Then the particle process $\mathbf{P}^{(r)}$ in the xy-plane comprising the orthogonal projections of the centers of such complete disks is "subclustered" relative to the standard Poisson process **P**, for clearly no pair of its particles can be closer than r apart. $\mathbf{P}^{(0)}$ is **P** itself. Simple argument shows that $\mathbf{P}^{(r)}$ has intensity $\{1 - \exp(-\pi\rho r^2)\}/\pi r^2$ per unit area, the maximum value of which for fixed r is $1/\pi r^2$, corresponding to quarter-coverage by the complete disks (maximal close packing gives 0.91-coverage). Increasing the lamina thickness is tantamount to increasing ρ. Equating z decreasing with time increasing, this may be regarded as a *sequential* partial packing by nonoverlapping disks.

Higher average packing densities may be achieved by rendering transparent any disk that is overlapped by any already recorded disk, as z decreases, so removing their overlapping effect on all lower (or later) disks; looked at in another and simpler way, disks are placed sequentially and uniformly into the spaces between already placed disks. This is the planar version of the one-dimensional "car-parking" problem, but has so far proved theoretically intractable. Turning from disks to rectangles, Palasti's conjecture that the average limiting packing density for equal aligned rectangles sequentially randomly packed into a large area is the square of the corresponding car-parking density, is still unresolved [7]. Another bounded model is obtained from **P** within a bounded domain by imposing the stochastic condition that no pair of particles be closer than r apart.

All the above are "hard core" models in the sense that disks of radius $r/2$ centered at the particles are disjoint. Kelly and Ripley [15] have described an interesting "soft core" generalization, again essentially based upon an underlying **P**, in which close particle contacts are inhibited rather than prohibited. It has a Markov-type property for the probability of occurrence of a particle at an arbitrary point, given the remainder of the realization.

There is currently high activity in England in this general area, much of it concerned with the statistical problem of fitting such models to planar particle data (locations of settlements, trees, etc.). The general method is novel in that a number m of independent realizations of the candidate model particle process are computer simulated. Key "significant" (but, inevitably, arbitrary) characteristics, often beyond the reach of a theoretical treatment, are measured for each of the m realizations *and* for the sample data. Statistical significance is derived from the consequent *ranking* of the data vis-à-vis the realizations [33]. [13] is a recent review of the area. Of course, sub- and superclustering is but one facet of the by now substantial theory of stochastic point processes [16], a reasonable amount of which extends from the line to the plane.

9.4. *Fractals.* To escape UR or Poisson bases for homogeneous stochastic planar images, it seems one must turn first to Gaussian stochastic processes. I can do no better than to refer you to the remarkable recent book [18] by Mandelbrot in which he develops the theory of random *fractal* curves and surfaces, and includes many computer-simulated illustrations. They possess the eminently natural property of being stochastically invariant (or "self-similar") under changes of scale. Consequently, they are irregular to the eye, with noninteger (fractal) dimensions! They are most appropriate models for such things as coastlines, topography, and lung structure.

10. SQUARE LATTICE ANALOGS

A natural question, especially in present company, is how much of the above theory carries over when $\mathcal{R}^2$ is replaced by the square lattice $\{(i, j): i, j \text{ integers}\}$. We are here much less interested in approximating by sufficiently fine lattices [8], than in true analogs. Suffering the heavy losses of continuous orientation and continuous scale change, one is reduced from Euclidean (and even homothetic) transformations to the rather limited translations by vectors with integer components, the corresponding invariant measure being simple counting measure. In this sense, a uniform random translate of a mobile finite particle array hitting a fixed finite particle array may be defined, leading to the hit model $\{X_1, \ldots, X_n \uparrow X_0\}$. Note that in

$$\{\underbrace{X, \ldots, X}_{n} \uparrow X_0\},$$

the locations of the random arrays may coincide with positive probability, leading to the notion of array *multiplicity*. Thus, in the Poisson limit $\mathbf{P}\{X \uparrow X_0\}$ the number of X occupying each permissible location has a Poisson distributed multiplicity. More useful may be the case where multiplicity at each location is either 0 or 1, decided by either sequential placement forbidding coincidences, or by a Bernoulli process of independent trials, one trial for each lattice point. These models are well known when the arrays are simply singletons, and the n-figure theory of Section 8 corresponds in a sense to the indicator method of calculating expected numbers of configurations of given type in a region: if at each lattice point the chance of a particle occurring is p, then the intensity of configurations of arbitrary geometric type is $p^i(1 - p)^j$ where i, j are the numbers of particles and voids in the configuration. The two standard indicator techniques of Section 6 easily extend.

Only A out of the three geometric quantities A, B, C seems to have a satisfactory analog, with (15) becoming

$$\sum_{012}\sum A_{012} = A_0A_1A_2.$$

A finite lattice array might be defined as convex if all lattice points in its geometric convex hull belong to the array, but this does not seem to lead to useful analogs of the fundamental IG formulas for convex domains in Sections 5 and 6.

The story is different for *rectangular* arrays $T_i(a_i, b_i)$ (a_i, b_i integers). Arbitrary intersections of rectangular arrays are rectangular arrays. For the hit model $\{T_1, \ldots, T_n \uparrow T_0\}$, it may be proved that

$$P(0\cdots n) = \{\prod_0^n a_i - \prod_0^n (a_i - 1)\}\{\prod_0^n b_i - \prod_0^n (b_i - 1)\}/\prod_0^n (a_0 + a_i - 1)(b_0 + b_i - 1)$$

and

$$E(A_{0\cdots n}|0\ldots n) = \prod_0^n a_ib_i/\prod_1^n (a_0 + a_i - 1)(b_0 + b_i - 1),$$

which, as the forms of these expressions indicate, stem by independence from the corresponding results for random segment arrays on a linear lattice.

Highly developed exceptions to this generally gloomy picture, largely on account of their importance in statistical mechanics, are percolation theory on lattices [36] and Markov-type particle processes on lattices [32].

REFERENCES

"Introductory"

1. M. G. Kendall, and P. A. P. Moran, *Geometrical Probability*, Hafner, New York, 1963.
2. P. A. P. Moran, A note on recent research in geometrical probability, *J. Appl. Prob.* **3**, 1966, 453–463.
3. P. A. P. Moran, A second note on recent research in geometrical probability, *Advan. Appl. Prob.* **1**, 1969, 73–89.
4. D. V. Little, A third note on recent research in geometrical probability, *Advan. Appl. Prob.* **6**, 1974, 103–130.
5. A. Baddeley, A fourth note on recent research in geometrical probability, *Advan. Appl. Prob.* **9**, 1977, 824–860.
6. L. A. Santaló, *Integral Geometry and Geometric Probability*, Vol. 1, Encyclopedia of Mathematics and Its Applications, Addison–Wesley, Reading, Mass., 1976.

Other

7. Y. Akeda and M. Hori, On random sequential packing in two and three dimensions, *Biometrika* **63**, 1976, 361–366.
8. G. Bernroider, The foundation of computational geometry: theory and application of the point-lattice-concept within modern structure analysis, in *Geometrical Probability and Biological Structures: Buffon's 200th Anniversary* (R. E. Miles and J. Serra, Eds.), pp. 153–170, Lecture Notes in Biomathematics No. 23, Springer-Verlag, Berlin. 1978.
9. W. Blaschke, *Vorlesungen über Integralgeometrie*, Chelsea, New York, 1949.

10. G. Buffon, Essai d'arithmétique morale. Supplément à l'Historie Naturelle, Vol. 4, Paris, 1777.
11. D. R. Cox, *Renewal Theory*, Methuen, London, 1962.
12. M. W. Crofton, Probability, in *Encyclopaedia Britannica*, Vol. 19, 9th ed., pp. 768–788, 1885.
13. P. J. Diggle, On parameter estimation and goodness-of-fit testing for spatial point patterns, *Biometrics* **35**, 1979, 87–101.
14. W. Feller, *An Introduction to Probability Theory and Its Applications*, Vol. I, 2nd ed., Wiley, New York, 1957.
15. F. P. Kelly and B. D. Ripley, A note on Strauss's model for clustering, *Biometrika* **63**, 1976, 357–360.
16. P. A. W. Lewis, (Ed.), *Stochastic Point Processes: Statistical Analysis, Theory and Applications*, Wiley, New York, 1972.
17. Z. A. Lomnicki and S. K. Zaremba, A further instance of the Central Limit Theorem for dependent random variables, *Math. Z.* **66**, 1957, 490–494.
18. B. B. Mandelbrot, *Fractals: Form, Chance and Dimension*, Freeman, San Francisco, 1977.
19. B. Matérn, *Spatial Variation*, Meddelanden från Statens Skogsforskningsinstitut, Band 49, Nr. 5, 1960.
20. R. E. Miles, Random polygons determined by random lines in a plane, *Proc. Nat. Acad. Sci. USA* **52**, 1964, 901–907; II, **52**, 1964, 1157–1160.
21. R. E. Miles, The asymptotic values of certain coverage probabilities, *Biometrika* **56**, 1969, 661–680.
22. R. E. Miles, On the homogeneous planar Poisson point process, *Math. Biosci.* **6**, 1970, 85–127.
23. R. E. Miles, Poisson flats in Euclidean spaces. II. Homogeneous Poisson flats and the Complementary Theorem, *Advan. Appl. Prob.* **3**, 1971, 1–43.
24. R. E. Miles, The random division of space, *Advan. Appl. Prob. Suppl.* **4**, 1972, 243–266.
25. R. E. Miles, The various aggregates of random polygons determined by random lines in a plane, *Advan. Math.* **10**, 1973, 256–290.
26. R. E. Miles, A synopsis of "Poisson flats in Euclidean spaces," pp. 202–227, in *Stochastic Geometry* (E. F. Harding and D. G. Kendall, Eds.), Wiley, London, 1974.
27. R. E. Miles, The fundamental formula of Blaschke in integral geometry and geometrical probability, and its iteration, for domains with fixed orientations, *Aust. J. Statist.* **16**, 1974, 111–118.
28. R. E. Miles and J. Serra (Eds.), *Geometrical Probability and Biological Structures: Buffon's 200th Anniversary*, Lecture Notes in Biomathematics No. 23, Springer-Verlag, Berlin, 1978.
29. L. Nachbin, *The Haar Integral*. Van Nostrand, Princeton, N. J., 1965.
30. J. Neyman and E. L. Scott, Processes of clustering and applications, in *Stochastic Point Processes: Statistical Analysis, Theory and Applications* (P. A. W. Lewis, Ed.), Vol. I pp. 646–681, Wiley, New York, 1972.
31. E. Parzen, *Modern Probability Theory and its Applications*, Wiley, New York, 1960.
32. C. J. Preston, *Gibbs States on Countable Sets*. Tracts in Math. No. 68. Cambridge Univ. Press, London/New York, 1974.
33. B. D. Ripley, Modelling spatial patterns. *J. Roy. Statist. Soc. Ser. B.* **39**, 1977, 172–212.
34. H. E. Robbins, On the measure of a random set, *Ann. Math. Statist.* **15**, 1944, 70–74; **16**, 1945, 342–347.
35. F. D. K. Roberts, A Monte Carlo solution of a 2-dimensional unstructured cluster problem, *Biometrika* **54**, 1967, 625–628.
36. V. I. S. Shante and S. Kilpatrick, An introduction to percolation theory, *Advan. Phys.* **20**, 1971, 325–357.
37. F. Streit, On multiple integral geometric integrals and their applications to probability theory, *Canad. J. Math.* **22**, 1970, 151–163.
38. I. Todhunter, *A History of the Mathematical Theory of Probability from the time of Pascal to that of Laplace*, Chelsea, New York, 1949.

Stochastic Image Models Generated by Random Tessellations of the Plane *

J. W. MODESTINO, R. W. FRIES,† AND A. L. VICKERS

Electrical, Computer, and Systems Engineering Department, Rensselaer Polytechnic Institute, Troy, New York 12181

A useful class of two-dimensional (2-D) random fields is described which can be generated by random tessellations of the plane. The random tessellations are in turn generated by marked point processes evolving according to a spatial parameter. Gray levels are assigned within elementary disjoint regions generated by the tessellations to have specified correlation properties with gray levels in contiguous regions. A complete second-order statistical description of the resulting class of random fields is provided. This includes not only autocorrelation functions and power spectral densities but joint probability density functions. Several applications are discussed including the modeling of real-world imagery possessing inherent edge structure.

1. INTRODUCTION

In a number of image processing applications it is important to have available a realistic and conveniently parameterized stochastic model for the class of images of interest. Examples include image enhancement, image coding, texture discrimination, and edge detection in noisy digitized images. This latter application provided the initial motivation for the work described here. It is clearly important in this case to have available a stochastic model for edge structure in two-dimensional (2-D) imagery data. This is also the case in several other applications to be described.

The ubiquitous 2-D Gaussian random field [1–4] has often been proposed as an appropriate model for real-world imagery. This has been particularly the case in image coding applications (cf. [5, 6]). Unfortunately, the 2-D Gaussian random field, except under pathological assumptions on the covariance, cannot account for the predominant and pronounced edge structure present in real-world imagery. It is of some interest then to develop alternative and more appropriate 2-D stochastic models for image data exhibiting inherent edge structure.

In this paper a useful class of 2-D random fields appropriate for this purpose are described and several of their more important properties discussed. This

* This work was supported in part by ONR under Contract N00014-75-C-0281 and in part by USAF RADC under Contract AF30602-78-C-0083.

† R. W. Fries is now with PAR Technology Corp., Rome, N.Y. 13440.

ISBN 0-12-597320-9

class of random fields is modeled as a marked point process [7] evolving according to a spatial parameter. According to this model the plane is randomly partitioned or tessellated into a number of disjoint geometric regions by an appropriately defined field of random lines which form the boundaries of these regions. The density of these random lines, or edges, is defined in terms of a rate parameter λ. Gray levels are then assigned within elementary regions to possess a specified correlation coefficient ρ with gray levels in contiguous regions. We describe several schemes for tessellating the plane into elementary geometrical regions where the line process generating the tessellation is modeled in terms of a stationary renewal point process possessing a gamma distributed interarrival distribution with characteristic parameter ν.

Given a particular tessellation scheme, the random fields are then completely defined in terms of the three parameters λ, ρ, and ν. The parameter λ represents the "edge busyness" associated with an image while ρ is indicative, at least on an ensemble basis, of the "edge contrast." For ρ large (in magnitude) and negative there is an abrupt almost black-to-white or white-to-black transition across an edge boundary. If $\rho > 0$, on the other hand, the transition across an edge boundary is much more gradual. Finally, the parameter ν provides a measure of the degree of randomness or "homogeneity" of the resulting mosaic appearance. These properties are closely related to real-world image properties.

This class of random fields has proved a useful and conveniently parameterized stochastic image model in a number of important applications. These include: the development of a class of edge detectors based upon 2-D least mean-square filtering concepts, image enhancement based upon 2-D stochastic homomorphic filtering, a stochastic texture model in the development of statistically optimum texture discriminators, and a stochastic source model for image coding applications. Several of these applications will be discussed.

2. PRELIMINARIES

We consider an image as a family of random variables $\{f_{\mathbf{x}}(\omega), \mathbf{x} \in R^2\}$, or a random field, defined on some fixed but unspecified probability space $(\Omega, \mathcal{A}, P)$. For convenience we suppress the functional dependence upon the underlying probability space and consistently write $f(\mathbf{x})$ for $f_{\mathbf{x}}(\omega)$. The covariance function of the random field then becomes[1]

$$R_{ff}(\mathbf{x}, \mathbf{y}) = E\{f(\mathbf{x})f(\mathbf{y})\}; \qquad \mathbf{x}, \mathbf{y} \in R^2, \tag{1}$$

where $E\{\cdot\}$ represents the expectation operator. If a random field $\{f(\mathbf{x}), \mathbf{x} \in R^2\}$ possesses a covariance function invariant under all Euclidean motions it will be called *homogeneous and isotropic* (cf. [2] for definitions). In this case the covariance function of the field evaluated at two points can depend only upon the Euclidean distance between these two points so that

$$E\{f(\mathbf{x} + \mathbf{u})f(\mathbf{x})\} = R_{ff}(\|\mathbf{u}\|), \tag{2}$$

where $\mathbf{u}^T = (u_1, u_2)$ is an element of R^2 and $\|\mathbf{u}\|$ represents the ordinary Euclidean

[1] We assume the field is of second order (i.e., variances exist) and possesses zero mean.

norm defined in terms of an inner product $\langle \cdot, \cdot \rangle$ according to

$$\|\mathbf{u}\|^2 = \langle \mathbf{u}, \mathbf{u} \rangle = u_1^2 + u_2^2. \tag{3}$$

By construction, the 2-D random fields to be described here are of this category. Furthermore, they have been explicitly constructed so that the joint probability density function (p.d.f.) of the field evaluated at two points likewise depends only upon the Euclidean distance between these points. More specifically, define the random variables $f_1 = f(\mathbf{x})$ and $f_2 = f(\mathbf{x} + \mathbf{u})$. The joint p.d.f. associated with these two random variables, parameterized by the spatial coordinates, then satisfies

$$p\{f_1, f_2; \mathbf{x}, \mathbf{x} + \mathbf{u}\} = p\{f_1, f_2; \|\mathbf{u}\|\}, \tag{4}$$

which is the 2-D concept of stationarity [8] or invariance which will be most useful for our purposes.

The corresponding power spectral density function is given by

$$S_{ff}(\boldsymbol{\omega}) = \int_{R^2} R_{ff}(\|\mathbf{u}\|) \exp\{-j\langle \boldsymbol{\omega}, \mathbf{u} \rangle\} d\mathbf{u}, \tag{5}$$

where $\boldsymbol{\omega}^T = (\omega_1, \omega_2)$ represents a 2-D spatial frequency vector and $d\mathbf{u}$ is the differential volume element in R^2. This expression can be evaluated up to functional form with the aid of a theorem of Bochner [9] with the result

$$S_{ff}(\boldsymbol{\omega}) = S(\Omega) = 2\pi \int_0^\infty \lambda R_{ff}(\lambda) J_0(\lambda\Omega) d\lambda, \tag{6}$$

where $\Omega \triangleq \|\boldsymbol{\omega}\| = (\omega_1^2 + \omega_2^2)^{\frac{1}{2}}$ represents radial frequency. Here $J_0(\cdot)$ denotes the ordinary Bessel function of the first kind of order zero. The quantities $S_{ff}(\cdot)$ and $R_{ff}(\cdot)$ are then related through a Hankel transform [10, 11].

3. CONSTRUCTION PROCEDURE

In the present section we describe several construction procedures for homogeneous and isotropic random fields generated by random tessellations of the plane. Relevant second-order properties are discussed in the next section. We begin with the case where the plane is tessellated into random *rectangular* regions.

Rectangular tessellations. A fundamental role in the construction of this class of processes will be played by the integer-valued random field[2] $\{N(\mathbf{x}), \mathbf{x} \geq 0\}$ which provides a 2-D generalization of a counting process [12]. In particular, suppose the vector $\tilde{\mathbf{x}}$ is obtained from $\mathbf{x}$ according to $\tilde{\mathbf{x}} = \mathbf{A}x$, where $\mathbf{A}$ is the unitary matrix

$$\mathbf{A} = \begin{bmatrix} \cos\theta & \sin\theta \\ -\sin\theta & \cos\theta \end{bmatrix}, \tag{7}$$

defined for some $\theta \epsilon [-\pi, \pi]$. This transformation results in a rotation of the Cartesian coordinate axes (x_1, x_2) by θ radians as illustrated in Fig. 1.

[2] By the notation $\mathbf{x} \geq 0$ we mean that $\mathbf{x}^T = (x_1, x_2)$ is such that $x_i \geq 0$, $i = 1, 2$.

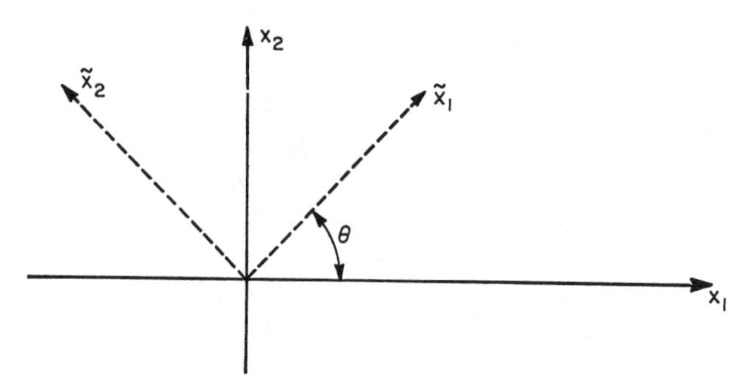

FIG. 1. Rotation of Cartesian coordinate axes.

Consider now the integer-valued random field defined by

$$N(\mathbf{x}) = N_1(\tilde{x}_1) + N_2(\tilde{x}_2); \qquad \mathbf{x} \geq 0, \tag{8}$$

where $\theta \epsilon [-\pi, \pi]$ is chosen according to some p.d.f. $p(\theta)$ and $\{N_i(l), l \geq 0\}$, $i = 1, 2$, are mutually independent 1-D counting processes. That is, $N_i(l)$ represents the number of events which have occurred in the interval $[0, l]$. We will be particularly concerned with the case where $\{N_i(l), l \geq 0\}$, $i = 1, 2$, are renewal point processes defined in terms of their interarrival distribution.

The random field $\{N(\mathbf{x}), \mathbf{x} \geq 0\}$ in (8) then assumes constant integer values on nonoverlapping rectangles whose sides are parallel to the transformed coordinate axes $(\tilde{x}_1, \tilde{x}_2)$ and whose locations are determined by the event times of the corresponding point processes $\{N_i(l), l \geq 0\}$, $i = 1, 2$. Consider now the random field $\{f(\mathbf{x}), \mathbf{x} \geq 0\}$ which undergoes transitions at the boundaries of these elementary rectangles. The gray level assumed throughout any elementary rectangle is zero-mean Gaussian[3] with variance σ^2 and correlation coefficient ρ with the gray levels in contiguous rectangles. More specifically, let $X_{i,j}$ represent the amplitude or gray level assumed by the random field after i transitions in the $\tilde{x}_1$ direction and j transitions in the $\tilde{x}_2$ direction. The sequence $\{X_{i,j}\}$ is assumed generated recursively according to

$$X_{i,j} = \rho X_{i-1,j} + \rho X_{i,j-1} - \rho^2 X_{i-1,j-1} + W_{i,j}; \qquad i, j \geq 1, \tag{9}$$

where $|\rho| \leq 1$, and $\{W_{i,j}\}$ is a 2-D sequence of independent and identically distributed (i.i.d.) zero-mean Gaussian variates with common variance $\sigma_w^2 = \sigma^2(1 - \rho^2)^2$. The initial values $X_{k,0}$, $X_{0,l}$, $k, l \geq 0$ are jointly distributed zero-mean Gaussian variates with common variance σ^2 and covariance properties chosen to result in stationary conditions. An alternative interpretation of the sequence $\{X_{i,j}\}$ is as the output of a separable 2-D recursive filter excited by a white noise field. It is easily seen that

$$E\{X_{i,j}X_{i+k_1,j+k_2}\} = \sigma^2\rho^{k_1+k_2}; \qquad k_1, k_2 \geq 0. \tag{10}$$

Typical computer-generated realizations of the resulting random field are

[3] For definiteness we assume Gaussian statistics. This assumption is not critical to the development which follows and is easily removed.

illustrated in Fig. 2 for selected values of ρ when $p(\theta)$ is uniform over $[-\pi, \pi]$ and $\{N_i(l), l \geq 0\}$, $i = 1, 2$, are Poisson with intensities $\lambda_1 = \lambda_2 = \lambda$. The displayed images here and throughout this paper are square arrays consisting of 256 elements or samples on a side. In Fig. 2, λ is measured in normalized units of events per sample distance so that there are on average 256λ transitions along each of the orthogonal axes. Similarly, in Fig. 3 we illustrate realizations of the resulting random field when the point processes $\{N_i(l), l \geq 0\}$, $i = 1, 2$, undergo jumps of unit height at equally spaced intervals $l = 1/\lambda$. The starting positions ϵ_i, $i = 1, 2$, will be assumed uniformly distributed over the interval $[0, l]$.

The preceding two examples are special cases of the situation where the point processes $\{N_i(l), l \geq 0\}$, $i = 1, 2$, are stationary renewal processes [13, 14] with gamma distributed interarrival times. This class of random fields represents a 2-D generalization of the class of 1-D processes described in [15]. In particular, we assume the common interarrival distribution of the two mutually independent

FIG. 2. Selected realizations of random field generated by Poisson partitions.

FIG. 3. Selected realizations of random field generated by periodic partitions.

point processes $\{N_i(l),\ l \geq 0\}$, $i = 1, 2$, possesses p.d.f.

$$f(x) = \frac{x^{\nu-1}}{\Gamma(\nu)\beta^{\nu}} \exp\{-x/\beta\}, \tag{11}$$

where $\nu = 1, 2, \ldots$, and $\beta = 1/\lambda\nu$ for fixed $\lambda > 0$. For example, if $\nu = 1$ we have the exponential distribution

$$f(x) = \lambda e^{-\lambda x}; \qquad x \geq 0, \tag{12}$$

associated with the Poisson process, while in the limit $\nu \to \infty$ we have

$$f(x) = \delta(x - 1/\lambda); \qquad x \geq 0, \tag{13}$$

corresponding to the case of periodic partitions as illustrated in Fig. 3.

In Fig. 4 we illustrate selected realizations of the resulting random field for several values of ν all with $\lambda = 0.05$ and $\rho = 0.0$. Clearly the parameter ν provides a measure of the degree of randomness or "homogeneity" of the structure. For small ν the random field $\{f(\mathbf{x}),\ \mathbf{x} \in R^2\}$ appears as a random rectangular mosaic. As ν increases, individual realizations rapidly approach a more periodic mosaic in appearance. The parameters λ, ρ, and ν then completely describe this class of 2-D random fields.

Although this class of 2-D random fields provides a useful image model in selected applications, the rectangular mosaic exhibited by individual realizations is not entirely consistent with edge structure in real-world imagery. That is, we would expect the edge structure to exhibit a much more random edge orientation. An alternative approach then is to randomly partition or tessellate the plane into more complex geometric regions. In what follows we describe two possible alternatives.

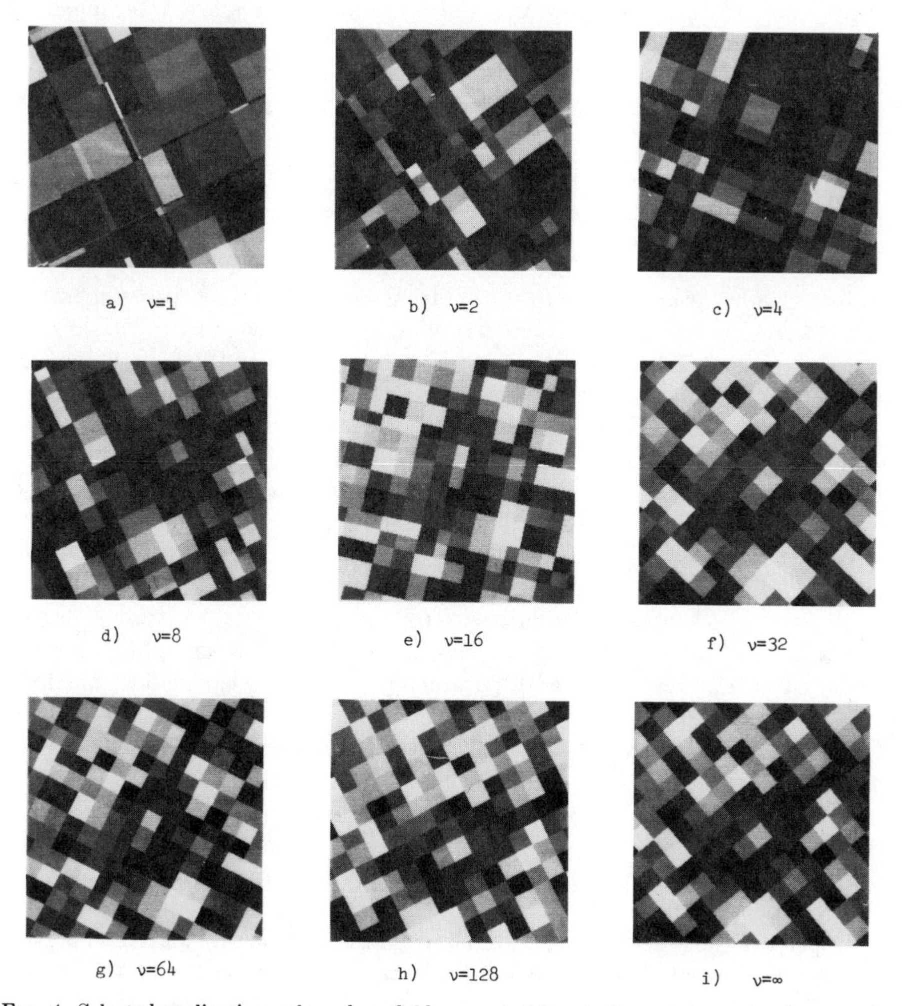

FIG. 4. Selected realizations of random field generated by stationary renewal point processes possessing gamma distributed interarrival distribution and with $\lambda = 0.05$ and $\rho = 0.0$.

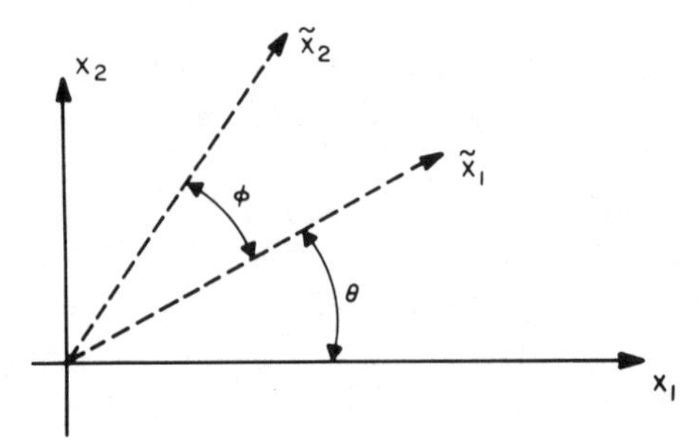

FIG. 5. Nonisometric transformation of Cartesian coordinate frame.

Parallelogram tessellations. Again the plane is partitioned by two mutually independent renewal point processes evolving along appropriately defined coordinate axes. In this case, however, these coordinate axes are determined by a nonunitary or nonisometric transformation of the Cartesian coordinate frame. More specifically, we suppose the vector $\tilde{\mathbf{x}}$ is obtained from $\mathbf{x}$ according to the linear transformation $\tilde{\mathbf{x}} = \mathbf{A}\mathbf{x}$ where now $\mathbf{A}$ is the 2×2 matrix defined for ϕ, $\theta \in [-\pi, \pi]$ according to

$$\mathbf{A} = \frac{1}{\sin \phi} \begin{bmatrix} \sin (\theta + \phi) & -\cos (\theta + \phi) \\ -\sin \theta & \cos \theta \end{bmatrix}. \tag{14}$$

This transformation has an interpretation as a distance-preserving rotation of the Cartesian coordinate frame by θ radians followed by a non-distance-preserving scaling. The new coordinate axes $(\tilde{x}_1, \tilde{x}_2)$ are illustrated in Fig. 5. In what follows we assume that the angle ϕ is fixed while θ is chosen uniformly on $[-\pi, \pi]$. The point processes $\{N_i(l), l \geq 0\}$ now evolving along the respective coordinate axes $\tilde{x}_i$, $i = 1, 2$, result in a tessellation of the plane into elementary regions comprised of disjoint parallelograms whose sides are parallel to the new coordinate axes. Gray levels are then assigned within these elementary regions as described previously.

In Fig. 6 we illustrate typical realizations of the resulting 2-D random field for selected values of ϕ all with $\lambda = 0.025$ and $\rho = 0.0$. The point processes generating the random field in this case are Poisson corresponding to gamma distributed interarrival times with parameter $\nu = 1$. Typical realizations for the case $\nu = \infty$, corresponding to periodic partitions, are illustrated in Fig. 7. This class of random fields results in a distinctive herringbone or tweed mosaic.

Polygonal tessellations. Consider the tessellation of the plane R^2 by a field of random sensed lines. More specifically, an arbitrary sensed line can be described in terms of the 3-tuple (r, θ, ζ). Here r represents the perpendicular or radial distance to the line in question, $\theta \in [-\pi, \pi]$ represents the orientation of this radial vector, and finally ζ is a binary random variable assuming values ± 1 which specifies the sense or direction imparted to this line segment. The pertinent geometry is illustrated in Fig. 8 for the case $\zeta = 1$. By virtue of the direction imposed on this line segment the plane is partitioned into two disjoint regions, R (right of line) and L (left of line) such that $R \cup L = R^2$.

Now consider the field of lines generated by the sequence $\{r_i, \theta_i, \zeta_i\}$. Here the sequence $\{r_i\}$ represents the "event times" associated with a homogeneous Poisson process $\{N(r), r \geq 0\}$ with intensity λ events/unit distance evolving according to the radial parameter r. The sequence $\{\theta_i\}$ is i.i.d. and uniform on $[-\pi, \pi]$ while $\{\zeta_i\}$ is also i.i.d. assuming the values ± 1 with equal probability.

The field of random lines so generated results in a partition of the plane into disjoint polygonal regions. Gray levels are assigned as described in [16] to result in correlation coefficient ρ with gray levels in contiguous regions. Typical realizations of the resulting random field are illustrated in Fig. 9 for selected values of

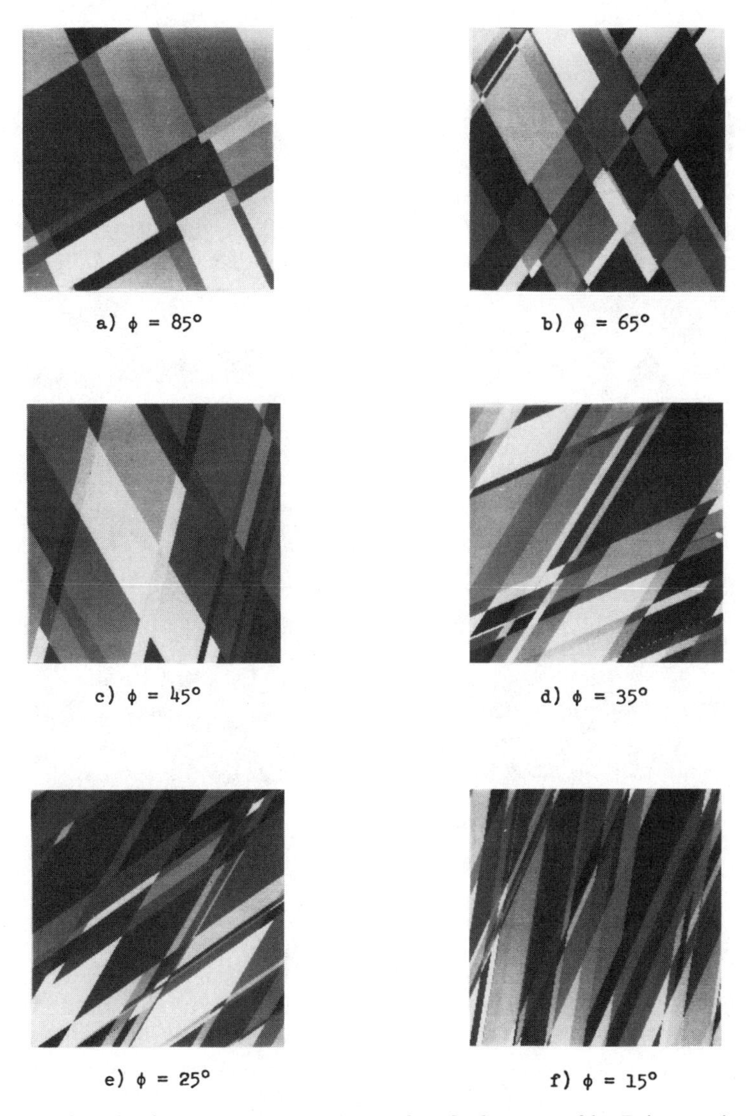

FIG. 6. Selected realizations of nonrectangular random field generated by Poisson point processes with $\lambda = 0.025$ and $\rho = 0.0$.

FIG. 7. Selected realizations of nonrectangular random field generated by periodic point processes with $\lambda = 0.025$ and $\rho = 0.0$.

$\lambda_e = \lambda/\pi$ and ρ. The quantity λ_e represents the average edge density along any randomly chosen line segment.[4] This random field is again described in terms of the two parameters λ_e, or equivalently λ, and ρ. This class of 2-D random fields can be extended to include more general point processes $\{N(r),\ r \geq 0\}$ controlling the radial evolution; for example, stationary renewal processes with gamma

[4] Similarly, in the case of rectangular tessellations it is easily shown that the average edge density along any randomly chosen line segment is $\lambda_e = 4\lambda/\pi$.

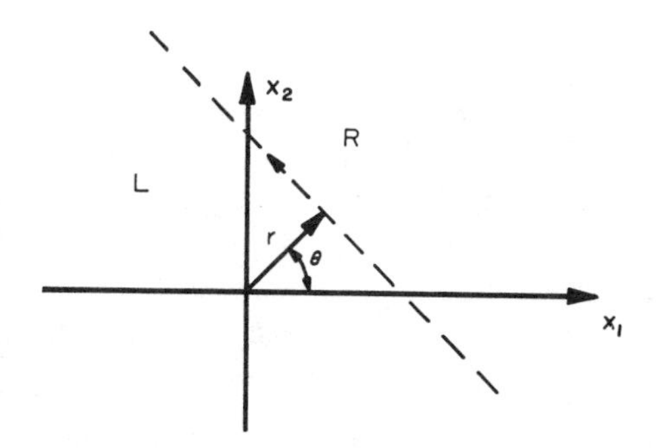

FIG. 8. Parameterization of directed line segment.

distributed interarrival times. Unfortunately, the analysis of the resulting processes becomes quite complicated and as a result we will not pursue this generalization here.

FIG. 9. Selected realizations of random field generated by polygonal partitions.

4. SECOND-ORDER PROPERTIES

We turn now to the second-order properties of the class of 2-D random fields described in the preceding section. In the interests of brevity the treatment will be condensed and will make extensive use of results reported elsewhere.

Rectangular tessellations. As a first step in the development of the covariance function, assume that the random orientation $\theta \in [-\pi, \pi]$ has been chosen and that k transitions have occurred[5] between the two points $\mathbf{x}$ and $\mathbf{x} + \mathbf{u}$ where we assume for the moment $\mathbf{u} \geq 0$. It follows from (10) that

$$E\{f(\mathbf{x} + \mathbf{u})f(\mathbf{x}) \,|\, \theta, k\} = \sigma^2 \rho^k; \qquad k = 0, 1, 2, \ldots. \tag{14}$$

The conditioning upon k is easily removed according to

$$E\{f(\mathbf{x} + \mathbf{u})f(\mathbf{x}) \,|\, \theta\} = \sum_{k=0}^{\infty} E\{f(\mathbf{x} + \mathbf{u})f(\mathbf{x}) \,|\, \theta, k\} p_{k|\theta}(\mathbf{u}), \tag{15}$$

where $p_{k|\theta}(\mathbf{u})$ is the probability of k transitions between $\mathbf{x}$ and $\mathbf{x} + \mathbf{u}$ given that θ is acting. We exploit the stationary renewal properties of the point processes $\{N_i(l), l \geq 0\}$, $i = 1, 2$, in writing this probability as a function only of the displacement $\mathbf{u}$. In particular, $p_{k|\theta}(\mathbf{u})$ can be evaluated according to

$$p_{k|\theta}(\mathbf{u}) = \sum_{j=0}^{k} q_{k-j|\theta}^{(1)}(\tilde{u}_1) q_{j|\theta}^{(2)}(\tilde{u}_2); \qquad k = 0, 1, \ldots, \tag{16}$$

where $q_{j|\theta}^{(i)}(\tilde{u}_i)$ is the probability that $\{N_i(l), l \geq 0\}$ has undergone j transitions in the interval $\tilde{u}_i$, $i = 1, 2$, which depends upon $\mathbf{u}^T = (u_1, u_2)$ and θ according to

$$\tilde{u}_1 = u_1 \cos\theta + u_2 \sin\theta, \tag{17a}$$

and

$$\tilde{u}_2 = u_2 \cos\theta - u_1 \sin\theta. \tag{17b}$$

Substituting (14) and (16) into (15) we obtain

$$E\{f(\mathbf{x} + \mathbf{u})f(\mathbf{x}) \,|\, \theta\} = \sigma^2 \sum_{k=0}^{\infty} \rho^k \sum_{j=0}^{k} q_{k-j|\theta}^{(1)}(\tilde{u}_1) q_{j|\theta}^{(2)}(\tilde{u}_2), \tag{18}$$

and by simple rearrangement of the double summation in this last expression we find

$$\begin{aligned} E\{f(\mathbf{x} + \mathbf{u})f(\mathbf{x}) \,|\, \theta\} &= \sigma^2 \sum_{j=0}^{\infty} \sum_{k=j}^{\infty} \rho^{k-j} q_{k-j|\theta}^{(1)}(\tilde{u}_1) \rho^j q_{j|\theta}^{(2)}(\tilde{u}_2) \\ &= \sigma^2 \Big[\sum_{m=0}^{\infty} \rho^m q_{m|\theta}^{(1)}(\tilde{u}_1)\Big] \cdot \Big[\sum_{n=0}^{\infty} \rho^n q_{n|\theta}^{(2)}(\tilde{u}_2)\Big]. \end{aligned} \tag{19}$$

[5] By this we mean that $k = k_1 + k_2$, where k_i, $i = 1, 2$, represents the number of transitions along each of the orthogonal axes which have now been rotated by θ radians.

Assuming a uniform distribution for θ, it follows that the covariance function becomes

$$R_{ff}(\mathbf{x}+\mathbf{u},\mathbf{x}) = \frac{1}{2\pi}\int_{-\pi}^{\pi} E\{f(\mathbf{x}+\mathbf{u})f(\mathbf{x})\,|\,\theta\}\,d\theta, \tag{20}$$

with the integrand given by (19). While not immediately apparent, it is easily shown that this last expression depends only upon $\|\mathbf{u}\|$ so that the resulting random field is indeed homogeneous and isotropic.

While explicit evaluation of (20) is in general quite cumbersome, it can be evaluated in special cases. For example, in the Poisson case $\nu = 1$ it can be shown [18] that

$$R_{ff}(\|\mathbf{u}\|) = \frac{2\sigma^2}{\pi}\int_0^{\pi/2} \exp\{-2^{\frac{1}{2}}(1-\rho)\lambda\|\mathbf{u}\|\cos(\theta-\pi/4)\}\,d\theta, \tag{21}$$

while the corresponding power spectral density computed according to (6) becomes

$$S_{ff}(\Omega) = \frac{8(1-\rho)\lambda\sigma^2}{\Omega^2+2(1-\rho)^2\lambda^2}\left[\frac{1}{\Omega^2+(1-\rho)^2\lambda^2}\right]^{\frac{1}{2}}. \tag{22}$$

Typical covariance surfaces together with intensity plots of the corresponding power spectral density in the case of periodic partitions (i.e., $\nu = \infty$) are illustrated in Fig. 10. The autocorrelation functions are plotted as a function of the normalized spatial variable[6] $\|\mathbf{u}\|/l$ over the range $0 \le \|\mathbf{u}\|/l \le 3$, while the power spectral density is plotted as a function of the normalized spatial frequency variable $\Omega/2\pi\lambda$ over the range $0 \le \Omega/2\pi\lambda \le 5$. Additional details can be found in [18]. Explicit evaluation of these quantities for the general case of gamma distributed interarrival times is provided in [19].

Similarly, the conditional joint probability of $f_1 = f(\mathbf{x})$ and $f_2 = f(\mathbf{x}+\mathbf{u})$ given both the random angle θ and the number of transitions k between $\mathbf{x}$ and $\mathbf{x}+\mathbf{u}$ is easily shown to be given by[7]

$$\begin{aligned} p\{f_1, f_2; \mathbf{x}, \mathbf{x}+\mathbf{u}\,|\,\theta, k\} &= \frac{1}{2\pi\sigma^2(1-\rho^{2k})^{\frac{1}{2}}}\exp\left\{-\frac{f_1^2-2\rho^k f_1 f_2+f_2^2}{2\sigma^2(1-\rho^{2k})}\right\}; \qquad k>0 \\ &= \frac{1}{(2\pi)^{\frac{1}{2}}\sigma}\exp\left\{-\frac{f_1^2}{2\sigma^2}\right\}\delta(f_1-f_2); \qquad k=0. \end{aligned} \tag{23}$$

Note that this quantity is independent of $\mathbf{x}$, $\mathbf{x}+\mathbf{u}$, and θ; we will make use of this observation later.

[6] Here $l = 1/\lambda$ with λ the common rate parameter of the two mutually independent point processes which provide a rectangular partition of the plane.

[7] It is at this point that the Gaussian assumption is crucial.

a) Autocorrelation Function
ρ= - 0.9

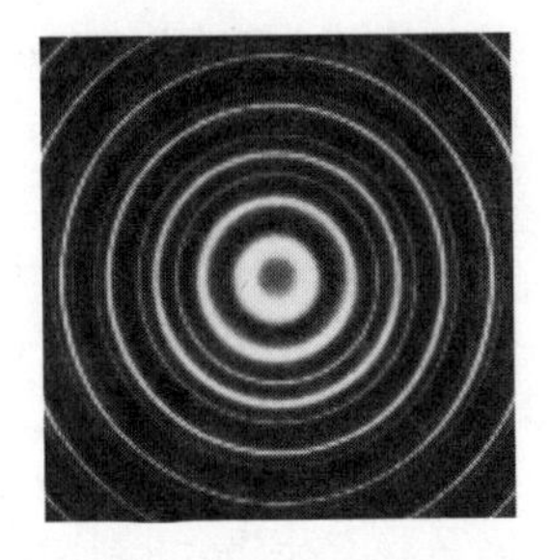

b) Power Spectral Density
ρ= - 0.9

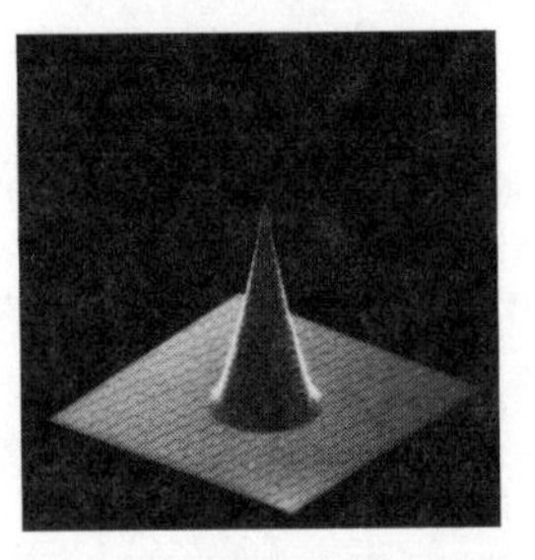

c) Autocorrelation Function
ρ= 0.0

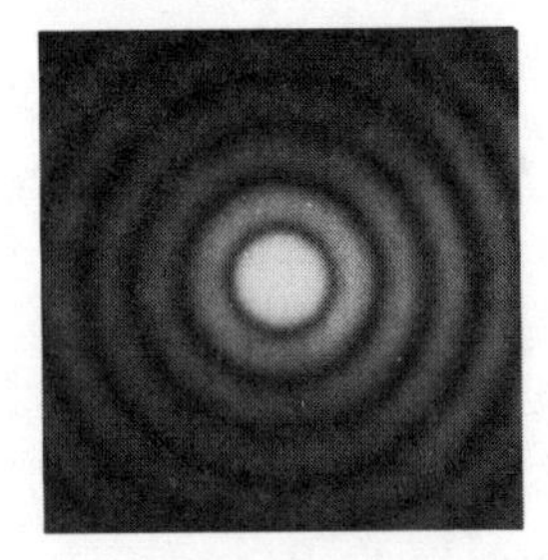

d) Power Spectral Density
ρ= 0.0

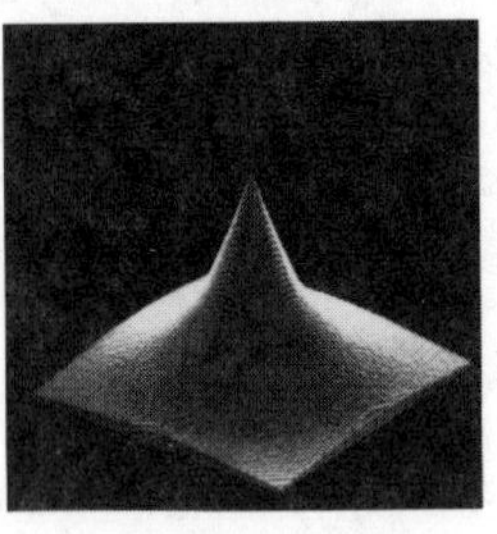

e) Autocorrelation Function
ρ= 0.5

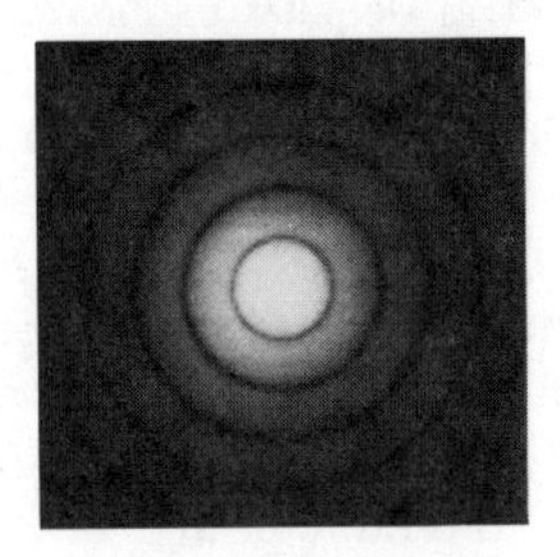

f) Power Spectral Density
ρ= 0.5

Fig. 10. Autocorrelation function and power spectral density of 2-D random checkerboard process generated by periodic partitions.

The conditioning upon k in this case is easily removed according to

$$p\{f_1, f_2; \mathbf{x}, \mathbf{x} + \mathbf{u} | \theta\} = \sum_{k=0}^{\infty} p\{f_1, f_2; \mathbf{x}, \mathbf{x} + \mathbf{u} | \theta, k\} p_{k|\theta}(\mathbf{u})$$

$$= \sum_{k=0}^{\infty} h_k(f_1, f_2) p_{k|\theta}(\mathbf{u}), \tag{24}$$

where $p_{k|\theta}(\mathbf{u})$ has been defined previously as the probability of k transitions between $\mathbf{x}$ and $\mathbf{x} + \mathbf{u}$ given that θ is acting. We have used $h_k(f_1, f_2)$ in the second expression of (24) in order to emphasize the functional independence of the spatial parameters $\mathbf{x}$ and $\mathbf{x} + \mathbf{u}$ and the rotation angle θ.

Again under the assumption of uniform distribution for θ, the joint p.d.f. can be evaluated as

$$p\{f_1, f_2; \mathbf{x}, \mathbf{x}+\mathbf{u}\} = \frac{1}{2\pi}\int_{-\pi}^{\pi} p\{f_1, f_2; \mathbf{x}, \mathbf{x}+\mathbf{u} \mid \theta\}d\theta$$

$$= \sum_{k=0}^{\infty} h_k(f_1, f_2)p_k(\|\mathbf{u}\|), \quad (25)$$

where

$$p_k(\|\mathbf{u}\|) \triangleq \frac{1}{2\pi}\int_{-\pi}^{\pi} p_{k|\theta}(\mathbf{u})d\theta, \quad (26)$$

and we have made explicit use of the fact that the integral on the right-hand side of this last expression depends only upon the Euclidean distance $\|\mathbf{u}\|$. It follows that (4) is indeed satisfied and hence the 2-D random field is homogeneous and isotropic through all second-order statistics.

To complete the evaluation of the joint p.d.f. $p\{f_1, f_2; \|\mathbf{u}\|\}$ it remains to provide explicit evaluation of $p_k(\|\mathbf{u}\|)$ in (26). This has proved rather cumbersome in general, although quite tractable in several important special cases. For example, again in the case $\nu = 1$ corresponding to Poisson partitions, it can be shown [19] that

$$p_k(\|\mathbf{u}\|) = \frac{4[2^{\frac{1}{2}}\lambda\|\mathbf{u}\|]^k}{\pi k!}\int_0^{\pi/4} \cos^k\theta \exp\{-2^{\frac{1}{2}}\lambda\|\mathbf{u}\|\cos\theta\}d\theta;$$

$$k = 0, 1, \ldots, \quad (27)$$

which does not seem capable of further simplification. At any rate, this expression is easily evaluated by numerical integration. Substitution into (25) then yields explicit evaluation of $p\{f_1, f_2; \|\mathbf{u}\|\}$. Actually, for evaluation and display purposes, it proves convenient to consider a normalized version of this joint p.d.f. defined according to[8]

$$p_0\{f_1, f_2; \|\mathbf{u}\|\} = \sigma^2 p\{\sigma f_1, \sigma f_2; \|\mathbf{u}\|\}, \quad (28)$$

which is plotted in Fig. 11 as a function of f_1, f_2 for selected values of ρ and the *normalized displacement* $d' \triangleq \lambda_e\|\mathbf{u}\|$. Here the point $f_1 = f_1 = 0$ appears in the center and the plots cover the range $-3 \leq f_i \leq 3$, $i = 1, 2$. Note the high concentration of discrete probability mass along the diagonal $f_1 = f_2$ for small values of d'. This is a direct result of the high probability of $\mathbf{x}$ and $\mathbf{x} + \mathbf{u}$ falling in the same rectangular regions and thus resulting in identical values for $f_1 = f(\mathbf{x})$ and $f_2 = f(\mathbf{x} + \mathbf{u})$. This probability diminishes for increasing d'. Indeed, as indicated in Fig. 11, this "ridge line" along the diagonal has virtually disappeared for $d' = 8$. The off-diagonal probability mass visible for $\rho = -0.9$ is a direct

[8] The net effect of this normalization is that the f_1, f_2 axes can be considered normalized to the standard deviation σ.

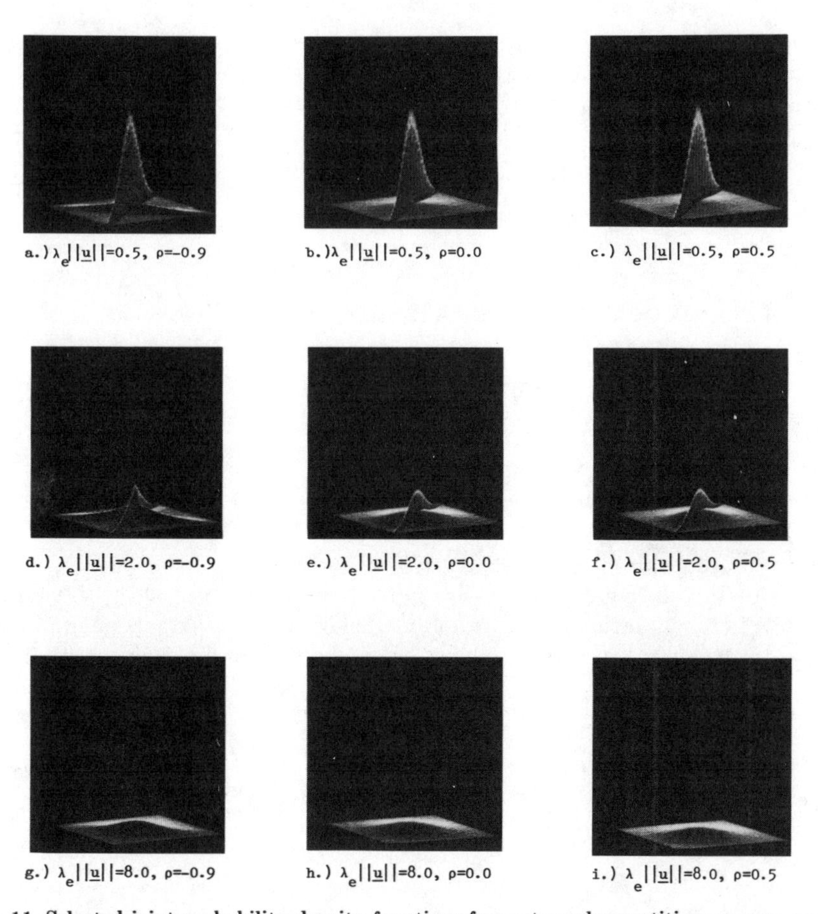

FIG. 11. Selected joint probability density functions for rectangular partition process, $\nu = 1$.

result of the negative correlation while for $\rho = 0.5$, as expected, there is visible probability mass distributed along the main diagonal. For $\rho = 0$, of course, this distribution is circularly symmetric about the origin. These observations are more apparent in Fig. 12 which illustrates intensity plots of the logarithms of the corresponding p.d.f.'s in Fig. 11. Note, in particular, the almost identical circularly symmetric distributions which result for large d' independent of the value of ρ.

Nonrectangular tessellations. Corresponding second-order properties of 2-D random fields generated by nonrectangular tessellations, although somewhat more complicated than in the rectangular case, have been determined as reported in [16, 19], to which the reader is referred to for details. For example, in the case of polygonal tessellations it is shown in [16] that, under the assumption of a Poisson line process generating the partitions, the autocorrelation function is given by

$$R_{ff}(\|\mathbf{u}\|) = \sigma^2 e^{-\lambda_e \|\mathbf{u}\|}\{I_0(\lambda_e\|\mathbf{u}\|) + 2 \sum_{k=1}^{\infty} \rho^k I_k(\lambda_e\|\mathbf{u}\|)\}, \tag{29}$$

where $I_k(\cdot)$ is the modified Bessel function of the first kind of order k. Similarly, the corresponding power spectral density is evaluated according to

$$S_{ff}(\Omega) = \frac{2\sigma^2(1-\rho^2)}{\lambda_e^2} \int_0^{\pi} \left[\frac{1-\cos\phi}{1-2\rho\cos\phi+\rho^2}\right] \frac{d\phi}{[(\Omega/\lambda_e)^2+(1-\cos\phi)^2]^{3/2}}. \qquad (30)$$

These quantities are illustrated in Fig. 13 for various values of ρ. One notable characteristic of this random field is that the power spectral density behaves as $(\Omega/\lambda_e)^{-3/2}$ for small values of (Ω/λ_e), i.e., $S_{ff}(\Omega)$ has a singularity at the origin except for $\rho = -1$. This high concentration of energy at low spatial frequencies is a direct result of the construction procedure which allows relatively large correlations between gray levels in regions relatively far apart. We feel that this characteristic is typical of selected image processing applications and as a result it was purposely built into the construction procedure.

Similarly, for the case of polygonal tessellations generated by Poisson point processes, the joint p.d.f. is easily seen to be given by (25) with the sum extended

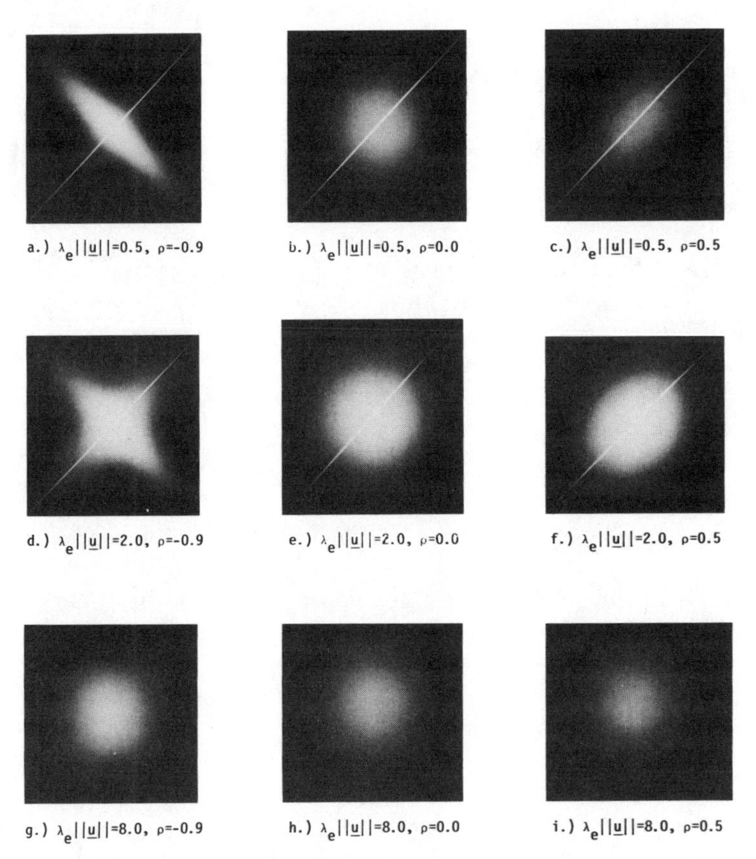

FIG. 12. Intensity plots of logarithm of selected joint probability density functions for rectangular partition process, $\nu = 1$.

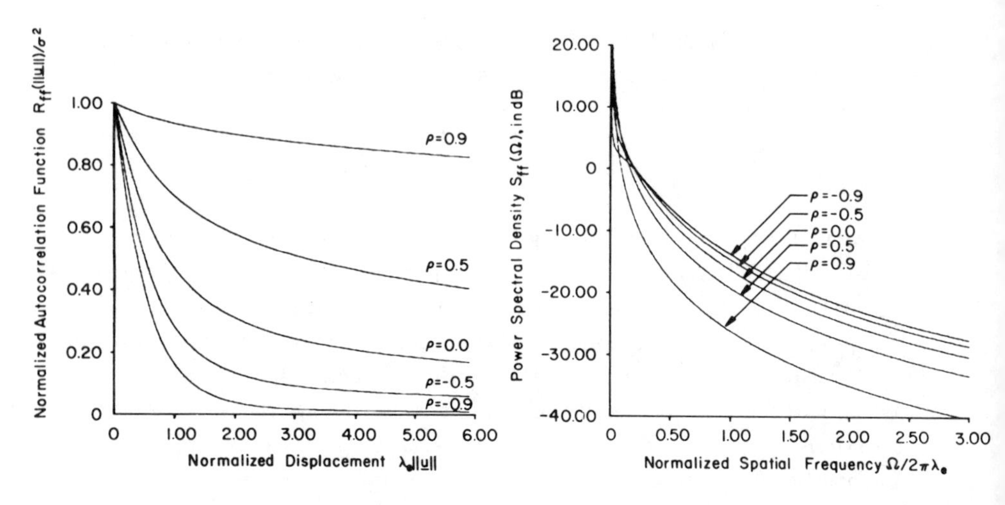

FIG. 13. Autocorrelation function and power spectral density of 2-D random field generated by polygonal partitions.

over both positive and negative values of k and now

$$p_k(\|\mathbf{u}\|) = [\lambda_e\|\mathbf{u}\|/2]^{|k|} \exp\{-\lambda_e\|\mathbf{u}\|\} \sum_{l=0}^{\infty} \frac{[\lambda_e\|\mathbf{u}\|/2]^{2l}}{(l + |k|)!l!};$$

$$k = 0, \pm 1, \pm 2, \ldots. \quad (31)$$

In Fig. 14 we provide intensity plots of the logarithm of $p_0\{f_1, f_2; \|\mathbf{u}\|\}$ as a function of f_1 and f_2 for selected values of ρ and $d' = \lambda_e\|\mathbf{u}\|$. An interesting observation to be drawn here is the persistence of the diagonal "ridge line" with increasing values of d'. This is, of course, a direct result of the construction procedure which allows return to the same gray level at distant spatial locations with relatively high probability.

5. APPLICATIONS

We consider now some selected applications of the 2-D random fields described in the preceding.

Edge detection. This problem is treated in some detail in [17]. We assume that the true edge structure in an image is described by the random field $f(\mathbf{x})$ modeled as one of the previously developed 2-D random fields. In many applications, the observed image is a noise-corrupted version of $f(\mathbf{x})$ described by

$$g(\mathbf{x}) = f(\mathbf{x}) + n(\mathbf{x}), \quad (32)$$

where $n(\mathbf{x})$ is a zero-mean homogeneous and isotropic noise field possessing noise spectral density $S_{nn}(r) = \sigma_n^2$, i.e., a white noise field. Here one assumes that the noise field $n(\mathbf{x})$ represents any additive noise or spurious detail not considered part of the essential contours or edges represented by the random field $f(\mathbf{x})$.

In [17] the problem of edge detection was posed as a 2-D Wiener filtering problem. More specifically, if $l(\mathbf{x})$ represents the output of some desired operation on $f(\mathbf{x})$ then design the imaging system with optical transfer function (OTF) $H_0(\boldsymbol{\omega})$ whose output $\hat{l}(\mathbf{x})$ in response to $g(\mathbf{x})$ at its input minimizes the mean-square error

$$I_e = E\{[l(\mathbf{x}) - \hat{l}(\mathbf{x})]^2\}. \tag{33}$$

Assuming the desired operation possesses OTF $H_d(\boldsymbol{\omega}) = \|\boldsymbol{\omega}\|^2 \exp\{-\frac{1}{2}\|\boldsymbol{\omega}\|^2\}$ (cf. [17] for justification) the optimum Wiener filter is isotropic with OTF

$$H_0(\Omega) = \frac{\Omega^2 e^{-\Omega^2/2} S_{ff}(\Omega)}{S_{ff}(\Omega) + \sigma_n^2}; \qquad \Omega \geq 0. \tag{34}$$

For example, if $\{f(\mathbf{x}), \mathbf{x} \in R^2\}$ is the rectangular 2-D random field with Poisson partitions then the power spectral density $S_{ff}(\Omega)$ is given by (22). The resulting Wiener filter is then completely defined in terms of the three parameters λ, ρ, and $\zeta \triangleq \sigma^2/\sigma_n^2$ which represents the signal-to-noise ratio (SNR). Typical results

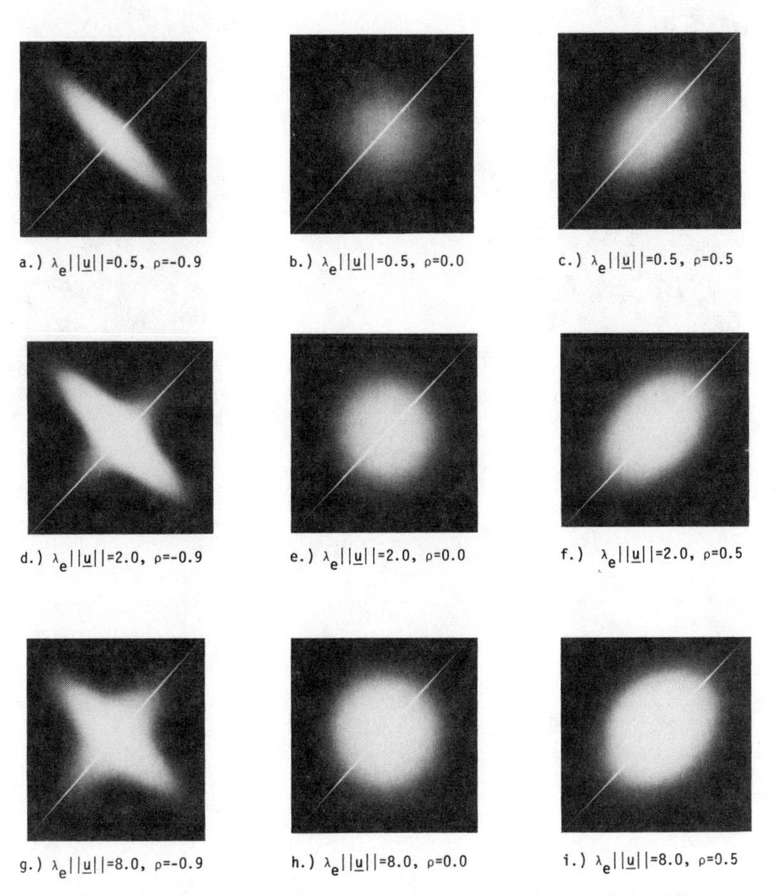

FIG. 14. Selected joint probability density functions for polygonal partition process.

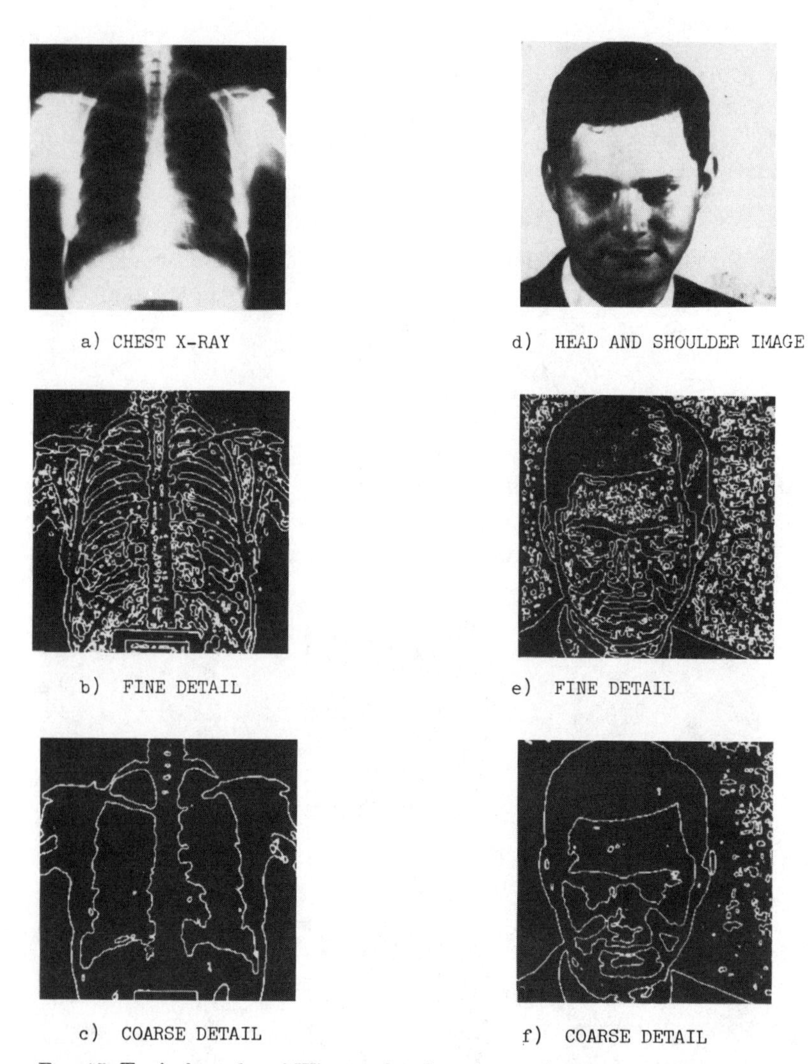

FIG. 15. Typical results of Wiener edge detector applied to real-world images.

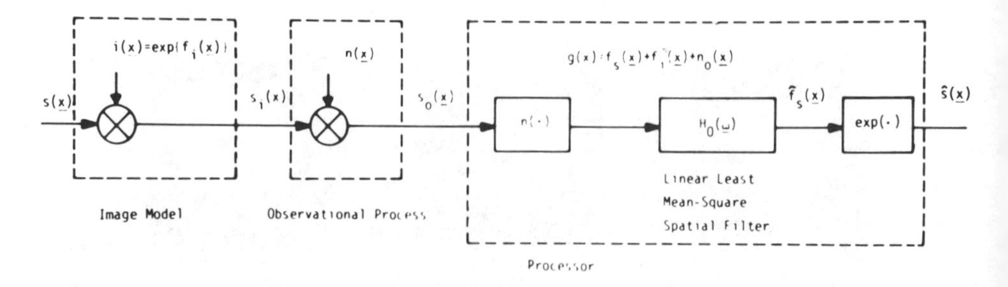

FIG. 16. Homomorphic filtering of degraded images.

are illustrated in Fig. 15 employing a digital implementation of the optimum Weiner filter. Additional results appear in [20].

Image enhancement. This class of 2-D random fields has found application as a stochastic model for spatially varying illumination in homomorphic filtering of

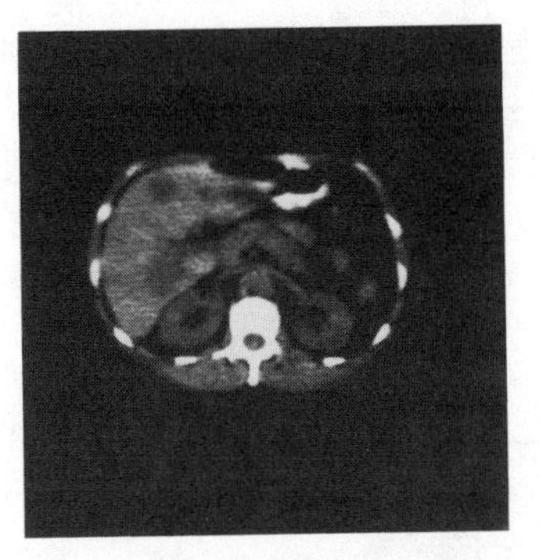

a) ORIGINAL

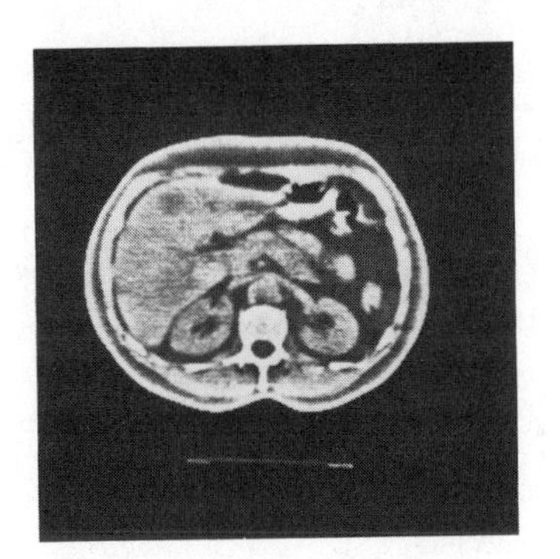

b) PROCESSED VERSION 1

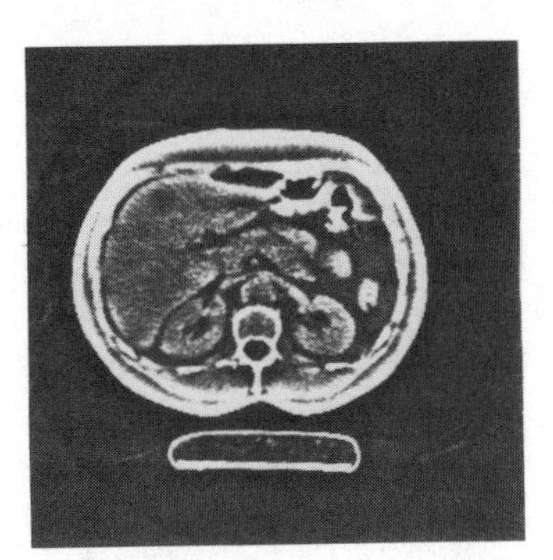

c) PROCESSED VERSION 2

FIG. 17. Typical results of homomorphic filtering of transaxial tomography image.

a.) Original; NW, $\rho=0.0$, $\lambda=0.16$; NE, $\rho=0.5$, $\lambda=0.32$; S, $\rho=0.0$, $\lambda=0.32$

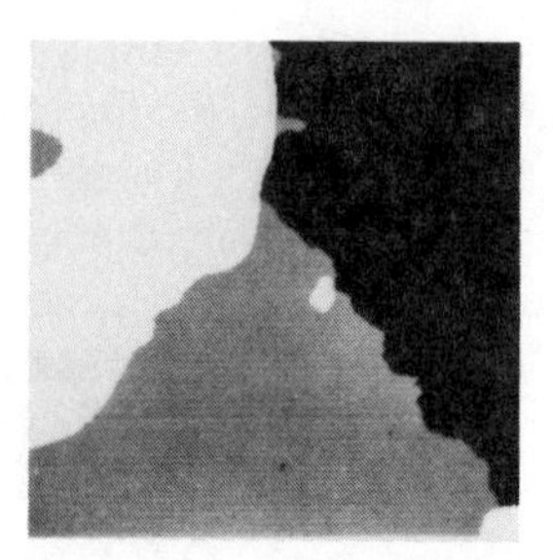

b.) Log-Likelihood Discriminator

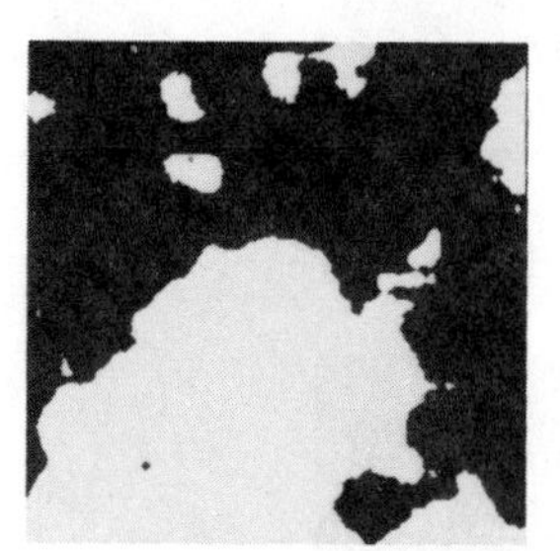

c.) Correlation Discriminator

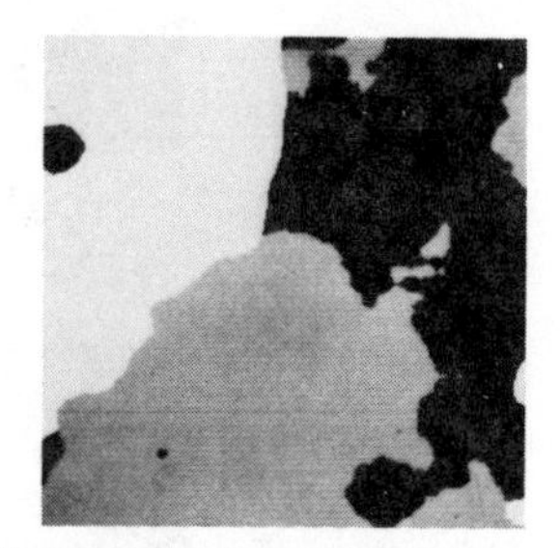

d.) Correlation / Edge Discriminator

FIG. 18. Illustration of texture discrimination results.

images. More specifically, consider the formation of an image as indicated in Fig. 16. Here the observed image $s_i(\mathbf{x})$ is the product of the true image $s(\mathbf{x})$ and the illumination function $i(\mathbf{x}) = \exp\{f_i(\mathbf{x})\}$, where $f_i(\mathbf{x})$ is a 2-D random field with polygonal partitions as described in the preceding section. The true image $s(\mathbf{x})$ will be similarly modeled although the edge density will be assumed much higher than that of the illumination process. Finally, we assume that the observational or recording process introduces the multiplicative white noise field $n(\mathbf{x})$. The image available for processing is then

$$s_0(\mathbf{x}) = s(\mathbf{x}) \cdot i(\mathbf{x}) \cdot n(\mathbf{x}). \tag{34}$$

As indicated in Fig. 16, by a simple application of homomorphic filtering concepts [21, 22] it is possible to design a linear least mean-square spatial filter to provide an estimate $\hat{s}(\mathbf{x})$ of the true image field. This approach attempts to minimize the effects of both the nonconstant illumination and the multiplicative noise. Typical results are indicated in Fig. 17 for a transaxial tomography image. Additional results can be found in [23].

Texture discrimination. This class of 2-D random fields has also been used as a stochastic texture model leading to a class of texture discrimination algorithms which approximate the statistically optimum maximum likelihood classifier. The details are provided in [24]. Typical performance of this texture discrimination scheme is illustrated in Fig. 18. Here Fig. 18a illustrates a source image which consists of realizations of three distinct rectangular Poisson tessellation processes for various parameter choices. The NW and NE corners have $\lambda = 0.16$, $\rho = 0.0$, and $\lambda = 0.32$, $\rho = 0.5$, respectively. These values were chosen to result in identical second-moment properties. As a result these two fields cannot be discriminated on the basis of autocorrelation functions and/or power spectral densities alone. The field in the S side has $\lambda = 0.32$ while $\rho = 0.0$. Since it possesses the same edge density as the field in the NE corner, these two textures cannot be discriminated on the basis of edge density alone.

As indicated in Fig. 18b, the log-likelihood discriminator does an excellent job of discriminating the three texture regions except in the vicinity of either texture or image boundaries. Included in Fig. 18 for comparison purposes is the performance of alternative more conventional texture discrimination schemes. In Fig. 18c we demonstrate the performance of a conventional correlation discriminator. This algorithm implements a threshold test on a least-squares estimate of the correlation of pixels separated by distance d which has been optimized in this case. The optimum threshold has been chosen empirically on the basis of histogram techniques. While this approach is useful in discriminating the texture in the S side from that in either the NW or NE corner, it cannot discriminate between the NW and NE regions due to the fact that they possess identical second-moment properties. As a partial remedy to this situation we have devised a discriminant that employs both correlation and edge density information. Using this correlation/edge density discriminator some degree of success has been

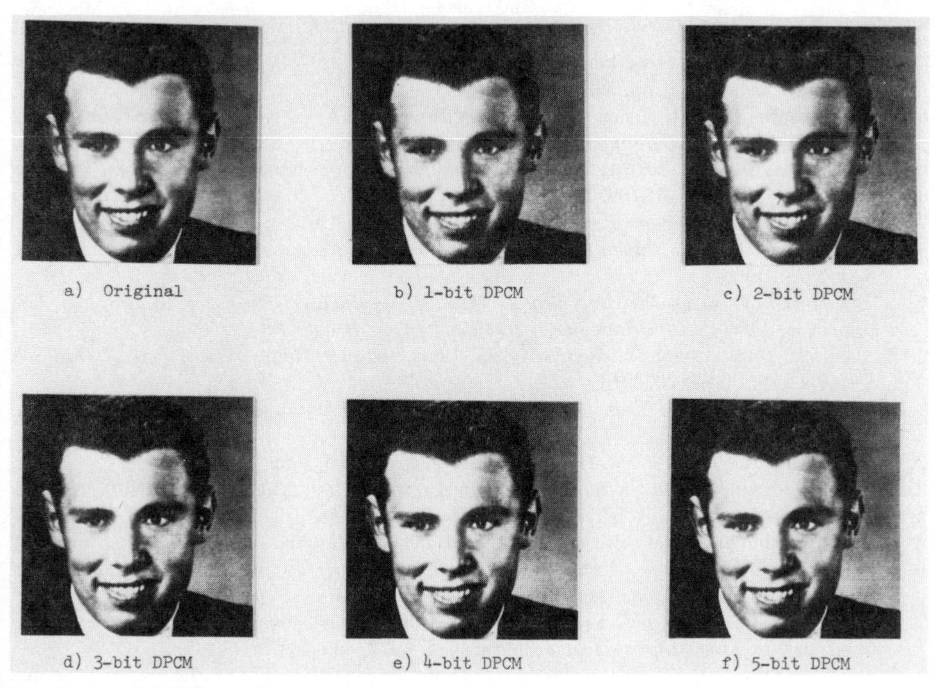

FIG. 19. 2-D DPCM encoding of face image using recursive predictor for $\lambda = 0.05$ and $\rho = -0.9$.

achieved in discriminating between the NW and NE regions as illustrated by the results in Fig. 18d. The results are, however, generally inferior to the performance of the log-likelihood discriminator.

Image coding. A 2-D differential pulse code modulation (DPCM) encoder for images has been developed using the rectangular Poisson tessellation process as an image model as an alternative to the conventional autoregressive (AR) modeling assumptions. This encoder performs better at 1 bit/pixel quantization than previous designs [25] based upon the AR model. The details are provided in [19]. In Fig. 19 we illustrate typical performance of this encoder on a head-and-shoulders image. Even at 1 bit/pixel it has achieved accurate reproduction of edges without excessive granular noise in regions of constant intensity.

6. SUMMARY AND CONCLUSIONS

A class of 2-D homogeneous and isotropic random fields has been described which we feel provides a useful model for real-world imagery possessing pronounced edge structure. This model is conveniently parameterized by several physically meaningful quantities. Several of the more important properties of this class of 2-D random fields have been discussed and some applications described.

REFERENCES

1. E. Wong, Homogeneous Gauss–Markov random fields, *Ann. Math. Statist.* **40**, 1969, 1625–1634.
2. E. Wong, Two-dimensional random fields and the representation of images, *SIAM J. Appl. Math.* **16**, 1968, 756–770.
3. E. Wong, *Stochastic Processes in Information and Dynamical Systems*, Chap. 7, McGraw–Hill, New York, 1971.
4. A. M. Yaglom, Second-order homogeneous random fields, In *Proc. 4th Berkeley Symp. Math. Stat. and Prob.*, Vol. 2, pp. 593–620, 1961.
5. D. J. Sakrison and V. R. Algazi, Comparison of line-by-line and two-dimensional encoding of random images, *IEEE Trans. Inform. Theory* **IT-17**, July 1971, 386–398.
6. J. B. O'Neal, Jr., and T. Raj Natarajan, Coding isotropic images, *IEEE Trans. Inform. Theory* **IT-23**, Nov. 1977, 697–707.
7. D. L. Synder, *Random Point Processes*, Wiley, New York, 1975.
8. A. Papoulis, *Probability, Random Variables and Stochastic Processes*, Chap. 11, McGraw–Hill, New York, 1965.
9. S. Bochner, *Lectures on Fourier Integrals*, Annals, of Math. Studies, No. 42, pp. 235–238, Princeton Univ. Press, Princeton, N.J., 1959.
10. A. Papoulis, Optical systems, singularity functions, complex Hankel transforms, *J. Opt. Soc. Soc. Amer.* **57**, 1967, 207–213.
11. A. Papoulis, *Systems and Transforms with Applications in Optics*, McGraw–Hill, New York, 1968.
12. E. Parzen, *Stochastic Processes*, Holden–Day, San Francisco, 1962.
13. E. Cinlar, *Introduction to Stochastic Processes*, Prentice–Hall, Englewood Cliffs, N.J., 1975.
14. W. Feller, *An Introduction to Probability and Its Applications*, Vol. 2, Wiley, New York, 1971.
15. J. W. Modestino and R. W. Fries, A generalization of the random telegraph wave, submitted.
16. J. W. Modestino and R. W. Fries, Construction and properties of a useful two-dimensional random field, *IEEE Trans. Inform. Theory*, in press.
17. J. W. Modestino and R. W. Fries, Edge detection in noisy images using recursive digital filtering, *Computer Graphics Image Processing* **6**, 1977, 409–433.
18. J. W. Modestino, R. W. Fries, and D. G. Daut, A generalization of the two-dimensional random checkerboard process, *J. Opt. Soc. Amer.* **69**, 1979, 897–906.

19. R. W. Fries, *Theory and Applications of a Class of Two-Dimensional Random Fields*, Ph.D. thesis, Electrical and Systems Engineering Dept., RPI, Troy, N.Y., in preparation.
20. R. W. Fries and J. W. Modestino, An empirical study of selected approaches to the detection of edges in noisy digitized images, submitted.
21. A. V. Oppenheim, R. W. Schafer, and T. G. Stockham, Jr., Nonlinear filtering of multiplied and convolved signals, *Proc. IEEE* **56**, Aug. 1968, 1264–1291.
22. A. V. Oppenheim and R. W. Schafer, *Digital Signal Processing*, Chap. 10, Prentice–Hall Englewood Cliffs, N.J., 1975.
23. R. W. Fries and J. W. Modestino, Image enhancement by stochastic homomorphic filtering, *IEEE Trans. Acoust., Speech, Signal Processing*, in press.
24. J. W. Modestino, R. W. Fries, and A. L. Vickers, Texture discrimination based upon an assumed stochastic texture model, submitted.
25. J. W. Modestino and D. G. Daut, Combined source-channel coding of images, *IEEE Trans. Commun.* **COM–27, 1979,** 1644–1659.

Long Crested Wave Models

BRUCE SCHACHTER*

General Electric Co., P.O. Box 2500, Daytona Beach, Florida 32015

This paper examines two long crested wave models. First, a traditional model is reviewed. It is based upon sums of a large number of sinusoids. It has found applications in oceanography, geology, and to some extent also in image analysis. Then, a new model is presented. It uses sums of three or fewer long crested narrow-band noise waveforms. It is designed specifically for the purpose of image analysis and synthesis. Results of experiments in the computer generation of textures are presented. This is the first texture model to be implemented in hardware in a real-time image generation system.

1. INTRODUCTION

Random phenomena in the plane may be studied under the framework of the theory of random functions, with a two-dimensional parameter space. We will investigate the spatial variation within scalar fields by building mathematical models of the variation, with enough details to make simulations possible. Realizations of such models may be called random fields. The scalar parameter of a field may represent gray level, color intensity, elevation, etc. We will usually regard these fields as gray level images, but sometimes will refer to them as surfaces when it helps our intuition. Under our primary interpretation, realizations of our models will be called textures.

2. THE STANDARD MODEL

A texture t may be formed from a background gray level μ, modulated by a zero mean function g. An image formed from a single long crested sinusoid is defined by

$$t(x, y) = \mu + g(x, y) = \mu + B \cos (ux + vy + \phi), \tag{1}$$

where B is the amplitude of the modulation function, u and v are the x and y components of frequency; the phase shift ϕ is uniformly distributed in the interval $[0, 2\pi)$ (Fig. 1). The frequency of the wave in a direction perpendicular to the crest front is given by $\omega = (u^2 + v^2)^{\frac{1}{2}}$.

Consider a pair of long crested sinusoids of equal amplitude (Fig. 2). The surface formed by their sum is given by

$$\begin{aligned} g(x, y) &= B \cos (u_1 x + v_1 y) + B \cos (u_2 x + v_2 y), \\ &= 2B \cos (u_a x + v_a y) \cos (u_b x + v_b y) \end{aligned}$$

* Present address: Westinghouse Defense and Electronics Systems Center, Box 746, Mail Stop 451, Baltimore, Maryland 21203.

ISBN 0-12-597320-9

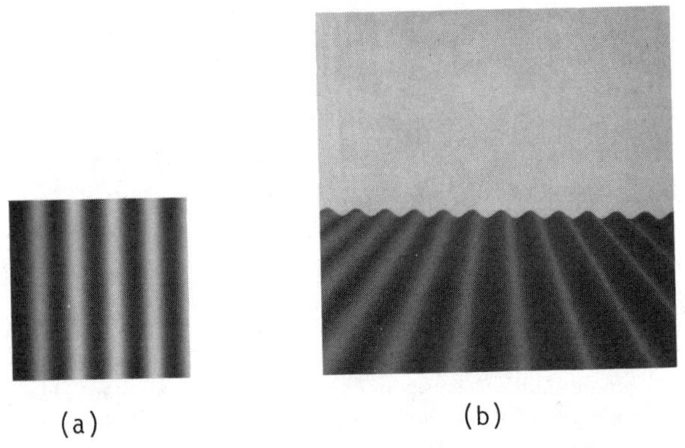

FIG. 1. Long crested sinusoid: (a) texture, (b) height field.

where

$$u_a = \tfrac{1}{2}(u_1 - u_2), \qquad v_a = \tfrac{1}{2}(v_1 - v_2), \tag{2}$$
$$u_b = \tfrac{1}{2}(u_1 + u_2), \qquad v_b = \tfrac{1}{2}(v_1 + v_2).$$

The pattern of maxima and minima resulting from two orthogonally intersecting waveforms is shown in Fig. 4. Notice that the contours of zero modulation intersect the crests and troughs of the individual waveforms at an angle of $\frac{1}{4}\pi$.

Now consider the image formed by three intersecting long crested sinusoids of equal amplitude, $\pi/3$ apart (Fig. 3):

$$g(x, y) = B[\cos(u_1x + v_1y + \phi_1) + \cos(u_2x + v_2y + \phi_2) + \cos(u_3x + v_3y + \phi_3)]. \tag{3}$$

In this case, the resulting pattern of maxima and minima depends upon the relative phases of the waveforms (Fig. 6). The pattern shown in Fig. 6a has a hexagonal structure, while the pattern shown in Fig. 6b has a triangular structure.

This representation may be extended to sums of a very large number of waves

$$g(x, y) = \sum_i B_i \cos(u_ix + v_iy + \phi_i). \tag{4}$$

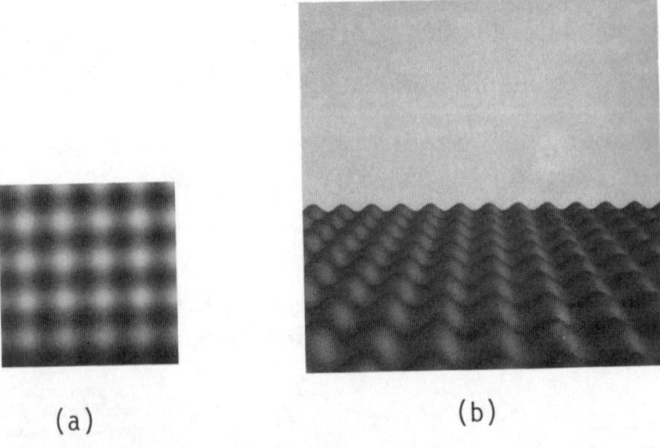

FIG. 2. Two intersecting long crested sinusoids, 90° apart: (a) texture, (b) height field.

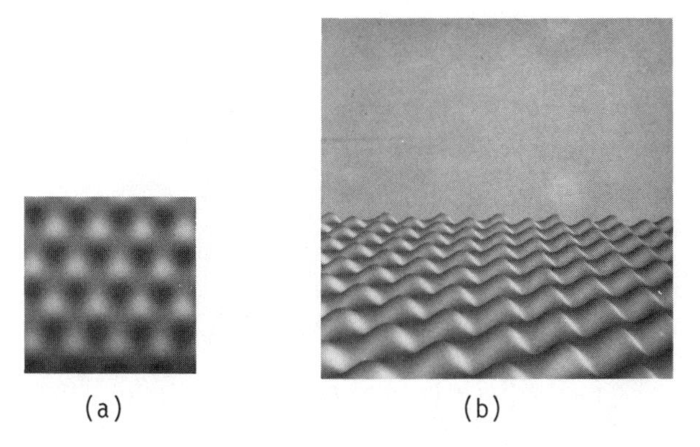

FIG. 3. Three intersecting long crested sinusoids, 60° apart: (a) texture, (b) height field.

The usual assumptions for this model are that the energy spectrum S is dense in frequency space and the amplitudes B_i are random variables such that for any finite surface element $dudv$

$$\tfrac{1}{2}B_i^2 = S(u, v)dudv. \tag{5}$$

We will briefly note this model's more basic attributes as derived by Longuet-Higgins [4–6, 14] in a famous series of papers on ocean wave modeling.

The (p, q) moment of the spectral density of $g(x, y)$ is given by

$$m_{pq} = \iint_{-\infty}^{\infty} S(u, v)u^p v^q dudv. \tag{6}$$

The mean square magnitude of $g(x, y)$ per unit surface area is

$$m_{00} = E(g^2(x, y)) = \sum_{i=1}^{\infty} \tfrac{1}{2}B_i^2 = \iint_{-\infty}^{\infty} S(u, v)dudv = \sigma_g^2. \tag{7}$$

The probability distribution of g is Gaussian (i.e., $N(0, \sigma_g^2)$) since g is the sum of a large number of random variables, each with an expected value of zero.

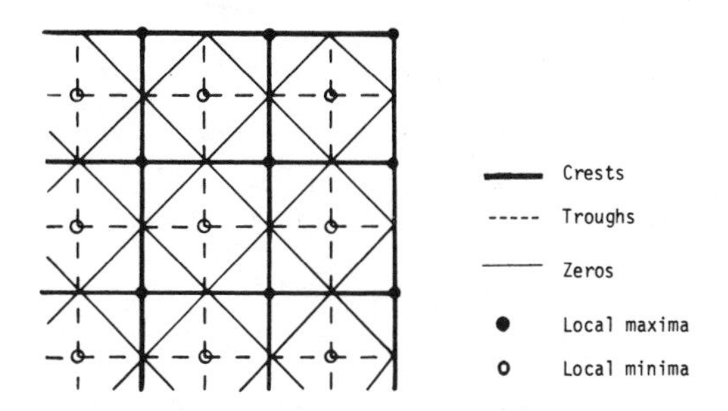

FIG. 4. Pattern resulting from two intersecting long crested sinusoids, 90° apart

FIG. 5. A texture produced by three intersecting long crested sinusoids, shown in perspective.

The correlation function of g is given by

$$\rho(x, y) = \frac{1}{\sigma_g^2} \{E(g(x_1, y_1)g(x_2, y_2))\} = \frac{1}{\sigma_g^2} \iint_{-\infty}^{\infty} S(u, v) \cos(ux + vy) du dv. \quad (8)$$

When $(p + q)$ is even,

$$\frac{m_{pq}}{m_{00}} = (-1)^{\frac{1}{2}(p+q)} \frac{\partial^{(p+q)}}{\partial x^p \partial y^q} \rho(0, 0). \quad (9)$$

Suppose that we drop a transect onto $g(x, y)$, at an angle ψ to the x-axis. The one-dimensional spectrum along this transect is denoted by

$$S_\psi(\hat{u}) = \int_{-\infty}^{\infty} S(u, v) d\hat{v}, \quad (10)$$

where $\hat{u} = u \cos \psi + v \sin \psi$; $\hat{v} = -u \sin \psi + v \cos \psi$. The moments along this

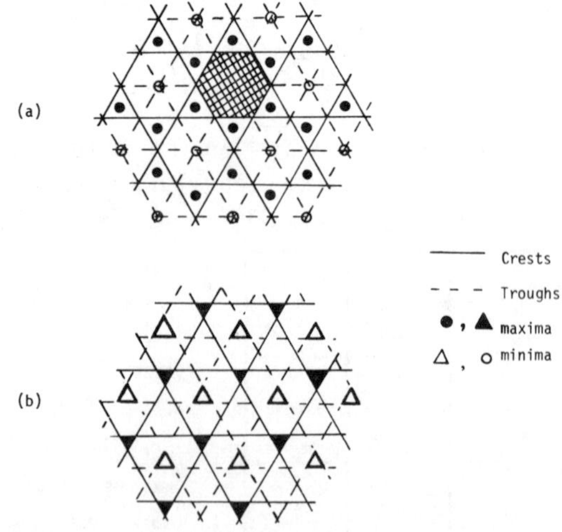

FIG. 6. Two basic patterns of maxima and minima formed by three intersecting long crested sinusoids. A third pattern is formed by interchanging the crests and troughs in (a).

line are given by

$$m_n(\psi) = \iint_{-\infty}^{\infty} S(u, v)(u \cos \psi + v \sin \psi)^n du dv = \int_{-\infty}^{\infty} S_\psi(\hat{u})\hat{u}^n d\hat{u},$$

$$m_0(\psi) = m_{00},$$

$$m_1(\psi) = m_{10} \cos \psi + m_{01} \sin \psi,$$

$$m_2(\psi) = m_{20} \cos^2 \psi + 2m_{11} \cos \psi \sin \psi + m_{02} \sin^2 \psi. \tag{11}$$

For the special case of isotropy, we may write instead

$$M_n = \iint_{-\infty}^{\infty} S(u, v)\omega^n du dv = \int_0^{\infty} \int_0^{2\pi} S(\omega)\omega^n d\omega d\psi$$

$$= 2\pi \int_0^{\infty} S(\omega)\omega^{n+1} d\omega. \tag{12}$$

For an isotropic spectrum, the odd moments vanish and $M_0 = m_0(\psi)$, $M_2 = 2m_2(\psi)$, $M_4 = (8/3)m_4(\psi)$.

For an anisotropic surface, the degree of anisotropy may be quantized in terms of the surface's moments. There are two special values of ψ for which $m_2(\psi)$ takes on the maximum and minimum values

$$m_{2,\max};\, m_{2,\min} = \tfrac{1}{2}\{(m_{20} + m_{02}) \pm ((m_{20} - m_{02})^2 + 4m_{11}^2)^{\frac{1}{2}}\}. \tag{13}$$

These maximum and minimum values always occur at right angles to each other. Let $\psi_{\max}$ denote the angle of the maximum

$$\tan (2\psi_{\max}) = \frac{2m_{11}}{m_{20} - m_{02}}. \tag{14}$$

The direction corresponding to this angle is called the principal direction of the wave field. The r.m.s. frequency $\bar{\omega}$, in the principal direction, is given by

$$\bar{\omega} = \left(\frac{m_{2,\max}}{m_{00}}\right)^{\frac{1}{2}}. \tag{15}$$

For an isotropic field, this equation reduces to

$$\bar{\omega} = |\rho''(0)|^{\frac{1}{2}}. \tag{16}$$

This is an extremely important result. It characterizes the "roughness" of a surface in terms of its correlation function.

The most important measure of an anisotropic surface is its "long-crestedness," denoted by $(1/\gamma)$:

$$(1/\gamma) = \left(\frac{m_{2,\max}}{m_{2,\min}}\right)^{\frac{1}{2}}. \tag{17}$$

For an isotropic field, clearly $\gamma = 1$.

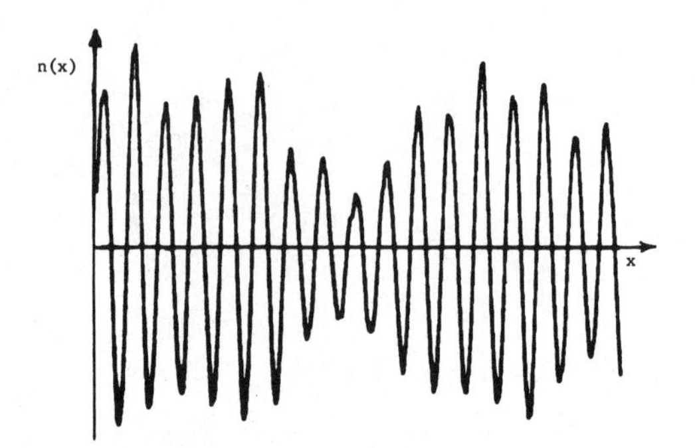

FIG. 7. Typical narrow-band noise waveform.

A spectrum of particular interest is the "ring" spectrum. All of the energy of a ring spectrum originates from waves of the same frequency, but possibly different directions. For this case, the following relation holds:

$$(m_{40} + 2m_{22} + m_{04})m_{00} - (m_{20} + m_{02})^2 = 0. \tag{18}$$

When the field is also isotropic, it is impossible for the shape of the spectrum in frequency space to be a true ring. However, it may be a very thin annulus having central frequency ω_c and thickness ∇. For this case

$$M_n \simeq \omega_c{}^n M_0, \tag{19}$$

$$\nabla \simeq \frac{(M_0 M_4 - M_2{}^2)^{\frac{1}{2}}}{M_2}. \tag{20}$$

Longuet-Higgins' model appears to offer a good description of many real world patterns whose statistics fit a single normal distribution (as opposed to a mixture of normals). However, there are a couple of factors which suggest that it is not appropriate as a generative pattern model.

(i) The assumption that a surface is formed by sums of long crested sinusoids has a strong theoretical basis for ocean wave modeling, but is not physically plausible for most other patterns.

(ii) The model is not practical for image generation, since it requires the sum of a large number of waveforms.

We will suggest a new model, designed to overcome these difficulties.

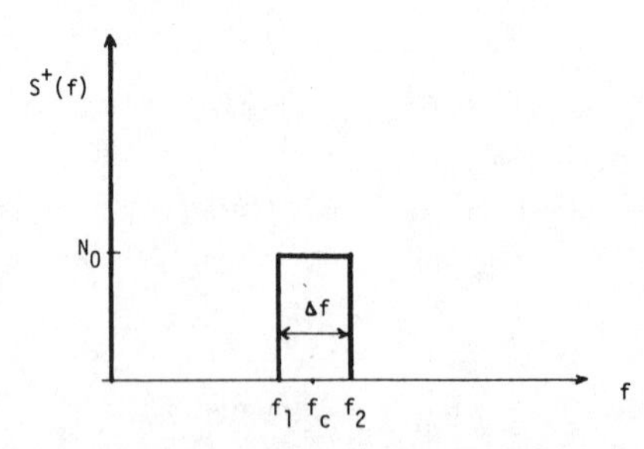

FIG. 8. Power spectrum of narrow-band white noise process.

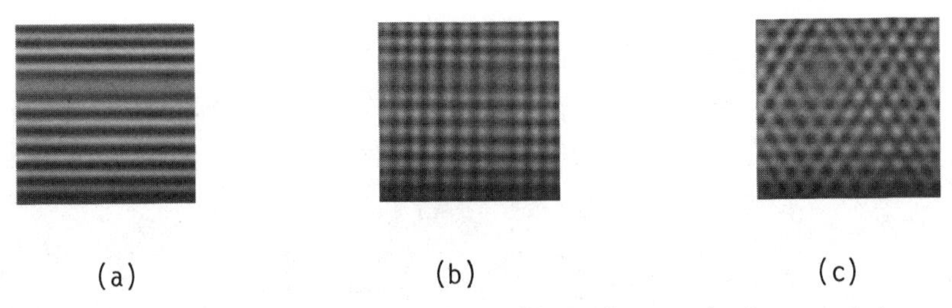

FIG. 9. (a) Long crested narrow-band waveform, (b) two intersecting long crested narrow-band waveforms 90° apart, (c) three intersecting long crested narrow-band waveforms 60° apart.

3. NARROW-BAND NOISE MODEL

A waveform is said to be "narrow-band" if the significant region of its energy spectrum is confined to a narrow bandwidth $\Delta\omega = 2\pi\Delta f$. A typical narrow-band noise wave is shown in Fig. 7. It appears to be more or less a sinusoid with a slowly varying envelope V and a slowly varying phase ϕ:

$$n(x) = v(x) \cos (\omega_c x + \phi(x)), \tag{21}$$

where ω_c is the center frequency.

We may approximate this function by its corresponding Fourier expansion. Since the random process $\{n(x), 0 < x < \infty\}$ is a real random process, the imaginary terms of the Fourier transform need not be considered.

$$n(x) = \sum_{i=1}^{\infty} C_i \cos (\omega_i x + \phi_i), \tag{22}$$

where C_i is the amplitude and ϕ_i is the phase of the ith harmonic relative to an arbitrary origin (Fig. 13).

In this representation,

$$S^+(f_i)\Delta f = \tfrac{1}{2}C_i^2, \tag{23}$$

where $S^+(f)$ denotes the one-sided (positive frequencies only) noise power spectrum. S^+ shows how the variance (or power) of the random process is distributed over frequency.

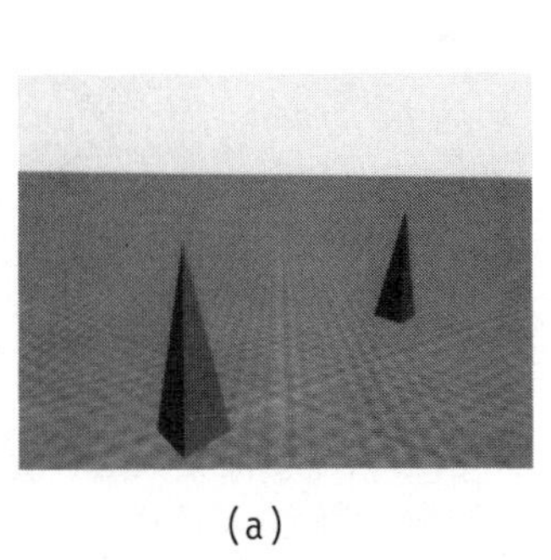

FIG. 10. Three intersecting long crested narrow-band noise waveforms shown in perspective: (a) as a texture, (b) as a height field.

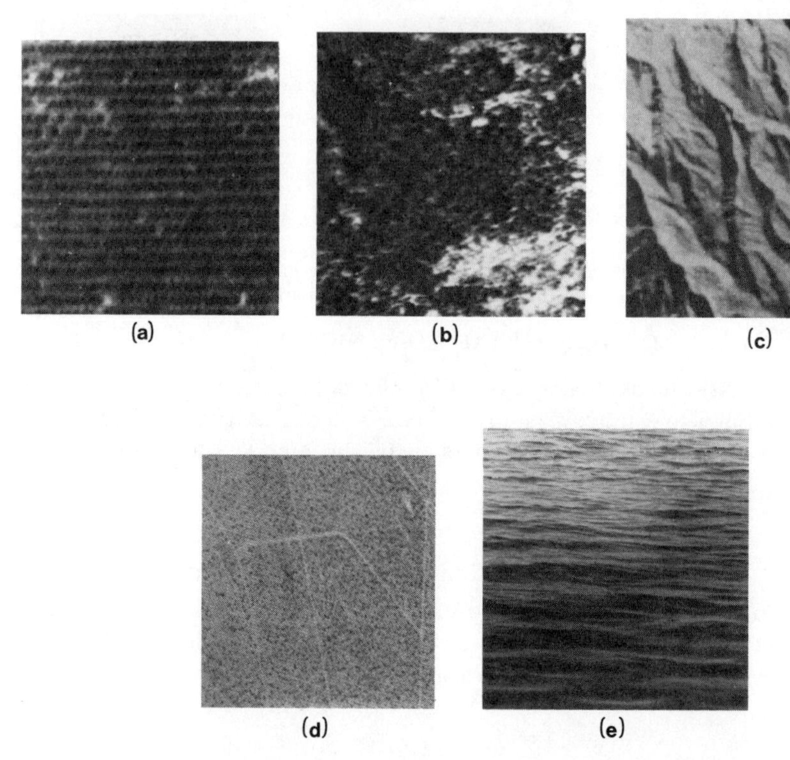

FIG. 11. Aerial photographs: (a) orchard, (b) deciduous forest, (c) rocky terrain, (d) desert, (e) water (in perspective).

The mean square magnitude of the function n is given by the integral over all frequencies of the spectrum

$$N = E[n^2(x)] = \sum_{i=1}^{\infty} \tfrac{1}{2}C_i^2 = \int_{f_1}^{f_2} S^+(f)df = \sigma_n^2. \tag{24}$$

The power spectrum of a narrow-band white noise process is shown in Fig. 8. For this case $N = N_0\Delta f$. The correlation function of n is given by the Fourier transform of its power spectral density (normalized power spectrum)

$$\begin{aligned}\rho(x) &= \frac{1}{\sigma_n^2} E[n(x_1)n(x_2)] \\ &= \frac{1}{\sigma_n^2}\int_{f_1}^{f_2} S^+(f)\cos(2\pi fx)df = N_0\frac{\sin(\omega_2 x) - \sin(\omega_1 x)}{2\pi x\sigma_n^2}. \end{aligned} \tag{25}$$

Since a narrow-band waveform may be represented by a sum of random variables, each with zero mean, the height distribution is Gaussian, i.e., $N(0, \sigma_n^2)$. This has an immediate advantage over the model of Section 2, which is Gaussian only for large sums of waveforms.

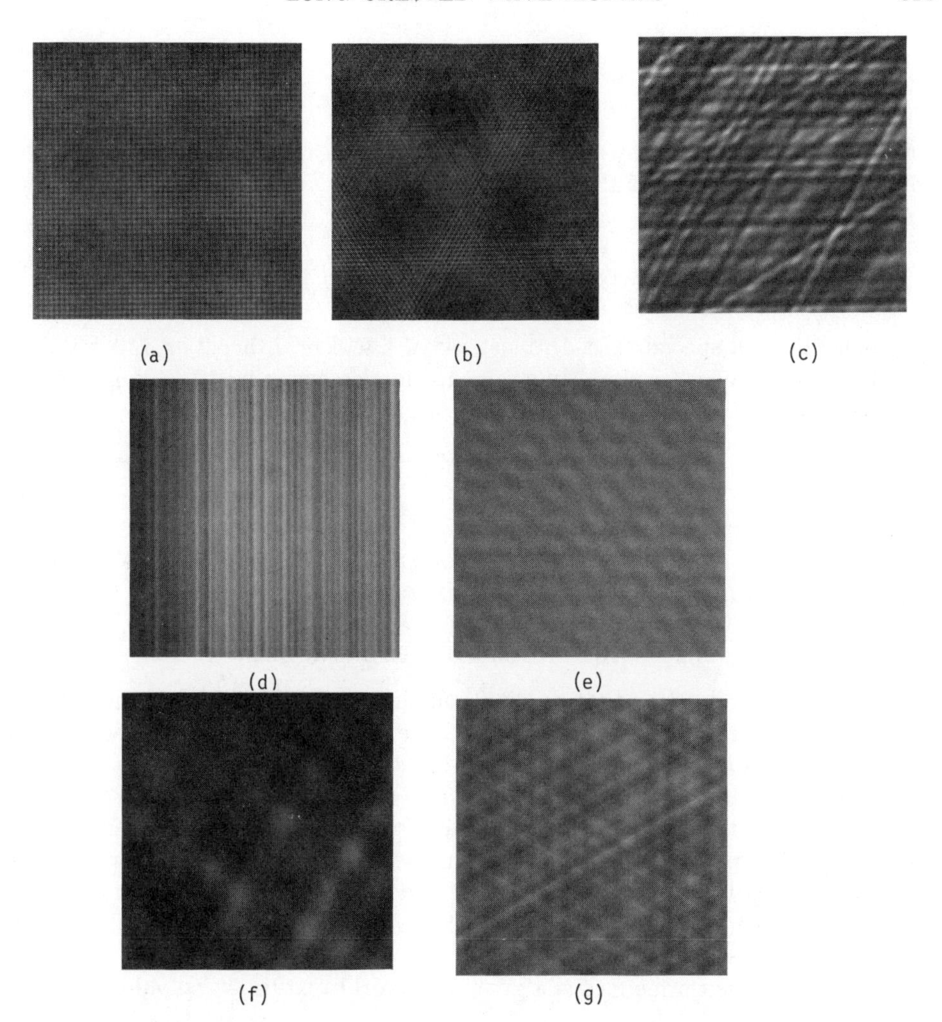

FIG. 12. Simulated textures: (a) orchard, (b) forest, (c) rocky terrain, (d) plowed field, (e) water, (f) tidal flat, (g) desert.

An image formed from sums of long crested narrow-band noise waveforms is defined by

$$t(x, y) = \mu + g(x, y) = \mu + (1/k) \sum_{i=1}^{k} B_i n(u_i x + v_i y + \phi_i). \tag{26}$$

Since each point of the image is the sum of k independent random variables,

$$t \sim N[\mu, \sigma_n^2 \sum_{i=1}^{k} (B_i/k)^2].$$

Several realizations of this model are given in Figs. 9 and 10.

We will increase the flexibility of this model by introducing multiplicative random noise. Let

$$m(x) = [(1 - \alpha) + \alpha R_x] n(x); \qquad 0 < \alpha < 1, \tag{27}$$

where $R \in [0, 1]$ is a uniformly distributed random variable, α is the parameter governing the degree of multiplicative randomness, and x is now discrete. Our model now becomes

$$t(x, y) = \mu + g(x, y) = \mu + (1/k) \sum_{i=1}^{k} B_i m(u_i x + v_i y + \phi_i) \tag{28}$$

of which (26) is a special case. In most cases, k will be either 2 or 3. A value of 2 will give the image an underlying square lattice structure. A value of 3 will produce a microstructure that drifts between hexagonal and triangular. Let

$$B = \begin{pmatrix} B_1 \\ \vdots \\ B_k \end{pmatrix}; \qquad \Theta = \begin{pmatrix} \theta_1 \\ \vdots \\ \theta_k \end{pmatrix},$$

where $\theta_i = \tan^{-1}(v_i/u_i)$. The narrow-band noise model has seven independent parameters: $\{\mu, B, \Theta, k, \omega_c, \Delta\omega, \alpha\}$.

4. FITTING THE MODEL TO REAL DATA

It should be made clear that no single model can depict all categories of textures [27]. The real world is infinitely variable. The model introduced in Section 3 describes certain members of the class of textures whose gray level statistics fit a single normal distribution. Realizations of this model have a "hilly" surface. A high multiplicative random noise parameter (i.e., letting $\alpha \to 1$) will give these hills a ragged appearance. Other major categories of generative texture models include the random mosaics and bombing models [28]. The problem of choosing the best type of model to depict a given real world texture is often difficult and will not be covered here.

If we suppose that a model fits a given real world texture, we should in theory

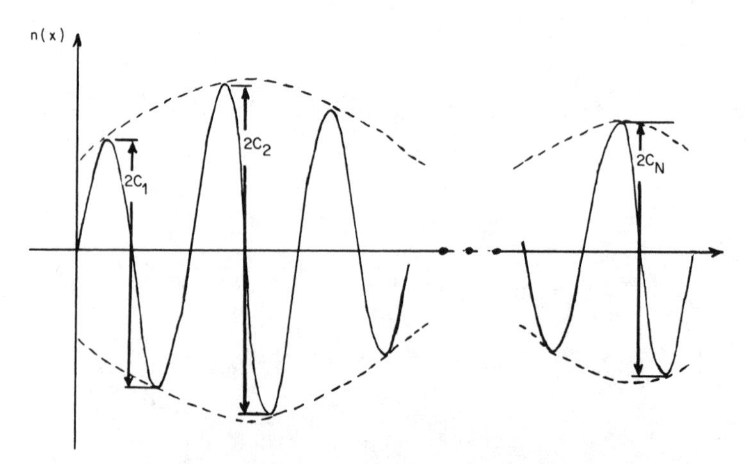

FIG. 13. A narrow-band noise waveform of N harmonics.

be able to extract the model parameters from the texture. The model presented in Section 3 has seven parameters which must be determined. Some are more difficult to obtain than others.

Parameters Θ and k usually may be determined by examination. For example, the aerial photograph of the orchard shown in Fig. 11a has a square lattice structure. This suggests that we set $\Theta = (0, \pi/2)$. A dense forest (Fig. 11b) will require a tightly packed structure. A hexagonal cell structure is the most efficient partitioning of the plane, suggesting that we set $k = 3$, $\Theta = (0, \pi/3, 2\pi/3)$. The rocky surface shown in Fig. 11c is highly anisotropic. We will orient the crests of the waveforms along the direction of the fault lines, giving $k = 3$, $\Theta = (0, \pi/6, \pi/3)$. A plowed field obviously requires a value of $k = 1$.

The mean and variance of a texture will give us model parameters μ and B.

Information about the power spectrum may be obtained from the Fourier transform of the correlation function. However, sample spectra are often so erratic that they are useless for estimation purposes. The basic reason why Fourier analysis breaks down when applied to random textures is that it is based upon the assumption of fixed amplitudes, frequencies, and phases. Random textures on the other hand are characterized by random changes of frequency, amplitude, and phase. The sample spectrum must be regarded as the realization of a random field. It is not a consistent estimator, in the sense that it does not converge toward the true spectrum as the sample size increases [13]. Sophisticated statistical techniques are available for overcoming these difficulties [13]. Instead of using them, we chose to arrive at the bandwidth and parameter α in an interactive manner. A photograph of the texture being simulated was displayed on the lower half of a CRT. Successive versions of the simulation were displayed on the upper half. The parameters α and $\Delta\omega$ were adjusted until these images appeared similar. Further research is needed to devise better techniques for obtaining these parameters. Some examples of our simulations are shown in Fig. 12.

The narrow-band noise model has been implemented in hardware in a real-time image generation system. Functions $m(x)$ are stored in core memory. They are called to modulate the background color of image surfaces as specified by the data base. Anti-aliasing operations are incorporated into the hardware.

5. GEOMETRICAL PROPERTIES

Let $g(x, y)$ be a real-valued "sufficiently smooth" random field. By sufficiently smooth, we will mean that the following regularity conditions are satisfied [16]:

(i) The sample function of $g(x, y)$ almost surely has continuous partial derivatives up to second order with finite variance in a compact subset A of the plane.

(ii) The number of points $(x, y) \in A$ for which $g(x, y) = u$ and either $(\partial g/\partial x)(x, y) = 0$ or $(\partial g/\partial y)(x, y) = 0$ is finite.

(iii) There is no point (x, y) on the boundary of A for which $g(x,y) = u$ and either $(\partial g/\partial x)(x, y) = 0$ or $(\partial g/\partial y)(x, y) = 0$.

(iv) There is no point $(x, y) \in A$ for which $g(x, y) = u$, $(\partial g/\partial x)(x, y) = 0$ and $(\partial g/\partial y)(x, y) = 0$.

(v) There is no point $(x, y) \in A$ for which $g(x, y) = u$ and either $(\partial g/\partial x)(x, y) = 0 = (\partial^2 g/\partial x^2)(x, y)$ or $(\partial g/\partial y)(x, y) = 0 = (\partial^2 g/\partial y^2)(x, y)$.

For the case when $g(x, y)$ is homogeneous and Gaussian, the last three conditions are automatically fulfilled [16]. An example of a Gaussian field not meeting the regularity conditions is any realization of the narrow-band noise model with $\alpha > 0$.

For a fixed real threshold u and a compact subset A of the plane, let

$$E(A, u) = \{(x, y) \in A : g(x, y) = u\}$$

define the *excursion set* of the field. The boundary of $E(A, u)$ is called a *level curve* of the field. It is composed of contour lines of height u. Let $l(\text{l.c.}(A, u))$ denote the length of the level curve.

Suppose that we create a new random field $w(x, y)$. All points in $g(x, y)$ having values above the threshold u will be set to 1 and the rest to 0. Thus $w(x, y)$ is a binary image defined by

$$\begin{aligned} w(x, y) &= 1 \qquad \text{if } g(x, y) \geq u \\ &= 0 \qquad \text{otherwise.} \end{aligned}$$

If the pdf of the original field is Guassian, i.e., $N(0, \sigma_g^2)$, then

$$P(w(x, y) = 1) = \int_u^\infty \frac{1}{(2\pi)^{\frac{1}{2}}\sigma_g} \exp\left(\frac{-u^2}{2\sigma_g^2}\right) du = 1 - \Phi\left(\frac{u}{\sigma_g}\right), \tag{29}$$

where

$$\Phi(\S) = \frac{1}{(2\pi)^{\frac{1}{2}}} \int_{-\infty}^{\S} \exp\left(\frac{-\S^2}{2}\right) d\S.$$

The first moment of the thresholded image is simply

$$E(w(x, y)) = 1 \cdot P(g(x, y) = 1) + 0 \cdot P(g(x, y) = 0) = 1 - \Phi\left(\frac{u}{\sigma_g}\right). \tag{30}$$

Similarly, the second moment is given by

$$E(w(x, y))^2 = 1 - \Phi\left(\frac{u}{\sigma_g}\right). \tag{31}$$

Thus [8]

$$\sigma_w^2 = E(w(x, y))^2 - E^2(w(x, y)) = \Phi\left(\frac{u}{\sigma_g}\right)\left(1 - \Phi\left(\frac{u}{\sigma_g}\right)\right). \tag{32}$$

The mean perimeter per unit area of $w(x, y)$ is just the length of the level curve of $g(x, y)$ for a level u. For Longuet-Higgins' model, this value is given by [14]

$$E(l\{\text{l.c.}(A, u)\}) = \frac{1}{\pi}\left(\frac{m_{20} + m_{02}}{m_{00}}\right)^{\frac{1}{2}} \frac{\Xi(1 - \gamma^2)^{\frac{1}{2}}}{(1 + \gamma^2)^{\frac{1}{2}}} \exp(-u^2/2m_{00}), \tag{33}$$

where $\Xi(\cdot)$ is the Legendre elliptical integral of the first kind. This is essentially the same result as given by Switzer [26].

Panda [8] determines the following lower bound on the average number of connected components per unit area in the excursion set for the isotropic version of Longuet-Higgins' model:

$$\frac{m_{02}}{(2\pi m_{00})^{\frac{3}{2}}}\, u\cdot\exp\left(\frac{-u^2}{2m_{00}}\right), \qquad \text{for} \qquad u > 0. \tag{34}$$

His results are essentially the same as those given by Adler [16] for the more general case of a "sufficiently smooth" Gaussian field.

The number of local maxima of $g(x, y)$ above u in A is denoted by $M(A, u)$. It is apparent that

$$P(Z(A) \leq u) = P(M(A, u) = 0) \geq 1 - E(M(A, u)),$$
$$\text{where} \qquad Z(A) = \max_{(x,y)\in A}\, [g(x, y)]. \tag{35}$$

If we know $E(M(A, u))$, then we have a lower bound for the distribution of $Z(A)$. Hasofer [22] shows that this lower bound is extremely close for high levels of u. When the spectrum of a random field has most of its energy concentrated in a small area of frequency space, the distribution of $Z(A)$ is better investigated by replacing $g(x, y)$ by its envelope $V(x, y)$, which has the property that

$$|g(x, y)| \leq |V(x, y)| \qquad \text{for all } x,\, y, \tag{36}$$

and $g(x, y) = V(x, y)$ at local maxima. Longuet-Higgins [4] has used this approach to get an approximate value for the maximum amplitude in a train of N waves, denoted by $C_{\max}$ (see Fig. 13):

$$E(C_{\max}) = \sigma_N((\log N)^{\frac{1}{2}} + 0.28861(\log N)^{-\frac{1}{2}} + O((\log N)^{-\frac{3}{2}})). \tag{37}$$

This equation gives good results for $N > 10$. For a field of two intersecting narrow-band trains of N waves, the expected maximum is $2E(C_{\max})$.

It can be shown that for any sufficiently smooth Gaussian long crested wave system, $g(x, y)$ approximates an elliptical paraboloid near a high maximum. Therefore, level curves near local maxima will be nearly elliptical.

Longuet-Higgins [5] shows that for any stationary random field, resulting from the intersection of long crested waves, the average density of local maxima plus local minima is equal to the average density of saddle points. Consider the example of two intersecting sinusoids. When a crest/trough from one system intersects a crest/trough from the other system, there is a local maximum/minimum. When a crest from one system intersects a trough from the other system, there is a saddle point (see Fig. 4). On this surface, the density of local maxima is equal to the density of local minima. However, this is not always the case. Consider the pattern of three intersecting sinusoids shown in Fig. 6a. There are twice as many local maxima as minima.

Let D_{ma} and D_{mi} denote the average density of local maxima and minima. For any long crested wave model, in which the phases of component waves are distributed uniformly between 0 and 2π, and the probability of a surface point being positive is the same as that of it being negative, $D_{ma} = D_{mi}$. There are twice as many saddle points as either local maxima or minima, since one saddle point falls between each local maximum–minimum pair. Thus the general equation

$$D_{ma} + D_{mi} = D_{sa} \tag{38}$$

is satisfied.

For two intersecting sinusoids or narrow band waves 90° apart, clearly

$$D_{ma} = (\omega_c/2\pi)^2. \tag{39}$$

For an isotropic version of Longuet-Higgins' model [14],

$$D_{ma} = \frac{1}{8(3)^{\frac{1}{2}}\pi} \frac{M_4}{M_2}. \tag{40}$$

When the spectrum falls within a thin annulus of center frequency ω_c, the above relation reduces to

$$D_{ma} = 0.907\,(\omega_c/2\pi)^2. \tag{41}$$

Thus, for this case, there are only slightly fewer local maxima per unit area than for two orthogonally intersecting sinusoids.

6. DISCUSSION

A new long crested wave model was presented for depicting certain types of Gaussian textures. The model appears more suited for image analysis and synthesis than the traditional long crested wave model developed by Longuet-Higgins.

This new model is particularly useful for real-time image generation. A number of random functions may be called from computer memory to modulate the background color of image surfaces. Perspective and smoothing transforms can be handled in hardware.

REFERENCES

1. W. M. Bunker and N. E. Ferris, *Computer Image Generation—Imagery Improvement: Circles. Contours, and Texture,* Air Force Human Resource Lab., TR 77-66, Air Force Systems Command, Brooks Air Force Base, Texas.
2. W. Freiberger and U. Grenander, Surface patterns in theoretical geography, *Comput. Geosci.* **3** (4), 1977, 547–578.
3. U. Grenander, Dynamical models of geomorphological patterns, *J. Math. Geol.* **7** (2), 1975, 267–278.
4. M. S. Longuet-Higgins, On the statistical distribution of the heights of sea waves, *J. Marine Res.* **11** (3), 1952, 245–266.
5. M. S. Longuet-Higgins, The statistical analysis of a random moving surface, *Phil. Trans. Roy. Soc. London, Ser. A* **249**, Feb. 1957, 321–387.
6. M. S. Longuet-Higgins, The statistical distribution of the curvature of a random Gaussian surface, *Proc. Cambridge Phil. Soc.* **54** (4), 1958, 439–453.
7. D. P. Panda, *Statistical Analysis of Some Edge Operators,* TR 558, Computer Science Center University of Maryland, College Park, July 1977.
8. D. P. Panda, Statistical properties of thresholded images, *Computer Graphics Image Processing* **8**, 1978, 334–354.

9. R. Schmidt, *The USC–Image Processing Institute Data Base*, USCIPI TR 780, University of Southern California, Los Angeles, Oct. 1977.
10. P. Swerling, Statistical properties of the contours of random surfaces, *IRE Trans. Inform. Theory* **8**(3), July 1962, 315–321.
11. S. O. Rice, The mathematical analysis of random noise, *Bell System Tech. J.* **23**, 1944, 282–332; **24**, 1945, 46–52.
12. L. Rayleigh, On the resultant of a large number of vibrations of the same pitch and arbitrary phase, *Phil. Mag.* **10**, 1880, 73–78.
13. G. Jenkins and D. Watts, *Spectral Analysis*, Holden Day, San Francisco, 1968.
14. M. S. Longuet-Higgins, Statistical properties of an isotropic random surface, *Phil. Trans. Roy. Soc. London Ser. A* **250**, Oct. 1957, 157–171.
15. L. J. Cote, *Two-dimensional Spectral Analysis*, Purdue University Dept. of Statistics, TR 83, July 1966.
16. R. J. Adler and A. M. Hasofer, Level crossings for random fields, *Ann. Probability* **4**(1), 1976, 1–12.
17. R. J. Adler, Excursions above high levels by Gaussian random fields, *Stochastic Processes Their Appl.* **5**, 1977, 21–25.
18. A. M. Hasofer, The mean number of maxima above high levels in Gaussian random fields, *J. Appl. Probability* **13**, 1976, 377–379.
19. V. P. Nosko, Local structure of Gaussian fields in the vicinity of high level shines, *Soviet Math. Dokl.* **10**, 1969, 1481–1484.
20. V. P. Nosko, The characteristics of excursions of Gaussian homogeneous fields above a high level, in *Proc. USSR–Japan Symp. on Prob., Novosibirsk, 1969*.
21. R. J. Adler, A spectral moment estimation problem in two dimensions, *Biometrika* **64**(2), 1977, 367–373.
22. A. M. Hasofer, Upcrossings of random fields, *Suppl. Advances in Appl. Probability* **10**, 1978, 14–21.
23. R. J. Adler, On generalizing the notion of upcrossings to random fields, *Advances in Appl. Probability* **8**, 1976, 789–805.
24. G. Lindgren, Local maxima of Gaussian fields, *Ark. Mat.* **10**, 1972, 195–218.
25. R. J. Adler, Excursions above a fixed level by n-dimensional random fields. *J. Appl. Probability* **13**, 1976, 276–289.
26. P. Switzer, *Geometrical Measures of the Smoothness of Random Functions*, TR 62, Stanford University Statistics Dept., 1974.
27. B. Schachter, Texture measures, submitted for publication.
28. B. Schachter and N. Ahuja, Random pattern generation processes, *Computer Graphics Image Processing* **10**, 1979, 95–114.

The Boolean Model and Random Sets *

J. SERRA

Centre de Morphologie Mathématique, Ecole des Mines de Paris,
35 Rue Saint-Honoré, 77305 Fontainebleau, France

We propose to present the main theorems which govern random set theory by studying one particular random set, namely the Boolean model. After defining the Boolean model X (union of almost surely compact random sets centered at Poisson points in $\mathbf{R}^n$), the probability $Q(B)$ that a given compact set B misses X is calculated. Several laws are derived from $Q(B)$ (covariance, law of the first contact, specific numbers, etc.). We then go back to the basic theoretical problems raised by such an approach:

— What are the morphological mappings which transform one random set into another?

— What pieces of information are sufficient to characterize a random set?

— What is the general expression for indefinitely divisible random sets?

Answers are given using G. Matheron and G. Choquet's theorems. In the last part we show how Boolean sets may be handled in view of constructing more sophisticated models (tessellations, n-phased sets, hierarchical sets, etc.). Examples are given.

NOTATION

1. *Sets*

$\mathbf{R}^1$, $\mathbf{R}^2$, $\mathbf{R}^3$	Euclidean space; of dimension 1, 2, or 3
x, y	A point in $\mathbf{R}^n$, or equivalently, the vector ($\mathbf{0}x$, $\mathbf{0}y$)
$h = x - y$	Vector having point x for origin and y for extremity
X	Set of points constituting the object to be studied
∂X	Boundary of set X
B	Set of points of the structuring element
$\check{B}$	Set of points x such that $-x \in B$
X_x, B_x	Translate of X, B by vector x
λB	Set similar to B (ratio λ)
ϕ	Empty set
X^c	Complement of set X, i.e., set of the points of $\mathbf{R}^n$ which do not belong to X
$\Psi(X)$	Transform of X by the transformation Ψ ($\Psi(X)$ is a set, not a number)
$\mathcal{K}$(resp. $\mathcal{F}$)	Set of the compact (resp. closed) sets of $\mathbf{R}^n$
RACS	Random closed set
$\mathcal{C}(a)$	Regular lattice of points with spacing a

ISBN 0-12-597320-9

2. *Logical Operations*

$X \cup Y$	Union of points belonging to X or to Y
$X \cap Y$	Intersection, i.e., set of points belonging to X and to Y
$y \in B$	Point y belongs to set B
$B_x \subset X$	B_x is included in X (if $y \in B_x$, then $y \in X$)
$B_x \Uparrow X$	B_x hits X (in other words: $B_x \cap X \neq \phi$)
$X \ominus B$	Eroded set of X by $\check{B}$: i.e., set of the points x such that $\check{B}_x$ is included in X; $X \ominus B = \cap_{y \in B} X_y$ (Minkowski subtraction)
$X \oplus B$	Dilated set of X by $\check{B}$: i.e., set of the points x such that $\check{B}_x$ hits X; $X \oplus B = \cup_{y \in B} X_y = \cup_{x \in X} B_x = \cup_{x \in X, y \in B} x + y$ (Minkowski addition)
X_B	Morphological opening set of X by B: X_B is the union of all the translates B_y contained in X, we have $X_B = (X \ominus \check{B}) \oplus B$
X^B	Morphological closing of set X by B, i.e., complement of opening of the complement of X: $X^B = ([X^c]_B)^c$

3. *Measures*

Mes X	Lebesgue measure of X (length in $\mathbf{R}^1$, area in $\mathbf{R}^2$, volume in $\mathbf{R}^3$); $\overline{\text{Mes } X}$ = mathematical expectation of Mes X
$Q(B_x)$	Probability that $B_x \subset X$, where X is a RACS; when B_x is reduced to point $\{x\}$, $Q(B_x) = q_x$ = porosity at x; when X is stationary $Q(B_x) = Q(B)$ is independent of X
$A(X)$, $U(X)$	Area, perimeter of X (in $\mathbf{R}^2$)
$V(X)$, $S(X)$, $M(X)$	Volume, surface area and integral of the mean curvature of X, in $\mathbf{R}^3$
$\bar{A}$, $\bar{U}$, $\bar{V}$, $\bar{S}$, $\bar{M}$	The corresponding mathematical expectation when X is a RACS
$N_A(X)$ (resp. $N_V(X)$)	Euler–Poincaré constant (or connectivity number) per unit area (resp. volume) for a stationnary RACS; N_A is the number of particles, minus their holes, per unit area

A COUNTERPOINT

The quantitative description of random sets has to be carried out at two different levels. First of all, we have to define them properly, i.e., provide them with adequate axiomatics, and then derive their main mathematical properties (characterization of random sets by their Choquet functionals, infinite divisibility, etc.). However, a purely mathematical approach would not be sufficient, and must be complemented with a forthright description of the random sets, i.e., by effectively giving recipes for the construction of random sets possessing desirable morphological properties. For the sake of pedagogy, it is better to start with this second aspect, and to concentrate on one particular model. We propose the Boolean model, since it is especially interesting in itself, and lends itself to many attractive derivations. Moreover, the model, and its derivations, are essentially due to the Centre de Morphologie Mathematique of Fontainebleau. (In fact, as often happens with good ideas, the Boolean model has been "rediscovered" a number of times since the publication of the original work. We will try to be scrupulous in giving the right references at each step of development. When we quote two references together, the older one corresponds to the original work, the other to a more accessible version.)

The counterpoint between the Boolean model and the general properties of random sets results in the following plan:

Boolean model	General underlying problems
1 Construction	Random sets: definition and basic properties
2 Functional moments	Infinite divisibility
3 Convex primary grains	Semi-Markov RACS
4 Connectivity number (in $\mathbf{R}^2$)	Digitization
5 Specified Boolean models	Estimation problems
6 Derived models	The rose of the models

1. CONSTRUCTION OF THE BOOLEAN SETS

The first outline of the Boolean model appears in the literature with Solomon [35] and B. Matern [18]; the latter took a disk with a constant radius as primary grain, and calculated the covariance of the model in this particular case. A few years later, G. Matheron [19] gave the general definition of the model, that we adopt here, and calculated the key formula (3).

The definition of the Poisson point process in $\mathbf{R}^n$ is well known. This random set of points is characterized by the following two properties:

(a) If B and B' are two sets such that $B \cap B' = \phi$, the numbers $N(B)$ and $N(B')$ of points falling in B and B' are two independent random variables.

(b) The elementary volume dv contains one point with probability $\theta(dv)$ and no points with probability $1 - \theta(dv)$. The measure θ is called the density of the process. Here we will take θ = constant, because it leads to more geometrically interpretable results.

Suppose we take a realization of a Poisson process of constant density θ, and consider each point as the germ of a crystalline growth. If two crystals meet each other, we suppose that they are not disturbed in their growth, which stops independently for each component. Let us transpose this description in terms of random sets. The I points of the Poisson realization are at the points x_i $(i \in I)$ in $\mathbf{R}^n$. The elementary grain is an almost surely compact random set X'; we pick out various realizations X'_i of X' from its space of definition, and implant each X'_i at the corresponding point x_i. The different X'_i are thus independent of each

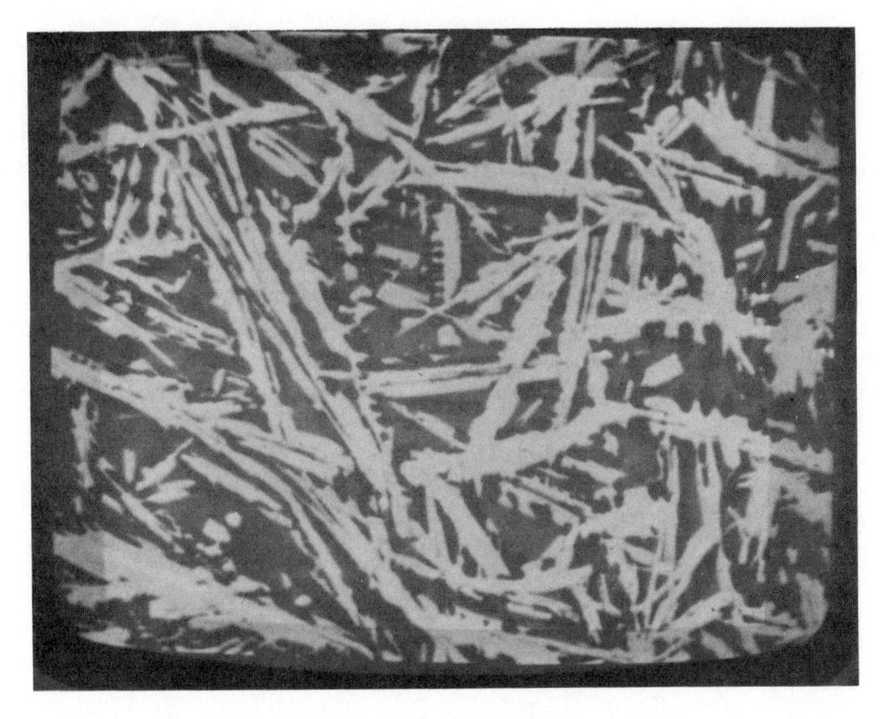

FIG. 1. Ferrite crystals in an iron ore sinter.

other. We shall call the realization X of a Boolean model, the *union* of the various X'_i after implantation at the points x_i:

$$X = \bigcup_{i \in I} X'_i \tag{1}$$

The Boolean model is extremely flexible: it is a first step, where one admits only negligible interactions between the particles X'_i. Figure 1 represents a typical Boolean structure.

1.* RANDOM SETS: DEFINITION AND BASIC PROPERTIES

We have just introduced a random closed set X (in brief, a RACS) via the technique which allows us to construct it (in the following, we use the same symbol X for denoting a RACS and its realization). However, just like a random variable, a RACS is mathematically defined from a collection of events, namely the relationships "K misses X," where K describes the class of compact sets. These events are governed by the classical axioms of probability, which require first a σ-algebra, call it σ_f, and then a probability P on the measure space $(\mathfrak{F}, \sigma_f)$, where $\mathfrak{F}$ denotes the set of closed sets in $\mathbf{R}^n$. σ_f is generated by the countable unions of the events with their complements. To define P we associate with any event V of σ_f the probability that this relation V is true.

Here, a few topological comments are necessary. To handle the random sets correctly, we must be able to express how a sequence $\{X_i\}$ of sets tends toward a limit X. On the other hand, we have to restrict the class $\mathcal{P}(\mathbf{R}^n)$ of all the possible parts of $\mathbf{R}^n$. Indeed Euclidian space is too rich for our purpose. For example a set such as "all the points with irrational coordinates in the plane" has absolutely no physical meaning. In order to make this simplification, define the distance ρ, from point x to set X, as follows:

$$\rho(x, X) = \inf_{y \in X} d(x, y), \qquad x \in \mathbf{R}^n, \qquad X \in \mathcal{P}(\mathbf{R}^n),$$

where d is the Euclidean distance. With respect to ρ, there is no difference between a set X and its topological closure $\bar{X}$ (i.e., X plus its boundary), since $\rho(x, X) = \rho(x, \bar{X})$, $\forall x$, $\forall X$. In other words, all the notions derived from ρ will not be related to the parts $\mathcal{P}(\mathbf{R}^n)$ of $\mathbf{R}^n$, but to the equivalence classes of sets which admit the same closure. For example, the points with irrational coordinates of $\mathbf{R}^2$, and the whole plane itself, will be considered as *identical*. Hence, from now on it suffices to concentrate upon the class of the closed sets of $\mathbf{R}^n$. The distance ρ generates a topology, called the intersection topology. Matheron [20, 21] and Kendall [14] exhaustively studied it in a more general frame than $\mathbf{R}^n$ (for a simpler presentation see also Serra [32, Chap. III]). By definition a sequence $\{X_i\}$, $X_i \in \mathfrak{F}$, converges toward a limit $X \in \mathfrak{F}$ if and only if, for any $x \in \mathbf{R}^n$, the sequence $\rho\{x, X_i\}$ converges towards $\rho\{x, X\}$ in $\overline{\mathbf{R}}_+$. From this standpoint we can derive all the basic topological notions, such as neighborhoods, continuity, semi-continuity, etc. We quote only one result, for it will be useful below. An increasing mapping Ψ from $\mathfrak{F}$ into itself (or more generally from $\mathfrak{F} \times \mathfrak{F} \to \mathfrak{F}$) is upper semi-continuous (u.s.c.) if and only if $X_i \downarrow X$ in $\mathfrak{F}$ implies $\Psi(X_i) \downarrow \Psi(X)$ in $\mathfrak{F}$. ($X_i \downarrow X$ means $X_{i+1} \subset X_i$ and $X = \bigcap_i X_i$; Ψ is said to be increasing when $X \subset Y \Rightarrow \Psi(X) \subset \Psi(Y)$.)

We now go back to the random sets, and study what they become after set transformations. It is not obvious that, if X is a RACS and Ψ an arbitrary set transformation, the transform $\Psi(X)$ is still a RACS. However, if Ψ is a *semi-continuous* mapping from $\mathfrak{F} \to \mathfrak{F}$ (or $\mathfrak{F} \times \mathfrak{F} \to \mathfrak{F}$), the resulting set $\Psi(X)$ is always a RACS. Hence, $X \cup X'$, $X \cap X'$, ∂X, X^c, $X \oplus K$, $X \ominus K$, and the finite iterations of these transformations provide RACS (Note that a mapping may be semi-continuous and *not* increasing, e.g., the boundary mapping $X \to \partial X$). Similarly, the volume, $V(X)$, in $\mathbf{R}^3$, and the area $A(X)$, in $\mathbf{R}^2$ are semi-continuous mappings $\mathbf{R}^3 \to \mathbf{R}^+$ (or $\mathbf{R}^2 \to \mathbf{R}^+$) and provide random variables.

It is well known that the probability distribution associated with an ordinary random variable is completely determined if the corresponding distribution function is given. There is a similar result for RACS. If X is a RACS and P the associated probability on σ_f define

$$Q(B) = P\{B \subset X^c\} \tag{2}$$

to be the probability that X misses a given compact set $B \in \mathcal{K}$. That is, we obtain a function Q on $\mathcal{K}$, called the functional moment, associated with the probability P. Conversely, the probability P is completely determined if the function Q on $\mathcal{K}$ is given (Matheron–Kendall's theorem [24, 14]). It is interesting to find the necessary and sufficient conditions for a given function Q to be associated with a RACS:

(i) One must have $0 \leq Q(B) \leq 1$ (Q is probability) and $Q(\phi) = 1$, since the empty set misses all the other ones.

(ii) If $B_i \downarrow B$ in $\mathcal{K}$, we must have $Q(B_i) \uparrow Q(B)$.

(iii) Let $S_n(B_0; B_1 \ldots B_n)$ denote the probability that X misses the compact set B_0, but hits the other compact sets $B_1, \ldots, B_n$. These functions are obtained by the following recurrence formula:

$$S_1(B_0; B_1) = Q(B_0) - Q(B_0 \cup B_1),$$

$$S_n(B_0; B_1 \ldots B_n) = S_{n-1}(B_0; B_1 \ldots B_{n-1}) - S_{n-1}(B_0 \cup B_n; B_1 \ldots B_{n-1}).$$

These functions, which are probabilities, must be ≥ 0 for any integer n and any compact sets $B_0, B_1, \ldots, B_n$.

(The last two prerequisites make the quantity $1 - Q$ an alternating Choquet capacity of infinite order.) The three requirements (i), (ii), (iii) must obviously be satisfied by the function Q. Choquet [5] proved that they are also sufficient. His basic theorem orients us toward the calculations to perform in order to characterize a RACS. The morphological interpretation of (2) is clear. When B is centered at point x, $Q(B)$ is nothing but the probability that x belongs to the pores of the dilate $X \oplus \check{B}$, i.e., the porosity $q_B(x)$ of $X \oplus \check{B}$ at point x.

2. THE FUNCTIONAL MOMENT OF THE BOOLEAN MODEL

We now apply Matheron–Kendall's theorem, which has just been stated, to the case of the Boolean model. By definition, the primary grain X' is known, and the question is to express the $Q(B)$'s of X as functions of those of X'. The

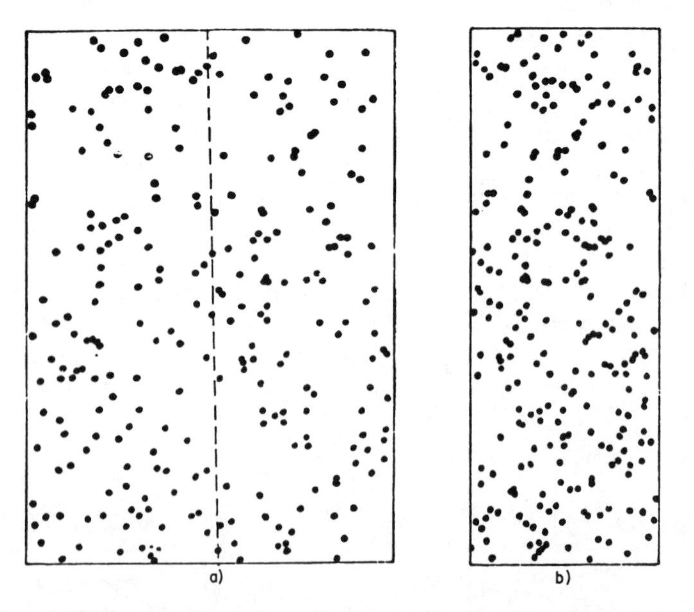

FIG. 2. The superposition of two Boolean realizations is again Boolean.

RACS X' being a.s. compact, the integral of $1 - q_B(x)$ over the space exists, and equals the average measure (volume or area) $\overline{\text{Mes}}\ (X' \oplus \check{B})$. The functional moment $Q(B)$ of X, after Booleanization, admits the very simple expression

$$P(B \subset X^c) = Q(B) = \exp(-\theta\, \overline{\text{Mes}}\ (X' \oplus \check{B})) \qquad \forall B \in \mathcal{K} \tag{3}$$

(Matheron [19], Serra [32]). Relation (3) is the *fundamental* formula of the model. It links the functionals of X to those of X', and, according to Matheron–Kendall's theorem, completely determines X. From it, one derives a series of important properties and formulas.

a. *Set Properties [19, 21, 32]*

α. We see from (3) that $Q(B)$ does not depend on the location of B. Hence the Boolean model is *stationary*. One can also prove that it is *ergodic*, i.e. the spatial averages for one realization tend toward the corresponding $Q(B)$. Thus, we can speak of porosity, of specific surface, covariance, etc., without referring to a particular portion of the space.

β. *X is stable for dilation.* We derived from (3) that the dilate of X by K (K a deterministic compact set) is still a Boolean model with primary grain $X' \oplus \check{K}$.

γ. *The cross-sections of X are Boolean,* since (3) does not depend on the fact that B belongs to a subspace of R^n. Similarly, if we cut a thick slice of X, limited by two parallel planes, of normal ω, and if we project the slice on a plane normal to ω, the projection set is still Boolean.

δ. The Boolean model is *infinitely divisible* (and this basic property in fact implies all the other ones of this section). "Infinitely divisible" means the following: if one picks out two realizations of the model, and superimposes them, taking the set union of phases X, then the result is again a model of the same family. For example, look at Fig. 2. Figure 2a shows a realization of a circular Boolean model; Fig. 2b has been obtained by reflecting the left side of Fig. 2a onto its right side, producing a Boolean realization with a double density.

ϵ. *Domain of attraction.* Just as the normal law appears as the limit of an average of independent random variables, the Boolean model also turns out to be the final term of an infinite union of *other* random sets. One says that it has a certain do-

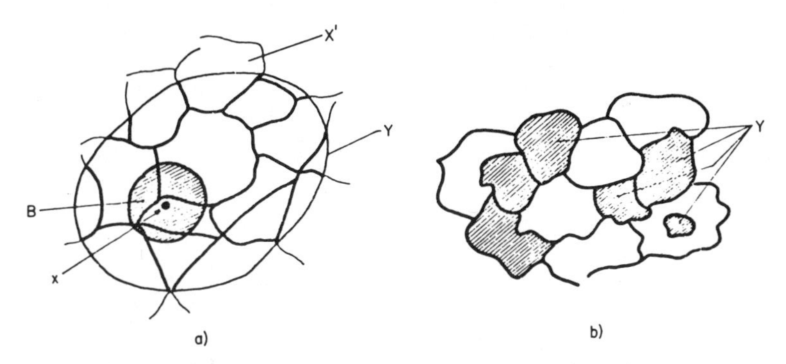

FIG. 3. An arbitrary mosaïc Y, whose infinite union tends toward a Boolean model.

main of attraction. The following result (Delfiner [7], Serra [32]) illustrates this point. Let X be a random partition of the space, arbitrary but stationary and ergodic. From X, construct a set Y by assigning each class X' of the partition to the phase "grain" with a probability p, independently from one class to another (Fig. 3). Now superimpose n realizations of Y, and let Y_n denote the intersection of the grains of the n realizations. Assuming that $np \to \theta$ $(0 < \theta < \infty)$ as $n \to \infty$ (if not, $Y\downarrow$ is trivially equal to ϕ!), and denoting by $Q_n(B)$ the functional moments of Y_n, we have

$$\lim_{n\to\infty} Q_n(B) = \exp(-\theta(\overline{\text{Mes}}\, X')^{-1} \cdot \overline{\text{Mes}}\, (X' \oplus B)),$$

i.e., according to Matheron–Kendall's theorem, a Boolean model of density $\theta(\overline{\text{Mes}}\, X')^{-1}$ and with primary grains the class X' of the initial partition.

ζ. The number of primary grains hitting B follows a *Poisson distribution* of parameter $\theta\, \overline{\text{Mes}}\, (X' \oplus \check{B})$. This is a direct consequence of the infinite divisibility. Stoyan [36] enlarged this result to the case of a density $\theta(x)$ variable over the space.

b. Applications of the Fundamental Formula

α. *Porosity.* Reduce the structuring element B to one point; then

$$B = \{\cdot\} \Rightarrow Q(B) = q = \exp(-\theta\, \overline{\text{Mes}}\, X').$$

β. *Covariance.* Take for B two points that are vector h apart

$$B = \{\cdot \overset{h}{\leftrightarrow} \cdot\} \Rightarrow Q(B) = C_{00}(h) = q^2 e^{\theta K(h)},$$

where $K(h) = \overline{\text{Mes}}\, (X' \cap X'_h)$ is said to be the geometric covariogram of X. The covariance $C_{00}(h)$ represents the probability that the two points of B lie in the pores. For the covariance $C_{11}(h)$ of the grains (i.e., $P(B \in X)$) we have

$$C_{11}(h) = 1 - 2q + C_{00}(h) = 1 - 2q + q^2 e^{\theta K(h)}.$$

γ. *Law of the first contact.* Suppose a random point x in the pores is chosen uniformly, and let R be its smallest distance to any grain. R is called the first contact distance; denoting by $F(r)$ the distribution function of the random variable R, we obtain

$$1 - F(r) = \frac{Q(Br)}{q}, \tag{5}$$

where B_r is the ball of radius r (or the disk in $\mathbf{R}^2$).

δ. *Specific surface and perimeter.* To avoid some pathological anomalies, such as fractal sets, we assume X' to be regular enough (a finite union of convex sets, for example). Then the specific surface of X, i.e., the surface S_V of ∂X per unit volume, in $\mathbf{R}^3$, and the specific perimeter U_A of X, in $\mathbf{R}^2$, are the derivative of

$Q(Br)$ in r for $r = 0$, that is,

$$S_V = \theta.\bar{S}(X').e^{-\theta\bar{V}(X')} \quad \text{in } \mathbf{R}^3, \qquad U_A = \theta.\bar{U}(X').e^{-\theta\bar{A}(X')} \quad \text{in } \mathbf{R}^2. \tag{6}$$

Before going on with the Boolean model for more specific cases, we now resume the analysis of the general properties suggested by this section.

2.* INFINITE DIVISIBILITY

The key notion we met in the preceding section is that of *infinite divisibility*. We say that a RACS X is infinitely divisible with respect to the union, if for any integer $n > 0$, X is equivalent to the union $\cup Y_i$ of n independent RACS Y_i, $i = 1 \ldots n$, equivalent to each other. This property depends only on the functional Q. More precisely, a function Q on $\mathcal{K}$ is associated with an infinitely divisible RACS without fixed points, if and only if there exists an alternating capacity of infinite order φ satisfying $\varphi(\phi) = 0$ and $Q(B) = \exp\{-\varphi(B)\}$ (Matheron [21]).

This theorem by Matheron opens the door to the Boolean sets, but not only to them. Imagine, for example, the union, in $\mathbf{R}^3$, of a Boolean set and trajectories of Brownian motions. The result is infinitely divisible, but not reducible to Boolean structures. For random variables, a classical theorem of P. Levy states that a infinitely divisible variable is a sum of independent Gaussian and Poisson variables. It seems that the equivalent of such a theorem does not exist for random sets, although the Boolean sets and the Poisson flats are in fact the major representatives of this type of models.

Beyond the infinite divisibility is another more demanding structure, namely, that of the RACS *stable with respect to union*. A RACS X belongs to this category when, for any integer n, a positive constant λ_n can be found, such that the union $X_1 \cup X_2 \ldots \cup X_n$ of n independent RACS equivalent to X is itself equivalent to $\lambda_n X$. Obviously a stable RACS is necessarily infinitely divisible, but the converse is false. The following theorem (Matheron [21]) characterizes stable RACS. A RACS X without fixed points is stable with respect to union if and only if its functional $Q(B) = \exp(-\varphi(B))$ for a capacity of infinite order φ satisfying $\varphi(\phi) = 0$ and homogeneous of degree $\alpha > 0$, i.e.,

$$\varphi(\lambda B) = \lambda^\alpha \varphi(B) \qquad (\lambda > 0, B \in \mathcal{K})$$

The Poisson lines (in $\mathbf{R}^2$) or planes (in $\mathbf{R}^3$), and the Brownian trajectories (in $\mathbf{R}^3$), are stable RACS. Indeed it results from their definition that the stable RACS are self-similar, and model the sets described by B. Mandelbrot [17].

3. CONVEX PRIMARY GRAINS

We go back to the Boolean thread of ideas. For the results we derived from (3) until now, we did not need to make explicit the quantity $\overline{\text{Mes}}\ (X' \oplus \check{B})$. If we want to go further in the analysis, we now must try to exploit it. There is a particular, but particularly important case, where $\overline{\text{Mes}}\ (X' \oplus B)$ admits a simple expression; it is when both X' and B are convex. Then Steiner's formula is applicable (J. Steiner (1840), reedited in Miles and Serra [28], Blaschke [3]) and provides several new fruitful results. Consequently, we now assume the compact random set X' to be almost surely convex. For the sake of simplicity we also assume that the various X' which constitute X are uniformly oriented. This latter hypothesis is not compulsory, but allows us to reduce the notation without substantially modifying the meaning of the results. (For more detailed expressions, see Matheron [19] and Serra [32].)

a. *Stereology*. Start from a Boolean set X_3 in $\mathbf{R}^3$, with a Poisson density θ_3, and primary grain X'_3. On any test plane Π (resp. line Δ), X_3 induces a Boolean set $X_2 = X_3 \cap \Pi$ of density θ_2 (resp. $X_1 = X_3 \cap \Delta$, density θ_1). The three densities are linked by the relationships

$$\theta_3 = \frac{2\pi\theta_2}{\bar{M}(X'_3)} = \frac{4\theta_1}{\bar{S}(X'_3)}. \tag{7}$$

The primary grains X'_2 and X'_1 induced on Π and Δ, respectively, are related to X'_3 by the following formulas:

$$\bar{V}(X'_3) = \frac{1}{2\pi}\cdot\bar{A}(X'_2)\cdot\bar{M}(X'_3) = \tfrac{1}{4}\bar{L}(X'_1)\cdot\bar{S}(X'_3), \tag{8}$$

$$\bar{S}(X'_3) = \frac{2}{\pi^2}\cdot\bar{U}(X'_2)\cdot\bar{M}(X'_3). \tag{9}$$

(Formulas (7), (8), (9) can be generalized to the nonconvex case; Serra [32].)

b. *Convex erosions*. Consider a 2-D Boolean set X_2, with parameters (θ_2, X'_2), and take for B's, first the segments of length l and second the families λB of sets similar to a given plane convex set B, having nonzero area. According to Steiner's formula (Minkowski [29], Watson [37]) we must write

$$Q(l) = \exp(-\theta_2[\bar{A}(X'_2) + (l/\pi)\bar{U}(X'_2)]), \tag{10}$$

$$Q(\lambda B) = \exp(-\theta_2[\bar{A}(X'_2) + (\lambda/2\pi)U(B)\bar{U}(X'_2) + \lambda^2 A(B)]). \tag{11}$$

We immediately derive from (10) the pore traverse length distribution $G(l)$,

$$1 - G(l) = \frac{Q'(l)}{Q'(0)} = \exp(-\theta_2(\bar{U}(X'_2)/\pi)\cdot l),$$

and from (11) and (5), the law $F(r)$ of the first contact;

$$1 - F(r) = \exp(-\theta_2[r\bar{U}(X') + \pi r^2]).$$

Now we can perfectly well consider X_2 as the plane cross section of a Boolean set X_3 in space. What knowledge of X_3 can we get from X_2? Obviously not an exhaustive one (take, for example, the union of X_3 and a 3-D Poisson point process Z; X_3 and $X_3 \cup Z$ induce the same X_2). However, using (7), (8), and (9), we can interpret the functionals (10) and (11) in terms of $\mathbf{R}^3$, i.e.,

$$Q(l) = \exp(-\theta_3[\bar{V}(X'_3) + (l/4)\bar{S}(X'_3)]), \tag{12}$$

$$Q(\lambda B) = \exp(-\theta_3[\bar{V}(X'_3) + (\lambda/8)U(B)\bar{S}(X'_3) + (\lambda^2/2\pi)A(B)\bar{M}(X'_3)]). \tag{13}$$

Comparison of (11) and (13) is instructive. In $\mathbf{R}^2$ we see that when B describes the class of compact convex sets, $Q(B)$ involves only three parameters of X, namely θ_2, $\bar{A}(X'_2)$, and $\bar{U}(X'_2)$, which can easily be estimated from (11). However, these three parameters do not exhaustively characterize X, since B only spans the class of compact *convex* sets. In other words, the morphological notions related to the convexity of B, such as size distributions, can easily be studied, but others, angularities, for example, cannot. In $\mathbf{R}^3$ we have a similar result, and X is described by the four parameters θ_3, $\bar{V}(X')$, $\bar{S}(X')$, and $\bar{M}(X')$. Unfortunately, if the structuring element B moved in $\mathbf{R}^3$ is planar, it does not allow us to access all the parameters, but only their linear combinations given by $\log Q(\lambda B)$ in (13), which provide only the three terms $\theta_3\bar{V}(X'_3)$, $\theta_3\bar{S}(X'_3)$, and $\theta_3\bar{M}(X'_3)$. In order to go further, we can alternatively specify the primary grain more completely as we will do in Section 5, or calculate the two-dimensional $Q(\lambda B)$ for the projection of the thick sections of X_3, as defined in Section 2.a.γ. For example, if d is the thickness of the section, and B is a polygon with n sides, we have:

$$Q(\lambda B) = \exp\left(-\theta_3 \left\{ \bar{V}(X'_3) + \frac{\lambda}{8}[U(B) + d(n-2)]\bar{S}(X'_3) + \frac{\lambda^2}{4}[2A(B) + dU(B)]\bar{M}(X'_3) + \lambda^3 dA(B) \right\} \right) \tag{14}$$

which enables us to estimate the four characteristics of X_3 (Serra [32]; see also Miles [25]).

3.* SEMI-MARKOV RACS

The underlying concept which governs Boolean textures with convex primary grains is the semi-Markov property. As soon as the chord distributions for the pores are exponential (10) we can suspect the model to be somehow Markov. Nevertheless the grain chords are *not* exponential, and thus X never induces a 1-D Markov process on any linear cross section. Such an asymmetry leads us to the semi-Markov property (Matheron [19, 24]). We say that two compact sets B and B' are *separated* by a compact set C if any segment (x, x') joining a point $x \in B$ to a point $x' \in B'$ hits C. Then a RACS X is said to be semi-Markov if for any B and B' separated by C the RACS $X \cap B$ and $X \cap B'$ are conditionally independent, given $C \cap X = \phi$. It can be shown that X is semi-Markov if and only if its functional Q satisfies the relationship

$$Q(B \cup B' \cup C)\cdot Q(C) = Q(B \cup C)\cdot Q(B' \cup C), \tag{15}$$

where B and B' are separated by C. In particular, if the union $B \cup B'$ is compact and convex, then they are separated by their intersection $B \cap B'$ and the semi-Markov property implies

$$Q(B \cup B')\cdot Q(B \cap B') = Q(B)\cdot Q(B'). \tag{16}$$

Moreover, if X is indefinitely divisible, the converse is true, and (16) implies that X is semi-Markov. In particular, if X is stationary and isotropic (i.e., Q

invariant under translations and rotations), a classical theorem of integral geometry (Hadwiger [11]) shows that $\log Q(B)$, for B a convex compact set, is given by the formula

$$\log Q(B) = \bigcap_{i=0}^{n} \beta_i w_i(B),$$

where the β_i are nonnegative coefficients and where $w_i(B)$ denotes the Minkowski functional $n^0 i$ of B. Remember that the w_i's are proportional, in $\mathbf{R}^2$, to the area, the perimeter, and 1, and in $\mathbf{R}^3$ to the volume, the surface, the norm, and 1; that explains the structure of (10)–(14), and shows that the Boolean model with primary grains is semi-Markov. One can rediscover this result intuitively in the following way. Intersect X by the straight line Δ, and pick up one point x in the pores of $X \cap \Delta$. x separates Δ into two half-lines Δ_1 and Δ_2. A primary grain X' of X which hits Δ_1 cannot hit Δ_2 since X' is convex; moreover the various primary grains are independently located, thus any event on $X \cap \Delta_1$ is independent of any other one on $X \cap \Delta_2$, given $x \in X^c$. Q.E.D.

The characterization relation (16) points out that it is interesting to combine the three properties of stationarity, infinite divisibility, and semi-Markovianness. Indeed, according Matheron [21], any RACS of $\mathbf{R}^3$ satisfying these three conditions is equivalent to the union of three stationary independent RACS, $X_1 \ldots X_3$. X_1 is a Boolean set with convex grains; X_2 and X_3 are the unions of cylinders the bases of which are 2-D and 1-D Boolean sets with convex grains. Note that if these two latter Boolean models have points as primary grains, the associated cylinders become Poisson lines and planes, respectively. In other words, the Poisson plane and line networks and the Boolean models are the two prototypes from which any stationary, infinitely divisible semi-Markov set may be constructed (the same comment applies to the RACS in $\mathbf{R}^2$). If now we replace infinite divisibility by the stability condition, then we restrict the possible class of sets. Indeed the only RACS which are stationary, stable, and semi-Markov are the Poisson flat networks (planes and lines in $\mathbf{R}^3$, lines in $\mathbf{R}^2$).

We should like to end this section by a remark on the use, and misuse, of Markov processes in picture processing. The fundamental difference between the line and the plane, or the space, is that the first one is ordered, and the other ones are not. Markov chains and processes are strongly based on this notion of an order relation (the future is independent of the past, given conditions x, y, etc.). Of course, one can define order relations in $\mathbf{R}^2$ (TV scanning, for example), but unfortunately there are too many ways to do this, and all of them give priority to particular directions. Moreover, when using the formalism of Markov chains (instead of continuous processes) for a phenomenon which in fact spreads out in the continuous space, one imposes a digital spacing, a, which parasitizes the information. (If two successive pixels are independent for a spacing a, are they still independent when $a' = a/2$?). Finally, the fact that the linear sections of a picture are Markov does not imply anything about the "Markov" structure of the 2-D phenomenon. One can easily construct counterexamples (Serra [32, Chap. IX]) of plane textures whose intersection by any line is Markov and which nevertheless exhibit splendid clusters of particles in $\mathbf{R}^2$!

On the contrary, the semi-Markov concept does not require any order relation in $\mathbf{R}^2$, and thus is isotropic. Defined in the continuous space, it can lend itself to digital sampling, and defined in $\mathbf{R}^2$ or $\mathbf{R}^3$, it is connected with spatial textures directly, and not via line sections. Its natural field of application is textures presenting a limited number of phases, which can thus be modeled by sets. In this frame, it presents more flexibility than the classical Markov concept, since one phase only is required to be Markov.

4. CONNECTIVITY NUMBER

We now come back to the Boolean model in $\mathbf{R}^2$, still assuming X' to be isotropic and convex. Several parameters concerning connectivity can be defined, such as the number of connected components (Roach [30], Ahuja [1]). The most interesting one is probably the specific connectivity number $N_A(X)$, also called the Euler–Poincaré constant. It has the double advantage of being a local parameter (i.e., statistically accessible from elementary square or triangle samples) and having a stereological interpretation: if X_3 is a stationary isotropic RACS (not necessarily Boolean) and X_2 a plane section of X_3, then $2\pi N_A(X_2)$ equals the sum of the mean curvature of boundary ∂X_3 per unit volume. Morphologically speaking, $N_A(X)$ is nothing but the number of connected components of X per unit volume, minus their holes.

A direct proof based on the normal rotation along the boundary ∂X leads to the following result (Serra [33], Miles [25]):

$$N_A = \theta e^{-\theta \bar{A}(X')} \left[1 - \frac{\theta}{4\pi} \bar{U}^2(X') \right]. \tag{17}$$

In practice, we work on digital images. In order to establish the link with $\mathbf{R}^2$, we assume that a regular square or hexagonal lattice of points $\mathcal{C}(a)$ with spacing a is superimposed on X_2, generating the binary digital image $X \cap \mathcal{C}(a)$ (Fig. 4).

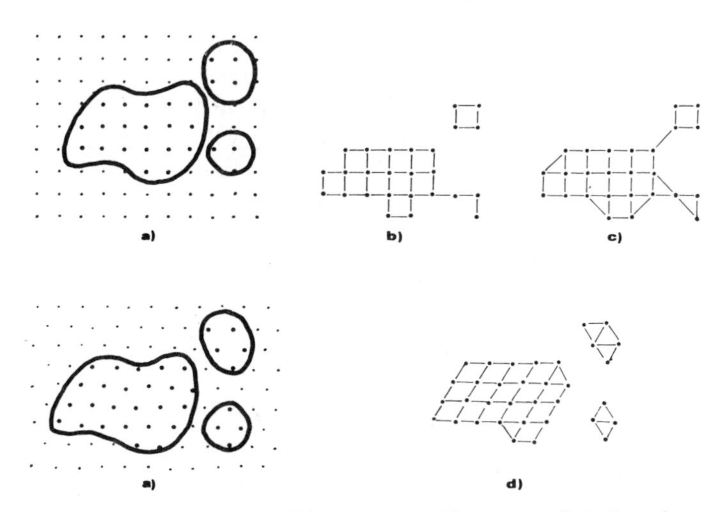

FIG. 4. A Euclidean set and its square and hexagonal digital versions.

Hoping that for a small enough spacing, a, the particles and the holes of X coincide with those of $X \cap \mathcal{C}(a)$, we calculate the latter using the notorious Euler relation for planar graphs. Denoting by

$$\begin{matrix} 0 & 0 \\ 0 & 0 \end{matrix}\,, \quad \begin{matrix} & 0 & \\ 0 & & 0 \end{matrix}\,, \quad 0\ 0,$$

etc., the elementary squares, triangles, edges, etc. made up of consecutive points of the pores of $X \cap \mathcal{C}(a)$, one can prove that the specific connectivity number is

— For the square lattice [10, 31]

$$N'_A = -\lim_{a \to 0} \frac{Q\left\{\begin{matrix} 0 & 0 \\ 0 & 0 \end{matrix}\right\} - Q\left\{0\ 0\right\} - Q\left\{\begin{matrix} 0 \\ 0 \end{matrix}\right\} - q}{a^2}.$$

— And for the hexagonal lattice [32]

$$N''_A = -\lim_{a \to 0} \frac{Q\left\{\begin{matrix} 0 & & 0 \\ & 0 & \end{matrix}\right\} + Q\left\{\begin{matrix} & 0 & \\ 0 & & 0 \end{matrix}\right\} - Q\left\{0\ 0\right\} - Q\left\{\begin{matrix} & 0 \\ 0 & \end{matrix}\right\} - Q\left\{\begin{matrix} 0 & \\ & 0 \end{matrix}\right\} + q}{a^2 (3/2)^{\frac{1}{2}}}.$$

Using the fundamental relation (3) these become

$$N'_A = \theta e^{-\theta \bar{A}(X')} \left[1 - \frac{\theta}{\pi^2} \bar{U}^2(X')\right], \tag{18}$$

$$N''_A = \theta e^{-\theta \bar{A}(X')} \left[1 - \frac{\theta 3^{\frac{1}{2}}}{2\pi^2} \bar{U}^2(X')\right]. \tag{19}$$

Strangely, the three numbers N_A, N'_A, and N''_A, which we could hope to be identical, at least at the limit, are different no matter how small a is. Indeed the differences are not negligible at all; for example, in the usual case where X' is a disk with a fixed radius r, we can express N_A, N'_A, and N''_A as functions of the porosity $q = e^{-\theta \pi r^2}$, i.e.,

$$\frac{N_A}{\theta} = q(1 + \log q), \qquad \frac{N'_A}{\theta} = q\left(1 + \frac{4}{\pi} \log q\right),$$

$$\frac{N''_A}{\theta} = q\left(1 + \frac{2(3)^{\frac{1}{2}}}{\pi} \log q\right). \tag{20}$$

For a very common value of q, say 0.5, we get

$$\frac{N_A}{\theta} = 0.153, \qquad \frac{N'_A}{\theta} = 0.059, \qquad \frac{N''_A}{\theta} = 0.118,$$

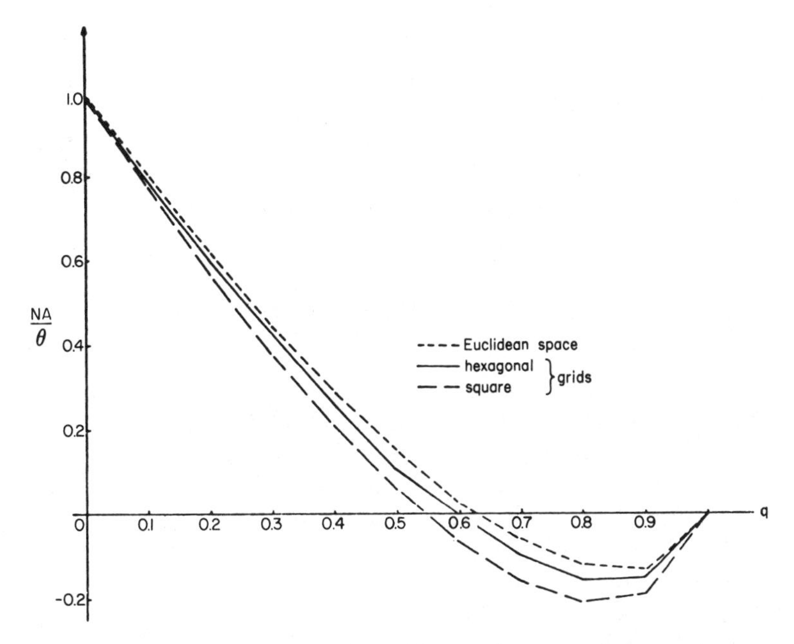

FIG. 5. Digitization causes a lattice-dependent bias on the connectivity number measurement.

which is not encouraging at all (the square lattice is particularly disastrous). We have plotted the three functions (20) in Fig. 5. Extrapolating beyond the disk to other convex shapes, we learn from these graphs that it is wise to try and estimate N_A only when the porosity q is rather large (although it is wiser still not to directly estimate N_A at all, but rather to calculate it from (17), after having estimated θ, $\bar{A}(X')$, and $\bar{U}^2(X')$ from (11)).

The same trouble also arises in $\mathbf{R}^3$. The Euler–Poincaré constant

$$N_V(X) = \theta_3 e^{-\theta_3 V(X')} \left\{1 - \frac{\theta_3}{4\pi} \bar{M}(X') \cdot \bar{S}(X') \cdot + \frac{\pi^2 \theta_3{}^2}{96} \bar{S}^3(X')\right\} \tag{21}$$

(Miles [25]) does not coincide with its digital versions (Serra [32]). Here again, the right way to evaluate $N_V(X)$, is by estimating the four characteristic parameters from (14) (if thick sections are available) and using these estimates in (21).

4.* DIGITIZATION

In Section 4, the estimation of $N_A(X)$ confronted us with the ticklish problem of digitization. More generally, we shall access the functional moments $Q(B)$ via digital versions of the realization of X. Are we sure that the $Q(B)$'s (defined in $\mathbf{R}^2$) are digitally accessible, and that we do not run the risk met with N_A, N'_A, and N''_A? The two following general theorems, due to Serra [32], answer the question. (In this section, X no longer denotes a random set, but its actual realization, on which the digitization holds.)

a. Connectivity and digitization

Consider the intersection $X \cap \mathcal{C}(a)$ of a 2-D closed set X with a square lattice $\mathcal{C}(a)$ of spacing a. It is always possible to construct several planar graphs on $X \cap \mathcal{C}(a)$, i.e., to provide the set of points. $X \cap \mathcal{C}(a)$, called vertices, with a collection D of edges which do not cross each other. For example, one can define as an edge each pair of consecutive 1's either in the horizontal or in the vertical direction

$$\left(1 \quad 1 \rightarrow 1\text{–}1; \quad \begin{matrix} 1 \\ \\ 1 \end{matrix} \rightarrow \begin{matrix} 1 \\ | \\ 1 \end{matrix}\right);$$

see Fig. 4.b. Figure 4.c shows another possible and more sophisticated graph. Note that a similar treatment may be applied to the digital background $X^c \cap \mathcal{C}(a)$, for transforming it into a graph, and also to the hexagonal lattice intersections $X \cap \mathcal{C}(a)$ of Fig. 4.d. On all these graphs, one can easily define what a connected component is, and what a hole inside a connected component is (see [31] for an exhaustive study of the graphs derived from the square lattice, and [32] for the hexagonal case). We denote by $\hat{X}(a)$ and $\hat{X}^c{}_{(a)}$ the planar graphs associated with $X \cap \mathcal{C}(a)$ and $X^c \cap \mathcal{C}(a)$ in all three cases of Fig. 4.

Given a 2-D closed set X, is it possible to find a lattice spacing a that is small enough to reach all the particles and all the holes of X? In other words, can we establish a one-to-one and onto correspondence between the particles of X and of $\hat{X}(a)$, and between the particles of X^c and of $\hat{X}^c(a)$, for every position and orientation of the lattice $\mathcal{C}(a)$ in the plane $\mathbf{R}^2$? One can show that such an isomorphism does exist if and only if Set X is opened and closed by a disk $B(a2^{\frac{1}{2}})$ of radius $a2^{\frac{1}{2}}$ in square lattices, or $B(a)$ of radius a in hexagonal lattices

$$X = X_{B(a2^{\frac{1}{2}})} = X^{B(a2^{\frac{1}{2}})} \qquad \text{square lattice,}$$

$$X = X_{B(a)} = X^{B(a)} \qquad \text{hexagonal lattice.}$$

Remarks: α. If so, no particles and no holes are created by dilating and by eroding X by the disk $B(a2^{\frac{1}{2}})$ (square case) or $B(a)$ (hexagonal case).

β. The conditions of the above theorem are rather severe, and exclude, among others, all the sets presenting angularities, such as the realizations of Boolean models.

γ. The theorem not only governs connectivity number estimation, but more generally, gives the conditions under which the digital algorithms based on connectivity (skeletons, etc.) possess a Euclidean meaning.

δ. The theorem generalizes, to 2-D sets, the well-known result due to Shannon that the highest harmonic detectable in a signal has a frequency equal to one half the sampling frequency.

ϵ. Given a lattice spacing a, the hexagonal lattice tolerates a weaker hypothesis on the structure of the object than does the square lattice (or equivalently, it tolerates a larger spacing, given the same hypothesis).

b. Digitizable Transformations for Closed Sets

We have just restricted the class of sets X by imposing a condition on the transformation (i.e., to respect connectivity). We now change our approach, and wonder what the digitizable transformations acting on the largest possible set class (i.e., on the closed sets) are?

Associate with the square lattice $\mathcal{C}(a)$ the mosaïc of closed squares of side a centered at each point of the lattice. Let X be the closed set under study and B be a compact structuring element. We adopt for digital versions of X and B their smallest coverings $\hat{X}$ and $\hat{B}$ by squares of the mosaïc (Freeman coding). The digital dilation of $\hat{X}$ by $\hat{B}$ generates a set of $\mathbf{R}^2$ which tends toward $X \oplus B$ when $a \rightarrow 0$. Now, any mapping Ψ from the closed sets of $\mathbf{R}^2$ into themselves, which is increasing, u.s.c. and translation invariant is intersection of dilations, and thus is digitizable for the Freeman coding.

Remarks: α. This theorem does not generate all possible digitizable transformations on the class of the closed sets; nevertheless, it opens the door to the analysis of erosions, dilations, openings, closings, size distributions, and gray tone functions.

β. The theorem is also valid for the mapping which takes the set of closed sets onto a closed part of $\mathbf{R}^+$; thus the functional moments $Q(B)$ are digitizable.

γ. We might wonder whether a similar theorem is possible if we replace the increasing condition and the semi-continuity condition with one of continuity. Unfortunately, it is not, as is well known by experimenters who have tried to digitize the rotations of $\mathbf{R}^2$.

5. SPECIFIC BOOLEAN MODELS

Coming back to the Boolean model, we now present three more specific models, which rather often appear in applications. Note that in all these cases the 3-D density is accessible from 2-D sections.

a. Isovolume Primary Grains

Let X be a Boolean model of density θ_3 and primary grain X' such that $V(X')$ is constant and not random. Then from the pore covariance $C_{00}(h)$, accessible from line sections, we get

$$\log C_{00}(0) = \log q = \theta_3 V(X'),$$

$$\frac{1}{q^2}\int_{R^3} \log C_{00}(h)dh = \theta_3 V^2(X'),$$

which allows us to estimate θ_3 and $V(X')$. One can also test the hypothesis "$V(X')$ is constant" from more sophisticated measurements performed on 2-D sections [32].

b. Spherical Primary Grains

The 3-D primary grains are balls. Denote the distribution function of their radii by $F_3(r)$. The model is completely determined by the covariance $C_{00}(h)$.

We have

$$\frac{1}{2\pi r}\,[\log C_{00}(2r)]^{(ii)} = \theta_3[1 - F_3(r)] \tag{22}$$

and in particular when $r \to 0$

$$\frac{1}{\pi}\,[\log C_{00}(2r)]_{r=0}{}^{(iii)} = \theta_3$$

[Using the left-hand side to estimate θ_3 would result in a large estimation variance, so it seems wiser to open the model with a small ball of radius r_0, i.e., essentially replace θ_3 by $\theta_3[1 - F_3(r_0)]$ given by (22).]

c. *Poisson Polyhedron Primary Grains*

In a partition of the space by isotropic Poisson plane networks, pick out the polyhedron containing the origin, and take it as a primary grain X'. This compact random set depends on one parameter only, namely, the intensity ρ of the Poisson planes (Miles [26, 27]). The geometric covariogram $K(h)$ of X' is given by the formula

$$K(h) = \bar{V}(X \cap X_h) = \frac{6}{\pi^4\lambda^3}\,e^{-\pi\rho|h|}. \tag{23}$$

The two basic quantities used for estimation are $Q(l)$ of (10) and the covariance $C_{00}(h)$. Putting $\theta' = 6\theta/\pi^4\rho^3$, we obtain

$$Q(l) = qe^{-\pi\rho\theta' l} \qquad \text{with} \qquad q = e^{-\theta'}, \tag{24}$$

$$C_{00}(h) = q^2 e^{\theta'} e^{-\pi\rho|h|} = q^2 e^{\theta K(h)} \tag{25}$$

which allows us to estimate the two parameters θ and ρ of the model.

5.* ESTIMATION PROBLEMS

We do not claim to present here a review of estimation theory for random sets, since this theory does not (yet) exist. Rather, we shall devote this aspect of the counterpoint to one example, followed step by step.

The Boolean model with Poisson polyhedra was first proposed by Delfiner [7], and used to model porous media for the petroleum industry. More recently, Celeski, Jeulin and Schneider [4] successfully developed it as a model for mineralization associated with various coking coals. Curiously, in both cases, the primary "grains" represent the physical *pores* of the medium, as if the overlapping condition of the Boolean model could more easily be satisfied by air bubbles than by solid materials.

Consider, for example, the photograph of coke in Fig. 6, which is meant to be representative of a certain type of coke. Twenty-five similar fields were measured per section on each of twenty-one polished sections of this type of coke. The purpose of the study is to find condensed but rather exhaustive descriptors of

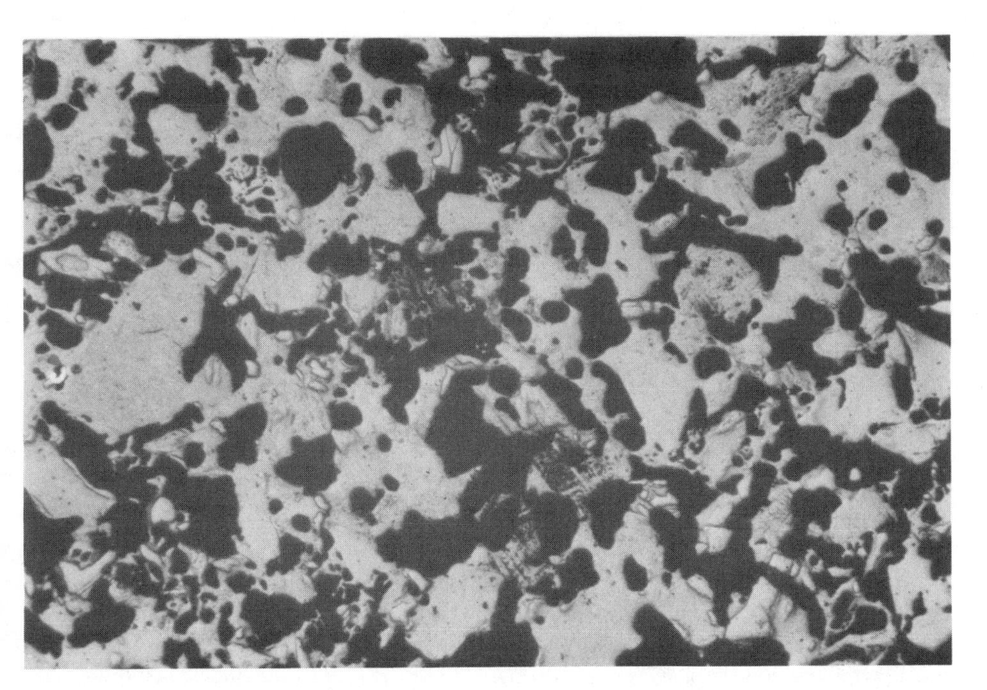

FIG. 6. Micrograph from a polished section of form coke. The carbon appears as white and the pores as dark. The latter will be considered as the grains when modeled by a Boolean set.

the structure (in order to follow how the coke changes when its components and its cooking conditions are modified). The following procedure describes a typical approach for estimating the Boolean models.

a. Using the Texture Analyzer (Serra and Klein [34]; Wild–Leitz Ltd., 1973) or a similar image analyzer, threshold the image into two complementary sets and compute estimates $Q^*(l)$ of the functional $Q(l)$ for the two complementary phases of the structure under study (or more, for multiphase structures). The edge effects must be corrected for by a suitable formula of Mathematical Morphology. Plot the quantities $\log Q^*(l)$ versus l, as in Fig. 7a.

b. Measure $Q(l)$ in different directions for detecting eventual anisotropies. In the present case, the structure is isotropic (i.e., $Q(B)$ does not depend on the orientation of B), and this has been tested for B's pairs of points, segments, and regular hexagons. When the structure is not isotropic, one must take the rotation average of $Q(B)$ for each B in order to apply (10)–(14). At the same time, the anisotropies will be described by applying techniques such as the rose of directions [32].

c. By means of geostatistics [23, 13], one can calculate the number of sections, and fields per section, in order to get a given estimation variance for the $Q(B)$'s.

d. The $Q^*(l)$ of the coke, in step a, exhibits a negative exponential behavior. One can suspect a Boolean structure with convex primary grains. Control this assumption by estimating $Q(\lambda B)$ of (11) for B a square or hexagon (here the

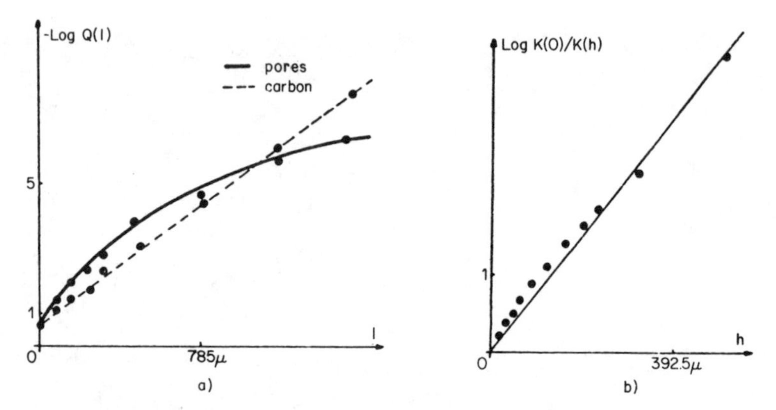

FIG. 7. (a) log of the linear erosion estimates for the carbon and the pores. (b) log of the geometric covariogram estimate, for the pore primary grain.

measurements were performed with B a hexagon, giving a parabolic log $Q^*(\lambda B)$ which corroborates the statement).

e. If we want to be rigorous, the next step is to check the semi-Markov and the infinite divisibility properties. The semi-Markov structure has already been shown, but not proved, by the exponential $Q^*(l)$ of the coke. To go further, it would be necessary to test whether (16) is true, for various shapes of B and B' (not realized in the present case). One checks the infinite divisibility by storing two disjoint fields of analysis and forming the union of the pore sets. On this union, compute log $Q^*(l)$ and log $Q^*(\lambda B)$ and check that they are still a straight line and a parabola, respectively (but with different parameters than formerly). This control was not carried out in the present case, and we assume that it would have been positive. If so, according to the Matheron's characterization theorem (Section 3*), the structure can be fitted with a union of Boolean sets with convex primary grains, and of cylinders. If they are cylinders, the ranges of the covariance, i.e., the quantities

$$\int_{\mathbf{R}^2} |h|^n C(|h|) d|h|,$$

are infinite for $n > 1$. This is obviously not the case here (see Fig. 7b). Therefore, the Boolean model with convex primary grains is acceptable.

f. It remains to estimate the characteristics of the primary grain X'. The next measurement is the covariance. Also, plotting the covariogram log $\theta K^*(h)$, estimated by the algorithm log [log $C^*(h)$ − 2 log q^*], we get the curve in Fig. 7b which is virtually a straight line. The primary grain is likely a Poisson polyhedron (the property is not characteristic but rather strong). We can confirm this assumption by noting that for a Poisson polyhedron X'

$$\bar{V}(X') = \frac{6}{\pi^4}(\rho)^{-3}, \qquad \bar{S}(X') = \frac{24}{\pi^3}(\rho)^{-2}, \qquad \bar{M}(X') = 3(\rho)^{-1}. \qquad (26)$$

The linear density ρ is estimated from the experimental data by a least-squares straight line fit to the points log $[K(0)/K(k)]^*$, and the parameter θ'(whence θ

follows immediatly) via two possible routes:

— from the experimental porosity, using $q = e^{-\theta'}$ and $\theta' = 6\theta/\pi^4\rho^3$.
— from the slope of $-\log^* Q(l)$, and (24).

The estimates θ^* and ρ^* and (26) provide estimates of $\theta\bar{V}$, $\theta\bar{S}$, and $\bar{M}$ that we can compare with those given by (13), fitted on the square or hexagonal erosion measurements $Q^*(\lambda B)$. The numerical results are the following: $\rho^* = 2.55\ \text{mm}^{-1}$, $\theta^* = 141\ \text{mm}^{-3}$ (from q^*), and $\theta^{**} = 144\ \text{mm}^{-3}$ (slope of $Q(l)$). We keep this second θ estimate, based on many more measurements than the other one. Then we have

Calculated	Parameters		
from	$\theta\bar{V}(X')$	$\theta\bar{S}(X')$	$\theta\bar{M}(X')$
ρ^* and θ^{**}	0.545	17.14 mm^{-1}	169 mm^{-2}
Q^* (λB) B hexagon	0.539	12.0 mm^{-1}	95 mm^{-2}

The Poisson polyhedron model is practically acceptable; as usual, the accuracy decreases with the dimension of the parameter.

g. *Remark.* In practice, one can say that a Boolean model with convex primary grain X' can be correctly estimated in general. This does not seem to be the case as soon as X' is no longer convex.

We end this section by briefly quoting a few studies where the Boolean model has successfully been estimated: Forest structures (Marbeau [16]); clay observed by scanning electron microscopy (Kolomenski and Koff [15]); dendritic metallic structures (Alberny *et al.* [2]).

6. DERIVED MODELS

The Boolean model is the seed for a considerable number of random sets. In what follows, we give only a short presentation of a few of them (more detailed information and examples may be found in Serra [32] and also in Jeulin [12]). We hope that these few derivations will suggest ideas to the reader for inventing his own models.

a. *Three-Phased Structures*

There are many ways of building up models for multiphased textures. Depending upon whether or not we wish to emphasize the dependence between two phases, we could use either the following metallographic or petrographic models.

α. Metallographic model. We developed this model in conjunction with Greco and Jeulin [9] in order to describe the morphology of sinter textures. It often happens, in metallography, that one type of crystal, say X_2, wipes out all others, which can survive only in places left by X_2 (i.e., in $X_2{}^c$). For example, assume that X_1 and X_2 are two independent Boolean sets, and that X_2 wipes out X_1. This leads to a three-phase texture X_2, X_3, X_4, with

$$X_2, \qquad X_3 = X_1 \cap X_2{}^c, \qquad X_4 = X_1{}^c \cap X_2{}^c = X_3{}^c \cap X_2{}^c.$$

If the primary grains X'_1 and X'_2 are convex the intercept histograms of $X_2{}^c$ and of X_3 should be negative exponentials (tests for the model result). An interesting piece of information for the three-phase structures is given by the co-occurrence matrix, which is equivalent here to the three covariances C_1, C_2, and C_3 of the three phases that are present. We have

$$C_2(h) = C_2(h) \qquad \text{(unchanged)},$$
$$C_3(h) = C_1(h)[1 - 2q_2 + C_2(h)],$$
$$C_4(h) = [1 - 2q_1 + C_1(h)][1 - 2q_2 + C_2(h)].$$

Figure 8a shows an example of such a structure.

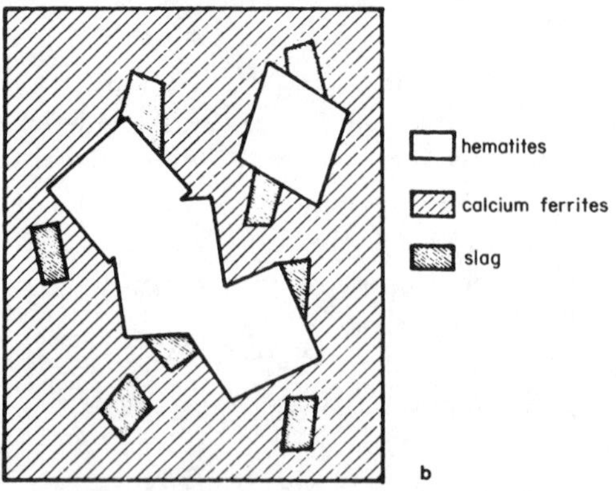

FIG. 8. (a) Plate-like calcium ferrite (gray) in contact with hematite (white) and with slag (dark). (b) Simulation of the corresponding model.

β. Petrographic model. In this second example, we emphasize the spatial distribution of one phase around the other. In cooperation with J. C. Fies, we studied the organization of clay in synthetic soils. The clay partly surrounds the quartz grains, and partly spreads out in the pores, looking like small spots. A Boolean model is chosen, whose primary grain is

— with probability p, a ball of quartz of radius r_1 surrounded by a spherical crown of clay of thickness $r_2 - 2$, which we call contour clay,
— with probability $1 - p$, a ball of clay of radius r'_2, which we call isolated clay.

We Booleanize and rule that when quartz and clay overlap, the quartz obliterates the clay. As in the former case, two phases exhibit intercept histograms which are negative exponential (the pores and the pore–clay union). The cooccurrence matrix is easily calculable, and constitutes the main piece of information on the model. The 3-D characteristics p, r_1, r_2, and r'_1 are derived from these covariances.

b. Dead Leaves Model

This other variation of the Boolean model that we now present is due to Matheron [22]. It has the dual advantage that it provides us with a tessellation of the space, as well as a model for nonoverlapping particles.

Dead leaves partition. When one looks at a cloudy sky, one only sees the lowest clouds, which hide those above. The dead leaves model is just a quantitative description of this type of superposition. Although the basic relation (27) is independent of the dimension of the space, we will take our realizations to be in $\mathbf{R}^2$ (Fig. 9a), and develop the model in this space.

Let X' be a primary grain in $\mathbf{R}^2$. Place the origin of time t in the distant past, at $-\infty$. From the origin to the present (i.e., $t = 0$) take the realization of Boolean sets, independent of each other, with identically distributed primary grains X' at time instants given by a constant spatial Poisson density θdt. The grains appearing between $-t$ and $-t + dt$ *hide* the portions of the former grains. At time

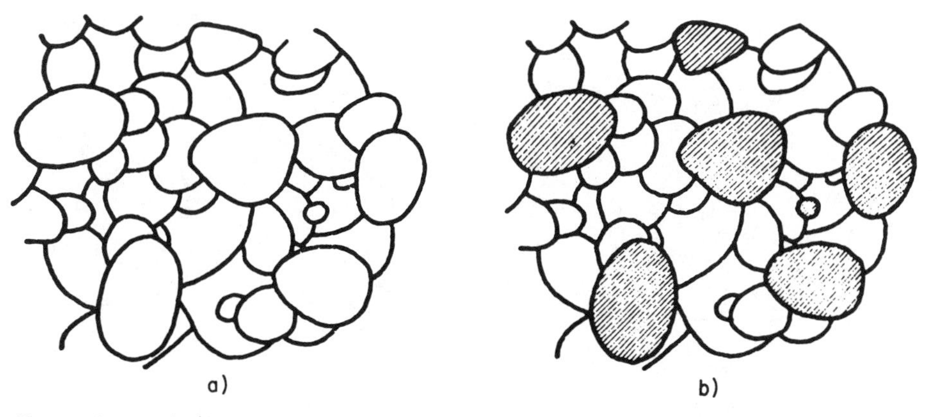

FIG. 9. (a) Realization of the dead leaves model. The remaining boundaries make up a closed set X. (b) The relief grains of Fig. 9a.

zero the plane is completely covered (since the origin of t is $-\infty$), giving a *random stationary partition* of the space. The probability $Q(B)$ that a given compact set B is included in one class of the partition is

$$Q(B) = \frac{\overline{\text{Mes}}\,(X' \ominus \check{B})}{\overline{\text{Mes}}\,(X' \oplus \check{B})}. \tag{27}$$

This relationship completely characterizes the partition, and opens the way to the classical measurements (size distribution, convex erosion, covariance, etc.).

Relief grains. From now on, the assumption of connexity of X' is very essential; however, our restriction to a 2-D isotropic analysis is only for pedagogical reasons. Imagine that the observer is now able to decide whether or not a given class of the final partition corresponds to an *entire* primary grain X_1 that has not been partially obscured by another (Fig. 9b). The union of all these classes generates a stationary random set X made up of disjoint compact convex grains.

Denote by A' and U' the random area and perimeter of the primary grain X', let $F(dA', dU')$ be the law of these two variables, and $\bar{\omega}(B; A', U')$ be the probability that $B \subset X'$, where X' is a primary grain with functionals A' and U' and located at the origin of $\mathbf{R}^2$. If B is a *connected* compact set (and only in this case), we have

$$Q(B) = \int_{\mathbf{R}^2} dx \int \frac{\bar{\omega}(B_{-x}; A', U')F(dA', dU')}{A' + \bar{A}' + U'\bar{U}'/2\pi}.$$

From this relation, we can derive the specific number N_A and the distribution $F_1(dA_1, dU_1)$ of the areas A_1 and the perimeters U_1 of the relief grains X_1:

$$N_A = \int \frac{F(dA', dU')}{A' + \bar{A}' + U'\bar{U}'/2\pi}, \qquad N_A F_1(dA_1, dU_1) = \frac{F(dA', dU')}{A' + \bar{A}' + U'\bar{U}'/2\pi}.$$

We see in the denominator the bias affecting the size distributions of the areas and perimeters; i.e., large values appear with a lower frequency in the relief grains X_1 than in the primary grains X'.

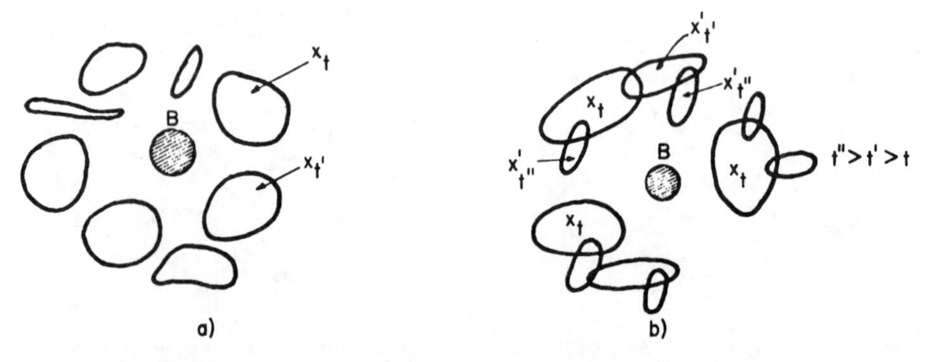

FIG. 10. Two types of hierarchical modes: (a) Separate grains; (b) Gypsum crystal model.

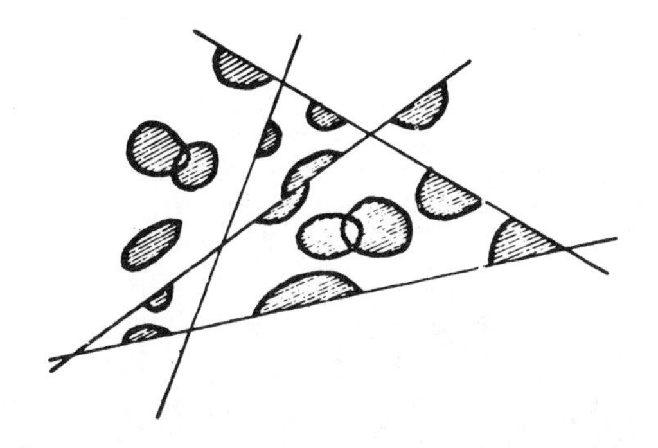

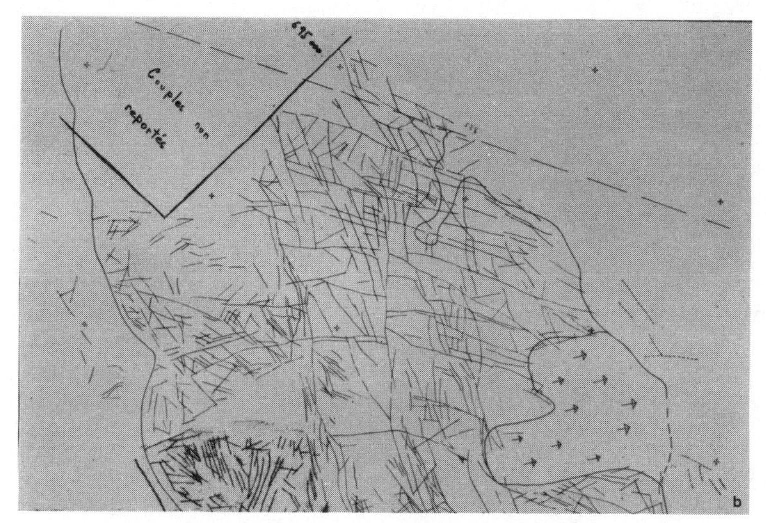

FIG. 11. Conrad and Jacquin's model, examples: (a) Simulation; (b) Real case.

c. *Hierarchical Models*

By hierarchical models (Serra [33]) we mean the following. During the time interval $(0, dt)$ we generate an initial Boolean model with density θdt and primary grain X'_0; at the instant dt we generate a second elementary process that will interact with the first, etc. The resulting set at time t is denoted by X_t. In both particular cases that we present below (α and β), the primary grains are assumed to be deterministic sets.

α. Any primary grain X'_{t+dt} appearing during the interval $[t, t + dt]$ is suppressed if it touches X_t; this will generate a random set with separate grains (Fig. 10a). Then, denoting by $Q_t(B)$ the probability that $B \subset X_t^c$, and by $Q'_t(B)$ its derivative with respect to t, we have

$$Q'_t(B) = -\theta \operatorname{Mes} (B \oplus \check{X}'_t) \cdot Q_t(X'_t). \tag{28}$$

β. Each primary grain X'_{t+dt} that does not touch X_t is suppressed. If it does touch X_t, we take its union with X_t for constructing X_{t+dt}. This will generate a random set with clusters of interpenetrating crystals, as in a gypsum crystal, for example. We now have

$$Q'_t(B) = -\theta \operatorname{Mes}(B \oplus \check{X}'_t)[1 - Q_t(X'_t)]. \tag{29}$$

The integral equations (28) and (29) cannot be solved in general, but only computed by numerical means. However, when the number of steps of the hierarchy is finite, they can effectively be calculated. One particularly interesting case happens when there are only two steps. Starting from a first Boolean model (θ_1, X'_1) we generate a second one (θ_2, X'_2) and allow them to interact in one of the two ways described above. The "separate grains" model now becomes a three-phased model (X_1, X_2, and the background) where X_1 and X_2 are *disjoint* from each other. The probability $Q(B)$ is

$$Q(B) = \exp\{-\theta_1 \overline{\operatorname{Mes}}(\check{X}'_1 \oplus B) - \theta_2 \overline{\operatorname{Mes}}(\check{X}'_2 \oplus B) \cdot \exp[-\theta_1 \overline{\operatorname{Mes}}(\check{X}'_1 \oplus X'_2)]\}.$$

(For the gypsum crystal type model, replace exp[ ] by 1 − exp[ ].

d. Boole-Poisson Model

Although Conrad and Jacquin [6] constructed the following model (Fig. 11) to describe aerial photographs of geological cracks, by using Poisson polygons and Boolean fissures, his "mixed" random set is extremely general. For simplicity, we will present it as it was originally. Start from an anisotropic Poisson line process in $\mathbf{R}^2$, with density ρ, and denote it as X_1. Intersect each polygon Π of the partition with a stationary random set X_2 in the following special way. For

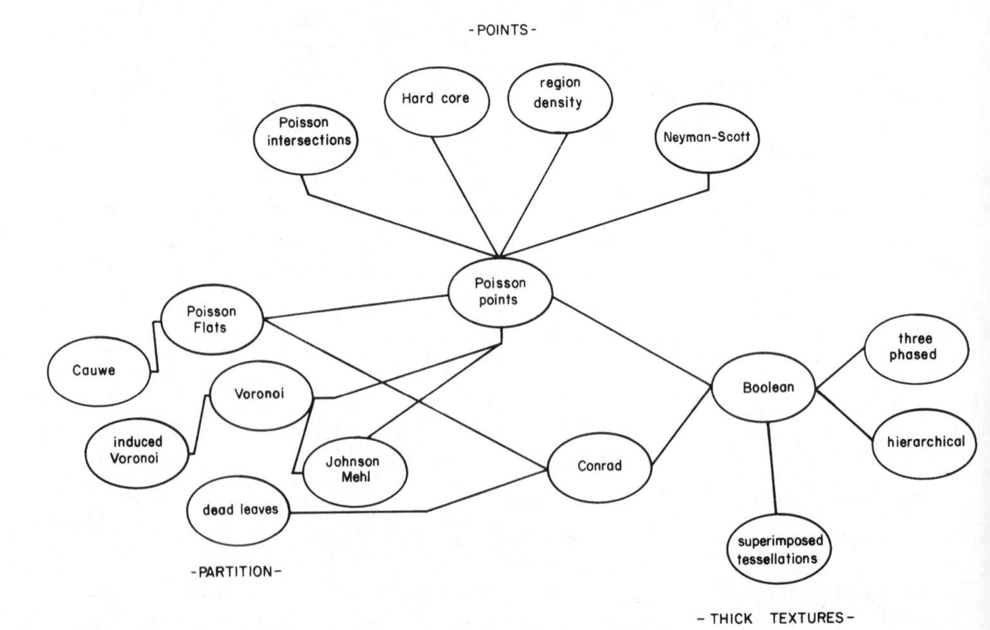

FIG. 12. The rose of the models.

each polygon of the given realization of X_1 use a *different* realization X_2 of X^*_2. The union of all these portions of X_2, together with X_1, makes up the set we shall call Y. If $Q_1(B)$, $Q_2(B)$, and $Q(B)$ denote the functional probabilities with X_1, X_2, and Y, respectively, we have

$$Q(B) = Q_1(B) \cdot Q_2(B) = \exp(-[\rho U(B) + \theta \bar{A}(X'_2 \oplus B)]),$$

where B is a *connected* compact set. Except for the covariance, all the basic classical measurements may be derived from this relation.

6.* THE ROSE OF MODELS

Will the reader accept that, as the ultimate step of our counterpoint, we propose a flower? Better than a long comment, it immediately visualizes (Fig. 12) the place of the Boolean model and its derivations among the main random sets, and their connections.

REFERENCES

1. N. Ahuja, Unpublished Ph.D. thesis, University of Maryland, 1979.
2. R. Alberny, J. Serra, and M. Turpin, Use of covariograms for dendrite arm spacing measurements, *Trans. AIME* **245**, 55–59 1969.
3. W. V. Blasckhe, Vorlesungen über integralgeometrie, Teubner, Leipzig, 1939.
4. J. C. Celeski, D. Jeulin, and M. Schneider, Etude quantitative du réseau poreux de trois cokes mulés, I.R.S.I.D. R.E. 433, 1977.
5. G. Choquet, Theory of capacities, *Ann. Inst. Fourier* **5**, 131–295, 1953.
6. F. Conrad and C. Jacquin, Representation d'un réseau bi-dimensionnel de failles par un modèle probabiliste. Int. report, Centre de Morphologie Mathématique, Fontainebleau, France, 1972.
7. P. Delfiner, Le schéma Booléen-Poissonien. Int. report, Centre de Morphologie Mathématique, Fontainebleau, France, 1970.
8. H. Giger and H. Hadwiger, Über Treff Zahlwahrscheinlichkeiten in Eikörperfeld, *Z. Wahrscheinlichkeitstheorie Verw. Geb.* **10**, 329–334, 1968.
9. A. Greco, D. Jeulin, and J. Serra, The use of the texture analyser to study sinter structure, *J. Micr.* **116**, pt. 2, 1979.
10. A. Haas, G. Matheron, and J. Serra, Morphologie mathématique et granulométries en place, Ann. Mines, XI 736–753 and XII 767–782, 1967.
11. H. Hadwiger, Vorslesungen über Inhalt, Oberfläche und Isoperimetrie, Springer, Berlin, 1957.
12. D. Jeulin, Multicomponent Random Models for the Description of Complex Microstructures, Fifth Congress for Stereology, Salzbourg, Austria, 1979.
13. A. Journel and Ch. Huijbregts, Mining Geostatistics, Academic Press, 1978.
14. D. G. Kendall, Foundations of a theory of random sets in stochastic geometry (E. F. Harding and D. G. Kendall, eds.), Wiley, 1974.
15. E. N. Kolomenski and G. L. Koff, in Ingenernaïa Gueloguia N° 3 (in Russian), 1979.
16. J. P. Marbeau, Etude structurale d'un peuplement en régénération. Int. report, Centre de Morphologie Mathématique, Fontainebleau, France, 1973.
17. B. B. Mandelbrot, Fractals: Form, chance, and dimension. W. H. Freeman, San Francisco-London, 1977.
18. B. Matern, Spatial variation. Medd. Statens skogsforskingsinstit, 1960.
19. G. Matheron, Eléments pour une théorie des milieux poreux, Masson, Paris, 1967.
20. G. Matheron, Théorie des ensembles aléatoires, Ecole des Mines, Paris, 1969.
21. G. Matheron, Random sets and integral geometry, Wiley, N.Y., 1975.
22. G. Matheron, Schéma booléen séquentiel de partitions aléatoires, Int. report, Centre de Morphologie Mathématique, Fontainebleau, France, 1968.

23. G. Matheron, Les variables régionalisées et leur estimation, Masson, Paris, 1965.
24. G. Matheron, Polyèdres poissoniens et ensembles semi-markoviens, Advances in Applied Prob., 1971.
25. R. E. Miles, On estimating aggregate and overall characteristics from thick sections by transmission microscopy, Fourth Congress for Stereology, NBS spec. pub. 431, 1976.
26. R. E. Miles, Random Polytopes, Ph.D. thesis, Cambridge, England, 1964.
27. R. E. Miles, Poisson Flats in Euclidean Spaces, Part. I, *Adv. Appl. Prob.* **1**, 211–237.
28. R. E. Miles and J. Serra (eds.), Geometrical Probabilities and Biological Structures. Springer Verlag, Lecture Notes in Biomathematics No. 23, 1978.
29. H. Minkowski, Volumen und Oberfläche, *Math. Ann.* **57**, 447–495, 1903.
30. S. A. Roach, The Theory of Random Clumping, Methuen, London, 1968.
31. A. Rosenfeld and A. C. Kak, Digital picture processing, Academic Press, N.Y., 1976.
32. J. Serra, Image Analysis and Mathematical Morphology, Academic Press, N.Y., 1980.
33. J. Serra, Hierarchical random set models, Int. report, Centre de Morphologie Mathématique, Fontainebleau, France, 1974.
34. J. Serra and J. C. Klein, The Texture Analyser, *J. Micr.* **95**, pt. 2, 349–356, 1972.
35. H. R. Solomon, Distribution of the Measure of a Random Two-Dimensional Set. *Ann. Math. Stat.* **24**, 650–656, 1953.
36. D. Stoyan, On some Qualitative Properties of the Boolean Model of Stochastic Geometry, 1978.
37. G. Watson. Texture analysis, *Geol. Soc. of Am. Mem.* **162**, 1975.

Scene Modeling: A Structural Basis for Image Description

JAY M. TENENBAUM,* MARTIN A. FISCHLER, AND HARRY G. BARROW*

Artificial Intelligence Center, SRI International, Menlo Park, California 94025

Conventional statistical approaches to image modeling are fundamentally limited because they take no account of the underlying physical structure of the scene nor of the image formation process. The image features being modeled are frequently artifacts of viewpoint and illumination that have no intrinsic significance for higher-level interpretation. This paper argues for a structural approach to modeling that explicitly relates image appearance to the scene characteristics from which it arose. After establishing the necessity for structural modeling in image analysis, a specific representation for scene structure is proposed and then a possible computational paradigm for recovering this description from an image is described.

1. INTRODUCTION

Current research on image modeling appears dominated by attempts to characterize, mathematically and statistically, spatial variations of brightness in gray-level imagery. The basic premise underlying much of this work is that one can extract invariant pictorial features (such as regions of homogeneous brightness) that correspond to semantically meaningful entities (such as surfaces of objects).

It is our position that this abstract mathematical approach to image modeling has fundamental limitations because it takes no account of the underlying physical structure of the scene nor of the image formation process. The image features being modeled are thus frequently artifacts of viewpoint and illumination that have no intrinsic significance for higher-level interpretation. To avoid artifacts, image appearance must be explicitly related to the scene structure from which it arose, primarily physical surface characteristics such as orientation, reflectance, color, and distance.

Models that represent image appearance in terms of physical scene structure will be called "structural models" to distinguish them from "statistical models" that restrict themselves to describing immediate image appearance.

The historical emphasis on statistical modeling, reflected in these proceedings, arose from two sources: The hope of finding simple solutions to image analysis problems which avoid the need to get involved with scene structure; and the fear that image formation is too complex a process to model deterministically. In this paper we will demonstrate that both the hope and fear are unfounded. We will first examine in detail the limitations of statistical models and build a strong case

* Present address: Fairchild, Artificial Intelligence Research Laboratory, 4001 Miranda Avenue, M/S 30-888, Palo Alto, California 94304.

ISBN 0-12-597320-9

for structural models. We will then propose a specific representation for scene structure, called "intrinsic imagery," and describe a computational paradigm for recovering this description from an input image.

2. STATISTICAL VERSUS STRUCTURAL COMPLEXITY

As a point of departure, it is important to reiterate that there are two distinct sources of image complexity: statistical and structural. Statistical complexity involves brightness variations that arise from truly random phenomena, such as noise or random dot textures [1]. Also included in this category are variations due to physical phenomena, such as atmospheric scatter, that are too complex to model in detail. Structural complexity, by contrast, refers to brightness variations that arise in a deterministic way from physical scene structure, such as surface albedo and orientation. While all real images contain some degree of statistical complexity, structural complexity is often dominant, particularly for images of three-dimensional scenes.

Statistical image models are clearly appropriate for describing statistical complexity, but structural complexity requires structural models.

3. NECESSITY FOR STRUCTURAL MODELS

The need for structural models derives from three fundamental flaws in current statistical approaches: They make invalid statistical assumptions, they use ad hoc tests of statistical significance, and they produce impoverished descriptions having limited utility.

First, statistical models generally assume that an image is composed of regions of homogeneous brightness (or some statistic thereof), with random noise superimposed. Implicit is the assumption that such regions correspond to homogeneous surfaces in a scene. Real surfaces do, in fact, generally have uniform reflectance. Image brightness, however, depends on many factors besides surface reflectance, notably incident illumination and surface orientation. Consequently, homogeneous surfaces frequently appear in images as regions with high brightness gradient and even discontinuities, due to shadows. The geometry of imaging imposes similar artifacts on spatial properties of surfaces, such as the shape, size, and density of texture elements. Figure 1 contains striking examples of both phenomena.

The second limitation is perhaps more fundamental. Even if statistical homogeneity were a valid model (as is approximately the case for large areas of sky in Fig. 1a), the problem of deciding what constitutes a statistically significant variation remains.

The traditional solution has been to attempt to remove known artifacts through a normalization process and then detect discontinuities using a threshold. While this philosophy is basically sound, implementations have failed in practice because they rely on ad hoc normalization schemes (such as histogram equalization). Normalization cannot be meaningfully accomplished without taking account of the causes of variation, which in turn depend on how the image is formed. Intensity variations and discontinuities resulting from illumination gradients, shadows, or a surface turning smoothly away from the illumination source are artifacts, no matter how large. Conversely, variations resulting from small step changes in surface reflectance or orientation can be very significant. Clearly, no purely statistical process can distinguish which image features correspond to significant scene events (i.e., surfaces and surface boundaries) and which do not (i.e., shadows and highlights).

The third limitation of statistical modeling is the inadequacy of the resulting description for subsequent levels of interpretation. Even if an image could be reliably partitioned into homogeneous regions corresponding to surfaces of objects, the two-dimensional shape features used for region descriptions would still limit recognition to known objects observed from standard viewpoints; a three-dimensional characterization of surface shape is needed to recognize unfamiliar views of known objects and to assimilate multiple views of previously unseen objects.

We find it significant that when a human looks at images of scenes like those in Fig. 1, what he sees, primarily, are the actual physical characteristics of three-dimensional surfaces, independent of artifacts of illumination and viewpoint. He tends to perceive color independent of illuminant, size independent of distance, and shape independent of orientation. Such constancies have been extensively validated in the psychological literature, and it is known that they do not depend on familiarity with scene content (see Fig. 2). There is considerable evidence, in fact, that these physical normalizations are performed by "hard-wired" neural circuitry, attesting to their fundamental importance; as Fig. 3 illustrates, it is only with great effort that they can be overridden to expose the raw image content.

In summary, we believe that the obfuscation of 3-D scene structure by 2-D image features is a fundamental limitation of current image analysis systems and an important reason why their performance is so inferior to that of the human visual system.

4. A PARADIGM FOR STRUCTURAL MODELING

Having argued for the necessity for structural modeling, we must now establish what scene characteristics to model, how to represent them in a computer, and how to recover a structural description from an image.

A. Intrinsic Characteristics and Their Representation

When an image is formed, whether by eye or camera, the light intensity at each point in the image is determined by three main factors at the corresponding point in the scene: the incident illumination, the surface reflectance, and the surface orientation. In the simple case of an ideally diffusing surface, for example, the image light intensity L is given by Lambert's law:

$$L = I^*R^* \cos i, \quad (1)$$

where I is incident illumination, R is surface reflectance (albedo), and i is angle of incidence with respect to the local surface normal.

In more complicated scenes, additional factors such as specularity, transparency, luminosity, distance, and so forth must be considered. We call these properties intrinsic characteristics to distinguish them from image features which have no physical significance.

A suitable representation for intrinsic characteristics, consistent with their acquisition and subsequent use, is a set of iconic arrays in registration with the original image array. Each array contains values for a particular characteristic of the surface element visible at the corresponding point in the sensed image. Each array also contains explicit indications of boundaries due to discontinuities in value or gradient. We call such arrays "intrinsic images." Figure 4 gives an example of one possible set of intrinsic images corresponding to a monochrome image of a simple scene.

A concrete example of intrinsic images and their usefulness in computer vision can be seen in Fig. 5, which summarizes experiments by Nitzan, Brain, and Duda with a scanning laser range finder [4]. This instrument directly measures the intrinsic properties of distance and apparent reflectance, based on the phase and amplitude of the signal received as a modulated laser beam is scanned over the scene.

FIG. 1. Examples refuting the assumption that homogeneous surface characteristics appear as homogeneous image features. (a) Varying incident illumination on different faces of the mountain transforms uniformly white snow into image regions with significantly varying hue and reflectance. (b) The significant shading gradients on the background hills are again due to variations in angle of incident illumination rather than any intrinsic change in surface color. The sharp two-dimensional texture gradient (foreground) is an artifact of perspective, not a characteristic of flowers! Despite the variations in image appearance, viewers correctly perceive the uniform reflectivity of snow and hills and the regular size and density of flowers.

FIG. 1—*Continued*

The left side of Fig. 5b is a range image of the scene in Fig. 5a, with brightness inversely related to range. Note the absence of surface markings on the top of the cart and on the gray-level test chart. The right side of Fig. 5b is a reflectance image, obtained by measuring returned amplitude and using the range to compensate for varying angles of incidence. Note the absence of shadows and shading gradients present in the original intensity image (Fig. 5a).

Because the range data are uncorrupted by reflectance variations and the amplitude data are unaffected by ambient lighting and shadows, it is easy to extract surfaces of uniform height (Fig. 5c) or reflectivity (Fig. 5e) and surface boundaries where range is discontinuous (Fig. 5d). Such tasks are difficult to perform reliably in gray-level imagery; but with pure range and amplitude data, even simple-minded techniques such as thresholding and region-growing work well. In intrinsic images, the assumptions underlying statistical image modeling are valid!

Instrumentation such as a laser range finder trivializes the problem of extracting intrinsic surface characteristics from sensory data. A key question is whether

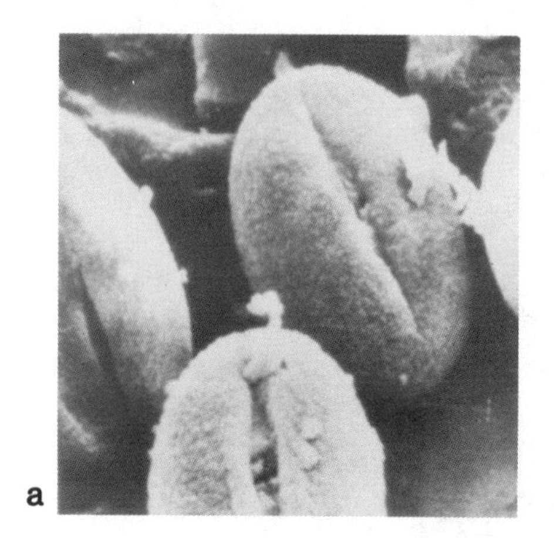

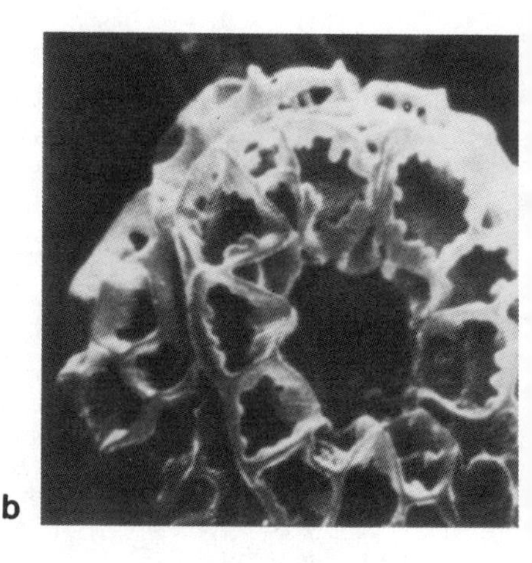

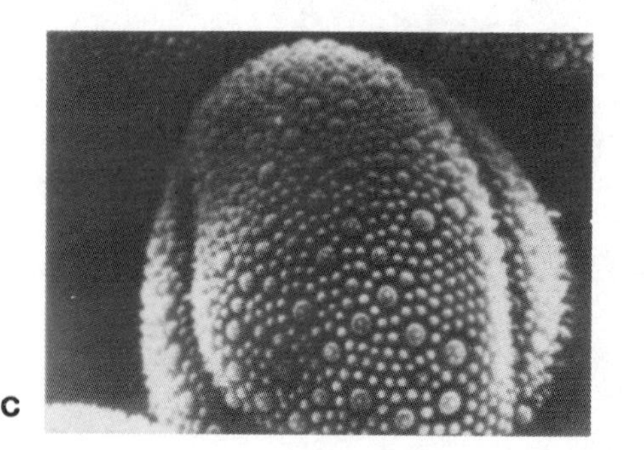

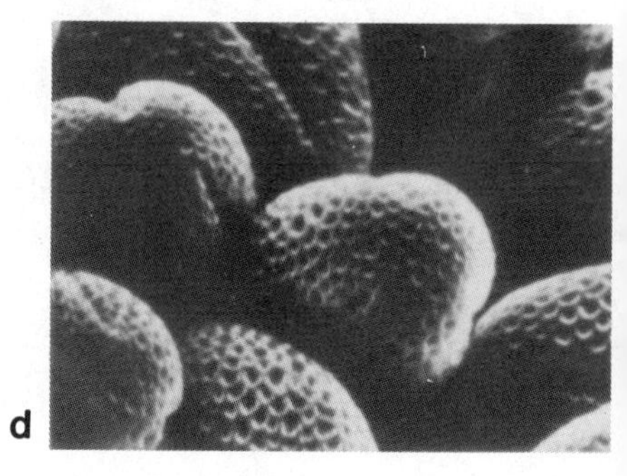

FIG. 2. Photomicrographs of pollen grains (Macleod [2]).

such information can be recovered from a single gray-level image, which is all that is available in many image analysis applications. While no definitive answer can yet be given, human performance and recent computer vision research both give cause for optimism. In following sections we discuss the computational problems involved in recovering intrinsic characteristics and briefly review promising solutions that have been proposed and demonstrated for the case of a single gray-level image. Research is also proceeding at SRI International and elsewhere for less deprived cases, such as where stereopsis [5], motion parallax [6], or a priori object models are available [7].

B. Recovery of Intrinsic Characteristics

The central problem in recovering intrinsic characteristics is that the desired information is confounded in the sensory data. Photometrically, the light intensity observed at each point in an image can result from any of an infinitude of illumination, reflectance, and orientation combinations at the corresponding scene point [see, for example, Eq. (1)]. Geometrically, each point in the image can correspond to any point along a ray in space (see Fig. 6). Recovery is thus an underconstrained problem that requires additional constraints for solution.

The necessary constraints follow from assumptions about the nature of the scene being viewed and the physics of the imaging process. In images of three-dimensional scenes, the brightness values are not independent but are constrained by various physical phenomena. Since surfaces are continuous in space, their characteristics (reflectance, orientation, range) are generally continuous across an image, except at surface boundaries. Incident illumination also usually varies smoothly over a scene, except at shadow boundaries. Elementary photometry tells us that where all intrinsic characteristics are continuous, image brightness is continuous; conversely, where one or more intrinsic characteristics are discontinuous, a brightness edge will usually result. The pattern of brightness variation in an image can provide important clues as to the local behavior of the intrinsic characteristics.

1. Shape from Shading

Horn showed that shading variations could be used to determine the three-dimensional shape of a Lambertian surface with uniform reflectance, viewed under distant point-source illumination [8, 9]. The basic approach can be grasped from Equation (2), which is obtained from Eq. (1) by taking logs and differentiating:

$$dL/L = dI/I + dR/R + d(\cos i)/\cos i. \tag{2}$$

Under Horn's assumptions, illumination and reflectance are both uniform, so that percentage changes in image brightness are directly proportional to percentage changes in (the cosine of) orientation, as follows:

$$dL/L = d(\cos i)/\cos i. \tag{3}$$

Horn showed that given suitable boundary conditions, Eq. (3) can be integrated to recover shape.

2. Lightness from Shading

A second example of exploiting photometry, also due to Horn [10], involves the recovery of surface lightness (a psychological term representing relative reflectance). This study assumed a planar surface, viewed under smoothly varying illumination. The surface was painted with a patchwork of contrasting regions, each of uniform reflectance. Within regions, since orientation and reflectance are both constant, variations in image brightness are directly proportional to variations in illumination, as given by Eq. (4a). Moreover, since illumination was assumed to vary smoothly, the brightness gradient within regions is small. At region boundaries, however, reflectance jumps discontinuously and dominates the small illumination gradient, as expressed in Eq. (4b). To recover surface lightness,

FIG. 3. Demonstration that human perception of surface boundaries is not critically dependent on image contrast. (a) Low-contrast interior boundary. (b) Masked cross sections of the interior boundary in Fig. 3a. (c) A subjective contour. In Fig. 3a, an intersection boundary is clearly perceived. Yet when that area of the image is viewed through a mask that exposes only local cross sections, no local contrast is visible (Fig. 3b). The boundary perceived in Fig. 3a is demanded by the integrity of the surfaces it joins. Subjective contour illusions, like the so-called sun illusion (Fig. 3c), appear to be an extreme example of this same phenomenon where an edge is clearly perceived despite the complete absence of local evidence. A plausible explanation is that the edge corresponds to the boundary of an occluding disk-shaped surface, whose presence is implied by the abrupt line endings [3].

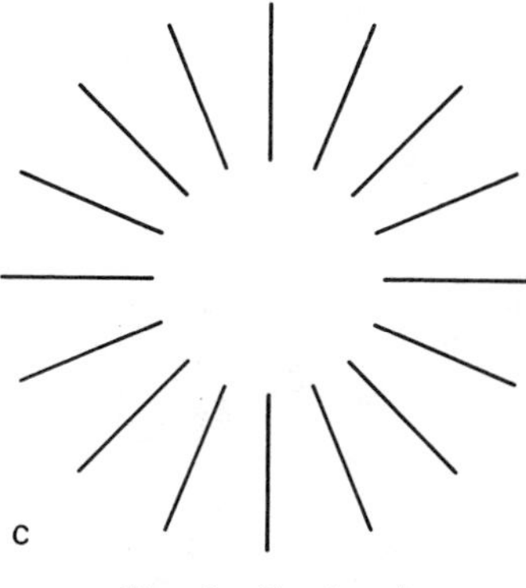

FIG. 3—*Continued*

Horn first spatially differentiated the input (log brightness) image to obtain a gradient image and then thresholded it to eliminate slow illumination variations. The remaining discontinuities, assumed to be reflectance jumps at the edges of regions [as in Eq. (4b)], were then reintegrated to recover the relative lightnesses of the various regions:

$$dL/L = dI/I \qquad \text{(within regions)}, \tag{4a}$$

$$dL/L = dR/R \qquad \text{(at boundaries)}. \tag{4b}$$

Surface color independent of illuminant can be estimated from lightness values recovered independently in three spectral bands, analogous to Land's retinex theory [11].

3. Shape From Contour

Horn's work emphasized photometric cues, but geometric cues to 3-D surface structure are at least as valuable (see Fig. 7).

The ability to perceive surface structure from line drawings is truly remarkable since, as Fig. 6 showed, each two-dimensional image curve can, in principle, correspond to an infinitude of possible three-dimensional space curves. However, people are not aware of this massive ambiguity. For example, when asked to provide a three-dimensional interpretation of an ellipse, the overwhelming response is a tilted circle, not some bizarrely twisting (or even discontinuous) curve that has the same image. As in Horn's work, some a priori assumptions about the scene must again be invoked. Recent research at SRI [12] and MIT [13, 14] suggests that humans resolve the projective ambiguity by perceiving the smoothest possible space curve corresponding to a given image curve. Mathematically, they seek the space curve having the most uniform curvature and the least torsion, as expressed by minimizing the terms

$$\int \left(\frac{dkb}{ds}\right)^2 ds = \int (\dot{k}^2 + k^2t^2)ds. \tag{5}$$

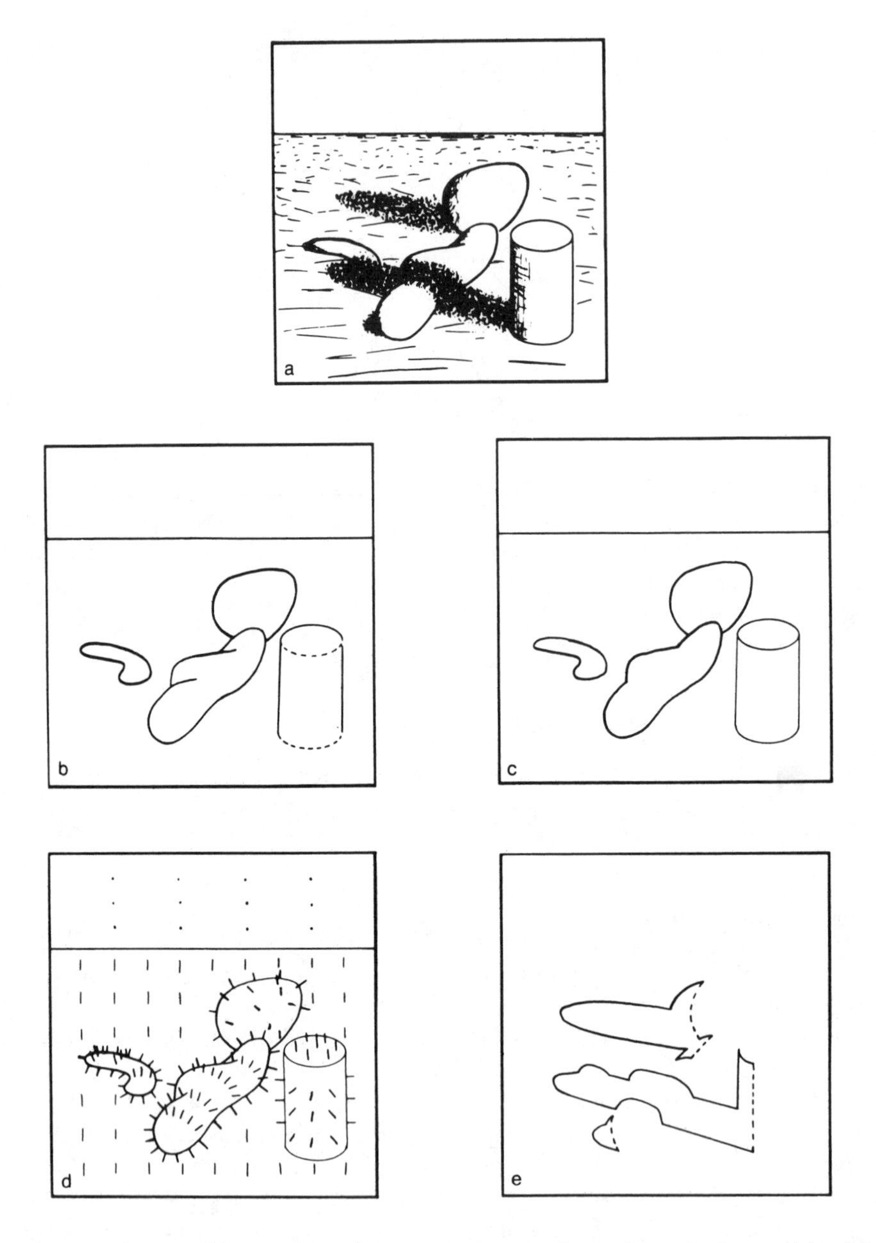

FIG. 4. A set of intrinsic images derived from a single monochrome intensity image: (a) original scene, (b) distance, (c) reflectance, (d) orientation (vectors), (e) illumination. The images are depicted as line drawings, but, in fact, would contain values at every point. The solid lines in the intrinsic images represent discontinuities in the scene characteristics; the dashed lines represent discontinuities in its derivative. In the input image, intensities correspond to the reflected light flux received from the visible points in the scene. The distance image gives the range along the line of sight from the center of projection to each visible point in the scene. The orientation image gives a vector representing the direction of the surface normal at each point. The reflectance image gives the albedo (the ratio of total reflected to total incident illumination) at each point.

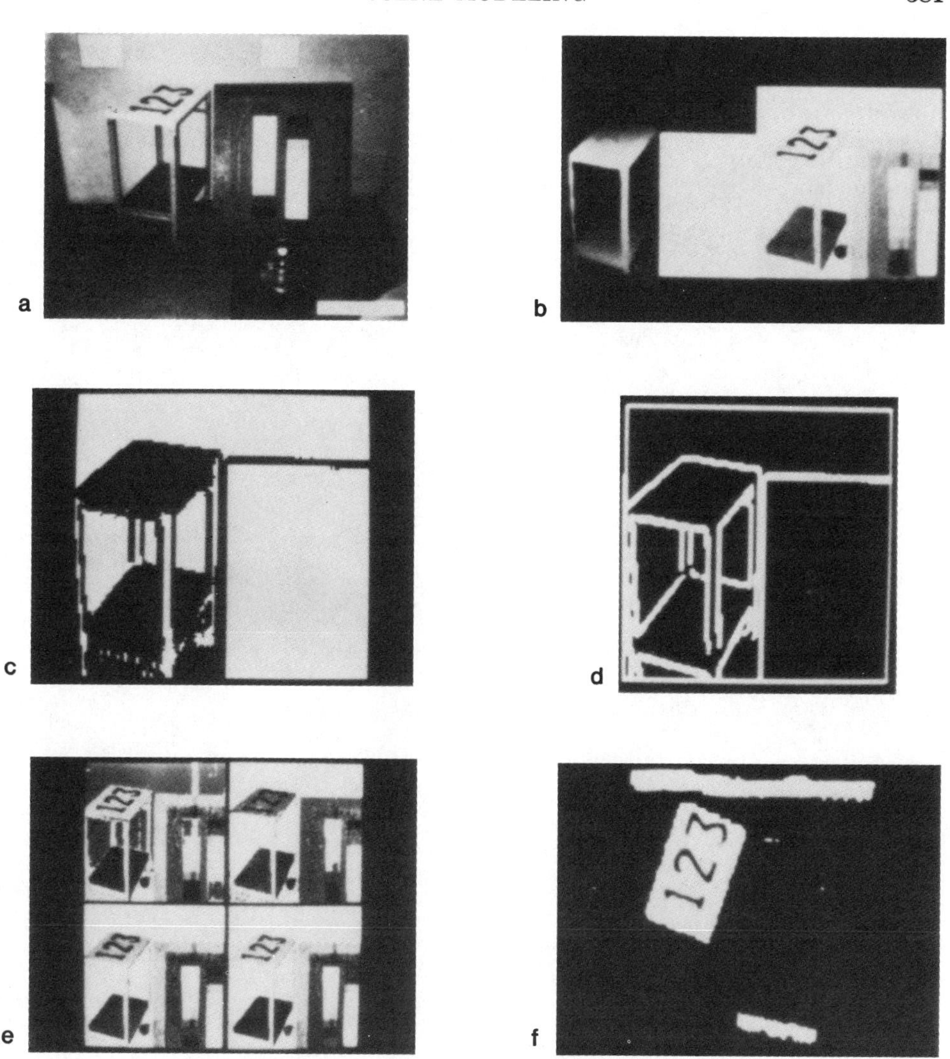

FIG. 5. Experiments with a laser range finder.

(Here k is the local differential curvature, $\dot{k}$ is its spatial derivative along the curve, B is the binormal, and t is the torsion.)

The smoothness assumption expressed by Eq. (5) has both an ecological and a statistical rationale. Ecologically, assumptions on smoothness of curves follow from the earlier assumptions on smoothness of surfaces, which, in turn, are rooted in assumptions that physical surfaces assume minimum energy configurations. Statistically, it is reasonable to assume that a scene is being viewed from a general position so that perceived smoothness is not an accident of viewpoint. In Fig. 6, for example, the discontinuous curve projects into an ellipse from only one viewpoint. Thus such a curve would be a highly improbable three-dimensional interpretation of an ellipse.

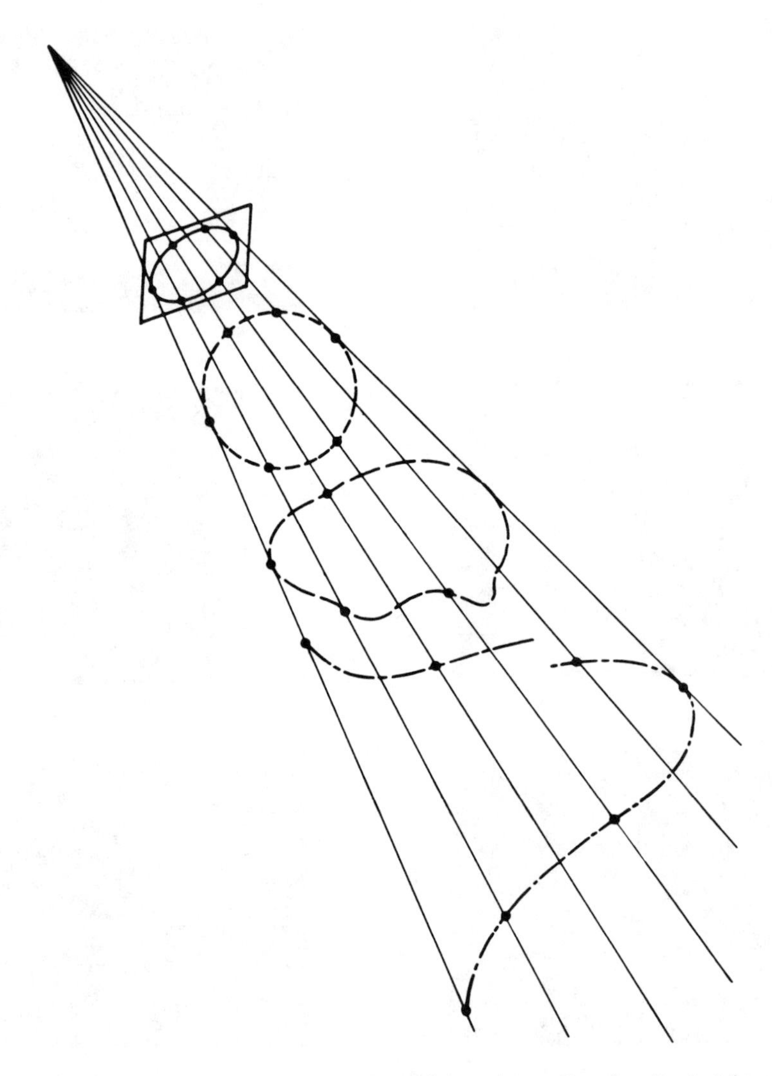

FIG. 6. Three-dimensional conformation of lines depicted in a line drawing is inherently ambiguous. All of the space curves in this figure project into an ellipse in the image plane, but they are not all equally likely interpretations.

A computer program has been written, based on Eq. (5), that can successfully determine three-dimensional space curves corresponding to simple image curves. Referring to Fig. 8, points along an image curve define rays in space along which the corresponding space curve points are constrained to lie. The program can adjust the distance associated with each space curve point by sliding it along the ray like a bead on a wire. An iterative optimization procedure determines the configuration of points that minimize the integral in Eq. (5). Optimization proceeds by independently adjusting each space curve point and observing the incremental change in local curvature and torsion. (Note that local perturbations

FIG. 7. Line drawing of a three-dimensional scene. Surface and boundary structure are distinctly perceived despite the ambiguity inherent in the imaging process.

have only local effects.) Witkin [13] used a similar approach to model the perception of planar orientation associated with simple closed curves.

The program produces correct 3-D interpretations for simple open and closed curves, such as interpreting an ellipse as a tilted circle and a trapezoid as a tilted rectangle. However, convergence is slow and somewhat dependent on the initial choice of z-values.

4. Shape from Texture

Texture gradient is a well-known geometric clue for inferring three-dimensional surface structure. The variations of size, density, eccentricity, and orientation of the texture elements in Fig. 9 are hardly random; they are predictable consequences of the foreshortening that occurs when a tilted surface is imaged under perspective projection. A recent thesis by Stevens [14] provides mathematical formulations for a number of texture depth cues, which previously have been described only qualitatively by psychologists. Another recent thesis by Kender establishes the principle that textured surfaces are perceived at orientations that maximize the regularity, homogeneity, and symmetry of the texture [15].

Underlying Kender's work on texture and our own work on line drawings is a fundamental assumption that the world is generally isotropic. This assumption, which we call "generalized isotropy," is reminiscent of the Gestalt notions of Prägnanz, but mathematically more precise.

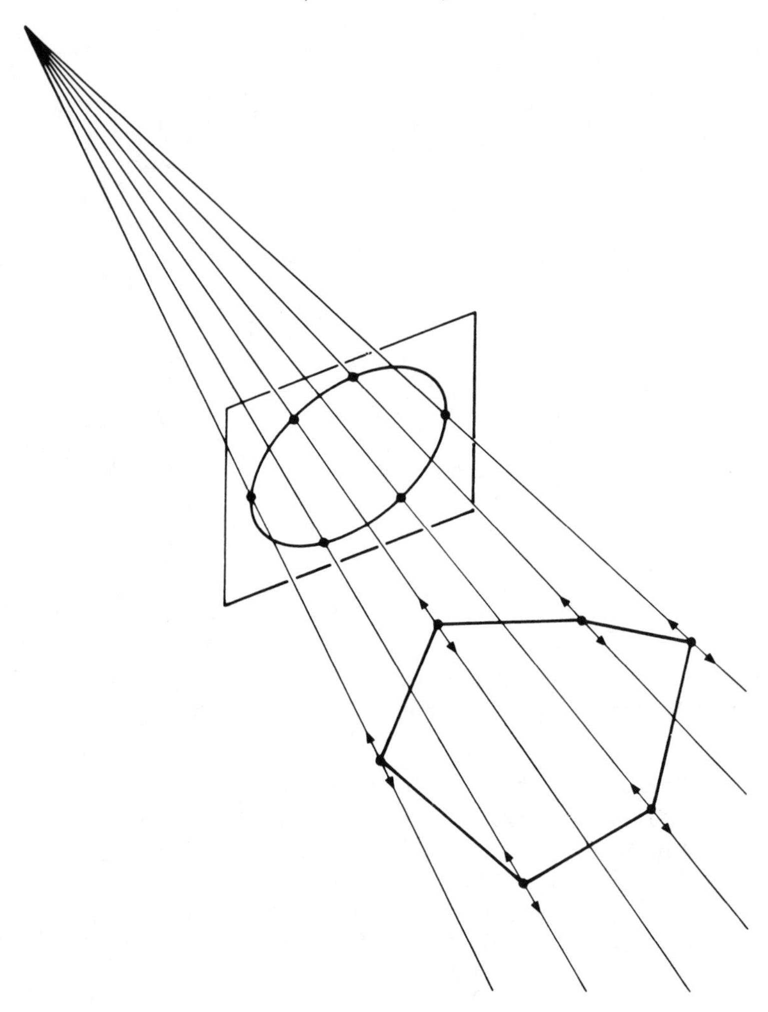

FIG. 8. An iterative procedure for determining the optimal space curve corresponding to a given line drawing. Projective rays constrain the three-dimensional position associated with each image point to one degree of freedom.

5. The Role of Edges

The previous sections described various ways in which surface characteristics could be inferred from image features, such as shading, contour, and texture. However, in each case the physical nature of the image feature being interpreted was known. For example, in determining shape from shading, Horn assumed that only surface orientation was changing; in determining lightness, he assumed that only reflectance was discontinuous; in determining shape from contour, we implicitly assumed that image curves corresponded to surface boundaries and were not shadows or lines painted on a flat surface. When a scene contains many occluding objects that may be varicolored or cast shadows, such simple assumptions are invalid. One is then faced with the problem of deciding what physical characteristic (or characteristics) is, in fact, responsible for an observed intensity variation and which characteristics are discontinuous across intensity edges.

The pattern of brightness variation on either side of an intensity edge can sometimes provide strong clues as to the type of scene event responsible (shadow or surface boundary), and thus to which intrinsic characteristics are actually discontinuous at that point. A simple example is the continuity of texture and high contrast at shadow edges, indicating a discontinuity in illumination. The interpretation of brightness edges as scene events is also important because knowing the type of scene event sometimes allows explicit values to be determined for some of the intrinsic characteristics. For example, at an extremal occluding boundary, where an object curves smoothly away from the viewer, the surface orientation can be inferred exactly at every point along the boundary. (The local surface normal is constrained to be normal to the edge element in the image and normal to the line of sight.) A test for extremal boundaries can be made by determining whether the observed brightness variation along an edge is consistent with assumed extremal orientations [3].

Having identified a number of edge and surface constraints, it was necessary to establish whether they were sufficient to permit the simultaneous recovery of surface reflectance and orientation from a single image. We defined the simplest domain in which confounding problems arose in a general way and proceeded to exhaustively catalog the physical interpretations corresponding to all possible image intensity patterns [3]. The domain consisted of smooth (no creases,

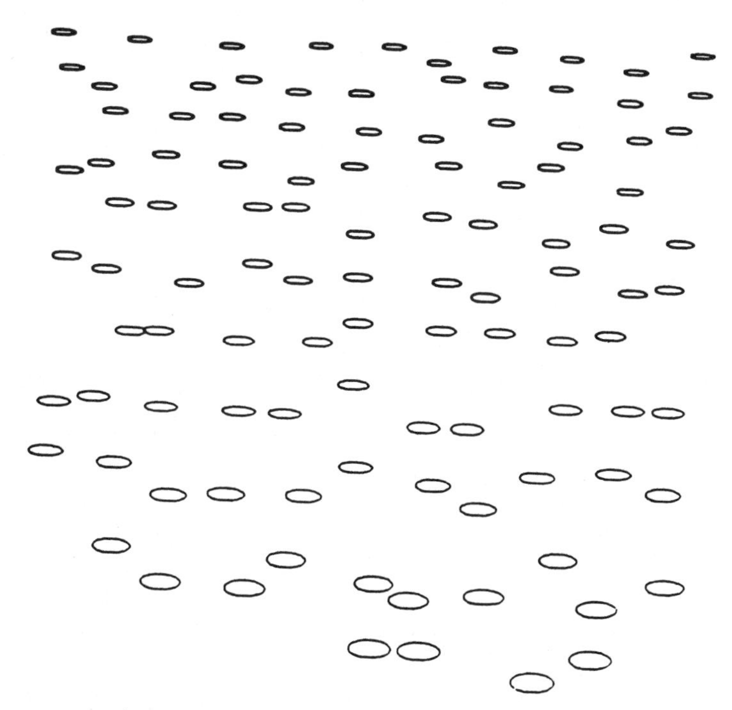

FIG. 9. Texture gradients on a tilted planar surface (Stevens [14]). The ellipses vary in width, eccentricity, and density, presenting a variety of texture clues.

folds), uniformly reflecting Lambertian surfaces, illuminated by a distant point source, and uniform, diffuse background light (approximating sun and sky). The resulting catalog is reproduced in Table 1. With but two exceptions, brightness discontinuities could be unambiguously interpreted as physical events. Furthermore, in most cases, values for one or more intrinsic characteristics were either determined or strongly constrained.

We concluded that the recovery problem was mathematically well posed, at least for the simple domain. The clues and constraints obtained by interpreting image discontinuities according to the catalog define a system of equations and inequalities, relating the intensity values to the intrinsic characteristics and boundary conditions at the discontinuities. In principle, this system can be solved, yielding feasible values for the intrinsic characteristics. In practice, however, the equations are highly nonlinear and boundary conditions may not always be determined reliably. This suggested an iterative numerical solution process, such as relaxation.

6. A Computational Model

A parallel computational model was proposed to illustrate how recovery might be performed. The basic model, reproduced in Fig. 10, can be regarded as a generalization of Horn's lightness model [10] and Marr and Poggio's cooperative stereopsis model [5], that simultaneously recovers both geometric and photometric attributes. In essence, it consisted of a stack of registered arrays representing the original intensity image (top) and the primary intrinsic image arrays. Processing was initialized by detecting intensity edges in the original image, interpreting

TABLE 1

The Nature of Edges[a]

Region Intensities LA	Region Intensities LB	Edge Type	Region Types	Intrinsic Edges Intrinsic Values D	N	R	I
Constant	Constant	Occluding sense unknown	A B shadowed	EDGE	EDGE	EDGE RA RB	IA IB
Constant	Varying	1 Shadow	A shadowed B illuminated		NB.S	RA RB	EDGE IA IB
		2 A occludes B	A shadowed B illuminated	EDGE DA<DB	EDGE NA	EDGE RA	EDGE IA
Varving	Varying	Inconsistent with domain					
Constant	Tangency	B occludes A	A shadowed B illuminated	EDGE DA>DB	EDGE NB	EDGE RA RB	EDGE IA IB
Varying	Tangency	B occludes A	A B illuminated	EDGE DA>DB	EDGE NB	EDGE RB	EDGE IB IA
Tangency	Tangency	Not seen from general position					

[a] LA and LB refer to variations of intensity along sides A and B of an edge. Intensities are either constant, varying, or varying in accordance with the assumed orientations along an extremal boundary, the so-called tangency condition.

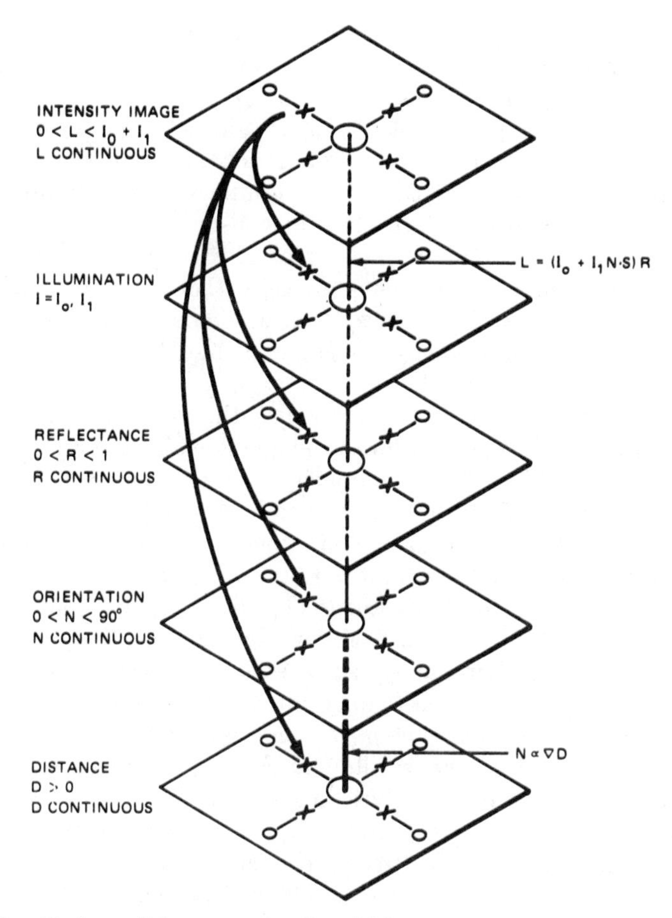

FIG. 10. A parallel computational model for recovering intrinsic images.

them according to the catalog of appearances, and then creating the appropriate edges in the intrinsic images (as implied by the descending arrows).

Parallel local operations (shown as circles) modified the values in each intrinsic image to make them consistent with intraimage continuity and limit constraints (for example, reflectance must be between 0 and 1). Simultaneously, a second set of processes (shown as vertical lines) operated to make the values at each point consistent with the corresponding intensity value, as required by the inter-image photometric constraint. A third set of processes (shown as *X*'s) operated to insert and delete edge elements, which inhibit continuity constraints locally. The constraint and edge modification processes operated continuously and interacted to recover accurate intrinsic scene characteristics and to perfect the initial edge interpretation.

The action was envisaged to resemble an analog computer: As the value in one image increased, corresponding values in other images increased or decreased to maintain consistency with the observed intensity at that point. Within each image, values tended to propagate in from boundary conditions established along edges. This resembles relaxation processes used in physics for determining temperature or potential over a region from boundary conditions.

How well does the proposed model work in the simplified domain? In theory exact recovery is generally possible for nonshadowed regions; where it is not possible, because of inadequate information in the original image, plausible estimates can usually be obtained. These results were significant, despite the simplicity of the domain, because they demonstrated for the first time the theoretical possibility of simultaneously recovering orientation, reflectance, and illumination from a single monochrome image, without recourse either to object prototypes or to primary depth cues, such as stereopsis, motion parallax, or texture gradient. (Such additional cues can, of course, be added to aid initialization in shadowed areas.)

We are currently implementing a version of the model and will soon test it on synthesized images of scenes from the simple domain. We are simultaneously studying the extensibility of the approach to more complex domains. We are convinced that the model can be extended to handle objects with creases, folds, painted-surface markings (texture), and other complicating features. Coping with more complex illuminations, however, appears difficult.

The theory of recovery outlined above depends heavily on analytic photometry and a precise lighting model, both for edge classification and for inferring shape from shading. Unfortunately, surfaces in real scenes have complex reflectance functions that are often strongly directional. Illumination patterns are equally complex, encompassing such phenomena as shading gradients from nearby or extended sources and secondary illumination by light reflected from nearby specular surfaces. The use of analytic photometry appears very questionable under these conditions. We are therefore working on an alternative theory of recovery that relies primarily on geometric cues, such as contour and texture gradient for gross shape determination, and uses photometry, only in a qualitative way, to indicate subtle refinements (e.g., bumps and dents).

5. DISCUSSION

In this paper we have argued that the explicit modeling of scene structure is a critical first step in the interpretation of images of three-dimensional scenes, and have provided a demonstration that, theoretically, such information can be obtained even from a single gray-level image. More generally, in almost any image analysis task, an understanding of the relationship between scene content and image appearance is necessary for meaningful interpretation. Such relationships invariably have an underlying physical basis.

We are not attempting to denigrate the role played by statistical techniques in image analysis. Clearly some images are inherently statistical and contain little structure. More generally, because images are noisy and ambiguous, their interpretation requires fitting a priori models to the observed data; interpretation is thus a proper subset of statistical decision theory. The problem with conventional statistical approaches is that in ignoring the physical nature of the scene and the imaging process, they are forced to base interpretation on ad hoc and often invalid assumptions. In the paradigm we are suggesting, structural scene models provide a rational basis for decision-making.

In conclusion, while both statistical and structural models play important roles in image analysis, whenever image variation can be accounted for in structural terms, there are compelling reasons for doing so.

ACKNOWLEDGMENTS

The work reported in this paper was performed under SRI's research program in computational vision, which is supported by ARPA, NSF, and NASA.

REFERENCES

1. B. Julesz, Experiments in the visual perception of texture, *Sci. Amer.* **232**, April 1975, 34–43.
2. I. D. G. Macleod, *A Study in Automatic Photo Interpretation*, Ph.D. thesis, Dept. of Engineering Physics, Australian National University, Canberra, Australia, 1970.
3. H. G. Barrow and J. M. Tenenbaum, Recovering intrinsic scene characteristics from images, in *Computer Vision Systems* (A. Hanson and E. Riseman, Eds), pp. 3–26, Academic Press, New York, 1978.
4. D. Nitzan, A. E. Brain, and R. O. Duda, The measurement and use of registered reflectance and range data in scene analysis, *Proc. IEEE* **65**, 1977, 206–220.
5. D. Marr and T. Poggio, Cooperative computation of stereo disparity, *Science* **194**, 1977, 283–287.
6. S. Ullman, *The Interpretation of Visual Motion*, MIT Press, Cambridge, Mass., 1979.
7. G. Falk, Interpretation of imperfect line data as a three-dimensional scene, *Artificial Intelligence* **4**, 1972, 101–144.
8. B. K. P. Horn, Obtaining shape from shading information, in *The Psychology of Computer Vision* (P. H. Winston, Ed.), McGraw–Hill, New York, 1975.
9. B. K. P. Horn, Understanding image intensities, *Artificial Intelligence* **8**, 1977, 201–231.
10. B. K. P. Horn, Determining lightness from an image, *Computer Graphics Image Processing* **3**, 1974, 277–299.
11. E. H. Land, The retinex theory of color vision, *Sci. Amer.* **237**, December 1977, 108–128.
12. H. G. Barrow and J. M. Tenenbaum, Recovery of 3-D shape information from image boundaries, in preparation.
13. A. Witkin, The minimum curvature assumption and perceived surface orientation, presentation at Optical Society of America, November 1978.
14. K. Stevens, *Surface Perception from Local Analysis of Texture and Contour*, Ph.D. dissertation, Electrical Engineering and Computer Science, Massachusetts Institute of Technology, Cambridge, February 1979.
15. J. R. Kender, Shape from texture: A computational paradigm, *Proc. ARPA Image Understanding Workshop*, pp. 134–138, Palo Alto, Calif., April 1979.

Pictorial Feature Extraction and Recognition via Image Modeling

JULIUS T. TOU

Center for Information Research, University of Florida, Gainesville, Florida 32611

1. INTRODUCTION

The field of pattern recognition and image processing by computer has made rapid advances in many directions. During the past five years a considerable amount of interest has been shown in computer-based image modeling for the description and recognition of visual patterns and scenes. A number of techniques for image modeling have been proposed in the literature. Some of them are reviewed and discussed in this volume. In this paper we intend to present the image modeling approaches for pictorial feature extraction and recognition which we have developed during the past six years. We will discuss three approaches: (a) two-dimensional time series model; (b) **G** matrix eigenvalue approach; and (c) pixel-vector clustering technique. Mathematical models are derived for feature extraction and recognition.

Research activities in computer-based picture processing have been centered around picture enhancement, interpretation, recognition, generation, and editing. Quantitative study of a picture is often concerned with four types of parameters which are of fundamental importance. They are contrast, color, shape, and texture. Contrast is a very important measure in picture processing, which often determines the quality of a picture. Color adds more useful discriminatory information to a picture. Shape is a much-used measure in recognizing the various objects contained in a picture. Texture provides very useful information for performing automatic interpretation and recognition of a picture by computer.

The contents of a picture may be categorized into three general types:

(1) identifiable objects with well-defined structural patterns,
(2) identifiable objects with fuzzy or diffused patterns,
(3) nonidentifiable objects.

These three types of pictures are illustrated in Fig. 1. To process a picture with identifiable objects, we may localize the objects and study their contrast, color, and shape for recognition. In the case of objects with fuzzy structures, identification of objects may require texture analysis. When we process pictures with nonidentifiable objects such as wood grain or forests, the study of the textures in

ISBN 0-12-597320-9

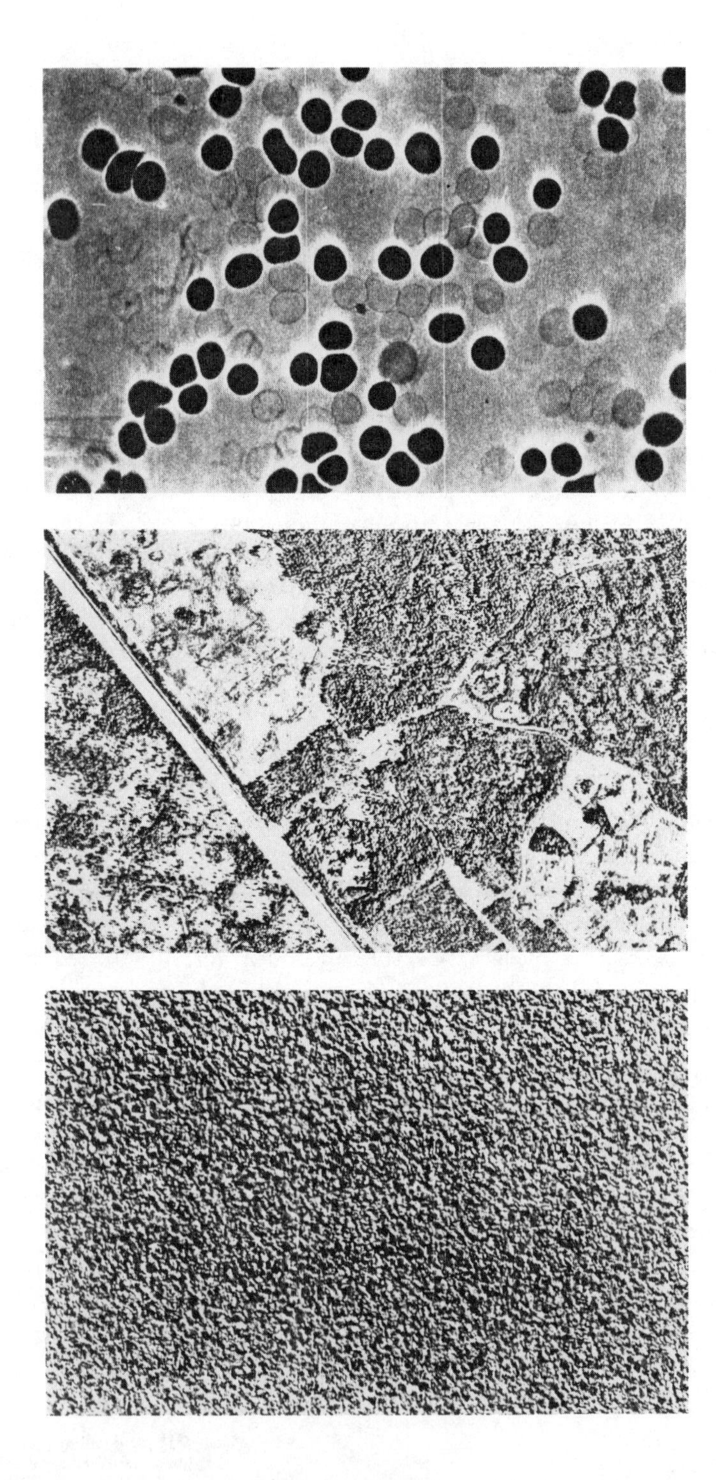

FIG. 1. Sample pictures.

the picture becomes indispensable. Under this circumstance, textural information provides important features for interpretation and recognition. This paper is concerned with the texture problem.

Texture information is very useful in automatic photointerpretation, earth resources exploration, and biomedical image processing. In photointerpretation, the most important task is to localize and detect discrete tactical targets such as armored vehicles, fortifications, artillery. In earth resources studies, texture information may be used in identifying and recognizing terrain and in determining and evaluating the properties of surface materials and natural vegetation from photographs. In cytopathology, important diagnostic clues are offered by the distribution, appearance, and granularity of chromatic material in the cell nucleus.

The general meaning of texture is well understood. Texture has been considered as a requisite for characterizing a surface. Texture analysis in pattern recognition studies is by no means a trivial problem. In [19], we proposed the following definition: Textures may be regarded as repetitive arrangements of a unit sub-pattern, as inhomogeneities in the gray scale, or as global properties of a picture or scene in a statistical sense. Based upon this notion, we suggested that textures in pattern recognition research may be studied either from the structural point of view or from the statistical point of view.

2. TWO-DIMENSIONAL TIME SERIES MODEL

On the basis of a statistical approach, the texture of a picture is treated as a set of statistical properties extracted from a large set of measurements made on the picture. In computer-based image processing, a picture is often characterized by pixels which are successive observations on the picture made by the computer. From the pixels we may derive a time series for modeling the textural properties of the picture. In this section we will first briefly review some basic notions of time series, then we will extend the time series concept to pictorial texture studies.

2.1. Time Series Modeling

Let a time series be characterized by $y_1, y_2, \ldots, y_i, \ldots$. The mean value of y_i is μ, and the independent white noise is $w_i = y_i - \hat{y}_i$ which is normally distributed with zero mean and variance σ^2, where $\hat{y}_i$ is the estimate of y_i. Then for

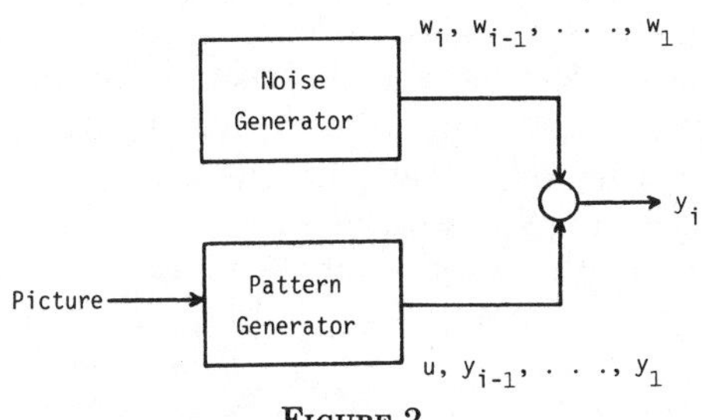

FIGURE 2

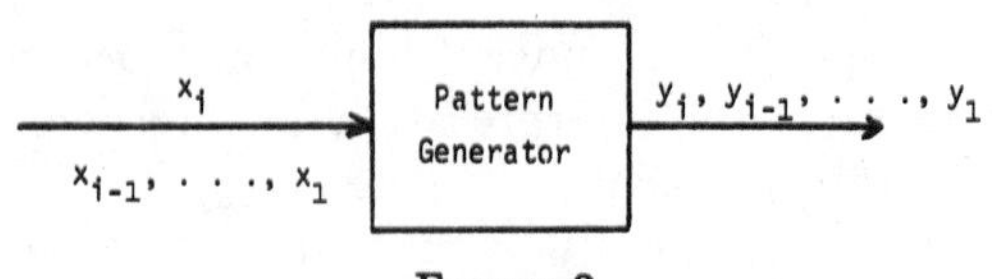

FIGURE 3

any y_i, we have a linear stochastic process

$$\begin{aligned} y_i &= \mu + w_i + \alpha_1 w_{i-1} + \alpha_2 w_{i-2} + \cdots \\ &= \sum_{k=0}^{\infty} \alpha_k w_{i-k} + \mu, \end{aligned} \tag{1}$$

where $\alpha_0 = 1$.

Alternatively, a linear stochastic process can be considered as a weighted sum of the past observations and the current white noise

$$\begin{aligned} y_i &= \beta_1 y_{i-1} + \beta_2 y_{i-2} + \cdots + w_i \\ &= \sum_{j=1}^{\infty} \beta_j y_{i-j} + w_i. \end{aligned} \tag{2}$$

From Eq. (2), we have autoregressive process of order p,

$$y_i = \sum_{j=1}^{p} \beta_j y_{i-j} + w_i. \tag{3}$$

From Eq. (1), we have the moving-average process of order q,

$$y_i = \sum_{k=0}^{q} \alpha_k w_{i-k} + \mu. \tag{4}$$

By combining Eqs. (3) and (4), we have the mixed autoregressive moving-average process,

$$y_i = \mu + \sum_{j=1}^{p} \beta_j y_{i-j} + \sum_{k=0}^{q} \alpha_k w_{i-k}. \tag{5}$$

The above analysis suggests a time series model for a textural picture as shown in Fig. 2. The model consists of a pattern generator and a noise generator. In response to a picture input, the pattern generator produces a sequence of measurements. The mean value of the current measurement is μ. Thus, the output of this model is y_i which is given by Eq. (5).

When we consider only the current white noise and p previous values of y_i, the model is the autoregressive process of order p, given by Eq. (3). When we consider only the current mean value of y_i and q previous values of the white noise, the model is the moving-average process of order q.

An alternative model may be derived on the basis of an input–output relationship, as illustrated in Fig. 3. The input–output relationship is given by

$$y_i + a_1 y_{i-1} + \cdots + a_p y_{i-p} = b_0 x_i + b_1 x_{i-1} + \cdots + b_q x_{i-q}. \tag{6}$$

Taking the z-transform yields

$$(1 + a_1 z^{-1} + \cdots + a_p z^{-p}) Y(z) = (b_0 + b_1 z^{-1} + \cdots + b_q z^{-q}) X(z). \tag{7}$$

The pulse transfer function of the pattern generator is

$$G(z) = (b_0 + \sum_{k=1}^{q} b_k z^{-k}) / (1 + \sum_{k=1}^{p} a_k z^{-k}). \tag{8}$$

Case (1). $b_k = 0$ for $k = 1, 2, \ldots, q$. Equation (8) reduces to

$$G(z) = b_0 / (1 + \sum_{k=1}^{p} a_k z^{-k}). \tag{9}$$

From Eq. (9) we obtain

$$(1 + \sum_{k=1}^{p} a_k z^{-k}) Y(z) = b_0 X(z) \tag{10}$$

which yields upon inverse z-transform

$$y_i = \sum_{k=1}^{p} (-a_k) y_{i-k} + b_0 x_i. \tag{11}$$

This is the autoregressive process of order p.

Case (2). $a_k = 0$ for $k = 1, 2, \ldots, p$. Equation (8) reduces to

$$G(z) = b_0 + \sum_{k=1}^{q} b_k z^{-k} \tag{12}$$

from which we have

$$Y(z) = (b_0 + \sum_{k=1}^{q} b_k z^{-k}) X(z). \tag{13}$$

Upon inverse z-transform, Eq. (13) yields

$$y_i = \sum_{k=0}^{q} b_k x_{i-k}. \tag{14}$$

This is the moving-average process of order q. In this case the mean value is zero, since zero initial condition is assumed.

In both the time series model and z-transform model we are looking for a pattern generator characterizing the pattern class on the basis of the observed samples $(y_n, y_{n-1}, \ldots, y_i, \ldots, y_1)$. The pattern generator is determined by the coefficients of the model.

The autoregressive model of order p is given by Eq. (3). Let the autocovariance be

$$\gamma_k = E\{y_i y_{i-k}\}. \tag{15}$$

Then it follows from Eq. (3) that

$$\gamma_k = \beta_1 \gamma_{k-1} + \beta_2 \gamma_{k-2} + \cdots + \beta_p y_{k-p}. \tag{16}$$

Let the autocorrelation function be

$$\rho_k = \gamma_k/\gamma_0. \tag{17}$$

Then we have

$$\rho_k = \beta_1\rho_{k-1} + \beta_2\rho_{k-2} + \cdots + \beta_k\rho_{k-p}. \tag{18}$$

By letting $k = 1, 2, \ldots, p$, we obtain the Yule–Walker equations,

$$\begin{aligned} \rho_1 &= \beta_1 + \beta_2\rho_1 + \cdots + \beta_p\rho_{p-1}, \\ \rho_2 &= \beta_1\rho_1 + \beta_2 + \cdots + \beta_p\rho_{p-2}, \\ &\vdots \\ \rho_p &= \beta_1\rho_{p-1} + \beta_2\rho_{p-2} + \cdots + \beta_p. \end{aligned} \tag{19}$$

This can be used to estimate the coefficients β_i after the autocorrelations are computed.

If we let $\phi_{\kappa i}$ be the ith coefficient in a kth order autoregressive process, the Yule–Walker equation for an autoregressive model of order k is

$$\begin{bmatrix} 1 & \rho_1 & \rho_2 & \cdots & \rho_{k-1} \\ \rho_1 & 1 & \rho_1 & \cdots & \rho_{k-2} \\ \vdots & \vdots & \vdots & & \vdots \\ \rho_{k-1} & \rho_{k-2} & \rho_{k-3} & \cdots & 1 \end{bmatrix} \begin{bmatrix} \phi_{k1} \\ \phi_{k2} \\ \vdots \\ \phi_{kk} \end{bmatrix} = \begin{bmatrix} \rho_1 \\ \rho_2 \\ \vdots \\ \rho_k \end{bmatrix}. \tag{20}$$

Among the coefficients ϕ_{ki}, the set ϕ_{kk} is of special interest, and ϕ_{kk} is often referred to as the partial autocorrelation function of log k. It follows from Eq. (20) that the autocorrelation functions are given by

$$\phi_{11} = \rho_1,$$

$$\phi_{22} = \frac{\begin{vmatrix} 1 & \rho_1 \\ \rho_1 & \rho_2 \end{vmatrix}}{\begin{vmatrix} 1 & \rho_1 \\ \rho_1 & 1 \end{vmatrix}},$$

$$\phi_{33} = \frac{\begin{vmatrix} 1 & \rho_1 & \rho_1 \\ \rho_1 & 1 & \rho_2 \\ \rho_2 & \rho_1 & \rho_3 \end{vmatrix}}{\begin{vmatrix} 1 & \rho_1 & \rho_2 \\ \rho_1 & 1 & \rho_1 \\ \rho_2 & \rho_1 & 1 \end{vmatrix}}, \qquad \text{etc.}$$

for $k = 1, 2, 3, \ldots$. It is noted that the last column in the numerator is $\rho_1, \rho_2, \rho_3, \ldots, \rho_k$. The partial autocorrelation function reveals some interesting properties of an autoregressive model. It can be shown that the partial autocorrelation function ϕ_{kk} for an autoregressive process of order p has the following properties:

$$\phi_{kk} \neq 0 \qquad \text{for } k \leq p,$$
$$\phi_{kk} = 0 \qquad \text{for } k > p.$$

Thus, the partial autocorrelation function of a pth order autoregressive process has a cutoff after log p.

The moving-average model of order q is given by Eq. (4). By letting $\tilde{y}_i = y_i - \mu$, we have

$$\tilde{y}_i = w_i + \alpha_1 w_i + \cdots + \alpha_q w_{i-q}. \tag{21}$$

Let the autocovariance be

$$\gamma_k = E\{\tilde{y}_i \tilde{y}_{i-k}\}.$$

Then

$$\gamma_0 = (1 + \alpha_1^2 + \alpha_2^2 + \cdots + \alpha_q^2)\sigma_w^2 = D_q \sigma_w^2, \tag{22}$$

where

$$\sigma_w^2 = E\{w_i w_i\} \tag{23}$$

and

$$E\{w_i w_j\} = 0 \qquad \text{for } i \neq j.$$

The autocovariance γ_k is

$$\begin{aligned} \gamma_k &= (\alpha_k + \alpha_1 \alpha_{k+1} + \cdots + \alpha_{q-k}\alpha_q)\sigma_w^2 \qquad &\text{for } k = 1, 2, \ldots, q \\ &= 0 &\text{for } k > q. \end{aligned} \tag{24}$$

Combining Eqs. (22) and (24), we obtain the autocorrelation function

$$\begin{aligned} \rho_k &= \frac{\alpha_k + \alpha_1 \alpha_{k+1} + \cdots + \alpha_{q-k}\alpha_q}{D_q} \qquad &\text{for } k = 1, 2, \ldots, q \\ &= 0 &\text{for } k > q \end{aligned} \tag{25}$$

From the above equation, by letting $k = 1, 2, \ldots, q$, we obtain q equations

$$\begin{aligned} \rho_1 D_q &= \alpha_1 + \alpha_1 \alpha_2 + \cdots + \alpha_{q-1}\alpha_q, \\ \rho_2 D_q &= \alpha_2 + \alpha_1 \alpha_3 + \cdots + \alpha_{q-2}\alpha_q, \\ &\vdots \\ \rho_q D_q &= 2\alpha_q. \end{aligned} \tag{26}$$

After we have estimated the autocorrelation functions ρ_k, these equations can be used to solve for the coefficients α_i.

2.2. 2-D Statistical Model

In computer-based picture processing studies, we deal with a two-dimensional array of data. As in the time series models, we can express an observation $z_{i,j}$ in a two-dimensional stochastical array as a weighted sum of the data and white noise associated with neighboring positions. A portion of the two-dimensional array is shown in Fig. 4, where z_{ij} denotes the value of the gray level at position (i, j), or the pixel intensity at that position.

Assuming zero mean for $z_{i,j}$, a linear stochastic process for $z_{i,j}$ in terms of white

$$\begin{matrix} & \vdots & \vdots & \vdots \\ \cdots & z_{i-2,j-2} & z_{i-2,j-1} & z_{i-2,j} \\ \cdots & z_{i-1,j-2} & z_{i-1,j-1} & z_{i-1,j} \\ \cdots & z_{i,j-2} & z_{i,j-1} & z_{i,j} \end{matrix}$$

FIGURE 4

noise $w_{i,j}$ is given by

$$z_{i,j} = w_{i,j} + \psi_{0,1}w_{i,j-1} + \psi_{1,0}w_{i-1,j} + \psi_{1,1}w_{i-1,j-1} + \cdots \tag{27}$$

We define the horizontal shift operator H by

$$H^n z_{i,j} = z_{i,j-n} \tag{28}$$

and the vertical shift operator V by

$$V^m z_{i,j} = z_{i-m,j}. \tag{29}$$

The horizontal difference operator $\nabla_h = 1 - H$,

$$\nabla_h z_{i,j} = (1 - H)z_{i,j} = z_{i,j} - z_{i,j-1}. \tag{30}$$

The vertical difference operator $\nabla_v = 1 - V$,

$$\nabla_v z_{i,j} = (1 - V)z_{i,j} = z_{i,j} - z_{i-1,j}. \tag{31}$$

These operators are commutative:

$$H^m V^n z_{i,j} = V^n H^m z_{i,j},$$

$$\nabla_h{}^m \nabla_v{}^n z_{i,j} = \nabla_v{}^n \nabla_h{}^m z_{i,j}.$$

Let

$$\Psi(H, V) = \sum_{m=0}^{\infty} \sum_{n=0}^{\infty} \psi_{n,m} H^m V^n \tag{32}$$

and $\psi_{0,0} = 1$; then Eq. (27) reduces to

$$z_{i,j} = \Psi(H, V)w_{i,j}, \tag{33}$$

where $\Psi(H, V)$ is referred to as the system function of the model.

Since the autocovariance at log (k, l) is given by

$$\gamma_{k,l} = E\{z_{i,j}z_{i+k,j+l}\} \tag{34}$$

from Eq. (33) we obtain

$$\begin{aligned} \gamma_{k,l} &= E\{\Psi(H, V)w_{i,j}\Psi(H, V)w_{i+k,j+l}\} \\ &= \sum_{p=0}^{\infty} \sum_{q=0}^{\infty} \sum_{m=0}^{\infty} \sum_{n=0}^{\infty} \psi_{p,q}\psi_{m,n}E\{w_{i-p,j-q}w_{i+k-m,j+l-n}\} \\ &= \sigma_w{}^2 \sum_{p=0}^{\infty} \sum_{q=0}^{\infty} \psi_{p,q}\psi_{p+k,q+l}, \end{aligned} \tag{35}$$

where σ_w^2 is the variance of $w_{i,j}$.

The autocovariance $\gamma_{k,l}$ of the model may be derived from its system function $\Psi(H, V)$. Let $\Gamma(H, V)$ be the 2-D autocovariance generating function given by

$$\Gamma(H, V) = \sum_{k=-\infty}^{\infty} \sum_{l=-\infty}^{\infty} \gamma_{k,l} V^k H^l \tag{36}$$

Then it follows from Eq. (35) that

$$\begin{aligned}\Gamma(H, V) &= \sigma_w^2 \sum_{k=-\infty}^{\infty} \sum_{l=-\infty}^{\infty} \sum_{p=0}^{\infty} \sum_{q=0}^{\infty} \psi_{p,q}\psi_{p+k,q+l} V^k H^l \\ &= \sigma_w^2 \sum_{p=0}^{\infty} \sum_{q=0}^{\infty} \sum_{m=0}^{\infty} \sum_{n=0}^{\infty} \psi_{p,q}\psi_{m,n} V^{m-p} H^{n-q} \\ &= \sigma_w^2 \Big(\sum_{m=0}^{\infty} \sum_{n=0}^{\infty} \psi_{m,n} V^m H^n\Big)\Big(\sum_{p=0}^{\infty} \sum_{q=0}^{\infty} \psi_{p,q} V^{-p} H^{-q}\Big) \\ &= \sigma_w^2 \Psi(H, V) \Psi(H^{-1}, V^{-1}).\end{aligned} \tag{37}$$

Following the reasoning for time series models we obtain the 2-D autoregressive model and moving-average model.

AR(m, n)—Autoregressive Model of Order (m, n)

$$\Phi(H, V) z_{i,j} = w_{i,j}, \tag{38}$$

where

$$\Phi(H, V) = \sum_{k=0}^{m} \sum_{l=0}^{n} \phi_{k,l} V^k H^l$$

and

$$\phi_{0,0} = 1.$$

MA(p, q)—Moving Average Model of Order (p, q)

$$z_{i,j} = \Theta(H, V) w_{i,j}, \tag{39}$$

where

$$\Theta(H, V) = \sum_{k=0}^{q} \sum_{l=0}^{p} \theta_{k,l} V^k H^l.$$

ARMA(m, n; p, q)—Mixed Autoregressive and Moving-Average Model

$$\Phi(H, V) z_{i,j} = \Theta(H, V) w_{i,j}. \tag{40}$$

Some useful properties of these three models are reflected in the autocorrelations and partial autocorrelations. These properties are summarized in Table 1, which may be used to identify possible models.

As an illustration for the determination of the model coefficients, we consider low-order models AR(1, 1) and MA(1, 1).

TABLE 1

Type of process	Autocorrelations	Partial autocorrelations
Autoregressive	Tailoff	Spikes at lags 1 through p, then cutoff
Moving-average	Spikes at lags 1 through q, then cutoff	Tailoff
Mixed ARMA	Irregular pattern at lags 1 through q, then tailoff	Tailoff

(*a*) *AR*(*1*, *1*).

$$\Phi(H, V)z_{i,j} = w_{i,j},$$
$$\Phi(H, V) = 1 - \phi_{1,0}V - \phi_{0,1}H - \phi_{1,1}HV.$$

The autocovariance is

$$\gamma_{k,l} = \phi_{1,0}\gamma_{k-1,l} + \phi_{0,1}\gamma_{k,l-1} + \phi_{1,1}\gamma_{k-1,l}.$$

The autocorrelation function is

$$\rho_{k,l} = \phi_{1,0}\rho_{k-1,l} + \phi_{0,1}\rho_{k,l-1} + \Phi_{1,1}\rho_{k-1,l-1}.$$

By letting k, $l = 0$ or 1, we obtain the Yule–Walker equation for this model

$$\begin{bmatrix} 1 & \rho_{1,1} & \rho_{0,1} \\ \rho_{1,1} & 1 & \rho_{1,0} \\ \rho_{0,1} & \rho_{1,0} & 1 \end{bmatrix} \begin{bmatrix} \phi_{1,0} \\ \phi_{0,1} \\ \phi_{1,1} \end{bmatrix} = \begin{bmatrix} \rho_{1,0} \\ \rho_{1,0} \\ \rho_{1,1} \end{bmatrix}.$$

The autocorrelation functions are determined from

$$\rho_{k,l} = \gamma_{k,l}/\gamma_{0,0},$$

where $\gamma_{k,l}$ is in turn determined from the observed values by using the relationship

$$\gamma_{k,l} = E\{z_{i,j}z_{i-k,j-l}\}.$$

From the Yule–Walker equation we may solve for the coefficients in terms of the autocorrelation functions.

(*b*) *MA*(*1*, *1*).

$$z_{i,j} = \Theta(H, V)w_{i,j},$$
$$\Theta(H, V) = 1 + \theta_{1,0}V + \theta_{0,1}H + \theta_{1,1}HV.$$

Since the autocovariance is given by $\gamma_{k,l} = E\{z_{i,j}z_{i-k,j-l}\}$, substitution yields

$$\gamma_{0,0} = (1 + \theta_{1,0}^2 + \theta_{0,1}^2 + \theta_{1,1}^2)\sigma_w^2,$$
$$\gamma_{1,0} = (\theta_{1,0} + \theta_{1,1}\theta_{0,1})\sigma_w^2,$$
$$\gamma_{0,1} = (\theta_{0,1} + \theta_{1,1}\theta_{1,0})\sigma_w^2,$$
$$\gamma_{1,1} = \theta_{1,1}\sigma_w^2.$$

The autocorrelation functions are

$$\rho_{1,0} = (\theta_{1,0} + \theta_{1,1}\theta_{0,1})/(1 + \theta_{1,0}^2 + \theta_{0,1}^2 + \theta_{1,1}^2),$$
$$\rho_{0,1} = (\theta_{0,1} + \theta_{1,1}\theta_{1,0})/(1 + \theta_{1,0}^2 + \theta_{0,1}^2 + \theta_{1,1}^2),$$
$$\rho_{1,1} = \theta_{1,1}/(1 + \theta_{1,0}^2 + \theta_{0,1}^2 + \theta_{1,1}^2).$$

From these equations we obtain the model coefficients.

2.3. Multiplicative Process

When the operator $\Psi(H, V)$ can be expressed as

$$\Psi(H, V) = \Psi_h(H)\Psi_v(V) \tag{41}$$

the model becomes a multiplicative process,

$$z_{i,j} = \Psi_h(H)\Psi_v(V)w_{i,j}. \tag{42}$$

This is true when the coefficients are multiplicative:

$$\psi_{m,n} = \psi_{m,0}\psi_{0,n}. \tag{43}$$

It can be shown that for a multiplicative process,

$$\rho_{i,j} = \rho_{i,0}\rho_{0,j}. \tag{44}$$

That is, the autocorrelation at lag (i, j) is equal to the product of the ith autocorrelation in the first column of the correlation matrix and the jth autocorrelation in the first row. Equation (44) may be used to test for the multiplicative property.

The autocovariance generating function $\Gamma(H, V)$ given in Eq. (37) reduces to

$$\Gamma(H, V) = \sigma_w^2\Psi_h(H)\Psi_h(H^{-1})\Psi_v(V)\Psi_v(V^{-1}) \tag{45}$$

for a multiplicative process. The autocovariance with horizontal lags, $\gamma_{0,k}$, is the coefficient of the term H^k in $\Gamma(H, V)$. Thus the generating function associated with $\gamma_{0,k}$ may be expressed as

$$\Gamma_h(H) = \sigma_w^2(\sum_{i=0}^{\infty} \psi_{vi}^2)\Psi_h(H)\Psi_h(H^{-1}). \tag{46}$$

Let A be the coefficient of H^j (or H^{-j}) in $\Psi_h(H)\Psi_h(H^{-1})$; then from Eq. (46) we have

$$\gamma_{0,j} = \sigma_w^2(\sum_{n=0}^{\infty} \psi_{vn}^2)A.$$

Similarly, by letting B be the coefficient of V^i (or V^{-i}) in $\Psi_v(V)\Psi_v(V^{-1})$, we have

$$\gamma_{i,0} = \sigma_w^2(\sum_{m=0}^{\infty} \psi_{hm}^2)B.$$

Now $\gamma_{i,j}$ is the coefficient of H^hV^i in $\Gamma(H, V)$. Since H^j comes from $\Psi_h(H)\Psi_h(H^{-1})$ with coefficient A and V^i comes from $\Psi_v(V)\Psi_v(V^{-1})$ with coefficient B, we have

$$\gamma_{i,j} = \sigma_w^2AB.$$

By making use of the fact that

$$\gamma_{0,0} = \sigma_w^2 \sum_{m=0}^{\infty} \psi_{hm}^2 \sum_{n=0}^{\infty} \psi_{vn}^2$$

we have

$$\frac{\gamma_{i,j}}{\gamma_{0,0}} = \frac{\sigma_w^2 AB}{\sigma_w^2 \sum_{m=0}^{\infty} \psi_{hm}^2 \sum_{n=0}^{\infty} \psi_{vn}^2}$$

$$= \frac{\gamma_{i,0}}{\gamma_{0,0}} \cdot \frac{\gamma_{0,j}}{\gamma_{0,0}}.$$

Hence,

$$\rho_{i,j} = \rho_{i,0}\rho_{0,j}.$$

This condition is used to verify that a linear stochastic process is multiplicative

2.4. *Equations for Model Building*

(a) *Mean*

$$\bar{z} = \frac{1}{n_1 n_2} \sum_{i=1}^{n_1} \sum_{j=1}^{n_2} z_{i,j},$$

where n_1 is number of rows of pixels and n_2 is number of columns of pixels.

(b) *Autocovariance*

$$\gamma_{k,l} = \frac{1}{(n_1 - k)(n_2 - l)} \sum_{i=1}^{n_1-k} \sum_{j=1}^{n_2-l} (z_{i,j} - \bar{z})(z_{i+k,j+l} - \bar{z})$$

where k is lag in row and l is lag in column.

(c) *Variance.*

$$\sigma_z^2 = \gamma_{0,0}$$

(d) *Autocorrelation.*

$$\rho_{k,l} = \frac{\gamma_{k,l}}{\gamma_{0,0}}$$

(e) *Partial autocorrelation function: In the horizontal direction,*

$$\phi_{k,k} = \rho_{0,1} \qquad \text{for } k = 1$$

$$= \Big(\rho_{0,k} - \sum_{j=1}^{k-1} \phi_{(k-1)j}\rho_{0,k-j}\Big) \Big/ \Big(1 - \sum_{j=1}^{k-1} \phi_{(k-1)}\rho_{0,j}\Big) \qquad \text{for } k = 2, 3, \ldots.$$

In the vertical direction,

$$\phi_{k,k} = \rho_{0,1} \qquad \text{for } k = 1$$

$$\phi_{k,k} = \Big(\rho_{k,0} - \sum_{j=1}^{k-1} \phi_{(k-1)j}\rho_{k-j,0}\Big) \Big/ \Big(1 - \sum_{j=1}^{k-1} \phi_{(k-1)j}\rho_{j,0}\Big) \qquad \text{for } k = 2, 3, \ldots,$$

where

$$\phi_{k-1,j} = \phi_{k-2,j} - \phi_{k-1,k-1}\phi_{k-2,k-j-1} \qquad \text{for } j = 1, 2, \ldots, k-1.$$

2.5. Applications to Pictorial Textures

For a given picture we obtain the array of digitized gray levels and build the statistical model for the two-dimensional array. Several steps are involved in model building:

Step 1. Generate a two-dimensional digitized array by a scanning device.
Step 2. Perform some preprocessing to enhance contrast and to reduce noise.
Step 3. Calculate the mean values.
Step 4. Calculate the autocovariances.
Step 5. Compute the variance of the data sample.
Step 6. Calculate the autocorrelations.
Step 7. Check stationarity. We note that the data in horizontal or vertical directions may be nonstationary, thus making the autocorrelations of the process undefined. Fortunately, the differences in many nonstationary process are stationary. Such a process is often called homogeneous nonstationary. In this case we difference the data until a stationary data series is apparent.
Step 8. Perform differencing in the direction that the two-dimensional data series is nonstationary.
Step 9. Determine the partial autocorrelation functions.
Step 10. Identify possible models with the aid of Table 1.
Step 11. Estimate parameters of the models. Having identified a tentative model, we may use the sample autocorrelations to obtain preliminary estimates of parameters. These parameters provide a useful approximation of the final model and are used as the starting values for the iterative procedure in computing maximum-likelihood estimates of the parameters. The preliminary estimates are obtained by solving the simultaneous equations relating the parameters and the autocorrelations. The procedure for an autoregressive process is straightforward, i.e., solution of the Yule–Walker equations. However, for moving-average and mixed processes, the relationship between autocorrelations and parameters is nonlinear and the determination of preliminary estimates is less tractable.

Having identified a tentative model with definite parameters for the pictorial data, we would like to choose the best estimate of the parameters. One of the methods is least-squares estimation by minimizing the sum of squared residuals. For example, in the general first-order model, the residual is given by

$$w_{i,j} = (1 + \phi_{1,0}V + \phi_{0,1}H + \phi_{1,1}HV)z_{i,j} - (\theta_{1,0}V - \theta_{0,1}H - \theta_{1,1}HV)w_{i,j}.$$

We find a set of coefficients which minimize the sum of squared residuals

$$s = \sum_{i=1}^{n} \sum_{j=1}^{n} (w_{i,j})^2,$$

where the size of the two-dimensional array is $n \times n$. This model will be the most suitable for the data.

Step. 12. Check the adequacy of the model. Once a suitable model has been identified, we would like to check the adequacy of the model for the data at hand. One method of doing this is to check the randomness of the residuals of a model. The randomness can be measured in terms of autocorrelation functions. If the

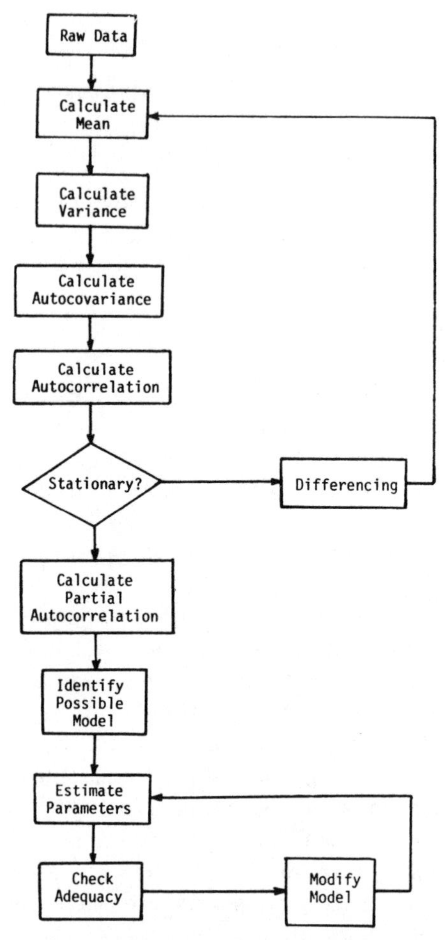

FIG. 5. Model building flow chart.

residuals are truly white noise, their estimated autocorrelation function should be close to zero except for $\rho_{0,0}$. The autocorrelation will suggest the direction in which the model should be modified.

Step 13. Make modifications in the model. If we find that the autocorrelation of the residuals is not insignificant, we know that $w_{i,j}$ is not pure white noise. It might be a stochastic process itself. Thus, for obtaining better results, we may have to use a model of higher order.

The above steps are summarized in a flow chart (Fig. 5).

2.6. Computational Results

Two texture samples used in our study are taken from Brodatz [4]: Sample 1 is wood grain and Sample 2 is pressed cork, as shown in Fig. 6. They are chosen for illustration because they are typical examples of the commonly found textures and they represent multiplicative and nonmultiplicative cases. The means and autocovariances are calculated.

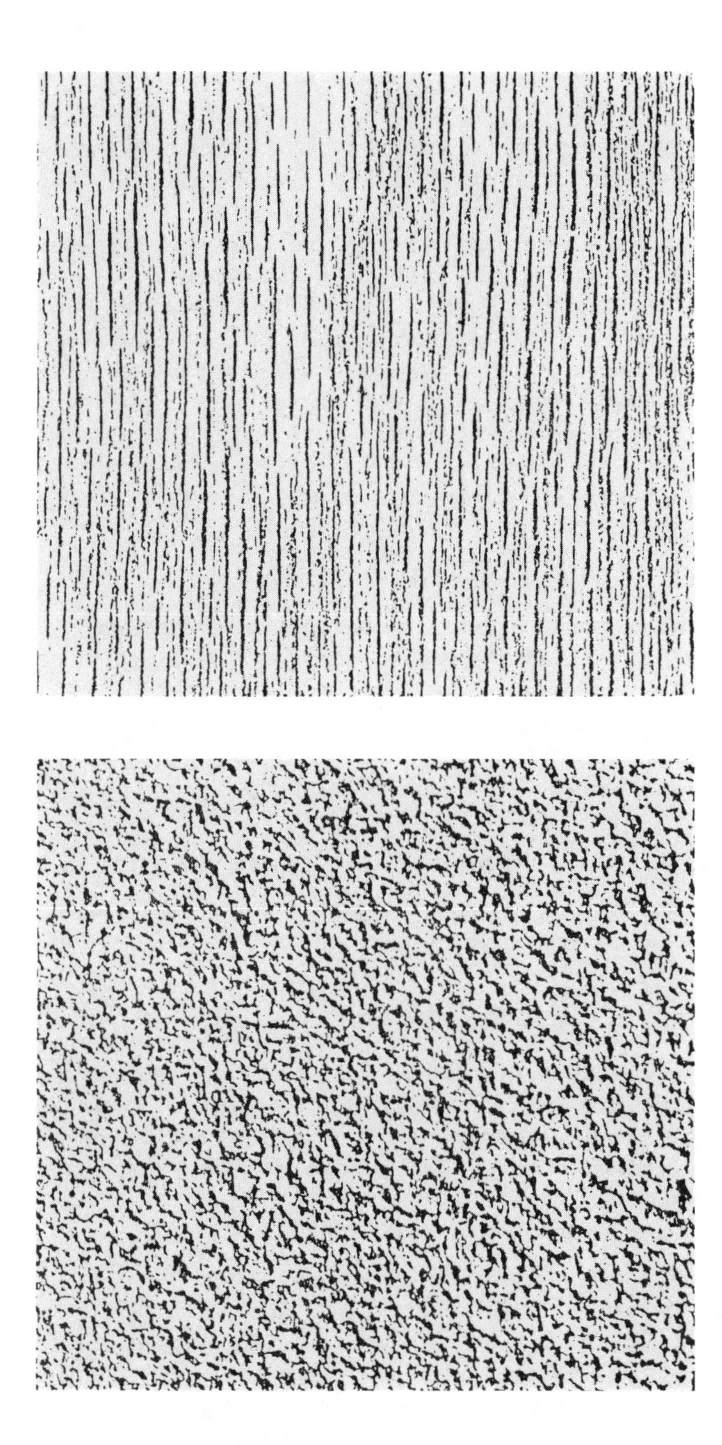

FIG. 6. Texture samples.

Texture	Mean	Autocovariance
Wood grain	4.842	3.608
Pressed cork	2.175	3.810

The autocorrelation functions of both Sample 1 and Sample 2 reveal that both two-dimensional data arrays are stationary. No differencing is needed in either the horizontal or vertical direction.

From the autocorrelation matrix for Sample 1, we note that the relationship $\rho_{i,j} = \rho_{i,0}\rho_{0,j}$ holds within the tolerance of estimation error. Hence this model is very likely a multiplicative process. The small values in the horizontal autocorrelation function and partial autocorrelation function suggest the possibility of a pure random noise or a first-order moving-average process in the horizontal direction. The slowly decaying vertical autocorrelation function suggests the possibility of taking vertical differences. After vertical differencing, the strong vertical correlation no longer exists. The vertical partial autocorrelation function is typical for a first-order autoregressive process. With the above considerations we choose the following tentative model for Sample 1:

$$(1 + \phi_{1,0}V)z_{i,j} = (1 + \theta_{0,1}H)w_{i,j}.$$

Model identification for Sample 2 was done in the same way. From the autocorrelation matrix we find that the multiplicative relationship does not hold. We noticed that the autocorrelation functions are essentially zero after lag 1. This suggests that we might have a first-order moving-average process. The rapidly decaying partial autocorrelations also indicate the property of a low-order moving-average process. Because of the strong possibility of nonmultiplicity, we choose the tentative model for Sample 2 as

$$z_{i,j} - \mu = (1 + \theta_{1,0}V + \theta_{0,1}H + \theta_{1,1}HV)w_{i,j}.$$

By minimizing the sum of square residuals, we find a set of coefficient values most suitable for the data. The tentative models with estimated coefficients are

$$\text{Sample 1:} \quad (1 - V)z_{i,j} = (1 + 0.4H)w_{i,j},$$
$$\text{Sample 2:} \quad z_{i,j} - 2.175 = (1 + 0.4H + 0.4V + 0.4HV)w_{i,j}.$$

Examination of the autocorrelation matrix for Sample 1 reveals that $w_{i,j}$ is not pure white noise. It might be a stochastic process itself. This suggests that we will have to use a model of higher order. On the other hand, the model for Sample 2 seems quite satisfactory so far as the randomness of residuals is concerned.

3. G MATRIX EIGENVALUE APPROACH

This section presents approaches to machine recognition and understanding of pictures via textural feature extraction. These approaches make use of **G** matrices, their eigenvalues, and the Karhunen–Loéve expansion. Visual texture patterns are a familiar concept to human beings. However, attempts to process textural information by computer have not been very successful. The major difficulty seems to lie in the lack of a general method to reduce texture information for machine implementation and interpretation. The texture analysis problem has been studied by two major approaches: the statistical approach and the structural approach. In the preceding section we have presented a statistical approach based upon time series modeling. This section will discuss the structural approach making use of **G** matrices.

In the structural approach, a texture is considered as consisting of subpatterns, which occur repeatedly within the picture according to some specified rules of arrangement. The textural information in a rectangular picture may be characterized by the distribution pattern of the intensity gradient between two pixels separated by a specified distance in a specified direction. We choose the gradient distribution method as the starting point to perform textural feature extraction. The gradient distribution method is often referred to as the gray level spatial dependence method. A major shortcoming of this method is that from the various arrays it generates a large number of function values which can be used as measurements for characterization of the texture picture. In our approach we extract features from these measurements by making use of Karhunen–Loéve expansion.

A gradient distribution pattern may be expressed in a matrix **G** whose elements g_{ij} are the joint probability density for point pairs with intensity gradient $(i - j)$. If K is the maximum number of gray levels, **G** is a $K \times K$ matrix. When the picture is characterized by discrete points and the specified separations are given in simple integers, the elements of a **G** matrix are determined by counting the number of times each intensity gradient occurs for a specified distance in a specified direction. This number is normalized by dividing it by the total number of intensity gradients. Thus, the **G** matrix is given by

$$\mathbf{G} = \begin{bmatrix} g_{00} & g_{01} & g_{02} & \cdots & g_{0(k-1)} \\ g_{10} & g_{11} & g_{12} & \cdots & g_{1(k-1)} \\ g_{20} & g_{21} & g_{22} & & g_{2(k-1)} \\ \vdots & \vdots & \vdots & & \vdots \\ g_{(k-1)0} & g_{(k-1)1} & g_{(k-2)2} & \cdots & g_{(k-1)(k-1)} \end{bmatrix} \tag{47}$$

in which

$$g_{ij} = \#(i, j)/N,$$

N being the total number of intensity gradients. The gradient distribution matrix is independent of the picture size. The **G** matrix is also referred to as a scatter matrix.

From the **G** matrices we may make a number of measurements on the textural information. However, these measurements contain redundant information, and they may not provide effective features for textural recognition. We propose that the **G** matrix method for textural feature measurement be augmented by optimal feature extraction via the Karhunan–Loéve expansion technique. The feature extraction process is optimal in the sense that the mean-square error committed by approximating the infinite series with a finite number of terms is a minimum.

A number of measurements have been proposed in the literature by Haralick *et al.* [8] for obtaining useful textural information from the **G** matrices. We have found that the following four measurements are more useful in our investigation:

(*a*) *Angular second moment.*

$$f_1 = \sum_{i=0}^{k-1} \sum_{j=0}^{k-1} (g_{ij})^2. \tag{48}$$

This quantity can be used to measure the homogeneity of the texture. For homogeneous textures there are few nonzero entries in the **G** matrix. For less homogeneous textures the **G** matrix has more entries with small values. Thus, the quantity f_1 will be large for homogeneous textures.

(*b*) *Contrast.*

$$f_2 = \sum_{i=0}^{k-1} \sum_{j=0}^{k-1} (i - j)^2 g_{ij}. \tag{49}$$

This quantity indicates the degree of spread of the matrix values. Thus, it can be used to measure the contrast in the picture. The quantity f_2 will be large for pictures with high contrast.

(*c*) *Entropy.*

$$f_3 = - \sum_{j=0}^{k-1} \sum_{j=0}^{k-1} g_{ij} \log g_{ij}. \tag{50}$$

This quantity also provides a measure of homogeneity.

(*d*) *Autocorrelation.*

$$f_4 = \left(\sum_{i=0}^{k-1} \sum_{j=0}^{k-1} (ijg_{ij}) - m_I m_J\right) / (\sigma_I \sigma_J), \tag{51}$$

where

$$m_I = \sum_{i=0}^{k-1} \sum_{j=0}^{k-1} i g_{ij},$$

$$m_J = \sum_{i=0}^{k-1} \sum_{j=0}^{k-1} j g_{ij},$$

$$\sigma_I = \left[\sum_{i=0}^{k-1} \sum_{j=0}^{k-1} i^2 g_{ij} - \left(\sum_{i=0}^{k-1} \sum_{j=0}^{k-1} i g_{ij}\right)^2\right]^{\frac{1}{2}},$$

$$\sigma_J = \left[\sum_{i=0}^{k-1} \sum_{j=0}^{k-1} j^2 g_{ij} - \left(\sum_{i=0}^{k-1} \sum_{j=0}^{k-1} j g_{ij}\right)^2\right]^{\frac{1}{2}}.$$

The quantity f_4 will indicate the gray level linear dependence in the picture. This quantity will be large for pictures with ordered patterns and containing less noise.

3.1. Textural Feature Extraction

Suppose a set of measurements from a textural picture is represented as a measurement vector

$$\mathbf{M}_i = \begin{bmatrix} m_i(1) \\ m_i(2) \\ \vdots \\ m_i(n) \end{bmatrix}, \qquad i = 1, 2, \ldots, N, \tag{52}$$

where n is the number of measurements, N is the number of sample pictures, and $m_i(k)$ denotes one of the above measurements from a $\mathbf{G}$ matrix characterizing the texture. The measurement vector $\mathbf{M}_i$ can be expressed as a linear combination of basis vectors $\mathbf{\Phi}_j$, $j = 1, 2, \ldots, n$,

$$\mathbf{M}_i = \sum_{j=1}^{n} \alpha_{ij} \mathbf{\Phi}_j, \tag{53}$$

where

$$\mathbf{\Phi}_j = \begin{bmatrix} \phi_j(1) \\ \phi_j(2) \\ \vdots \\ \phi_j(n) \end{bmatrix}, \qquad j = 1, 2, \ldots, n. \tag{54}$$

The basis vectors are eigenvectors of the autocorrelation matrix

$$\mathbf{S} = E_i\{\mathbf{M}_i\mathbf{M}_i'\}, \tag{55}$$

where $\mathbf{M}_i'$ is the transposed vector of $\mathbf{M}_i$. Thus,

$$\mathbf{S}\,\mathbf{\Phi}_j = \lambda_j - \mathbf{\Phi}_j \tag{56}$$

in which λ_j and $\mathbf{\Phi}_j$ are the jth eigenvalue and eigenvector of $\mathbf{S}$.

For two-class pattern recognition problems, let $M_i^{(1)}$ be the measurement vectors of pattern class 1, and $M_i^{(2)}$ be the measurement vectors of pattern class 2. Both $M_i^{(1)}$ and $M_i^{(2)}$ can be expanded by the same basis vector $\mathbf{\Phi}_j$ with $j = 1, 2, \ldots, n$:

$$M_i^{(k)} = \sum_{j=1}^{n} \alpha_{ij}^{(k)} \mathbf{\Phi}_j, \qquad i = 1, 2, \ldots, N, \qquad k = 1, 2, \tag{57}$$

where $\mathbf{\Phi}_j$ is the eigenvector of the matrix $\mathbf{B}$,

$$\mathbf{B} = p(w_i)E_i\{M_i^{(k)}M_i^{(k)\prime}\}, \qquad k = 1, 2, \tag{58}$$

which is the autocorrelation matrix multiplied by a priori probability for the pattern class under consideration. In view of the orthonormal condition of the eigenvectors, the coefficient $\alpha_{ij}^{(k)}$ is given by

$$\alpha_{ij}^{(k)} = \mathbf{\Phi}_j'\mathbf{M}_i^{(k)}. \tag{59}$$

The eigenvectors $\mathbf{\Phi}_j$ associated with larger eigenvalues λ_j will contain more important information of both classes. In picture recognition and understanding, we wish to extract features which represent the difference between two texture pattern classes. In order to extract these features, we use a normalization process.

Let us assume that the autocorrelation matrices of two textural pattern classes are $\mathbf{R}_1$ and $\mathbf{R}_2$, and the autocorrelation matrix of the mixture of both classes is

$$\mathbf{R}_0 = \mathbf{R}_1 + \mathbf{R}_2. \tag{60}$$

Since $\mathbf{R}_0$ is a symmetric matrix, there exists a transformation matrix $\mathbf{A}$, such that

$$\mathbf{A}\mathbf{R}_0\mathbf{A}' = \mathbf{I}\begin{bmatrix} \lambda_1 \\ \lambda_2 \\ \vdots \\ \lambda_n \end{bmatrix}, \tag{61}$$

where $\mathbf{I}$ is the identity matrix and λ_i is an eigenvalue of $\mathbf{R}_0$. Hence,

$$\mathbf{A}(\mathbf{R}_0 + c\mathbf{I})\mathbf{A}' = \mathbf{I}\begin{bmatrix} \lambda_1 + c \\ \lambda_2 + c \\ \vdots \\ \lambda_n + c \end{bmatrix} \tag{62}$$

in which $c = \epsilon - \lambda_i$, ϵ is a small positive number, and λ_i is the smallest negative eigenvalue of $\mathbf{R}_0$. The value of c is set equal to zero when all eigenvalues are nonnegative. Then, we can find a matrix $\mathbf{P}$ such that

$$\mathbf{P}(\mathbf{R}_0 + c\mathbf{I})\mathbf{P}' = \mathbf{I}, \tag{63}$$

where

$$\mathbf{P} = \begin{bmatrix} \mathbf{e}_1/(\lambda_1 + c)^{\frac{1}{2}} \\ \mathbf{e}_2/(\lambda_2 + c)^{\frac{1}{2}} \\ \vdots \\ \mathbf{e}_n/(\lambda_n + c)^{\frac{1}{2}} \end{bmatrix} \tag{64}$$

and $\mathbf{e}_j$ is an eigenvector of $\mathbf{R}_0$. Thus, the transformed autocorrelation matrices $\mathbf{S}_1$ and $\mathbf{S}_2$ satisfy

$$\mathbf{S}_1 + \mathbf{S}_2 = \mathbf{I}, \tag{65}$$

$$\mathbf{S}_1 = \mathbf{P}(\mathbf{R}_1 + \tfrac{1}{2}c\mathbf{I})\mathbf{P}', \tag{66}$$

$$\mathbf{S}_2 = \mathbf{P}(\mathbf{R}_2 + \tfrac{1}{2}c\mathbf{I})\mathbf{P}'. \tag{67}$$

Following the above transformation, the eigenvalues of class 1 are given by

$$\mathbf{S}_1 \mathbf{\Phi}_j^{(1)} = \lambda_j^{(1)} \mathbf{\Phi}_j^{(1)}. \tag{68}$$

The eigenvalues and eigenvectors for class 2 are given by

$$\mathbf{S}_2 \mathbf{\Phi}_j^{(2)} = \lambda_j^{(2)} \mathbf{\Phi}_j^{(2)}. \tag{69}$$

In view of Eq. (65), we have

$$\mathbf{S}_1 \mathbf{\Phi}_j^{(2)} = [1 - \lambda_j^{(2)}] \mathbf{\Phi}_j^{(2)}. \tag{70}$$

Hence

$$\mathbf{\Phi}_j^{(2)} = \mathbf{\Phi}_j^{(2)} \tag{71}$$

and

$$\lambda_j^{(1)} = 1 - \lambda_j^{(2)}. \tag{72}$$

After the normalizing transformation, both classes have the same set of eigenvectors and the corresponding eigenvalues are reversely ordered. The most important features of the effective basis vectors for class 1 are the least important features for class 2, and vice versa. Therefore, by applying this normalization process, we may extract major textural features from the measurements for classification and recognition.

If we consider n eigenvectors of matrix $\mathbf{S}_1$ or $\mathbf{S}_2$ to be basis vectors of the measurements of pictures in class 1 and class 2, then the measurements of pictures in both classes can be expanded as weighted sums of these basis vectors. The weights of these vectors will be used as features for classification. If we consider

$$\mathbf{\Phi} = (\boldsymbol{\phi}_1, \boldsymbol{\phi}_2, \ldots, \boldsymbol{\phi}_m), \qquad m < n, \tag{73}$$

to be the transformation matrix, then the features are the coefficients of the K–L expansion. That is

$$\mathbf{w}_i^{(1)} = \mathbf{\Phi}'\mathbf{M}_i^{(1)}, \tag{74}$$

$$\mathbf{w}_i^{(2)} = \mathbf{\Phi}'\mathbf{M}_i^{(2)}, \tag{75}$$

where $\mathbf{\Phi}$ is an $n \times m$ matrix, $\mathbf{M}_i$ is an n-vector, and $\mathbf{w}_i^{(1)}$ and $\mathbf{w}_i^{(2)}$ are feature vectors of m dimensions.

3.2. Textural Feature Extraction Experiments

Pictures of four sample classes used for our experimental study were taken from Brodatz:

(1) Class 1—pressed cork.
(2) Class 2—pigskin.
(3) Class 3—handmade paper.
(4) Class 4—pressed calf leather.

These are illustrated in Figure 7.

To evaluate the effectiveness of the method used to extract textural features, we choose these four classes because of their general resemblance. In order to save computation time for feature extraction, we reduce the digitized image to a smaller array by taking the average value of each window of 6×6 points to be the intensity of a new point. Then we apply an equal probability quantizing technique to transform the pictures into 10 gray levels, from 0 to 9.

The second step is to generate the **G** matrices and to calculate the four function values discussed previously. We generate **G** matrices with spatial distances $d = 1, 2, 3$, and directions $\theta = 0, 45, 90, 135°$, for each digitized image. We eliminate the direction effect by taking the average and the deviation of function values calculated from four matrices corresponding to four different directions. Thus, there are 24 measurements which are independent of the direction of each picture.

Since most of those measurements are somewhat correlated, the K–L expansion is utilized to extract a set of distinct textural features from the 24 measurements for each picture. Since textural features characterizing the differences between two sample classes are desired, the normalization process is applied during the K–L expansion. The eigenvalues and basis vectors for feature extraction are given below:

Class 1 vs class 2:

$$\lambda_1 = 0.999986, \qquad \lambda_2 = 0.999494, \qquad \lambda_3 = 0.000005,$$

$$\boldsymbol{\Phi}_1 = \begin{pmatrix} 0.633655 \\ 0.745025 \\ 0.122130 \\ -0.159945 \\ -0.048515 \\ 0.019009 \\ 0.008275 \\ 0.011730 \\ -0.000152 \\ -0.000145 \\ -0.000236 \\ 0.000064 \\ -0.000425 \\ 0.000408 \\ 0.000113 \\ 0.000008 \\ -0.000004 \\ 0.000064 \\ 0.000122 \\ -0.000238 \\ -0.000023 \\ -0.000050 \\ -0.000034 \\ -0.000017 \end{pmatrix}, \quad \boldsymbol{\Phi}_2 = \begin{pmatrix} 0.108022 \\ -0.118161 \\ 0.700400 \\ -0.471838 \\ 0.247122 \\ 0.439137 \\ 0.012630 \\ -0.084099 \\ 0.004538 \\ 0.000966 \\ 0.001519 \\ -0.002259 \\ 0.001319 \\ -0.001417 \\ -0.001021 \\ -0.000041 \\ 0.000162 \\ -0.000280 \\ -0.000589 \\ -0.000291 \\ 0.000213 \\ -0.000156 \\ -0.000182 \\ -0.000105 \end{pmatrix}, \quad \boldsymbol{\Phi}_3 = \begin{pmatrix} 0.753695 \\ 0.650287 \\ 0.047944 \\ 0.080062 \\ -0.003006 \\ 0.001786 \\ -0.018568 \\ 0.002237 \\ 0.000199 \\ -0.000125 \\ -0.000034 \\ -0.000011 \\ -0.000040 \\ 0.000048 \\ -0.000034 \\ -0.000001 \\ 0.000000 \\ 0.000000 \\ 0.000000 \\ 0.000000 \\ 0.000000 \\ 0.000000 \\ 0.000000 \\ 0.000000 \end{pmatrix}.$$

Class 1 vs class 3:

$$\lambda_1 = 0.999749, \qquad \lambda_2 = 0.998610, \qquad \lambda_3 = 0.000118,$$

$$\boldsymbol{\Phi}_1 = \begin{pmatrix} 0.656790 \\ -0.623534 \\ 0.364613 \\ -0.064394 \\ -0.007769 \\ -0.206460 \\ 0.003872 \\ -0.004134 \\ 0.000928 \\ 0.000831 \\ 0.000977 \\ -0.001096 \\ -0.001183 \\ -0.000124 \\ -0.000422 \\ -0.000121 \\ 0.000009 \\ -0.000782 \\ -0.000184 \\ 0.000245 \\ -0.000177 \\ -0.000028 \\ -0.000039 \\ -0.004260 \end{pmatrix}, \quad \boldsymbol{\Phi}_2 = \begin{pmatrix} -0.178025 \\ 0.172213 \\ 0.543756 \\ -0.085251 \\ 0.783248 \\ -0.132914 \\ -0.043362 \\ 0.048662 \\ -0.006245 \\ 0.001512 \\ -0.001485 \\ -0.000461 \\ 0.002757 \\ 0.000396 \\ -0.000187 \\ -0.000705 \\ -0.000235 \\ -0.000347 \\ -0.001860 \\ 0.001276 \\ -0.000254 \\ -0.000105 \\ -0.000091 \\ 0.006992 \end{pmatrix}, \quad \boldsymbol{\Phi}_3 = \begin{pmatrix} 0.686585 \\ 0.713686 \\ 0.009294 \\ -0.080308 \\ -0.003451 \\ 0.070428 \\ 0.066783 \\ 0.057309 \\ -0.000339 \\ -0.000144 \\ -0.000075 \\ -0.000414 \\ -0.000180 \\ -0.000108 \\ 0.000006 \\ -0.000003 \\ -0.000001 \\ 0.000004 \\ 0.000000 \\ 0.000001 \\ 0.000000 \\ 0.000000 \\ 0.000000 \\ 0.000787 \end{pmatrix}.$$

Class 1 vs class 4:

$\lambda_1 = 0.999789,\qquad \lambda_2 = 0.999505,\qquad \lambda_3 = 0.000142,$

$$\boldsymbol{\Phi}_1 = \begin{bmatrix} 0.604189\\ 0.473117\\ -0.467744\\ 0.355055\\ -0.202467\\ 0.154977\\ 0.023277\\ -0.023595\\ 0.011726\\ -0.003745\\ 0.000839\\ -0.001084\\ 0.000022\\ -0.001664\\ -0.000780\\ 0.000012\\ -0.000281\\ 0.000239\\ 0.000793\\ -0.000417\\ 0.000232\\ -0.000299\\ -0.000026\\ -0.000076 \end{bmatrix},\quad \boldsymbol{\Phi}_2 = \begin{bmatrix} 0.123707\\ 0.688781\\ 0.511049\\ -0.264041\\ 0.400955\\ 0.090082\\ -0.086543\\ -0.052706\\ 0.014822\\ 0.003123\\ 0.002533\\ -0.000562\\ 0.001104\\ -0.000072\\ 0.000584\\ -0.000117\\ -0.001150\\ -0.000591\\ -0.001109\\ -0.000425\\ 0.000026\\ 0.000388\\ -0.000146\\ -0.000144 \end{bmatrix},\quad \boldsymbol{\Phi}_3 = \begin{bmatrix} 0.674781\\ -0.502508\\ 0.431831\\ -0.050713\\ -0.003684\\ 0.320706\\ -0.015069\\ 0.003241\\ 0.001401\\ -0.000925\\ 0.000425\\ -0.000465\\ 0.000081\\ -0.000339\\ -0.000243\\ -0.000005\\ 0.000001\\ 0.000001\\ 0.000001\\ -0.000001\\ 0.000000\\ 0.000000\\ 0.000000\\ 0.000000 \end{bmatrix}.$$

Class 2 vs class 3:

$\lambda_1 = 0.999994,\qquad \lambda_2 = 0.999186,\qquad \lambda_3 = 0.000065,$

$$\boldsymbol{\Phi}_1 = \begin{bmatrix} 0.752889\\ 0.051171\\ -0.046389\\ 0.045228\\ 0.055395\\ 0.016778\\ -0.038060\\ 0.011584\\ -0.000375\\ -0.001728\\ 0.000116\\ -0.000598\\ -0.000257\\ -0.000312\\ -0.000174\\ -0.000072\\ -0.000075\\ -0.000158\\ 0.000176\\ 0.000168\\ 0.000005\\ 0.000012\\ 0.000006\\ -0.000899 \end{bmatrix},\quad \boldsymbol{\Phi}_2 = \begin{bmatrix} -0.105318\\ 0.118553\\ 0.207141\\ 0.750670\\ -0.272518\\ 0.411763\\ 0.339476\\ -0.095543\\ -0.000513\\ 0.013192\\ -0.000589\\ 0.004758\\ 0.003147\\ 0.003323\\ 0.000663\\ 0.000256\\ 0.000332\\ 0.001494\\ -0.000042\\ 0.000280\\ 0.000167\\ 0.000068\\ 0.000038\\ 0.005735 \end{bmatrix},\quad \boldsymbol{\Phi}_3 = \begin{bmatrix} -0.613864\\ 0.721326\\ -0.206584\\ -0.142289\\ -0.169709\\ -0.103941\\ -0.007299\\ 0.015916\\ 0.004866\\ -0.000580\\ -0.000138\\ 0.000052\\ 0.000032\\ 0.000452\\ -0.000015\\ -0.000008\\ -0.000003\\ 0.000001\\ -0.000001\\ 0.000000\\ 0.000000\\ 0.000000\\ 0.000000\\ -0.000765 \end{bmatrix}.$$

Class 2 vs class 4:

$$\lambda_1 = 0.999997, \qquad \lambda_2 = 0.999703, \qquad \lambda_3 = 0.000021,$$

$$\Phi_1 = \begin{bmatrix} 0.760414 \\ 0.646298 \\ -0.020592 \\ 0.053055 \\ 0.019727 \\ -0.007857 \\ -0.019045 \\ 0.004179 \\ -0.000246 \\ 0.000344 \\ 0.000181 \\ -0.000052 \\ -0.000014 \\ -0.000002 \\ 0.000008 \\ -0.000001 \\ -0.000084 \\ -0.000000 \\ 0.000019 \\ 0.000006 \\ 0.000008 \\ 0.000005 \\ 0.000022 \\ 0.000171 \end{bmatrix}, \quad \Phi_2 = \begin{bmatrix} -0.161119 \\ 0.156840 \\ 0.558736 \\ 0.707298 \\ -0.080589 \\ 0.359174 \\ 0.029385 \\ 0.024672 \\ -0.000776 \\ -0.001530 \\ -0.000336 \\ 0.001518 \\ -0.000322 \\ -0.000116 \\ 0.000024 \\ 0.000016 \\ 0.001305 \\ 0.000027 \\ -0.000856 \\ 0.000402 \\ 0.000069 \\ 0.000059 \\ -0.001424 \\ 0.001695 \end{bmatrix}, \quad \Phi_3 = \begin{bmatrix} -0.612732 \\ 0.799133 \\ -0.333712 \\ 0.010289 \\ 0.024560 \\ -0.084337 \\ 0.049972 \\ 0.001592 \\ 0.001797 \\ -0.001053 \\ 0.000259 \\ 0.000157 \\ 0.000006 \\ -0.000005 \\ 0.000005 \\ 0.000003 \\ 0.000002 \\ 0.000002 \\ 0.000000 \\ 0.000000 \\ 0.000000 \\ 0.000000 \\ 0.000484 \\ 0.000005 \end{bmatrix}.$$

Class 3 vs class 4:

$$\lambda_1 = 0.999969, \qquad \lambda_2 = 0.999394, \qquad \lambda_3 = 0.000061,$$

$$\Phi_1 = \begin{bmatrix} 0.658098 \\ 0.671880 \\ -0.240818 \\ 0.053583 \\ 0.232944 \\ -0.018415 \\ -0.002592 \\ -0.002206 \\ 0.000029 \\ 0.001725 \\ -0.000546 \\ 0.000102 \\ -0.000931 \\ 0.000004 \\ 0.000034 \\ -0.000016 \\ -0.000020 \\ 0.000115 \\ -0.000123 \\ 0.000028 \\ -0.000051 \\ -0.000095 \\ -0.000046 \\ -0.000932 \end{bmatrix}, \quad \Phi_2 = \begin{bmatrix} 0.191992 \\ -0.24463 \\ -0.267368 \\ 0.828840 \\ -0.308258 \\ 00.011936 \\ -0.215905 \\ -0.055604 \\ 0.002977 \\ 0.002731 \\ -0.002018 \\ -0.003250 \\ -0.002261 \\ 0.000035 \\ 0.000229 \\ 0.000026 \\ -0.000790 \\ 0.001268 \\ 0.000549 \\ 0.000031 \\ 0.000048 \\ 0.000048 \\ 0.000790 \\ 0.001917 \end{bmatrix}, \quad \Phi_3 = \begin{bmatrix} 0.677512 \\ -0.474742 \\ 0.526569 \\ -0.111193 \\ 0.037404 \\ 0.155987 \\ 0.002836 \\ -0.014784 \\ -0.000785 \\ 0.000076 \\ -0.000051 \\ 0.000085 \\ 0.000061 \\ -0.000013 \\ -0.000002 \\ -0.000003 \\ -0.000003 \\ -0.000002 \\ 0.000000 \\ 0.000000 \\ 0.000000 \\ 0.000000 \\ -0.000051 \\ 0.000074 \end{bmatrix}.$$

The feature vectors, $\mathbf{w}_i$, extracted from the textural patterns are tabulated below:

(a) For class 1 and class 2,

$$\mathbf{w}_1 = \begin{bmatrix} 1.159125 \\ 5.252234 \\ -0.284561 \end{bmatrix}, \qquad \mathbf{w}_2 = \begin{bmatrix} 0.551217 \\ 2.333391 \\ 0.212000 \end{bmatrix}.$$

(b) For class 1 and class 3,

$$\mathbf{w}_1 = \begin{bmatrix} -0.353544 \\ -0.163591 \\ 3.063955 \end{bmatrix}, \qquad \mathbf{w}_3 = \begin{bmatrix} -0.255401 \\ -0.086797 \\ 2.829094 \end{bmatrix}.$$

(c) For class 1 and class 4,

$$\mathbf{w}_1 = \begin{bmatrix} 4.749453 \\ 2.740117 \\ -3.853419 \end{bmatrix}, \qquad \mathbf{w}_4 = \begin{bmatrix} 5.060804 \\ 2.593646 \\ -2.833081 \end{bmatrix}.$$

(d) For class 2 and class 3,

$$\mathbf{w}_2 = \begin{bmatrix} -2.740342 \\ 3.556821 \\ -0.009156 \end{bmatrix} \qquad \mathbf{w}_3 = \begin{bmatrix} -3.834904 \\ 2.133827 \\ 0.213698 \end{bmatrix}.$$

(e) For class 2 and class 4,

$$\mathbf{w}_2 = \begin{bmatrix} -2.583534 \\ 0.275734 \\ -0.049975 \end{bmatrix} \qquad \mathbf{w}_4 = \begin{bmatrix} -4.009842 \\ 4.885375 \\ 0.389147 \end{bmatrix}.$$

(f) For class 3 and class 4,

$$\mathbf{w}_3 = \begin{bmatrix} 5.432416 \\ 3.094403 \\ -1.286592 \end{bmatrix}, \qquad \mathbf{w}_4 = \begin{bmatrix} 5.688379 \\ 3.248096 \\ -1.261984 \end{bmatrix}.$$

The procedure is implemented on a PDP-11/40 computer. It takes only a few seconds to extract the textural features from a digitized image.

3.3. Eigenvalues of the G *matrix*

In the study of aerial photographs for forestry research, we have proposed the use of eigenvalues of the **G** matrix as measurements of textural information.

FIG. 7. Four classes of sample pictures.

The eigenvalues for a sample picture are used as a measurement vector. By generalizing the **G**-matrix concept, we introduce the **E** matrix. If the elements of the **G** matrix are g_{ij}, then the elements of the **E** matrix are

$$e_{ij} = f(i, j)g_{ij}, \tag{76}$$

where $f(i, j)$ is an intensity weighting function. We use the eigenvalues of the **E** matrix to form a measurement vector.

The original number of gray levels is 256. We quantize it to 16 levels, from which we determine a 16 × 16 **E** matrix. In order to eliminate the variation due to the rotational effect of the picture, we calculate the eigenvalues for **E** matrices in four directions 0, 45, 90, 135°, and take the average. Let the measurement vectors in terms of the eigenvalues be

$$\begin{bmatrix} a_{1,1} \\ a_{2,1} \\ \vdots \\ a_{16,1} \end{bmatrix} \quad \begin{bmatrix} a_{1,2} \\ a_{2,2} \\ \vdots \\ a_{16,2} \end{bmatrix} \quad \begin{bmatrix} a_{1,3} \\ a_{2,3} \\ \vdots \\ a_{16,3} \end{bmatrix} \quad \begin{bmatrix} a_{1,4} \\ a_{2,4} \\ \vdots \\ a_{16,4} \end{bmatrix}$$

where $a_{i,j}$ are eigenvalues for the jth direction. These four directional measurement vectors are merged to form a measurement vector $(b_1, b_2, \ldots, b_{16})'$, in which

$$b_i = \tfrac{1}{4} \sum_{j=1}^{4} a_{ij}.$$

To apply this method, we scan the photograph by a square window. For each window area we determine the **G** matrices and the **E** matrices. The eigenvalues for each matrix upon directional averaging form a measurement vector for that window area. This method has been applied with some success to the identification of seven vegetation types from aerial photographs. The seven types are cypress, melaleuca, palmetto, pine, mixed stands, pasture, and Florida ash. The eigenvalues associated with the **E** matrix increase the separability of the vegetation from its background. Research is under way on the selection of the intensity weighting function $f(i, j)$ in Eq. (76).

4. PIXEL-VECTOR CLUSTERING TECHNIQUE

In this section we present a technique for extracting information patterns in a picture for image segmentation, classification, storage, transmission, and understanding. When a digital computer is employed for the determination of information patterns in pictures, the picture is usually scanned to generate a digital image consisting of a matrix of pixels. We consider the pixel matrix as an image of the picture. Each pixel is characterized by a property vector, which we will refer to as a pixel vector. Pixels with certain properties contain a type of information in the picture. Thus, information patterns in a picture may be automatically extracted by grouping pixel vectors into clusters. Pixel vectors offer a good medium for modeling images. When the clusters are formed, we will be able to segment the image according to cluster characteristics and to store in the database the coordinates of strategic pixels together with the global properties of each cluster. From the information patterns, objects in the original picture can be identified and a picture containing all the essential information of the original picture can be reconstructed. The information patterns can be transmitted to a remote station and can then be used for the regeneration of the picture.

4.1. Characterization of Pixel Vectors

A picture digitizer transforms the data in a picture into a digitized image which can be accessed by a digital computer. An optical picture can be represented mathematically by a real function of f on a picture plane P which is a simply connected subset of the real plane such that

$$f: P \rightarrow R \tag{77}$$

where R is the set of intensity values of the picture points. A picture plane may be quantized in two principal ways: the hexagonal grid and the rectangular grid. Hexagonal grids have the advantage of having six nearest-neighboring picture points for every picture point p. They have the drawback of being based on an uncommon, nonorthogonal coordinate system. Rectangular grids contain only four nearest-neighboring points for every picture point p. They offer the advantage of easy access to every picture point. The rectangular grid forms an orthogonal coordinate system. Thus, the picture plane may be represented by

$$I = I_x \times I_y, \tag{78}$$

where I_x and I_y are subsets of the integer set.

The intensities of a picture are quantized into n levels, where $n = 2^k$ to maximize storage efficiency within bit-oriented digital computers. We call a 2^k level picture a k-bit picture. An n-level digitized picture may be characterized by a mapping

$$\phi : I \to N \tag{79}$$

where N is the set $\{0, 1, 2, \ldots, n - 1\}$ of quantized intensity values. A digitized picture can also be represented as a matrix as mentioned earlier. The relative location of a pixel is specified by the location of the element in the matrix. The intensity of a pixel is given by the value of the corresponding element in the matrix. A picture digitizer performs a transformation from mapping f to a new mapping ϕ. Once a digitized image is obtained the picture data is then readily accessible by a digital computer.

The above discussion reviews the traditional characterization of a pixel by its intensity value and its location. In the pixel-vector clustering approach, we characterize a pixel by its location in the picture plane and by a property vector which contains the intensity value as an element of the vector. The other elements could be the intensity gradients in four neighboring directions, if a rectangular grid scheme is used, or in six neighboring directions, if a hexagonal grid scheme is used. Other characterizations may include intensity variation patterns, contextual information, or neighborhood statistics for the pixel under consideration.

When a multispectral scanner is used, each channel will produce an intensity representation for the pixel. If the scanner is designed to respond to K wavelength bands, the pixel may be characterized by a K-dimensional property vector. For instance, if the scanner responds to light in the 0.40 to 0.44, 0.58 to 0.62, 0.66 to 0.72, and 0.80 to 1.00 μm wavelength bands, each pixel of the scanned region can be characterized by a four-dimensional property vector [17]. These ranges are in the violet, green, red, and infrared bands, respectively. The elements of the property vector represent a level of violet, a level of green, a level of red, and a level of infrared, respectively. In general, we denote the pixel vector as a n-vector

$$\mathbf{x} = \begin{bmatrix} x_1 \\ x_2 \\ \vdots \\ x_n \end{bmatrix}. \tag{80}$$

Each element describes a specific property of the pixel. In our study of aerial photographs for forestry applications, we have used the intensity value and the intensity gradients for the characterization of pixel vectors. Research has been undertaken to develop a theoretical framework for the characterization of pixel vectors, to determine the dimensions of the vector which are needed for automatic

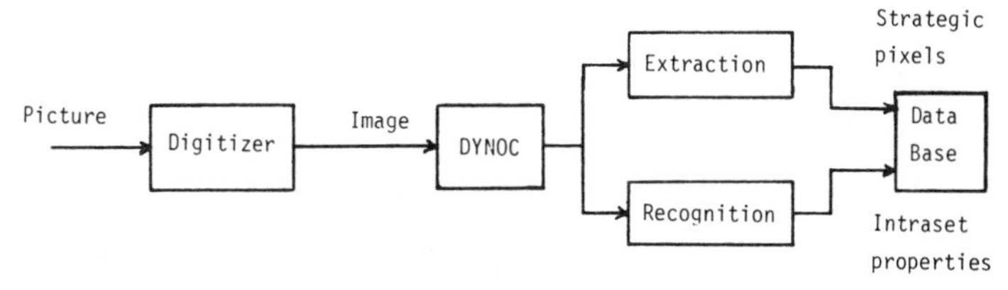

FIGURE 8

extraction of information patterns in an image, to evaluate the relative importance of the characterizing measurements, and to find strategic pixels.

After a picture has been characterized by appropriate pixel vectors, automatic

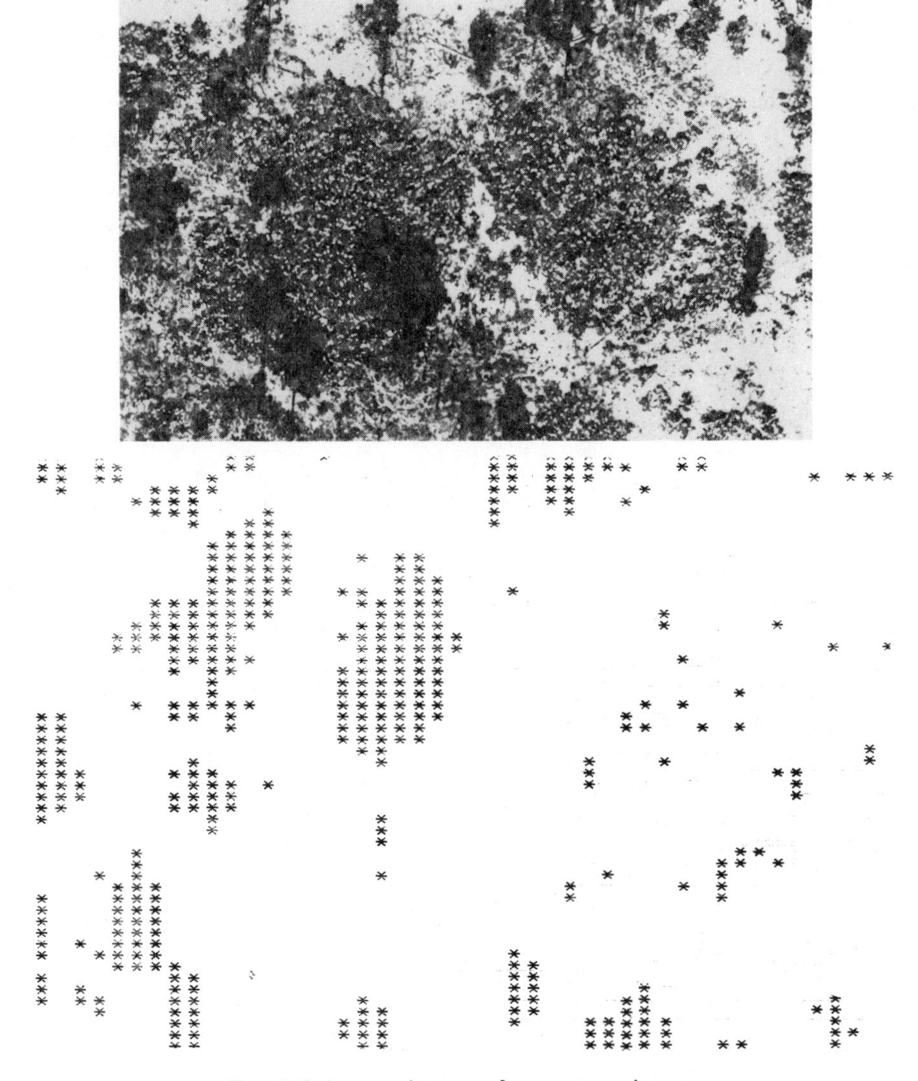

FIG. 9. Palmetto picture and computer printout.

clustering is performed on these vector points to extract information patterns. We have developed the DYNOC algorithm [13] for automatic clustering of pixel vectors. DYNOC is a dynamic optimal cluster-seeking algorithm which is more efficient than the commonly used clustering methods. The procedures involved in the pixel-vector clustering approach are summarized in the block diagram shown in Fig. 8.

4.2. Experimental Results on Vegetation Identification

The pixel-vector clustering technique has been applied to segmentation and identification of vegetation in aerial photographs. We have considered aerial photographs for four vegetation types: melaleuca, palmetto, pine, and pasture. The sizes of the four scenes which we have chosen are 25 mm $\times$ 20 mm, 25 mm $\times$ 20 mm, 25 mm $\times$ 8 mm, and 27 mm $\times$ 15 mm, respectively. The raster size of the pixels is 100 μm. The melaleuca scene contains melaleuca, shadow, and light areas. The palmetto scene contains the ground and two types of vegetation, one of which is palmetto. The pine scene contains pine, shadow, and ground. The pasture scene contains road and two different vegetation types, one of which is pasture. Our preliminary study indicates that this technique yields promising results. Shown in Fig. 9 are the palmetto photograph and the computer printout resulting from the pixel-vector clustering approach. More comprehensive results will be reported at a later date.

ACKNOWLEDGMENT

The work reported in this paper was supported in part by the National Science Foundation.

REFERENCES

1. M. R. Anderberg, *Cluster Analysis for Applications*, Academic Press, New York, 1973.
2. D. Ausherman, *Texture Discrimination within Digital Imagery*, Ph.D. dissertation, University of Missouri, December 1972.
3. J. E. P. Box and G. M. Jenkins, *Time Series Analysis*, Holden–Day, San Francisco, 1970.
4. P. Brodatz, *Textures*, Dover, New York, 1966.
5. L. Carlucci, A formal system for texture languages, *Pattern Recognition* **4**, January 1972.
6. Y. T. Chien and K. S. Fu, On the generalized Karhunen–Loéve expansion, *IEEE Trans. Inform. Theory*, **IT-13**, July 1967.
7. R. W. Conners and C. A. Harlow, Some theoretical considerations concerning textural analysis of radiographic images, in *Proceedings of the 1976 IEEE Conference on Decision and Control, 1976.*
8. R. M. Haralick, K. Shanmugam, and I. Dinstein, Texture features for image classification, *IEEE Trans. Systems Man Cybernet.* **SMC-3**, November 1973.
9. J. J. Hawkins, Textural properties for pattern recognition, in *Picture Processing and Psychopictorics* (B. S. Lipkin and A. Rosenfeld, Eds.), Academic Press, New York, 1970.
10. R. P. Kruger, W. B. Thompson, and A. F. Turner, Computer diagnosis of pneumoconiosis, *IEEE Trans. Systems Man Cybernet.* **SMC-4**, January 1974.
11. B. H. McCormick and S. N. Jayaramaurthy, Time series model for texture synthesis, *Internat. J. Computer Inform. Sci.* **3**, 1974.
12. A. Rosenfeld, A note on automatic detection of texture gradients, *IEEE Trans. Computers* **C-24**, October 1975.

13. J. T. Tou, DYNOC—A dynamic optimal cluster-seeking technique, *Internat. J. Computer Inform. Sci.* **8**, 1979.
14. J. T. Tou (Ed.), *Advances in Information Systems Science,* Vol. 3, Plenum, New York, 1971.
15. J. T. Tou and Y. S. Chang, An approach to texture pattern analysis and recognition. in *Proceedings of the 1976 IEEE Conference on Decision and Control, 1976.*
16. J. T. Tou and Y. S. Chang, Picture understanding by machine via textural feature extraction, in *Proceedings of 1977 IEEE Pattern Recognition and Image Processing Conference, 1977.*
17. J. T. Tou and R. C. Gonzalez, *Pattern Recognition Principles*, Addison–Wesley, Reading, Mass., 1974.
18. J. T. Tou and R. P. Heydorn, Some approaches to optimum feature selection, in *Computer and Information Sciences—II* (J. T. Tou, Ed.), Academic Press, New York, 1967.
19. J. T. Tou, D. B. Kao, and Y. S. Chang, Pictorial texture analysis and synthesis, in *Proceedings of the Third International Joint Conference on Pattern Recognition, 1976.*
20. J. S. Weszka, C. R. Dyer, and A. Rosenfeld, A comparative study of texture measures for terrain classification, *IEEE Trans. Systems Man Cybernet.* **SMC-6**, April 1976.

Finding Structure in Co-Occurrence Matrices for Texture Analysis *

STEVEN W. ZUCKER AND DEMETRI TERZOPOULOS

Computer Vision and Graphics Laboratory, Department of Electrical Engineering, McGill University, Montreal, Quebec, Canada

Co-occurrence matrices are a popular representation for the texture in images. They contain a count of the number of times that a given feature (e.g., a given gray level) occurs in a particular spatial relation to another given feature. However, because of the large number of spatial relations that are possible within an image, heuristic or interactive techniques have usually been employed to select the relation to use for each problem. In this paper we present a statistical approach to finding those spatial (or other) relations that best capture the structure of textures when the co-occurrence matrix representation is used. These matrices should thus be well suited for discriminations that are structurally based.

1. INTRODUCTION

Texture can be viewed as a global pattern arising from the repetition, either deterministically or randomly, of local subpatterns [11]. The structure resulting from this repetition is often important in discriminating between different textures. In this paper, we present a statistical approach to selecting a description from within a well-known class of representations that best reflects such structure. The class of representations is the co-occurrence matrix introduced by Haralick [6], which has been used for remote sensing, biomedical, and many other application areas (see, e.g., the recent survey [5]).

The co-occurrence matrix is essentially a two-dimensional histogram of the number of times that pairs of intensity values (or, more generally, arbitrary local features) occur in a given spatial relationship. Thus, it forms a summary of the subpatterns that could be formed by intensity pairs and the frequency with which they occur.

The success of this kind of representation is tied directly to the fidelity with which it captures the structure of the underlying texture. To understand the importance of selecting the spatial relationship properly, consider a textural pattern with a dominant banded (i.e., almost constant) structure in the horizontal direction, but with a random structure in the vertical direction. If the co-occurrence matrix were built using a spatial relationship between vertical

* This research was supported by MRC Grant MA-6154.

ISBN 0-12-597320-9

pairs of pixels (at some separation), then it would appear as if the underlying texture were perfectly random. On the other hand, if a horizontal spatial relationship were used, then it would appear as if the texture were highly structured. For a co-occurrence matrix computed as an average over several directions, the description would be somewhere between these two extremes.

In the above example, only one variable (orientation) was considered in defining the spatial relationship on which to base the co-occurrence description, and this variable was shown to be critical in capturing the structure of the texture. In more realistic situations, such dependencies are rarely that obvious. Furthermore, similar difficulties arise in attempting to choose a spatial relation on which to base first-order gray-level statistics [10] or generalized co-occurrence matrices [3]. In fact, for generalized co-occurrence matrices, the situation is further complicated by the wide choice of local features that can be used.

The specific problem to be addressed in this paper is the selection of spatial relationships for defining co-occurrence matrices such that these matrices maximally reflect the structure of the underlying texture. The model adopted for structure is statistical and is based on a (chi-square) measure of independence of the rows and columns of the co-occurrence matrix. Thus it provides quantitative support for the design of co-occurrence-based classification schemes, and supplements the interactive and heuristic (in particular, the exhaustive [10]) tools currently available.

The measure of independence is formulated by interpreting the co-occurrence matrix as a contingency table [7]. These notions are formally defined in the next two sections, and are followed by example applications to several Brodatz [2] textures and to the LANDSAT-1 images considered in [10]. The dependency measure clearly indicates the size of the structural units in the Brodatz textures. It further provides quantitative corroboration for Weszka, Dyer, and Rosenfeld's empirical observations about the merits of certain spatial relationships for the LANDSAT images. Finally, it leads to the specification of co-occurrence matrices upon which successful classifiers can be designed.

2. CO-OCCURRENCE MATRICES FOR TEXTURE CLASSIFICATION

In this section, we formally define co-occurrence matrices and list several feature functions that can be computed over them. Such features usually provide the basis for classification in practical problems.

Let f be a rectangular, discrete picture containing a finite number of gray levels. f is defined over the domain

$$\mathfrak{D} = \{(i, j) \,|\, i \in [0, n_i),\ j \in [0, n_j),\ i, j \in I\}$$

by the relation

$$f = \{((i, j), k) \,|\, (i, j) \in \mathfrak{D},\ k = f(i, j),\ k \in [0, n_g),\ k \in I\},$$

where I is the set of integers, n_i and n_j are the horizontal and vertical dimensions of f, and n_g is the number of gray levels in f.

The unnormalized co-occurrence matrix, Ψ, is a square matrix of dimension n_g and is a function of both the image f and a displacement vector $\bar{d} = [\Delta i, \Delta j]$

in the (i, j) plane; i.e.,

$$\Psi(f, \bar{d}) = \{\psi_{ij}(f, \bar{d})\}.$$

Its entries, ψ_{ij}, are the unnormalized frequencies of co-occurring gray levels in f which are separated by the spatial relation $\bar{d}$:

$$\psi_{ij}(f, \bar{d}) = \#\{((k_1, l_1), (k_2, l_2)) \mid (k_1, l_1), (k_2, l_2) \in \mathfrak{D}, \\ f(k_1, l_1) = i, f(k_2, l_2) = j, [k_2, l_2] - [k_1, l_1] = \bar{d}\},$$

where # denotes the number of elements in the set.

Symmetric co-occurrence matrices are generated by pooling frequencies of gray-level occurrences that are separated by both $\bar{d}$ and $-\bar{d}$. That is, the symmetric co-occurrence matrix, Φ, is defined by the relation

$$\Phi(f, \bar{d}) = \Psi(f, \bar{d}) + \Psi(f, -\bar{d}).$$

For the remainder of this paper, we shall deal with symmetric co-occurrence matrices exclusively.

It is often convenient to normalize co-occurrence matrices so that they approximate discrete joint probability densities of co-occurring gray levels. The appropriate normalization is accomplished by dividing each entry in the co-occurrence matrix by the total number of paired occurrences

$$\Phi_N = \frac{1}{N}\Phi, \qquad \text{where} \qquad N = \sum_i \sum_j \phi_{ij}.$$

Texture classification is usually accomplished by using certain characteristic features of co-occurrence matrices. That is, the values of a number of feature functions can be used to summarize the content of the matrices. Fourteen such functions were introduced by Haralick *et al.* [6], four of which appear to be the most widely used in practice. These functions are listed below since they are used for the classification experiments described in Section 5.

(1) Angular second moment: $\text{ASM} = \sum_i \sum_j \phi_{ij}^2;$

(2) Contrast: $\text{CON} = \sum_i \sum_j (i - j)^2 \phi_{ij};$

(3) Correlation: $\text{COR} = \sum_i \sum_j (ij\phi_{ij} - \mu_x\mu_y)/(\sigma_x\sigma_y);$

(4) Entropy: $\text{ENT} = -\sum_i \sum_j \phi_{ij} \log \phi_{ij}.$

ϕ_{ij} denotes the (i, j)th element of a normalized co-occurrence matrix Φ_N, μ_x, and σ_x are the mean and standard deviation of the marginal probability vector, obtained by summing over the rows of Φ_N, and μ_y and σ_y are the corresponding statistics for the column sums.

3. CONTINGENCY TABLES AND χ^2 SIGNIFICANCE TESTS

A new point of view toward co-occurrence matrices can be developed by interpreting intensity pairs (or local feature values) in an image as samples ob-

FIG. 1. A banded texture (gray levels 1 to 8 repeated every eight columns).

tained from a (two-dimensional) random process. The rows and columns of a co-occurrence matrix separate the samples into various classes based on observed intensities. The matrix thus tabulates the frequencies of samples belonging to each class.

Such tables have been used for some time in statistics, and are called "contingency" tables. They typically take the following form:

A \ B	$1 \cdots j \cdots n$	Row totals
1	$x_{11} \cdots x_{1j} \cdots x_{1n}$	r_1
$\vdots$	$\vdots \quad \vdots \quad \vdots$	$\vdots$
i	$x_{i1} \cdots x_{ij} \cdots x_{in}$	r_i
$\vdots$	$\vdots \quad \vdots \quad \vdots$	$\vdots$
m	$x_{m1} \cdots x_{mj} \cdots x_{mn}$	r_m
Column totals	$c_1 \cdots c_j \cdots c_n$	N

In the above table, x_{ij} is the number of times that variable A has been observed to fall into class i while B has fallen into class j. Furthermore, the various row and column totals are indicated.

If we interpret variables A and B as image pixels at either end of $\bar{d}$, and the classes into which they can fall as the gray levels $[0, n_g)$, the correspondence between co-occurrence matrices and contingency tables becomes clear. An unnormalized co-occurrence matrix, Ψ, is a square contingency table with $m = n = n_g$.

The importance of adopting this interpretation of co-occurrence matrices is that it allows the formulation of a precise statistical measure for the amount of textural structure that is contained in any particular matrix. For motivation, we present the following simple example. Consider a highly structured, one-dimensional, banded texture (Fig. 1) that repeats every eight columns. If we restrict the displacement vector $\bar{d}$ to lie in the horizontal direction, different

TABLE 1

Co-Occurrence Matrices, over Two Spatial Relationships, for the Banded Texture of Fig. 1[a]

Spatial relation: $\bar{d} = [3, 0]$

0.0000	0.0000	0.0000	0.0574	0.0000	0.0656	0.0000	0.0000
0.0000	0.0000	0.0000	0.0000	0.0656	0.0000	0.0574	0.0000
0.0000	0.0000	0.0000	0.0000	0.0000	0.0656	0.0000	0.0574
0.0574	0.0000	0.0000	0.0000	0.0000	0.0000	0.0656	0.0000
0.0000	0.0656	0.0000	0.0000	0.0000	0.0000	0.0000	0.0656
0.0656	0.0000	0.0656	0.0000	0.0000	0.0000	0.0000	0.0000
0.0000	0.0574	0.0000	0.0656	0.0000	0.0000	0.0000	0.0000
0.0000	0.0000	0.0574	0.0000	0.0656	0.0000	0.0000	0.0000

Spatial relation: $\bar{d} = [8, 0]$

0.1250	0.0000	0.0000	0.0000	0.0000	0.0000	0.0000	0.0000
0.0000	0.1250	0.0000	0.0000	0.0000	0.0000	0.0000	0.0000
0.0000	0.0000	0.1250	0.0000	0.0000	0.0000	0.0000	0.0000
0.0000	0.0000	0.0000	0.1250	0.0000	0.0000	0.0000	0.0000
0.0000	0.0000	0.0000	0.0000	0.1250	0.0000	0.0000	0.0000
0.0000	0.0000	0.0000	0.0000	0.0000	0.1250	0.0000	0.0000
0.0000	0.0000	0.0000	0.0000	0.0000	0.0000	0.1250	0.0000
0.0000	0.0000	0.0000	0.0000	0.0000	0.0000	0.0000	0.1250

[a] Note that the matrix becomes diagonal for distances that are integer multiples of eight pixels.

values of $|\bar{d}|$ would lead to co-occurrence matrices with very different entries. In particular, a choice of $|\bar{d}| = 8 \cdot n$, where n is an integer, would lead to a diagonal Φ (Table 1), conveying the banded structure; any other choice would miss the important regularity, thus falsely indicating a more random underlying texture. This example shows that different choices for $\bar{d}$ can lead to very different descriptions of the same textural pattern.

Our notion of structure, conveyed by co-occurrence matrices, is related to the strength of the statement that can be made about variable B given observations about A (and vice versa). If the texture is highly structured and the co-occurrence matrix reflects this structure, then observations about A should bias the probabilities of observing various classes for B. On the other hand, if the structure is not being captured, then observations about A will not influence the probabilities for B. In other words, A and B will be independent. The amount of structure conveyed by a co-occurrence matrix clearly depends on the choice of variables A and B; that is, it is a function of $\bar{d}$.

A quantitative measure of this structure can be obtained by hypothesizing that the variables A and B are independent, and then using a chi-square goodness of fit test to determine the degree to which this hypothesis can be rejected by the observed data (i.e., the image). For textures with a lot of structure, it should be rejected overwhelmingly. Operationally, the hypothesis translates into a statement about row/column independence in contingency tables.

In general, the chi-square test is employed to determine whether observed frequencies of occurrences in a set of randomly drawn samples appear to have been drawn from an assumed distribution. The test involves calculation of the quantity

$$\chi^2 = \sum_{i=1}^{k} \frac{(x_i - e_i)^2}{e_i}, \tag{1}$$

where the x_i and e_i represent observed and expected frequencies in the ith class, respectively. As the number of samples drawn approaches infinity the above distribution approaches that of a chi-square function with $k - 1$ degrees of freedom.

The chi-square test can still be applied in situations where the expected frequencies depend on unknown parameters, if maximum likelihood estimators are used to estimate these parameters ([7, Sect. 9.6]; but see also [1, Sect. 3.1.1]). Furthermore, it becomes necessary to subtract one degree of freedom for each parameter estimated.

To formulate the independence hypothesis consider an arbitrary contingency table. Let p_{ij} be the probability corresponding to the cell in the ith row and jth column, and let $p_{i\cdot}$ be the probability corresponding to the ith row and $p_{\cdot j}$ the probability corresponding to the jth column. The hypothesis that the two variables, A and B, are independent may now be written as

$$H_0\colon p_{ij} = p_{i\cdot}p_{\cdot j}, \qquad i = 1, \ldots, m, \; j = 1, \ldots, n.$$

If N denotes the total number of samples, i.e.,

$$N = \sum_{i=1}^{m} \sum_{j=1}^{n} x_{ij},$$

then the measure of compatibility between observed and expected frequencies is (from (1))

$$\chi^2 = \sum_{i=1}^{m} \sum_{j=1}^{n} \frac{(x_{ij} - Np_{ij})^2}{Np_{ij}}$$

$$= \sum_{i=1}^{m} \sum_{j=1}^{n} \frac{(x_{ij} - Np_{i\cdot}p_{\cdot j})^2}{Np_{i\cdot}p_{\cdot j}} \tag{2}$$

under hypothesis H_0. Since $p_{i\cdot}$ and $p_{\cdot j}$ are unknown, it is necessary to compute their maximum likelihood estimates. First, however, note that

$$\sum_{i=1}^{m} p_{i\cdot} = 1 \qquad \text{and} \qquad \sum_{j=1}^{n} p_{\cdot j} = 1. \tag{3}$$

Thus $(m - 1) + (n - 1) = m + n - 2$ parameters must be estimated. Therefore, the number of degrees of freedom for testing H_0 is

$$\begin{aligned}\nu &= (\text{number of cells}) - 1 - (\text{number of estimated parameters}) \\ &= (m - 1)(n - 1).\end{aligned} \tag{3.1}$$

To determine the maximum likelihood estimates of $p_{i\cdot}$ and $p_{\cdot j}$, note that the data samples are discrete and independent. Then the likelihood of the sample (i.e., the probability of obtaining the sample in the order of its occurrence) is

$$L = \prod_{i=1}^{m} \prod_{j=1}^{n} p_{ij}^{x_{ij}}$$

which, under H_0, becomes

$$\begin{aligned} L &= \prod_{i=1}^{m} \prod_{j=1}^{n} (p_{i\cdot} p_{\cdot j})^{x_{ij}} \\ &= \prod_{i=1}^{m} \prod_{j=1}^{n} p_{i\cdot}^{x_{ij}} \prod_{i=1}^{m} \prod_{j=1}^{n} p_{\cdot j}^{x_{ij}} \\ &= \prod_{i=1}^{m} p_{i\cdot}^{\sum_{j=1}^{n} x_{ij}} \prod_{j=1}^{n} p_{\cdot j}^{\sum_{i=1}^{m} x_{ij}} \\ &= \prod_{i=1}^{m} p_{i\cdot}^{r_i} \prod_{j=1}^{n} p_{\cdot j}^{c_j}, \end{aligned}$$

where

$$r_i = \sum_{j=1}^{n} x_{ij}$$

and

$$c_j = \sum_{i=1}^{m} x_{ij}$$

are the sums of the frequencies in the ith row and jth column, respectively. It is convenient to express one of the $p_{\cdot j}$, say $p_{\cdot n}$, in terms of the others by using relation (3).

Hence

$$L = \left(1 - \sum_{j=1}^{n-1} p_{\cdot j}\right)^{c_n} \prod_{i=1}^{m} p_{i\cdot}^{r_i} \prod_{j=1}^{n-1} p_{\cdot j}^{c_j}.$$

Taking logarithms, we have

$$\log L = c_n \log \left(1 - \sum_{j=1}^{n-1} p_{\cdot j}\right) + \sum_{i=1}^{m} r_i \log p_{i\cdot} + \sum_{j=1}^{n-1} c_j \log p_{\cdot j}.$$

A maximum likelihood estimate of $p_{\cdot j}$ may be obtained by differentiating with respect to $p_{\cdot j}$ and setting the derivative to zero:

$$\frac{\partial \log L}{\partial p_{\cdot j}} = -c_n / \left(1 - \sum_{j=1}^{n-1} p_{\cdot j}\right) + \frac{c_j}{p_{\cdot j}} = 0.$$

Now

$$1 - \sum_{j=1}^{n-1} p_{\cdot j} = p_{\cdot n}.$$

Thus

$$p_{\cdot j} = \frac{p_{\cdot n}}{c_n} c_j = \lambda c_j,$$

where λ is independent of j. Since this must be true for all $j = 1, \ldots, n$,

$$1 = \sum_{j=1}^{n} p_{\cdot j} = \lambda \sum_{j=1}^{n} c_j = \lambda N.$$

Therefore, $\lambda = 1/N$, and the maximum likelihood estimate of $p_{\cdot j}$ is

$$\hat{p}_{\cdot j} = \frac{c_j}{N}.$$

Similarly, the maximum likelihood estimate of $p_{i\cdot}$ can be shown to be

$$\hat{p}_{i\cdot} = \frac{r_i}{N}.$$

Replacing $p_{i\cdot}$ and $p_{\cdot j}$ in the expression for χ^2 (Eq. (2)) by their maximum likelihood estimates, we obtain

$$\chi^2 = \sum_{i=1}^{m} \sum_{j=1}^{n} \frac{(x_{ij} - (r_i c_j/N))^2}{r_i c_j/N}. \tag{4}$$

If N is sufficiently large and H_0 is true, then (4) will possess a chi-square distribution with $(m - 1)(n - 1)$ degrees of freedom.

A more computationally efficient expression for (4) may be obtained after some algebra:

$$\chi^2 = N\left(\sum_{i=1}^{m} \sum_{j=1}^{n} \frac{x_{ij}^2}{r_i c_j} - 1\right). \tag{5}$$

Computing (4) or (5) over a square contingency table yields a direct measure of the significance of H_0 with respect to that table. The level of significance is a function of the number of degrees of freedom and may be found by inspection of χ^2 tables [7] or by numerical integration of the χ^2 distribution. If χ^2 (adjusted by the proper number of degrees of freedom) exceeds a critical value, χ_0^2 (usually 0.05), then H_0 is rejected; otherwise H_0 is accepted. Or, for comparison purposes, χ^2 provides a continuous measure of structure within contingency tables.

4. AN ALGORITHM FOR SELECTING CO-OCCURRENCE MATRICES FOR TEXTURE CLASSIFICATION

Since we are interested in co-occurrence matrices that reflect the greatest amount of structure in the underlying texture, it is straightforward to devise algorithms to select the best matrix (or matrices) from a set of candidate matrices (these candidates may be obtained, for example, by using different spatial relations $\bar{d}$). One merely determines the goodness of fit of H_0 (by applying (4) or (5)) and selects the matrix (or matrices) yielding the highest value of χ^2.

As long as the co-occurrence matrices are complete (i.e., contain no zero entries) the theory presented in the previous section is applicable as stated. However, in the analysis of real texture pictures, one or more allowable gray levels may never occur. This causes entire rows and columns (due to symmetry) in co-occurrence matrices to have zero entries. When such matrices are interpreted as contingency tables, each degenerate row will result in one of the r_i having a value of zero, while each degenerate column will result in a zero c_j. Thus, zero maximum likelihood estimates would be obtained for one or more of the $p_{i\cdot}$ or $p_{\cdot j}$ parameters.

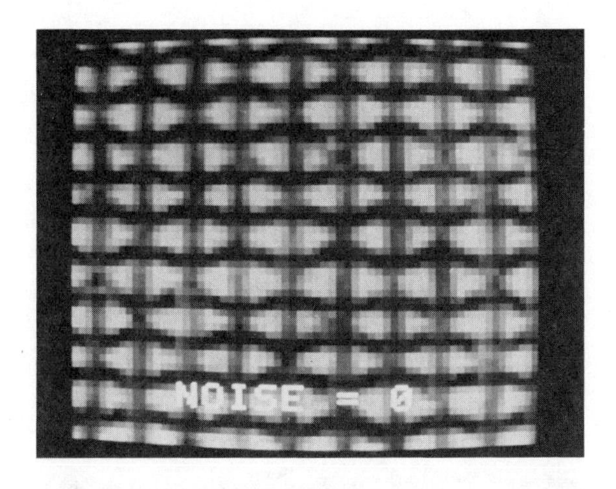

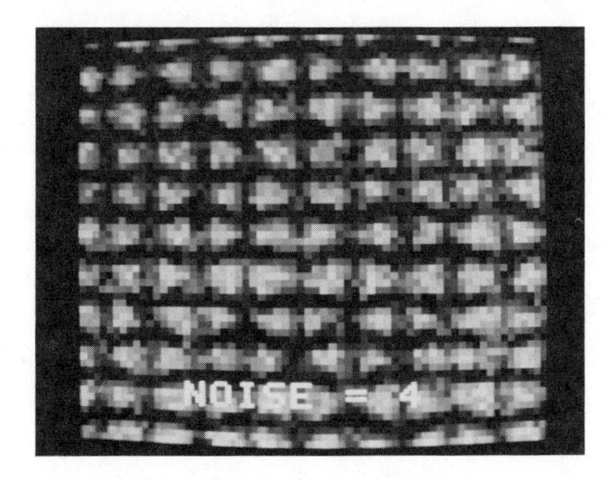

FIG. 2a. A texture that has been subjected to several magnitudes of additive noise in order to vary the structure.

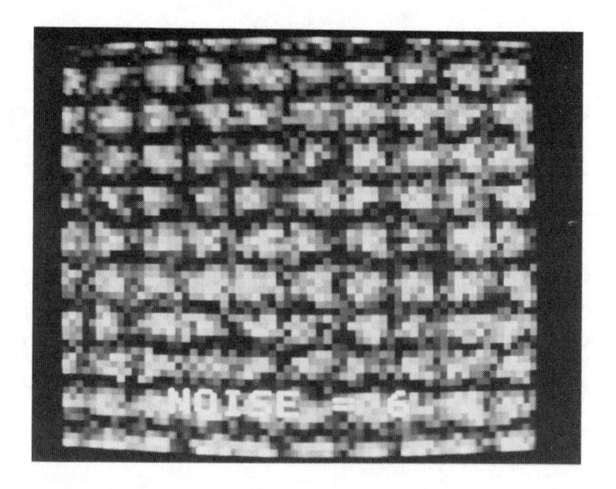

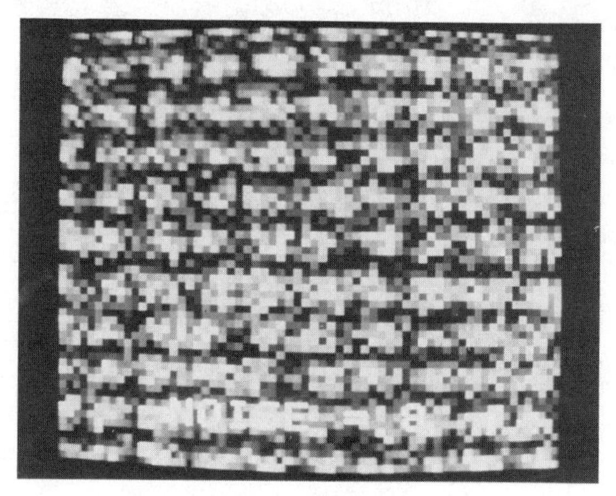

FIG. 2a—*Continued.*

The presence of nonestimable parameters requires that special action be taken in the application of (4) or (5), as well as adjustments in the number of degrees of freedom. Extensive discussion of the problems caused by incomplete contingency tables may be found in [1].

For our purposes, since each nonestimable parameter contains no information about a particular category (i.e., intensity), incomplete contingency tables may be collapsed into smaller, complete tables by eliminating all degenerate rows and columns. Equation (3.1) should then be used to determine the proper number of degrees of freedom for the complete table. That is, if ν is the result of (3.1) for the original incomplete table, $\mathfrak{Z}_e$ the number of cells with zero estimates, and $\mathfrak{Z}_p$ the number of nonestimable parameters, the adjusted number of degrees of freedom is given by

$$\nu' = \nu - \mathfrak{Z}_e + \mathfrak{Z}_p.$$

In actual practice, the proper number of degrees of freedom for any co-occur-

TABLE 2

χ^2 Values for Co-Occurrence Matrices Computed from the Images in Fig. 2a

Distance	Angles			
	0°	45°	90°	135°
Noise = 0				
2	4,781.5	1,100.4	1,261.2	1,200.4
3	5,046.8	2,659.0	2,199.7	2,601.1
4	4,082.2	1,097.8	738.2	1,114.0
5	3,537.9	923.5	1,853.3	896.9
6	5,466.1	6,112.5	5,749.3	6,794.2
7	6,048.3	2,267.1	1,826.3	2,323.1
8	3,688.7	738.4	738.1	830.7
9	3,055.4	1,874.4	1,698.4	1,873.6
10	4,029.7	986.8	844.2	925.9
11	3,668.5	763.3	2,278.2	627.8
12	4,761.5	2,533.1	7.934.4	2,739.0
13	10,492.9	1,487.2	2,217.7	1,472.0
14	7,745.0	564.9	779.6	597.4
Noise = 4				
2	1,636.4	500.7	501.7	559.3
3	1,605.7	1,159.0	621.2	1,066.0
4	1,476.8	580.8	361.1	639.6
5	1,543.1	483.4	820.0	451.4
6	1,983.1	2,455.8	2,082.4	2,675.5
7	2,136.2	993.0	902.0	1,073.4
8	1,568.3	440.6	368.9	345.4
9	1,255.2	963.8	571.9	819.9
10	1,372.8	494.7	372.5	428.2
11	1,202.4	404.5	910.8	376.9
12	2,037.8	1,090.9	2,781.8	1.246.3
13	4,142.1	707.7	1,068.9	741.8
14	2,910.2	403.4	364.3	380.3
Noise = 6				
2	972.9	407.8	366.9	377.7
3	934.6	599.6	555.3	700.9
4	1,002.0	394.6	418.7	387.6
5	843.6	519.7	635.2	454.2
6	1,073.6	1,065.7	1,013.1	1,209.3
7	1,141.2	669.1	572.2	633.1
8	920.6	292.8	346.2	315.3
9	778.2	582.0	447.3	484.0
10	791.2	326.1	343.9	377.9
11	853.5	439.6	552.8	418.4
12	1,119.9	715.4	1,233.5	749.2
13	1,352.0	492.4	585.8	530.7
14	1,303.5	316.6	344.6	313.3

TABLE 2—*Continued*

Distance	Angles			
	0°	45°	90°	135°
Noise = 8				
2	714.8	446.0	407.6	433.7
3	608.4	528.4	438.8	485.9
4	601.3	386.5	448.4	438.1
5	601.3	445.1	527.2	481.5
6	629.1	687.7	582.3	682.8
7	651.8	611.1	438.5	463.2
8	525.8	378.4	353.2	420.3
9	500.1	530.7	380.5	426.2
10	624.0	379.5	371.5	358.8
11	533.9	398.2	522.5	361.6
12	586.3	508.8	623.0	429.6
13	677.5	451.1	576.1	422.9
14	730.5	383.2	402.3	314.3

rence matrix is not required by our selection algorithm, because absolute levels of significance need not be computed. Rather, only relative comparisons of the magnitudes of χ^2 values (computed over candidate matrices) need be made; the degrees of freedom are constant and hence cancel out.

5. EXPERIMENTS

Two different kinds of experiments were performed. The first kind was designed to show that the χ^2 measure did indeed vary with the amount of structure in the texture pattern, while the second kind involved the classification of textures.

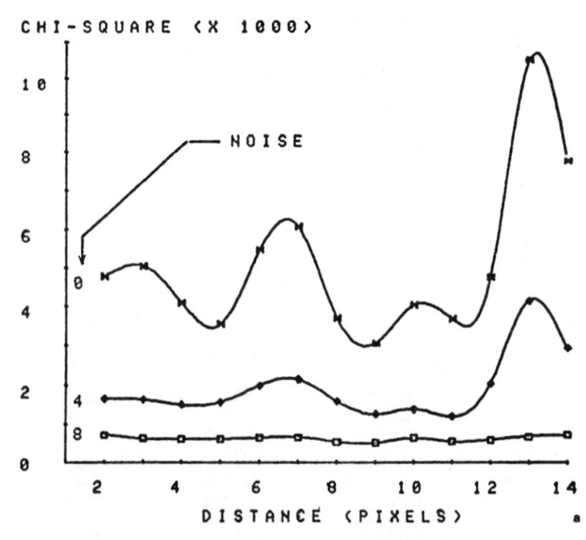

FIG. 2b. Graph of χ^2 vs distance (at 0°) for several magnitudes of additive noise. The curves were generated from selected entries in Table 2.

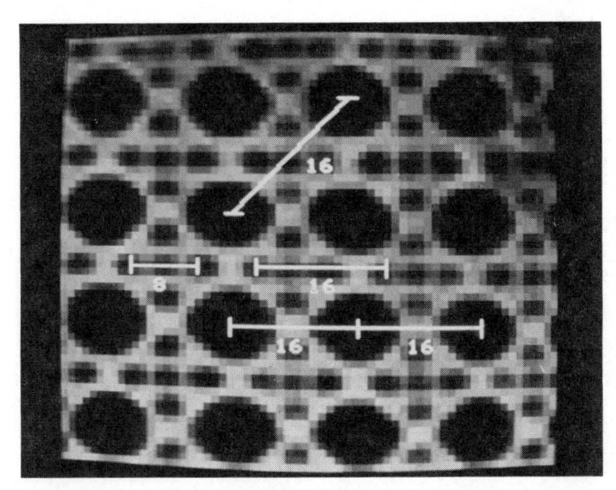

FIG. 3. A texture pattern on which the displacements (in pixels) between several major structural units are superimposed.

To show the relation between χ^2 values computed over co-occurrence matrices and the degree of structure present in the underlying texture, we used a number of Brodatz patterns. These texture samples were digitized into 256 × 256 images with 256 gray levels. The image intensities were then requantized into 16 equiprobable intensities in order to compensate for varying brightnesses and contrasts among the originals. The amount of structure in each image was varied by the pointwise addition of (uniform, zero mean) pseudorandom noise over different ranges (i.e., uniformly distributed in $[-N, N]$). χ^2 values for co-occurrence matrices constructed over several spatial relationships (various distances over the four angles, 0, 45, 90, and 135°) were computed. As expected, the χ^2 statistics decreased with increasing noise at all values of $\bar{d}$ for all images tested. A typical texture is illustrated in Fig. 2a and the resulting χ^2 values are shown in Table 2 and in Fig. 2b. The results confirm that the independence hypothesis, H_0, becomes more plausible as the underlying texture becomes more random.

Similar variations in the χ^2 measure can be observed for a single image as a function of the spatial relation $\bar{d}$. For values of $\bar{d}$ that capture texture structure very well, χ^2 will be high. To show this, consider the Brodatz texture in Fig. 3. Co-occurrence matrices were computed at various spatial relationships and the corresponding χ^2 statistics are shown in Table 3 and Fig 4. Maxima occur for $|\bar{d}| = 16$ and 32. A number of line segments of length 8 and 16 pixels have been superimposed on the pattern in Fig. 3 to show corresponding image scales and the distances between structural units.

A further property of the χ^2 measure is that it accurately reflects image magnification. That is, if a particular image is uniformly magnified, the maxima of the χ^2 measure will occur for displacements $\bar{d}$ that are correspondingly enlarged. For example, the image of Fig. 3 was magnified by a factor just less than 2, as shown in Fig. 5. Table 4 shows the corresponding set of χ^2 values, and Fig. 6 is a graph of some of them. Maximal χ^2 values clearly occur at a displacement $|\bar{d}| = 24$, almost twice the displacement obtained for the original image. In both images,

TABLE 3

χ^2 Values for Co-Occurrence Matrices Computed from the Texture of Fig. 3

Distance	Angles			
	0°	45°	90°	135°
2	22,458.4	8,320.1	14,298.8	8,731.5
4	7,295.5	7,983.4	3,788.8	7,159.5
6	8,583.4	5,165.5	4,920.6	4,959.8
8	9,718.9	4,649.2	4,325.6	5,217.7
10	7,030.4	4,986.1	3,949.4	4,615.4
12	6,022.6	2,478.0	4,160.5	2,857.6
14	22,041.3	13,154.4	21,233.0	9,864.8
16	53,866.1	22,583.3	40,782.3	34,186.0
18	12,413.4	2,069.3	7,000.1	4,332.1
20	3,796.6	2,858.9	2,329.0	3,066.1
22	4,820.5	2,674.4	2,809.5	2,903.3
24	4,963.5	2,694.2	2,558.3	3,400.9
26	3,413.6	1,608.0	2,028.5	2,215.3
28	5,338.6	2,658.9	6,942.3	1,097.8
30	20,597.7	14,588.6	23,054.7	8,588.1
32	25,696.2	5,582.3	13,537.1	12,547.6
34	6,747.6	778.5	2,489.9	2,041.3
36	1,950.7	1,401.9	1,320.9	1,299.0
38	2,724.4	2,078.3	1,950.6	2,146.6
40	2,478.9	1,454.2	1,244.9	2,236.6

the displacement between the major structural units in the texture is accurately reflected by the χ^2 measure.

Next, we present the results of several texture classification experiments. A

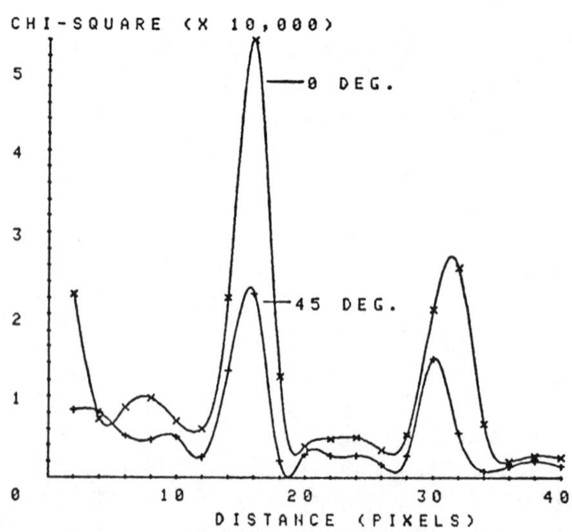

FIG. 4. Graph of χ^2 vs distance at two orientations. The curves were generated from selected entries in Table 3. Note how the displacements between structural units in Fig. 3 are reflected by the location of the maxima in the curves.

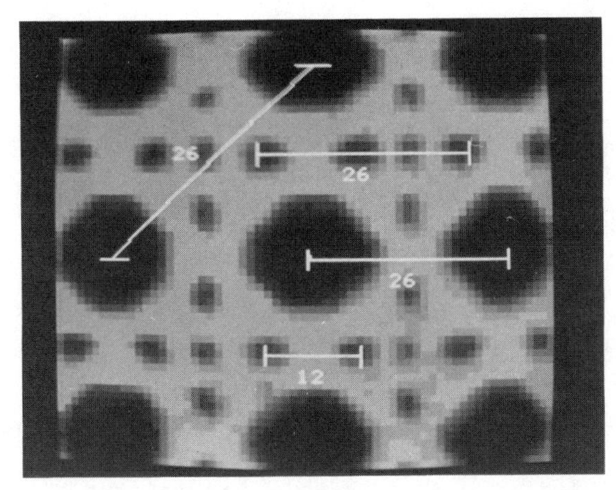

FIG. 5. The texture of Fig. 3 uniformly magnified. Displacements between major structural units are shown.

TABLE 4

χ^2 Values for Co-Occurrence Matrices Computed from the Texture of Fig. 5

Distance	Angles			
	0°	45°	90°	135°
8	3,452.6	5,724.2	3,987.1	5,379.1
10	4,177.5	3,223.3	5,131.1	3,205.7
12	4,889.5	4,073.3	5,333.7	3,959.5
14	3,884.0	2,798.1	4,709.8	2,486.6
16	2,328.3	2,442.9	3,923.9	2,394.3
18	3,375.6	4,037.4	3,012.7	3,271.7
20	8,574.5	1,435.3	5,015.7	1,382.5
22	18,968.3	6,518.4	12,719.5	7,777.2
24	42,427.4	19,848.5	32,862.7	21,826.0
26	33,877.6	26,246.5	44,901.0	16,567.0
28	12,053.4	8,033.9	17,494.5	4,676.1
30	3,879.4	1,159.3	6,604.2	714.6
32	1,420.6	1,432.8	1,886.5	1,192.8
34	1,805.3	2,117.5	2,062.5	1,407.9
36	2,508.5	1,737.3	2,815.1	1,291.4
38	2,280.8	1,709.5	2,795.2	1,156.9
40	1,222.8	1,341.1	2,509.8	948.9
42	1,223.9	1,257.8	1,785.3	1,145.7
44	3,489.6	1,046.8	1,405.2	449.9
46	8,484.3	737.2	3,918.2	1,730.2
48	17,021.4	3,948.2	9,356.4	6,447.5
50	26,934.0	11,206.3	22,176.3	10,927.8
52	12,280.2	8,308.6	25,231.2	6,049.5
54	5,179.8	2,527.7	10,557.4	2,089.6
56	1,261.7	236.3	3,844.9	756.4
58	620.4	662.4	924.3	1,294.4
60	1,064.8	893.5	1,164.0	1,519.2

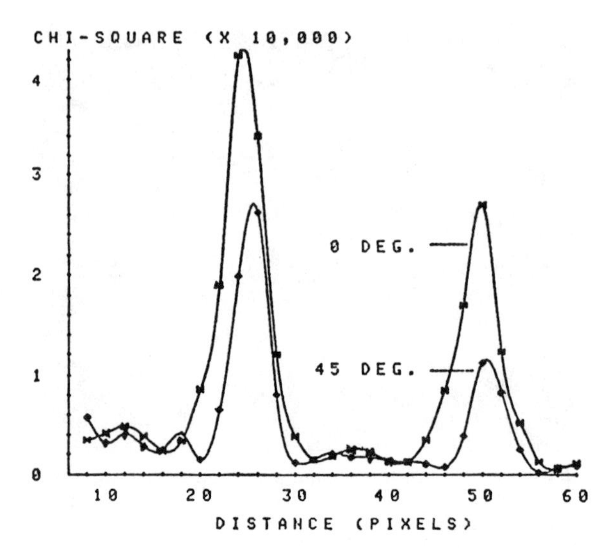

FIG. 6. Graph of selected χ^2 values from Table 4. Compared to Fig. 4, the maxima occur at larger distances due to the magnification of the structural units.

minimum Mahalanobis distance, linear discriminant classifier was used, the details of which are given in the Appendix.

The first experiment involved five Brodatz textures which are shown in Fig. 7. Twenty-five samples were taken from each texture (class) by extracting 64×64 windows over the equiprobability quantized images. Co-occurrence matrices were computed over several distances and orientations of 0, 45, 90, and 135°. χ^2 values were computed for each matrix and the distance yielding the maximum χ^2 measure, averaged over the four orientations, was chosen. The ASM, CON, and COR feature functions were computed over the four chosen matrices (at the above distance). The feature vectors (length 6) were generated by computing the mean and range of the values of each feature function. The classifier was trained on the 100 feature vectors (25 in each class). One hundred new windows were then selected and classified. The classification was 100% correct.

The second classification experiments involved the set of LANDSAT-1, Eastern Kentucky terrain images used by Weszka *et al.* [10] in their main study. The data set consists of three groups consisting of 60 images each. Each image is 64×64 pixels and has been histogram flattened to cover 64 gray levels (see Fig. 8). Co-occurrence matrices were, once again, computed for each window over several spatial relationships. For all of the 180 windows, $\bar{d} = [1, 0]$ and $\bar{d} = [0, 1]$ yielded matrices with the largest χ^2 values. Hence, these spatial relationships should be preferred in a classification of the above texture sample. Table 5 shows typical χ^2 results. It is interesting to note that these findings support Rosenfeld's observation that displacements of 1 pixel in the horizontal paired with size 1 in the vertical direction yielded the best classification results.

To further corroborate this observation, we trained our classifier on the basis of matrices selected by maximal χ^2 values. Feature vectors (length 3) were constructed for each of the 180 matrices using the ASM, CON, and ENT features.

The training set was then classified into three classes. In the final analysis, 85% of the samples were classified correctly. This is comparable to the best result obtained in [10] using pairs of features and a Fisher linear discriminant classifier.

6. DISCUSSION AND CONCLUSIONS

In this paper we have presented a statistical approach to selecting co-occurrence matrices for texture classification on the basis of how well they captured texture structure. Although the methodology was only demonstrated for matrices of intensity pair occurrence frequencies, its application to secondary images derived from more general local features is straightforward.

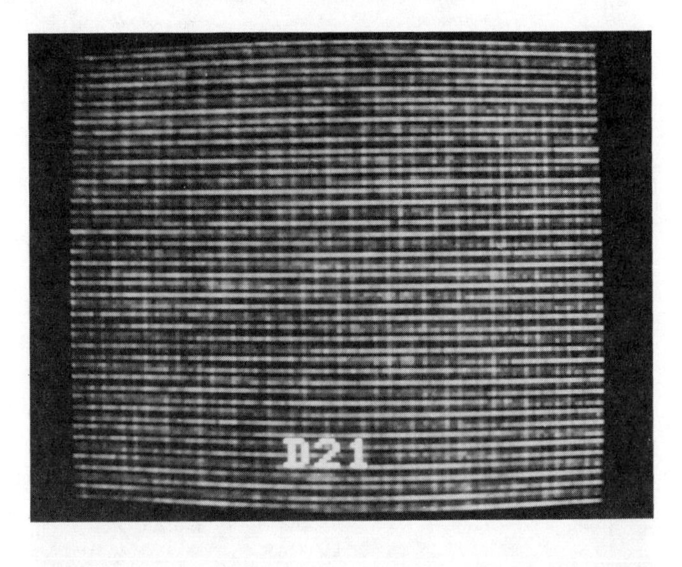

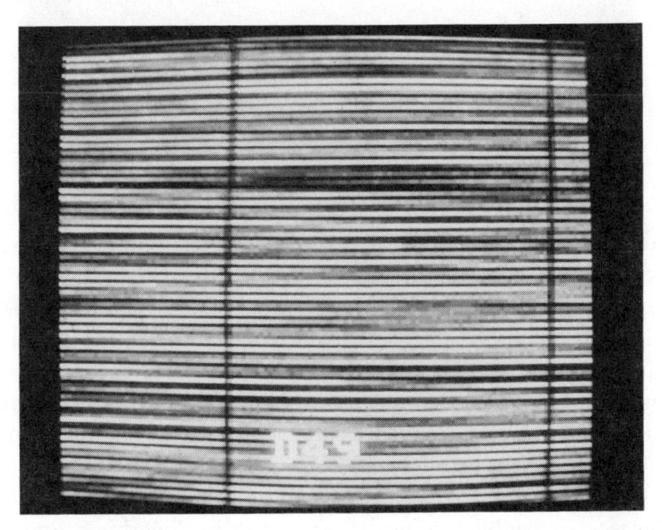

FIG. 7. The five Brodatz textures used in the first classification experiment.

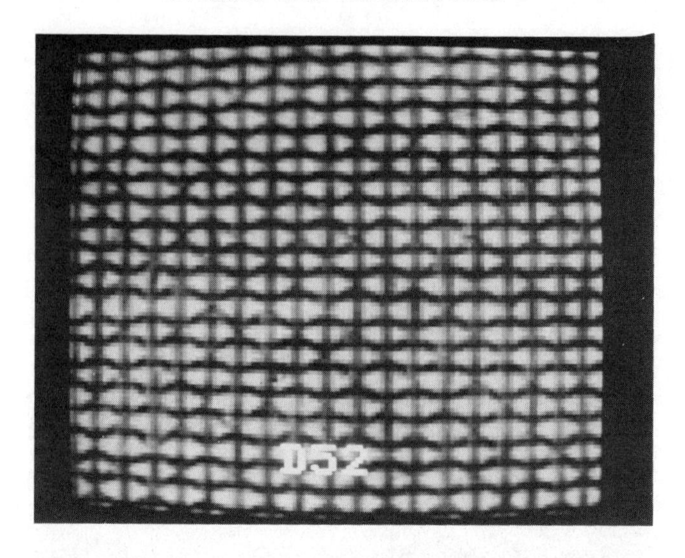

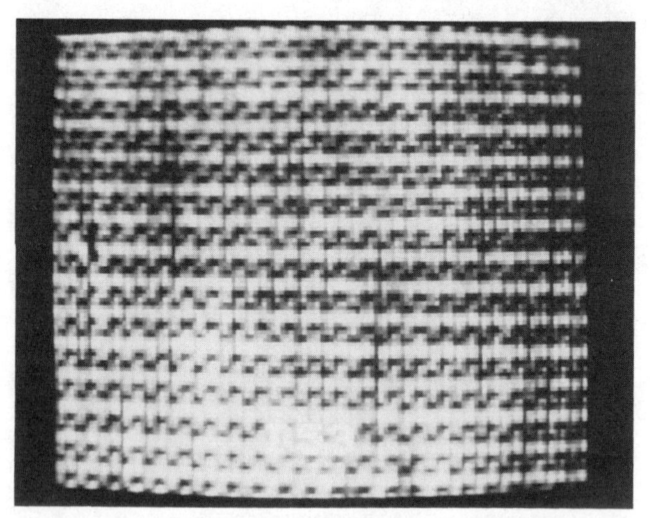

FIG. 7—*Continued.*

The measure of texture structure captured by co-occurrence matrices, a χ^2 statistic, was used to select such matrices for classification. This implies that the feature vectors for different samples in a particular classification experiment may be computed from matrices derived from different spatial relationships. An important variation on the technique presented here is to actually use the chosen $\bar{d}$ values themselves as features for classification. Such a feature should separate magnified images of identical textures very well, and should also reflect size variations between texture primitives.

The χ^2 measure of texture structure is only one possibility for quantifying the association between variables in contingency tables. Goodman and Kruskal [8] have characterized several other measures of association which should also be

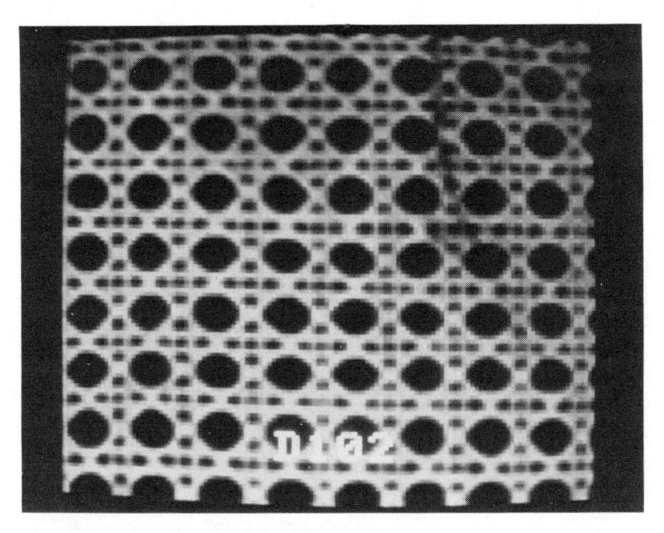

FIG. 7—*Continued.*

examined, both as features for texture classification and as measures of texture structure. Such measures may provide the beginnings of a formal bridge between statistical and structural models for texture.

APPENDIX: A LINEAR DISCRIMINANT CLASSIFIER

In this appendix we describe the classifier used in our texture experiments. It is a minimum Mahalanobis distance classifier that computes a set of linear discriminant functions.

Let ω_i, $i = 1, 2, \ldots, c$, be the set of classes and $\bar{X}$ be a sample (i.e., feature vector). The decision rule divides feature space into c decision regions $R_1, \ldots, R_c$. Let $g_i(\bar{X})$, $i = 1, 2, \ldots, c$, denote the set of discriminant functions. If $g_i(\bar{X}) > g_j(\bar{X})$ for all $i \neq j$ then $\bar{X}$ is in R_i and the decision rule assigns $\bar{X}$ to class ω_i.

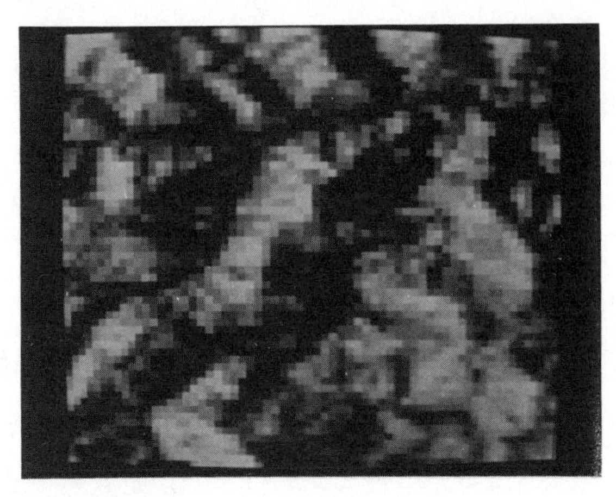

FIG. 8. LANDSAT terrain images used in the second classification experiment. A typical image is shown from each of the three groups.

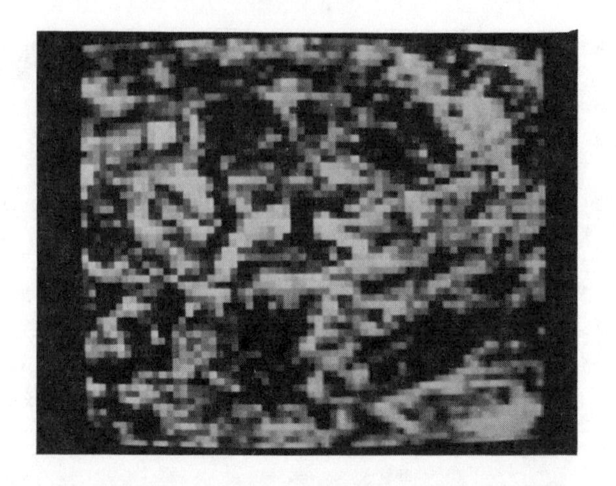

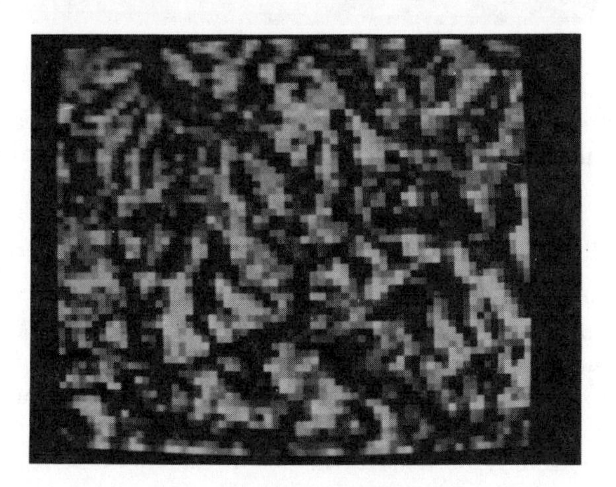

FIG. 8—*Continued.*

It can be shown [4] that minimum error rate classification can be obtained with the discriminant functions

$$g_i(\bar{X}) = \log p(\bar{X}\,|\,\omega_i) + \log P(\omega_i),$$

where $p(\bar{X}\,|\,\omega_i)$ is the likelihood of ω_i with respect to $\bar{X}$, and $P(\omega_i)$ is the a priori probability of samples falling into ω_i. To evaluate the functions, we assume that $p(\bar{X}\,|\,\omega_i)$ are multivariate normal densities. Furthermore, we assume that the covariance matrices are identical for all the classes (i.e., $\Sigma_i = \Sigma$). Thus, geometrically, the clusters for all classes are hyperellipses of equal size and shape, with the cluster of the ith class centered about mean vector $\overline{\mu_i}$. Under these conditions, the discriminant functions are linear and the resulting decision boundaries are hyperplanes.

Suppose that there are c classes and that the classifier is to be trained on a pool of N samples, with n_k, $k = 1, 2, \ldots, c$, samples belonging to each class. In addition, suppose that X_i is the ith feature of $\bar{X}$ and that $\bar{X}$ has length d. For each

TABLE 5

χ^2 Values for Co-Occurrence Matrices, Computed over Several Spatial Relationships for the LANDSAT Images[a]

Distance	Angles			
	0°	45°	90°	135°
1	8,795.4	6,737.6	11,994.6	6,998.7
2	5,026.1	4,641.0	5,577.7	4,267.7
3	4,246.3	3,897.1	4,904.9	4,315.7
4	4,103.0	4,009.3	4,497.7	4,177.0
5	4,023.4	3,894.6	4,438.9	4,089.9
6	4,079.3	4,066.3	4,172.4	4,242.1
7	4,003.6	3,879.2	4,106.7	4,006.4
8	3,805.0	3,926.4	4,146.1	4,201.5
1	7,268.6	5,067.1	6,973.1	4,949.7
2	4,660.9	4,131.5	3,985.6	4,002.5
3	4,112.1	3,744.2	4,257.4	3,974.1
4	4,056.8	4,035.2	4,356.5	4,037.5
5	3,970.7	3,950.1	4,143.4	4,039.1
6	4,152.3	3,968.3	4,488.4	4,013.9
7	4,041.1	4,043.4	4,018.1	3,928.8
8	3,958.3	4,043.3	3,892.7	4,121.6
1	13,628.7	9,618.6	17,446.3	11,182.7
2	7,282.4	5,560.0	7,814.8	6,275.2
3	5,858.3	4,450.6	5,486.4	4,861.8
4	4,988.2	4,329.8	4,616.3	4,390.2
5	4,473.5	4,309.3	4,305.2	4,265.8
6	4,419.1	4,185.8	4,073.6	4,229.1
7	3,904.4	4,293.5	4,185.6	4,085.2
8	3,863.1	4,300.1	4,458.9	4,186.0

[a] The tables were generated from the images in Fig. 8, in the order of their appearance.

class, $k = 1, 2, \ldots, c$, we compute the means

$$\mu_j{}^k = \frac{1}{n_k} \sum_{i=1}^{n_k} X_{ij}^k, \qquad j = 1, 2, \ldots, d,$$

and the matrix of cross products of deviations from the means

$$D^k = \{d_{jl}^k\} = \sum_{i=1}^{n} (X_{ij}^k - \mu_j{}^k)(X_{il}^k - \mu_l{}^k), \qquad j, l = 1, 2, \ldots, d.$$

The pooled covariance matrix is then

$$\Sigma = \{\sigma_{ij}\} = \sum_{k=1}^{c} D^k / ((\sum_{k=1}^{c} n_k) - c).$$

Let σ_{ij}^{-1} be the (i, j)th element of Σ^{-1}. The coefficients of the linear discriminant functions are then given by

$$W_i^k = \sum_{j=1}^{d} \sigma_{ij}^{-1} \mu_j^k, \qquad i = 1, 2, \ldots, d,$$

and the constant term by

$$W_0^k = -\tfrac{1}{2} \sum_{i=1}^{d} \sum_{j=1}^{d} \sigma_{ij}^{-1} \mu_i^k \mu_j^k.$$

Thus, the discriminant functions are

$$g_k(\bar{X}) = W_0^k + \sum_{i=1}^{d} W_i^k X_i^k.$$

The decision rule assigns $\bar{X}$ to the class yielding the largest discriminant function. The confidence in the classification is given by

$$P_L = 1 / \sum_{k=1}^{c} e^{(g_k - g_L)},$$

where L is the class whose discriminant function gives the largest value, g_L, for $\bar{X}$.

One may obtain a general measure of the usefulness of a set of discriminant functions obtained from a particular training set by computing a generalized Mahalanobis D^2 statistic. Let m_i, $i = 1, 2, \ldots, d$, be the common means for all c groups.

$$m_i = \sum_{k=1}^{c} n_k \mu_i^k / \sum_{k=1}^{c} n_k.$$

The statistic is given by

$$D^2 = \sum_{i=1}^{d} \sum_{j=1}^{d} \sigma_{ij}^{-1} \sum_{k=1}^{c} n_k (\mu_i^k - m_i)(\mu_j^k - m_j).$$

D^2 can be used as a chi square (under the assumption of normality), with $m(k - 1)$ degrees of freedom, to test the hypothesis that the d feature mean values are the same in all c groups (i.e., the groups are nonseparable).

REFERENCES

1. Y. M. M. Bishop, S. E. Feinberg, and P. W. Holland, *Discrete Multivariate Analysis: Theory and Practice*, MIT Press, Cambridge, Mass., 1976.
2. P. Brodatz, *Textures: A Photographic Album*, Dover, New York, 1966.
3. L. S. Davis, S. Johns, and J. K. Aggarwal, Texture analysis using generalized co-occurrence matrices, in *Proc. IEEE Conf. on Pattern Recognition and Image Processing, Chicago 1978.*
4. R. O. Duda and P. E. Hart, *Pattern Classification and Scene Analysis*, Wiley, New York, 1973.
5. R. M. Haralick, Statistical and structural approaches to texture, *Proc. IEEE* **67**, 1979, 786–804.
6. R. M. Haralick, K. Shanmugam, and I. Dinstein, Textural features for image classification, *IEEE Trans. Systems Man Cybernet.* **SMC-3**, 1973, 610–621.
7. P. G. Hoel, *Introduction to Mathematical Statistics*, 4th ed., Wiley, 1971.
8. L. A. Goodman and W. H. Kruskal, Measures of association for cross classifications, I–IV,

J. Amer. Statist. Assoc. **49**, 1954, 732–764; **54**, 1959, 123–163; **58**, 1963, 310–364; **67**, 1972, 415–421.

9. R. L. Plackett, *The Analysis of Categorical Data*, Griffin, London, 1974.
10. J. Weszka, C. R. Dyer, and A. Rosenfeld, A comparative study of texture measures for terrain classification, *IEEE Trans. Systems Man Cybernet.* **SMC-6**, 1976, 269–285.
11. S. W. Zucker, Toward a Model of Texture, *Computer Graphics Image Processing* **5**, 1976, 190–202.